Biotechnology: Pharmaceutical Aspects

Biotechnology: Pharmaceutical Aspects
Ronald T. Borchardt and C. Russell Middaugh, Series Editors

Volume I: *Pharmaceutical Profiling in Drug Discovery for Lead Selection*
R.T. Borchardt, E.H. Kerns, C.A. Lipinski, D.R. Thakker, B. Wang

Volume II: *Lypophilization of Biopharmaceuticals*
H.R. Costantino, M.J. Pikal

Volume III: *Methods for Structural Analysis of Protein Pharmaceuticals*
W. Jiskoot, D.J.A. Crommelin

Volume IV: *Optimizing the "Drug-Like" Properties of Leads in Drug Discovery*
R.T. Borchardt, E.H. Kerns, M.J. Hageman, D.R. Thakker, J.L. Stevens

Volume V: *Prodrugs: Challenges and Rewards, Parts 1 and 2*
V.J. Stella, R.T. Borchardt, M.J. Hageman, R. Oliyai, H. Maag, J.W. Tilley

Volume VI: *Solvent Systems and Their Selection in Pharmaceutics and Biopharmaceutics*
P. Augustijns, M.E. Brewster

Solvent Systems and Their Selection in Pharmaceutics and Biopharmaceutics

Patrick Augustijns
Catholic University of Leuven, Belgium

Marcus E. Brewster
Janssen Pharmaceutica N.V., Beerse, Belgium

Springer

AAPS PRESS

Patrick Augustijns
Laboratory for Pharmacotechnology
and Biopharmacy
Catholic University of Leuven
Belgium

Marcus E. Brewster
Janssen Pharmaceutica N.V.
Beerse, Belgium

Library of Congress Control Number: 2007924356

ISBN-13: 978-0-387-69149-7 e-ISBN-13: 978-0-387-69154-1

Printed on acid-free paper.

9 8 7 6 5 4 3 2 1

springer.com

Dedication

We would like to dedicate this volume to Prof. David Grant, who passed away on December 9, 2005. Prof. Grant held the William and Mildred Peters Endowed Chair in the College of Pharmacy, Department of Pharmaceutics at the University of Minnesota. Prof. Grant was an internationally recognized authority on solid-state properties of drugs. His research directly impacted the ability to make safe and effective pharmaceutical agents with reproducible and predictable biopharmaceutical performance. Prof. Grant was a prolific scientist with more than 200 scientific articles to his credit. He also gave back to the scientific community in many ways, including his participation on the editorial boards of various scientific publications including the *Journal of Pharmaceutical Sciences*, where he served as Associate Editor, as well as *Pharmaceutical Development and Technology* and *The AAPS Journal* (formerly *AAPS Pharm Sci*). Prof. Grant was often singled out by his peers for his excellent contributions to science; he received such awards as the Pharmaceutics Award in Excellence from the Pharmaceutical Research and Manufacturers Association Foundation as well as the 2004 Dale E. Wurster Research Award, the highest recognition in his discipline from the American Association of Pharmaceutical Scientists (AAPS). The legacy left by Prof. Grant is substantial and transformational. We are indebted to him on so many levels, including his contribution to this monograph.

Contents

Preface ix

1. Principles of Solubility 1
Yuchuan Gong, David J.W. Grant, and Harry G. Brittain

2. Ionic Equilibria and the pH Dependence of Solubility 29
Harry G. Brittain

3. Solvent Systems for Crystallization and Polymorph Selection 53
Jonathan M. Miller, Anthony C. Blackburn, Dainius Macikenas, Benjamin M. Collman, and Naír Rodríguez-Hornedo

4. Solubility Issues in Early Discovery and HTS 111
Li Di and Edward H. Kerns

5. Practical Aspects of Solubility Determination in Pharmaceutical Preformulation 137
Wei-Qin (Tony) Tong

6. Biorelevant Dissolution Media 151
Anette Müllertz

7. Selection of Solvent Systems for Membrane-, Cell-, and Tissue-Based Permeability Assessment 179
Françoise Ingels, Anna-Lena Ungell, and Patrick Augustijns

8. The Use of Solubilizing Excipients and Approaches to Generate Toxicology Vehicles for Contemporary Drug Pipelines 221
Marcus E. Brewster, Claire Mackie, Ann Lampo, Marc Noppe, and Thorsteinn Loftsson

9. Solubilizing Vehicles for Oral Formulation Development 257
Robert G. Strickley and Reza Oliyai

10. Solubilizing Systems for Parenteral Formulation Development—Small Molecules 309
James E. Kipp

11. Aqueous Solubilizing Systems for Parenteral Formulation Development—Proteins 341
Zahra Shahrokh

12. Non-Aqueous Systems for Formulation Development—Proteins 359
Cynthia L. Stevenson

13. Pharmaceutical Solvents as Vehicles for Topical Dosage Forms 403
Adrian Williams

14. Pharmaceutical Solvents for Pulmonary Drug Delivery 427
Erik Mogalian and Paul Brian Myrdal

Index 443

Preface

Solvent systems are integral to drug development and pharmaceutical technology. This single topic encompasses numerous allied subjects running the gamut from recrystallization solvents to biorelevant media. The goal of this contribution to the **Biotechnology: Pharmaceutical Aspects** series is to generate both a practical handbook as well as a reference allowing the reader to make effective and informed decisions concerning the use of solvents and solvent systems. To this end, the monograph was created by inviting recognized experts from a number of fields to author relevant sections. Specifically, 14 chapters have been designed to cover the theoretical background of solubility, the effect of ionic equilibria and pH on solubilization, the use of solvents to effect drug substance crystallization and polymorph selection, the use of solvent systems in high throughput screening and early discovery, solvent use in preformulation, the use of solvents in biorelevant dissolution and permeation experiments, solvents and their use as toxicology vehicles, solubilizing media and excipients in oral and parenteral formulation development, specialized vehicles for protein formulation, and solvent systems for topical and pulmonary drug administration. The chapters are organized such that useful decision criteria are included together with the scientific underpinning for their application. In addition, trends in the use of solvent systems and a balance of current views make this monograph useful, we hope, to both the novice and experienced researcher and to scientists at all developmental stages from early discovery to late pharmaceutical operations.

Patrick Augustijns
Catholic University of Leuven

Marcus E. Brewster
Janssen Pharmaceutica N.V.

I

Principles of Solubility

YUCHUAN GONG AND DAVID J.W. GRANT

Department of Pharmaceutics, College of Pharmacy, University of Minnesota, Minneapolis, MN

HARRY G. BRITTAIN

Center for Pharmaceutical Physics, Milford, NJ

Introduction

Solubility is defined as the maximum quantity of a substance that can be completely dissolved in a given amount of solvent, and represents a fundamental concept in fields of research such as chemistry, physics, food science, pharmaceutical, and biological sciences. The solubility of a substance becomes especially important in the pharmaceutical field because it often represents a major factor that controls the bioavailability of a drug substance. Moreover, solubility and solubility-related properties can also provide important information regarding the structure of drug substances, and in their range of possible intermolecular interactions. For these reasons, a comprehensive knowledge of solubility phenomena permits pharmaceutical scientists to develop an optimal understanding of a drug substance, to determine the ultimate form of the drug substance, and to yield information essential to the development and processing of its dosage forms.

In this chapter, the solubility phenomenon will be developed using fundamental theories. The basic thermodynamics of solubility reveals the relation between solubility, and the nature of the solute and the solvent, which facilitates an estimation of solubility using a limited amount of information. Solubility-related issues, such as the solubility of polymorphs, hydrates, solvates, and amorphous materials, are included in this chapter. In addition, dissolution rate phenomena will also be discussed, as these relate to the kinetics of solubility. A discussion of empirical methods for the measurement of solubility is outside the scope of this chapter, but is reviewed elsewhere (Grant and Higuchi, 1990; Grant and Brittain, 1995).

Units for the Expression of Solubility

A discussion of the thermodynamics and kinetics of solubility first requires a discussion of the method by which solubility is reported. The solubility of a substance may be defined in many different types of units, each of which represents an expression of the quantity of solute dissolved in a solution at a given temperature. Solutions are said to be *saturated* if the solvent has dissolved the maximal amount of solute permissible at a particular temperature, and clearly an *unsaturated* solution is one for which the concentration is less than the saturated concentration. Under certain conditions, metastable solutions that are *supersaturated* can be prepared, where the concentration exceeds that of a saturated solution. The most commonly encountered units in pharmaceutical applications are molarity, normality, molality, mole fraction, and weight or volume percentages.

The *molarity* (abbreviated by the symbol M) of a solution is defined as the number of moles of solute dissolved per liter of solution (often written as mol/L or mol/dm^3), where the number of moles equals the number of grams divided by its molecular weight. A fixed volume of solutions having the same molarity will contain the same number of moles of solute molecules. The use of molarity bypasses issues associated with the molecular weight and size of the solute, and facilitates the comparison of different solutions. However, one must exercise caution when using molarity to describe the concentrations of ionic substances in solution, because the stoichiometry of the solute may cause the solution to contain more moles of ions relative to the number of moles of dissolved solute. For example, a 1.0 M solution of sodium sulfate (Na_2SO_4) would be 1.0 M in sulfate ions and 2.0 M in sodium ions.

The *normality* (abbreviated by the symbol N) of a solution is defined as the number of equivalents of solute dissolved per liter of solution, and can be written as eq/L or eq/dm^3. Normality has the advantage of describing the solubility of the ionic compounds since it takes into account the number of moles of each ion in the solution liberated upon dissolution of a given number of moles of solute. The number of equivalents will equal the number of grams divided by the equivalent weight. For ionic substances, the equivalent weight equals the molecular weight divided by the number of ions in the compound. Equivalent weight of an ion is the ratio of its molecular (atomic) weight and its charge. Therefore, a molar solution of Na_2SO_4 is 2 N with respect to both the sodium and the sulfate ion. Since the volume of solution is temperature dependent, molarity and normality can not be used when the properties of solution, such as solubility, is to be studied over a wide range of temperature.

Molality is expressed as the number of moles of solute dissolved per kilogram of solvent, and is therefore independent of temperature since all of the quantities are expressed on a temperature-independent weight basis. The molality of a solution is useful in describing solubility-related phenomena at various temper atures, and as the concentration unit of colligative property studies. When the density of the solvent equals unity, or in the case of dilute aqueous solutions, the molarity and the molality of the solution would be equivalent.

Expressing solution concentrations in terms of the *mole fraction* provides the ratio of the number of moles of the component of interest to the total number of moles of solute and solvent in the solution. In a solution consisting of a single solute and a single solvent, the mole fraction of solvent, X_A, and solute, X_B, is expressed as:

$$X_A = \frac{n_A}{n_A + n_B} \tag{1}$$

$$X_B = \frac{n_B}{n_A + n_B} \tag{2}$$

where n_A and n_B are the number of moles of solvent and solute, respectively. Obviously the sum of the mole fraction of the two components must equal one:

$$X_A + X_B = 1 \tag{3}$$

Since mole fractions provide quantitative information of a mixture that can be readily translated down to the molecular level, this unit is most commonly used in thermodynamic studies of solubility behavior.

Volume fraction is frequently used to define the composition of mixed solvent systems, or to express the solubility of one solvent in another. However, since the volumes of solutions exhibit a dependence on temperature, the expression of concentrations in terms of volume fraction requires a simultaneous specification of the temperature. In addition, since volume defects may occur during the mixing of the solvents, and since these will alter the final obtained volume, defining the solubility of a solution in terms of volume fraction can lead to inaccuracies that can be avoided through the use of other concentration parameters.

The concept of *percentage* is widely used as a concentration parameter in pharmaceutical applications, and is expressed as the quantity of solute dissolved in 100 equivalent units of solution. The *weight percentage* (typically abbreviated as % w/w) is defined as the number of grams of solute dissolved in 100 grams of solution, while the *volume percentage* (typically abbreviated as % v/v) is defined as the number of milliliters of solute dissolved in 100 mL of solution. A frequently encountered unit, the *weight-volume percentage* (typically abbreviated as % w/v) expresses the number of grams of solute dissolved in 100 mL of solution. The choice of unit to be used depends strongly on the nature of solute and solvent, so the solubility of one liquid in another is most typically expressed in terms of the volume percentage. The use of weight or weight-volume percentages is certainly more appropriate to describe the concentration or solubility of a solid in its solution.

For very dilute solutions, solubility is often expressed in units of parts per million (ppm), which is defined as the quantity of solute dissolved in 1,000,000 equivalent units of solution. As long as the same unit is used for both solute and solvent, the concentration in parts per million is equivalent to the weight, volume, or weight-volume percentages multiplied by 10,000. The descriptive terms of solubility that is expressed in units of parts of solvent required for each part of solute can be found in each edition of the United States Pharmacopeia (Table 1).

Descriptive term	Parts of solvent required for 1 part of solute
Very soluble	Solubility < 1
Freely soluble	1 < Solubility < 10
Soluble	10 < Solubility < 30
Sparingly soluble	30 < Solubility < 100
Slightly soluble	100 < Solubility < 1,000
Very slightly soluble	1,000 < Solubility < 10,000
Practically insoluble, or Insoluble	Solubility > 10,000

Table 1. Descriptive terms of solubility.

Reproduced from:

United States Pharmacopeia, 25th edition. United States Pharmacopeial Convention; Rockville, MD; 2002, p. 2363.

Thermodynamics of Solubility

The *equilibrium solubility* of a substance is defined as the concentration of solute in its saturated solution, where the saturated solution exists in a state of equilibrium with pure solid solute. As solutes and solvents can be gaseous, liquid, or solid, there are nine possibilities for solutions, although liquid-gas, liquid-liquid, and liquid-solid are of particular interest for pharmaceutical applications. Among these, the most frequently encountered solubility behavior involves solid solutes dissolved in liquid solvent, so systems of this type will constitute the examples of the following discussions.

For the particular system of a saturated solution, the dissolved solute in the solution and the undissolved solute of the solid phase are in a state of dynamic equilibrium. Under those conditions, the rate of dissolution must equal the rate of precipitation and hence the concentration of the solute in the solution remains constant (as long as the same temperature is maintained).

For two phases in equilibrium, the chemical potential, μ_i, of the component in the two phases must be equal:

$$\mu_{\text{solute}} = \mu_{\text{solid}} \tag{4}$$

The chemical potential, also known as the molar free energy, can be represented by:

$$\mu = \mu^{\circ} + RT \ln a \tag{5}$$

where μ° is the chemical potential of the solute molecule in its reference state, and a is the activity of the solute in the solution. Since both the dissolved solute and the undissolved solid must refer back to the same standard state, it follows that the activities of the dissolved solute and that of the undissolved solid must be identical.

The activity of a component in a solution is defined as the product of its activity coefficient, γ, and its mole fraction, X:

$$a = \gamma X \tag{6}$$

For the solute **B** in a saturated solution:

$$a_{\text{solid}} = a_{\text{solute}} = \gamma_{\text{B}} X_{\text{B}} \tag{7}$$

or

$$X_{\text{B}} = \frac{a_{\text{solid}}}{\gamma_{\text{B}}} \tag{8}$$

According to equation (8), the solubility of a substance would be proportional to the activity of the undissolved solid, and inversely proportional to its activity coefficient. Although the activity of a substance in its standard state is defined as unity, the activity of the undissolved solid must depend on reference state. A hypothetical, supercooled liquid state of solute at the temperature of interest is commonly taken as the standard state, making the activity coefficient a more complicated term. The activity coefficient will depend on the nature of both the solute and solvent, as well as on the temperature of the solution.

Solubility in Ideal Solutions

In order to understand the thermodynamics of solubility, it is appropriate to begin with a simplified model of solution, namely that of an ideal solution. An ideal solution is defined as one where the activity coefficient of all components in the solution equals one. Under these stipulations, the activity of the dissolved solute, the activity of the solid, and the molar solubility of the dissolved solute would be equal.

$$a_{\text{solute}} = a_{\text{solid}} = X_{\text{B}} \tag{9}$$

As discussed above, the absolute activity of the solid depends on the chosen reference or standard state, and the usual practice is to take the supercooled liquid state of the pure solute at the temperature of solution as the standard state of unit activity. At temperatures lower than the melting point, the liquid state of the solute is less stable than its solid state, making the activity of the corresponding solid less than one.

An ideal solution requires that the scope of solute-solute, solvent-solvent, and solute-solvent intermolecular forces be all the same. Thus, the net energy change associated with breaking bonds between two solute molecules and two solvent molecules, and then forming new bonds between solute and solvent molecules must be zero. Moreover, the mixing process is ideal as well, so that the total volume of the solute/solvent system does not change during the mixing process.

$$\Delta U_{\text{mix}} = 0 \tag{10}$$

$$\Delta H_{\text{mix}} = 0 \tag{11}$$

$$\Delta V_{\text{mix}} = 0 \tag{12}$$

where ΔU_{mix} is the energy of mixing, ΔH_{mix} is the enthalpy of mixing, and ΔV_{mix} is the volume change of mixing. The ideal entropy of mixing, ΔS_{mix}, can be derived from pure statistical substitution

$$\Delta S_{\text{mix}} = -R(n_{\text{A}} \ln X_{\text{A}} + n_{\text{B}} \ln X_{\text{B}}) \tag{13}$$

where n_{A} and n_{B} are the number of moles of the solvent (A) and the solute (B), respectively. Because the mole fractions of the solvent and the solute, X_{A} and X_{B}, are less than unity, it follows that ΔS_{mix} is always positive. From this analysis, one can conclude that the mixing processes associated with an ideal solution would be thermodynamically favored.

The dissolution of a solid in a solvent can be considered as consisting of two steps. The first step would be, in effect, a melting of the solid at the absolute temperature (T) of the solution, and the second step would entail mixing of the liquidized solute with the solvent. The enthalpy of solution (ΔH_{s}) is therefore equal to the sum of the enthalpy of fusion ($\Delta H_{\text{f}}^{\text{T}}$) and the enthalpy of mixing (ΔH_{mix}). However, since the enthalpy of mixing must equal zero for an ideal solution, it follows that the enthalpy of solution must equal the enthalpy of fusion of the solid at the given temperature, T:

$$\Delta H_{\text{s}} = \Delta H_{\text{f}}^{\text{T}} \tag{14}$$

For those situations where the temperature of study is not the same as the melting point, then $\Delta H_{\text{f}}^{\text{T}} \neq \Delta H_{\text{f}}^{\text{m}}$, where now $\Delta H_{\text{f}}^{\text{m}}$ is the enthalpy of fusion at the melting point(T_{m}). If one makes the approximation that the enthalpy of fusion is constant over the temperature range in the vicinity of the melting point, then:

$$\Delta H_{\text{s}} = \Delta H_{\text{f}}^{\text{T}} \approx \Delta H_{\text{f}}^{\text{m}} \tag{15}$$

Applying the Clausius-Clapeyron equation to the solubility calculation yields:

$$\left(\frac{\partial \ln a}{\partial T} \right)_{\text{P}} = \frac{\Delta H_{\text{s}}}{RT^2} \tag{16}$$

Integration of equation (16) provides the relationship known as the van't Hoff equation, which expresses the temperature dependence of the solubility of a solid solute (identified as species B) in an ideal solution:

$$\ln X_{\text{B}} = \ln a_{\text{B}} = -\frac{\Delta H_{\text{s}}}{R} \left(\frac{1}{T} - \frac{1}{T_{\text{m}}} \right) \tag{17}$$

By combining equations (15) and (17), one finds that the molar solubility of the solute in an ideal solution (expressed in natural logarithmic form) is given by:

$$\ln X_B = \ln a_B = -\frac{\Delta H_f^m}{R}\left(\frac{1}{T} - \frac{1}{T_m}\right) \tag{18}$$

Since the solid solute and its corresponding molten solid must be in a state of equilibrium at the melting point, it follows that:

$$\Delta G_f^m = \Delta H_f^m - T_m \Delta S_f^m = 0 \tag{19}$$

where the enthalpy of fusion (ΔH_f^m) is equal to $T_m \Delta S_f^m$, where ΔS_f^m is the entropy of fusion at the melting temperature. Under these circumstances, equation (18) may also be written as:

$$\ln X_B = \ln a_B = -\frac{\Delta S_f^m}{R}\left(\frac{T_m}{T} - 1\right) \tag{20}$$

The enthalpy and entropy of fusion, and the melting temperature may all be measured through the use of differential scanning calorimetry (DSC), and therefore equations (18) and (20) provide a simple way to predict the solubility of a solute in an ideal solution.

To achieve a better prediction of the solubility of a solute, one must consider the temperature dependence of the enthalpy of fusion, which is described by the Kirchoff equation

$$\left(\frac{\partial \Delta H_f}{\partial T}\right)_P = \Delta C_p \tag{21}$$

where ΔC_p is the difference between the heat capacities of the supercooled liquid and that of the corresponding solid. Therefore:

$$\Delta H_f^T = \Delta H_f^m - \Delta C_p (T_m - T) \tag{22}$$

With the assumption that ΔC_p is independent of temperature, integration of equation (16) and the replacement of ΔH_s by ΔH_f^T, yields the Hildebrand equation

$$\ln X_B = \ln a_B = -\frac{\Delta H_f^m}{R}\left(\frac{1}{T} - \frac{1}{T_m}\right) + \frac{\Delta C_p}{R}\frac{T_m - T}{T} - \frac{\Delta C_p}{R}\ln\frac{T_m}{T} \tag{23}$$

Equation (23) provides a better prediction of the solubility of a solute in an ideal solution.

Prediction of solubility in an ideal solution can also be performed using the entropy approach developed by Hildebrand and Scott (Hildebrand and Scott, 1962). Assuming that $\Delta H_s \approx T\Delta S_f^m \approx T\Delta C_p$, they found that:

$$\ln X_B = \ln a_B = -\frac{\Delta S_f^m}{R}\ln\frac{T}{T_m} \tag{24}$$

Equation (24) is similar to equation (20), except that $\ln(X_B)$ is correlated to $\ln(T)$ instead of $1/T$. The solubility prediction using equation (24) was found to have a better tolerance for the non-ideality of the solution than that obtained using equation (20).

Several approaches have been used to predict the entropy of fusion required for the prediction of solubility. According to Walden's rule, the entropy of fusion (ΔS_f^m) is approximately equal to 13 cal/K·mol for most organic compounds (Walden, 1908). Use of this approximation reduces equation (20) to:

$$\ln X_B = \ln a_B = -\frac{\theta_m - 25}{298.15} \tag{25}$$

where θ_m is the melting point of the solute in degrees centigrade.

Yalkowsky proposed that the entropy of fusion of an organic compound is the sum of translational, rotational, and internal entropy changes when it is released from the crystal lattice (Yalkowsky, 1979):

$$\Delta S_f = \Delta S_{trans} + \Delta S_{rot} + \Delta S_{int} \tag{26}$$

while, the translational entropy change consists of the components associated with the expansion and change of position as the solid melts.

$$\Delta S_{trans} = \Delta S_{exp} + \Delta S_{pos} \tag{27}$$

Yalkowsky also proposed empirical values and limits for these components. Both the Walden and Yalkowsky models provide ways by which one can predict the entropy of fusion, and therefore predict the solubility of the solute in an ideal solution.

Over a small temperature range, the enthalpy of solution of a solid can be assumed to be independent of temperature. The van't Hoff equation shows that $\ln(X_B)$ increases with temperature, until the solid melts at $T = T_m$. At this condition, the solid forms a liquid in the absence of solvent, and since $X_B = 1$, the slope of the van't Hoff plot is equal to ($\Delta H_S/R$). The degree of ideality associated with a given solution may therefore be tested by evaluating the degree of linear correlation between $\ln(X_B)$ and $1/T$. Figure 1 shows the ideal behavior of naphthalene dissolved in benzene and xylene, which is due to the similar nature of the molecules involved, and the strength of intermolecular interactions such as polarity, polarizability, molecular volume, and hydrogen-bonding characteristics (Grant and Higuchi, 1990). On the other hand, the molecular properties of ethanol are very different from those of naphthalene. Thus one finds that for solutions of naphthalene in ethanol, $\ln(X_B)$ does not exhibit a linear dependence on $1/T$, which is taken as an indication of the non-ideal character of the solution.

Typically, one finds that the solubility that would be predicted assuming the model of an ideal solution is normally much higher than the solubility that is actually measured for a non-ideal solution.

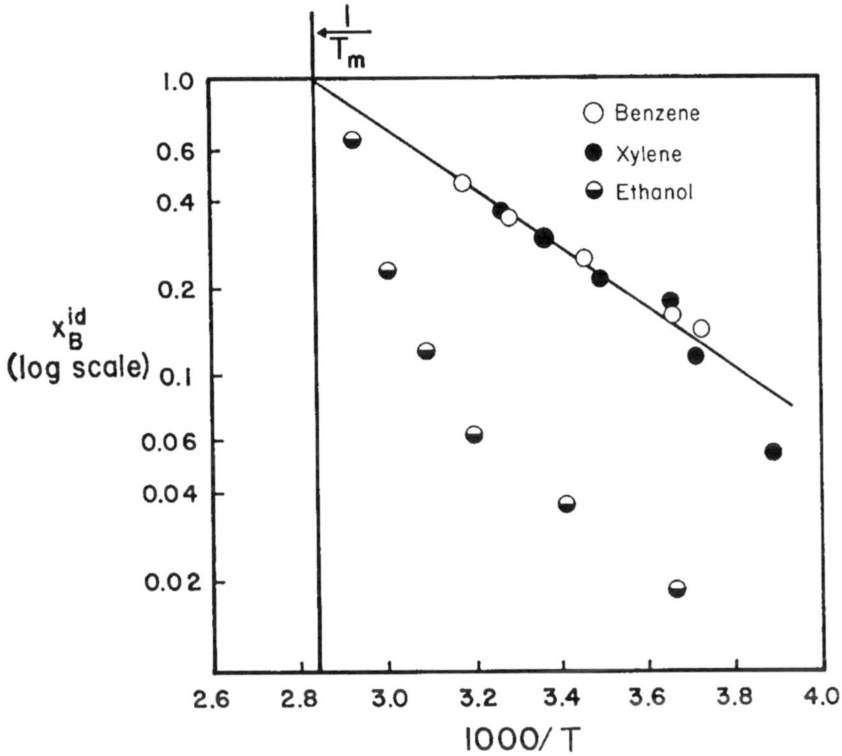

Figure 1. Van't Hoff plot of the molar solubility of naphthalene in benzene, xylene, and ethanol as a function of the reciprocal of the absolute temperature. The solid line corresponds to equation (17) for the ideal solubility of solid. Reproduced from DJW Grant, and T Higuchi, *Solubility Behavior of Organic Compounds*, John Wiley & Sons, New York, NY, 1990, p. 17.

Solubility in Regular Solutions

One rarely encounters ideal solutions in practice, and practically all solutions of pharmaceutical interest are non-ideal in character. For such non-ideal solutions, the activity coefficient (γ_B) of the solute does not equal one because the range of solute-solute, solvent-solvent, and solute-solvent interactions are significant. Therefore, one must consider the effect of the activity coefficient in order to predict the properties of non-ideal solutions:

$$X_B = \frac{a_B}{\gamma_B} \tag{28}$$

$$\ln X_B = \ln a_B - \ln \gamma_B \tag{29}$$

In equations (28) and (29), a_B is the activity of the dissolved solute and the undissolved solid, which may be evaluated using the hypothetical supercooled liquid as the standard state of unit activity. $\ln(a_B)$ may be expressed by equation

(17), as was the case for ideal solutions. Therefore:

$$\ln X_B = -\frac{\Delta H_s}{R}\left(\frac{1}{T} - \frac{1}{T_m}\right) - \ln \gamma_B \tag{30}$$

The value of the activity coefficient depends on many factors, and for non-ideal solutions the activity coefficient may be predicted from knowledge of the nature of the solute and the solvent.

For the sake of simplicity, the prediction of activity coefficients in regular solutions, the simplest non-ideal solution, will be discussed. For a regular solution, the energy of mixing and the enthalpy of mixing are not negligible because the intermolecular solute-solute, solvent-solvent, and solute-solvent interactions are different. However, the total volume is still assumed to be unchanged during mixing.

The activity coefficient in a regular solution can be estimated by considering the changes in intermolecular interaction energies that accompany the mixing of solute and solvent. For this purpose, the solution process may be divided into the three steps illustrated in Figure 2. The first step would consist of the removal of a solute molecule from its pure solute phase into the vapor phase, the second step would be the creation of a hole in the solvent for incorporation of the solute molecule, and the third step is the process where the free solute molecule fills the hole created in the solvent (Higuchi, 1949; Hildebrand and Scott, 1950; Martin, 1993).

To begin the analysis, the potential energy of solute-solute, solvent-solvent, and solute-solvent pairs is identified as w_{BB}, w_{AA}, and w_{AB}. In the first step, an energy equal to $2w_{BB}$ must be absorbed to break the solute-solute interaction between two adjacent solute molecules in the solid. After the solute molecule

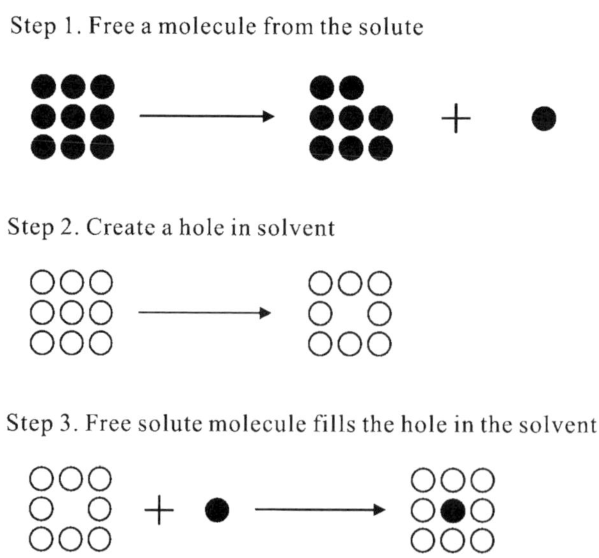

Figure 2. Hypothetical steps in solution process.

is removed to the vapor phase, the hole created in the solute closes, which releases an energy equal to w_{BB}, making the net energy change associated with liberation of a solute molecule equal to w_{BB}. In the second step, energy equal to w_{AA} is absorbed to separate a pair of solvent molecules, and to produce a hole in the solvent which the solute molecule may occupy. Finally, the solute molecule liberated from its solid phase is inserted in the hole in the solvent, forming two solute-solvent interactions and releasing an energy equal to $2w_{AB}$. The overall potential energy change, Δu, is therefore:

$$\Delta u = w_{AA} + w_{BB} - 2w_{AB} \tag{31}$$

Using this simplified model, Hildebrand and Wood (1933) proposed

$$\ln \gamma_B = (w_{AA} + w_{BB} - 2w_{AB}) \frac{V_B \Phi_A^2}{RT} \tag{32}$$

where V_B is the molar volume of the solute in the supercooled state, Φ_A is the volume fraction of the solvent in solution, R is the gas constant, and T is the absolute temperature of the solution.

The attractive interactions between pairs of solute and solvent molecules are assumed to be derived from van der Waals forces, so the solute-solvent interaction energy (w_{AB}) may be represented by the geometric mean of the solute-solute (w_{BB}) and the solvent-solvent (w_{AA}) interaction energies:

$$w_{AB} = \sqrt{w_{AA}\, w_{BB}} \tag{33}$$

Therefore, equation (32) becomes:

$$\ln \gamma_B = ((w_{AA})^{\frac{1}{2}} - (w_{BB})^{\frac{1}{2}})^2 \frac{V_B \Phi_A^2}{RT} \tag{34}$$

The square root of the interaction energy is defined as the solubility parameter, δ, and so equation (34) can be rewritten as:

$$\ln \gamma_B = (\delta_A - \delta_B)^2 \frac{V_B \Phi_A^2}{RT} \tag{35}$$

where δ_A and δ_B are the solubility parameters of the solvent and solute, respectively. In the case of a mixed solvent system, the total solubility parameter of the solvent mixture is given by:

$$\delta_A = \phi_1 \delta_1 + \phi_2 \delta_2 + \cdots \tag{36}$$

where δ_1 and δ_2 refer to the respective solvent parameters of pure solvents 1 and 2, and ϕ_1 and ϕ_2 are the respective volume fractions in the solvent mixture.

Introducing equation (35) into equation (30) yields the Hildebrand solubility equation describing regular solution behavior:

$$\ln X_B = -\frac{\Delta H_f}{R}\left(\frac{1}{T} - \frac{1}{T_m}\right) - (\delta_A - \delta_B)^2 \frac{V_B \Phi_A^2}{RT} \tag{37}$$

According to equation (37), if the difference between δ_A and δ_B is very small, then the second term approaches zero. The implication of this is that a regular solution would behave in an ideal manner when the solute and solvent have similar chemical properties. It may be seen that the Hildebrand solubility equation enables the prediction of solubility in regular solutions, as long as one has knowledge of the solubility parameters of both components in the solution.

Following the introduction of the Hildebrand model, the topic of solubility parameters has been extensively discussed (Hildebrand and Scott, 1962; Hildebrand et al., 1970; Kumar and Prausnitz, 1975; Barton, 1983), and values of δ can be found in these reference works. As a general rule, compounds having stronger London forces will be characterized by larger solubility parameters values.

Hildebrand and Scott (1950) proposed that the solubility parameters of similar molecules could be calculated using the enthalpy of vaporization (ΔH_v) and the molar volume of the liquid component (V_l) at the temperature of interest:

$$\delta = \left(\frac{\Delta H_v - RT}{V_l}\right)^{\frac{1}{2}} \tag{38}$$

Predictions of the solubility of non-polar solutes in non-polar solvents have been successfully achieved using the Hildebrand solubility equation (Davis et al., 1972). These solutions may be classified as regular solutions since the primary intermolecular interactions are London dispersion forces. However, the equation does not provide a good prediction of solubility for solutions involving polar components. When dipole-dipole, dipole-induced-dipole, charge transfer, and/or hydrogen-bonding interactions exist in the solution, $w_{AB} \neq \sqrt{w_{AA}w_{BB}}$, and with the presence of hydrogen bonding the entropy of mixing is no longer ideal. In addition, ΔV_{mix} will not equal zero if the dimensions of the solute and solvent molecules are very different.

Modifications to the Hildebrand solubility parameter model have been advanced in attempts to achieve better degrees of solubility prediction (Taft et al., 1969; Rohrschneider, 1973). Among these, the three-dimensional solubility parameter introduced by Hansen and Beerbower (1971) showed the most practical application. These workers calculated the total solubility parameter (δ_{total}) using three partial parameters, δ_D, δ_P, and δ_H:

$$\delta_{total}^2 = \delta_D^2 + \delta_P^2 + \delta_H^2 \tag{39}$$

where the parameters δ_D, δ_P, and δ_H account for dispersion, polar, and hydrogen-bonding interactions, respectively. Some of the values deduced for δ_D, δ_P, δ_H, and δ_{total} are listed in Table 2. Another modification of Hildebrand solubility parameter considered the effects of polar interaction and hydrogen bonding, and was found to yield good solubility predictions in many cases (Kumar and Prausnitz, 1975). However, the modified Hildebrand solubility equation can only be used empirically in predicting solubility in polar solvents, since the original assumptions associated with regular solutions do not apply in polar solvents (Grant and Higuchi, 1990).

In solvent systems where polar interactions exert a major role, the molecular and group-surface-area (MGSA) approach provides a better quality solubility prediction (Yalkowsky et al., 1972, 1976; Amidon et al., 1974, 1975). Instead of

Solvents	Solubility parameter $(cal/cm^3)^{1/2}$			
	δ_D	δ_P	δ_H	δ_{total}
n-Butane	6.9	0	0	6.9
n-Hexane	7.3	0	0	7.3
n-Octane	7.6	0	0	7.6
Diethyl ether	7.1	1.4	2.5	7.7
Cyclohexane	8.2	0	0.1	8.2
n-Butyl acetate	7.7	1.8	3.1	8.5
Carbon tetrachloride	8.7	0	0.3	8.7
Toluene	8.8	0.7	1.0	8.9
Ethyl acetate	7.7	2.6	3.5	8.9
Benzene	9.0	0	1.0	9.1
Chloroform	8.7	1.5	2.8	9.3
Acetone	7.6	5.1	3.4	9.8
Acetaldehyde	7.2	3.9	5.5	9.9
Carbon disulfide	10.0	0	0.3	10.0
Dioxane	9.3	0.9	3.6	10.0
1-Octanol	8.3	1.6	5.8	10.3
Nitrobenzene	9.8	4.2	2.0	10.9
1-Butanol	7.8	2.8	7.7	11.3
1-Propanol	7.8	3.3	8.5	12.0
Dimethylformamide	8.5	6.7	5.5	12.1
Ethanol	7.7	4.3	9.5	13.0
Dimethyl sulfoxide	9.0	8.0	5.0	13.0
Methanol	7.4	6.0	10.9	14.5
Propylene glycol	8.2	4.6	11.4	14.8
Ethylene glycol	8.3	5.4	12.7	16.1
Glycerin	8.5	5.9	14.3	17.7
Formamide	8.4	12.8	9.3	17.9
Water	7.6	7.8	20.7	23.4

Table 2. Solubility parameters for some common solvents.

Reproduced from:

Hansen C, and Beerbower A. Solubility Parameters. In: Standen A. *Kirk-Othmer Encyclopedia of Chemical Technology, 2nd ed. Supplement Volume.* New York, NY: Wiley; 1971. 889–910.

the potential energy term that was used in equation (32), a free energy model was used in the MGSA approach to represent the change of the interactions at mixing. The power of this approach is that changes in enthalpy and entropy are included:

$$\ln \gamma_B = (W_{AA} + W_{BB} - 2W_{AB})\frac{V_B\phi_A^2}{RT} \tag{40}$$

In equation (40), W is reversible work which represents the internal free energy. Yalkowsky et al. (1976) used the molar surface area (A) and the surface tension (σ) to replace the molar volume (V) and reversible work. Under those circumstances, equation (40) becomes:

$$\ln \gamma_B = \frac{\sigma_{AB} A_B}{kT} \tag{41}$$

where σ_A and σ_B are the surface energies of the pure liquids A and B, while σ_{AB} is the interfacial energy between the two liquids. The interfacial tension can be experimentally measured for substances of different polarity, and therefore equation (41) better predicts solubility in polar solvents.

Intermolecular Interactions in Non-Ideal Solutions

Prediction of solubility using the regular solution theory usually fails when the solute and solvent are polar in character. The dipole-dipole, dipole-induced-dipole, charge-transfer, and hydrogen bonding interactions that exist between solute and solvent molecules may reduce the free energy of the solution, and increase the solubility. In these solutions, the activity coefficient may be less than one, a fact that cannot be explained using regular solution theory. The range of dipole-dipole, dipole-induced-dipole, and hydrogen bonding interactions in polar solutions may also lead to molecular orientation, which would tend to decrease the entropy of mixing. Clearly the nature of the forces involved in solution, and the influence of the forces on solubility, are important in order to arrive at a better understanding of solubility behavior.

Coulombic interaction is a valence force between counterions, and in extreme situations a cation-anion pair might form a strong ion-dipole interaction in solution. Such interactions would tend to be major for ionic substances dissolved in non-polar solvent systems, but less so in polar solvents where the forces of solvation serve to disrupt ion pairs into individual solvated ions. These trends provide an insight into why salts tend to be soluble in polar solvents, but not in non-polar solvents.

Van der Waals forces represent important intermolecular interactions between nonelectrolyte substances, and can be categorized into dipole-dipole, dipole-induced-dipole, and induced-dipole-induced-dipole forces. Polar molecules, by definition, will have a permanent dipole moment, and will interact with the oppositely charged portions or other molecules having permanent dipole moments. The dipole-dipole interaction is known as the orientation effect, or as the Keesom force.

Molecules having delocalized electron systems or large molar volumes often are characterized by high degrees of polarizability. Their interaction with polar molecules can induce shifts in electron density that result in the transient presence of induced dipole moments, and the charged portions of the induced dipoles can form an attractive interaction with the neighboring polar molecule. This type of interaction is termed the dipole-induced-dipole force, the induction effect, or the Debye force. It is found that Keesom and Debye forces provide efficient molecular packing in crystals, accounting for the high stability, low thermodynamic activity, and the high melting point of many organic crystals. These attractive effects may yield substantial lattice energies for such crystals, and therefore tend to reduce their solubility in potential solvents.

All molecules, whether polar or nonpolar, are also attracted to each other by induced-dipole-induced-dipole interactions, which are known as dispersion forces, or London forces.

Nonpolar molecules can only interact by dispersion forces, while the interactions of polar molecules are often dominated by the Keesom forces. However, under certain circumstances it is still possible that dispersion forces might predominate over the other forces, even for polar molecules such as HCl. The Debye forces are often stronger than the London forces for highly polar molecules, and would predominate over Keesom forces for weakly polar molecules. Debye forces are selective, and important in explaining why certain nonpolar but polarizible molecules can still be soluble in polar solvents (Krishnan and Fredman, 1971).

Hydrogen atoms are small in size, and would be positively polarized in molecules where it is bound adjacent to an electronegative atom, A. Should another strongly electronegative atom, B, approach the hydrogen atom at a short distance, a strong interaction may develop that is termed a hydrogen bond. The strongest hydrogen bonds are formed when the electronegative atoms involved are fluorine, oxygen, or nitrogen, although chlorine and sulfur are known to form weak hydrogen bonds in some molecules.

The strengths of hydrogen bonds are similar in magnitude to those of van der Waals forces, but is also directional in the manner of a covalent bond. Hydrogen bonding tends to stabilize molecular pairs and reduces the enthalpy, but also tends to orient the molecules involved and decrease the entropy. The effect of hydrogen bonding on solubility is complicated, and the analysis must proceed on a case-by-case basis. Extensive intermolecular hydrogen bonding in a crystal would tend to decrease the free energy, with this stabilization effect reducing the activity of the solute, and tending to reduce the solubility. However, the hydrogen bonds formed between solute and solvent molecules would tend to reduce the activity coefficient, and this effect would lead to increased solubility.

Influence of Temperature on Solubility

For ideal solutions, the van't Hoff relation of equation (17), and the Hildebrand relation of equation (24), state that the $\ln(X_B)$ term is linearly dependent on $1/T$ and on $\ln(T)$. The enthalpy of solution is equal to the enthalpy of melting (*i.e.*, $\Delta H_s = \Delta H_f^T$), since the enthalpy of mixing is zero for an ideal solutions. Since ΔH_s for ideal solutions is always endothermic and positive, the solubility of an ideal solution would increase with increasing temperature.

In non-ideal solutions, however, the enthalpy of solution does not equal the enthalpy of melting because the enthalpy of mixing does not equal zero. Moreover, because the heat capacity of the solid is different from the heat capacity of the supercooled liquid, the ΔC_p term does not equal zero, and:

$$\Delta H_s = \Delta H_f^m - \Delta C_p(T_m - T) + \Delta H_{mix} \tag{42}$$

The strong solute-solvent interactions in solution may significantly reduce the free energy of the final solution compared to that of the pure solute and solvent. Despite the positive entropy of mixing, the enthalpy of mixing term may be negative, especially when the molecules in solution are oriented by the strong polar-polar, polar-induced-polar, and/or hydrogen-bonding interactions. Moreover, the second term in equation (42) may yield a negative contribution to the total enthalpy of solution. Therefore, the dissolution of a solute in a non-ideal solution might turn out to be an exothermic process, characterized by a negative ΔH_s. For those systems where ΔH_s is negative, it follows that the solubility would decrease with increasing temperature. The dissolution of carbon dioxide in water is characterized by a negative enthalpy of solution, and therefore carbonated waters go flat when their temperature is raised.

Grant et al. (1984) proposed an equation that better represents the temperature dependence of the molar solubility of polar organic compounds in water:

$$\ln X_B = -\frac{a}{R}\frac{1}{T} + \frac{b}{R}\ln T + c \tag{43}$$

In equation (43), *a*, *b*, and *c* are adjustable parameters, and this equation enables one to simulate the solubility of most solute-solvent combinations over a wide temperature range.

Solubility of Substances in Various Solid-State Forms

Many pharmaceutical solids are capable of existing in several different solid-state forms, such as polymorphs, solvatomorphs, and amorphous form (Brittain, 1999; Bernstein, 2002). *Polymorphism* is defined as the ability of a substance to exist in two or more crystalline phases that differ in the arrangement and/or conformation of the molecules in the crystal structure with the empirical formula of a polymorphic pair being identical. Polymorphism can arise from a different packing arrangement of molecules having the same conformation, or from the alternate assembly of different conformational states of the same molecule. *Solvatomorphism (pseudopolymorphism)* is defined as the ability of a substance to exist in two or more crystalline phases that differ in their empirical formulae with solvatomorphs being characterized by the presence of water (i.e., hydrates) or other solvent molecules (i.e., solvates) in the crystal structure. An *amorphous* solid is characterized by a disordered arrangement of molecules, where intermolecular forces impose short range order and where there is no long range order in the solid.

The different internal energies of these structural types are manifested in different magnitudes of lattice energy, and hence lead to the existence of different solubilities for the various forms. The solubility difference may be understood using the solution models that have been developed in the previous sections. For dissolution to take place, the solute-solvent attractive forces must be stronger than the solute-solute and solvent-solvent attractive interactions so that the latter may be overcome by the former. As always, the free energy change associated with the process determines the ultimate equilibrium solubility of the solute in the solution. Details of the internal structure of the various solid-state forms will determine the respective enthalpies of solution, and the differing enthalpies of solution associated with the various different solid-state forms will lead to the existence of differing solubilities. These phenomena will be considered using the basic thermodynamic theory.

Solubility of Polymorphic Substances

The attraction force between two neighboring molecules of a solute is determined by the interactions existing in the crystal structure. Consequently, the internal energy (U) of a particular polymorph is equal to the sum of the individual energies of interaction between each pair of neighboring molecules as these are dictated by the details of the crystal structure. At constant pressure (P), the enthalpy (H) of a polymorph is defined as:

$$H = U + P \times V \tag{44}$$

where V is the volume of the crystal. The stability of the polymorph is determined by its free energy (G):

$$G = H - T \times S \tag{45}$$

where S is the entropy of the polymorph. The polymorph with the lowest free energy is termed the thermodynamically stable form, and the polymorphs having higher free energies are termed the metastable forms. Following accepted nomenclature, Form-I will be identified as the stable crystal form, and Form-II will be identified as the metastable form.

The solubility of the most stable crystal form in a polymorphic system is termed the *equilibrium solubility*. While the measurement of equilibrium solubility at a given temperature is a routine practice in pharmaceutical research (Grant and Brittain, 1995), evaluation of the solubility of a metastable polymorph is frequently more complicated owing to the tendency of metastable forms to undergo a phase transformation to the more stable polymorph in the medium of measurement. It is therefore prudent to include a determination of the phase at the completion of any solubility measurement to verify exactly which polymorphic form has been the subject of the measurement.

Several indirect methods have been proposed to determine the solubility of metastable polymorphs. Milosovish (1964) deduced the relative solubilities of metastable and stable polymorphs based on the measurement of intrinsic dissolution rates. Ghosh and Grant (1995) proposed an extrapolation technique to determine the solubility of a crystalline solid that undergoes a phase change upon contact with a solvent medium. Brittain (1996) used the time evolution of light scattering from aqueous suspensions of anhydrous theophylline as a means to evaluated its solubility, and also to study its phase transformation into its monohydrate solvatomorph.

In many systems, measurement of the solubility of a metastable form can be directly obtained if there is an energy barrier between the metastable polymorph and the stable polymorph that prevents interconversion during the lifetime of the measurement. If the free energy difference of the polymorphs, which is the driving force of the phase transformation, does not overcome the activation energy barrier, the metastable polymorph may stay unchanged for a sufficiently long period of time to permit a direct determination of solubility to be made.

Solubility of Solvatomorphic Substances

Solvatomorphs are formed when solvent molecules become incorporated into a crystalline solid, and occupy regular positions in the crystal lattice. In other cases, the crystal structure may contain channels having repetitive sites of hydrophilicity or hydrophobicity, and solvent molecules can become attached to those sites. Hydrates are those solvatomorphs where water molecules constitute an integral part of the crystal structure, and are typically contained in a defined ratio. Hydrates will be specifically discussed since those solvatomorphs are often of highest interest for pharmaceutical applications, but the results of the discussion apply equally well to solvatomorphs containing solvent molecules other than water.

In the presence of water, hydrated and anhydrous crystals can be considered as being in equilibrium:

$$\mathrm{A(solid)} + \mathrm{mH_2O} \overset{K_h}{\longleftrightarrow} \mathrm{A \cdot mH_2O(solid)} \tag{46}$$

where A(solid) and $\mathrm{A \cdot mH_2O(solid)}$ refer to the anhydrous and hydrated phase, respectively, **m** is the stoichiometry of the hydrate, and K_h is the equilibrium constant of hydration:

$$K_h = \frac{a[\mathrm{A \cdot mH_2O(solid)}]}{a[\mathrm{A(solid)}]\,a[\mathrm{H_2O}]^{\mathrm{m}}} \tag{47}$$

Equation (47) indicates that the activity ratio of the hydrated and anhydrous crystals depend on the activity of water. When $a[\mathrm{H_2O}]^{\mathrm{m}}$ is greater than $\{a[\mathrm{A \cdot mH_2O(solid)}]/K_h \cdot a[\mathrm{A(solid)}]\}^{1/\mathrm{m}}$, the hydrated form is more stable than the anhydrous form (Zhu and Grant, 1996). Obviously when the value of

$\{a[\text{A} \cdot \text{mH}_2\text{O(solid)}]/K_h \cdot a[\text{A(solid)}]\}^{1/m}$, exceeds that of $a[\text{H}_2\text{O}]^m$, the anhydrate form would be more stable. The addition of a miscible cosolvent would reduce the water activity, and would move the position of equilibrium toward that of the anhydrous form.

As a rule of thumb, hydrated crystalline forms are usually less soluble in water than are the corresponding anhydrate crystalline forms (Grant, 1990), and thus solid solvates are usually less soluble in the solvating solvent than the original solid. However, the solubility of a solvate in a solvent that is miscible with the solvating solvent is higher than the corresponding non-solvated form. This phenomenon arises because the negative energy change of mixing associated with the solvents provides an additional contribution to the negative free energy of solution.

When $a[\text{H}_2\text{O}]$ equals zero, then $a[\text{A} \cdot \text{mH}_2\text{O(solid)}]/a[\text{A(solid)}]$ also equals zero. The consequence of this is that, thermodynamically speaking, the hydrated form is only stable in the presence of water. For this reason, the solubility of a hydrate crystal form can only be measured in water, as the solubility of a solvate can only be measured in the solvating liquid corresponding to the included solvate molecule. Similar to the metastable polymorphs, however, a solvate may be temporarily stable in absence of the solvating liquid due to a high energy barrier of desolvation.

Solubility of the Amorphous Form

As described above, amorphous solids are disordered in nature, and contain only short range order between the constituent molecules. Amorphous solids lack the stabilizing influence of lattice energy, and therefore are thermodynamically less stable than any of the corresponding crystalline forms of the substance. Since the amorphous form represents the most highly energetic solid state form of a material (Hancock and Zografi, 1996), it follows that amorphous materials exhibit the highest degree of solubility for a given substance.

In some instances, the relative solubilities of the amorphous and crystalline forms of a substance can be estimated using the same methodologies as would be used in the measurement of the solubility of polymorphic materials. Using a theoretical approach, Hancock and Parks (2000) proposed that the solubility advantage of the amorphous drug to its most stable crystalline form was about 16-fold to 1600-fold. The maximum concentration measured during the course of dissolution of the amorphous form was taken to represent the solubility of the amorphous form. However, the empirical data were less than that predicted, suggesting that the amorphous substances partially converted to a crystalline form during the lifetime of the solubility measurement. It is probably true that amorphous materials cannot achieve their maximum theoretical solubility under practical experimental conditions owing to phase transformations.

Sato et al. (1981) measured the solubility of amorphous substance by adding a nucleation inhibitor, but the measured solubility could have been affected by the presence of the nucleation inhibitor.

Dissolution Phenomena: Kinetics of Solubility

Systemic absorption of a drug substance from a particulate form takes place after the compound enters the dissolved state. If the dissolution rate of the substance is less than the diffusion rate to the site of absorption and the absorption rate itself, then the dissolution process will be the rate-determining step. This situation is characteristic of drug substances that have low degrees of aqueous solubility, and therefore low dissolution rates, and it has become an established tenet in pharmaceutics that one method to improve the dissolution rate of a relatively insoluble substance is to reduce the particle size of its component particles. As discussed above, the solubilities of polymorphs, solvatomorphs, and amorphous forms are different, and these differences may lead to differences in the dissolution rate, which in turn could lead to differences in bioavailability.

The mechanism of dissolution was proposed by Nernst (1904) using a film-model theory. Under the influence of non-reactive chemical forces, a solid particle immersed in a liquid experiences two consecutive processes. The first of these is solvation of the solid at the solid-liquid interface, which causes the formation of a thin stagnant layer of saturated solution around the particle. The second step in the dissolution process consists of diffusion of dissolved molecules from this boundary layer into the bulk fluid. In principle, one may control the dissolution through manipulation of the saturated solution at the surface. For example, one might generate a thin layer of saturated solution at the solid surface by a surface reaction with a high energy barrier (Mooney et al., 1981), but this application is not commonly employed in pharmaceutical applications.

In the majority of dissolution phenomena, the solvation step is almost instantaneous. The diffusion process is much slower and, therefore constitutes the rate limiting step. Noyes and Whitney (1897) developed an equation based on Fick's second law of diffusion to describe dissolution within the scope of their model, and report the relation:

$$\frac{dC}{dt} = \frac{DS}{h}(C_s - C) \tag{48}$$

where dC/dt is the rate of drug dissolution at time t, D is the diffusion coefficient, S is the surface area of the particle, h is the thickness of the stagnant layer, C_s is the concentration of the drug in the stagnant layer (usually taken as the equilibrium solubility), and C is the concentration of the drug in the bulk solvent. According to the Stock-Einstein equation for the small particles, the diffusion coefficient, D, is related to the viscosity of the liquid medium:

$$D = \frac{kT}{6\pi\eta r} \tag{49}$$

where k is the Boltzmann constant, T is the temperature, η is the viscosity of the solvent, and r is the radius of the particle.

According to the Noyes-Whitney equation (48), the dissolution rate of a drug substance is directly proportional to its equilibrium solubility. However, the nature of the dissolving solid and the dissolution medium also exert strong influences on the dissolution rate. For example, metastable polymorphs will exhibit faster dissolution rates than would the thermodynamically stable polymorph, and amorphous materials will dissolve faster than any corresponding crystalline forms. Temperature may affect both the solubility and the diffusion coefficient, and in many cases the dissolution rate will increase with increasing temperature. Consequently, as was the case for solubility determinations, evaluation of drug dissolution must be conducted at a fixed and reported temperature.

The effect of particle size and dissolution rate has been known since the pioneering work of Noyes and Whitney (1897), and Hixson and Crowell (1931) subsequently derived a highly useful equation that expresses the rate of dissolution based on the cube root of the weight of the particles. When the Hixson-Crowell model is applied to micronized particles, for which the thickness of the aqueous diffusion layer around the dissolving particles is comparable to or larger than the radius of the particle, the change in particle radius with time is given by:

$$r^2 = r_0^2 - \frac{2DC_s t}{\rho} \tag{50}$$

where r_0 is the initial radius of the particle, r is the radius of the particle at time equal to t, D is the diffusion coefficient of the molecules dissolving from the particle, C_S is the equilibrium solubility of the substance, and ρ is the density of the solution.

A very useful relation is obtained for the time, T, which would be required to achieve complete dissolution of the particle, or the condition where $r^2 = 0$:

$$\mathrm{T} = \frac{\rho r_0^2}{2DC_s} \tag{51}$$

For most aqueous solutions, D is typically equal to 5×10^{-6} cm^2/sec, ρ is approximately equal to 1.0 g/mL, so the calculation of equation (51) can be performed if the equilibrium solubility of particles having a known initial particle size is known. Consider a substance whose equilibrium solubility is 1.0 mg/mL. For a particle whose initial diameter equals 10 μm, the time to achieve complete dissolution would be predicted to be 25 seconds (0.42 minutes). For the same substance, if the initial diameter instead equaled 50 μm, then the time to achieve complete dissolution would be predicted to be 625 seconds (10.4 minutes). For 100 μm particles of this substance, the time to achieve complete dissolution is calculated to be 2500 seconds (41.7 minutes). The relationship between particle size and the time required to completely dissolve particles of various sizes as defined in equation (51) has been illustrated in Figure 3.

This effect of particle size on dissolution rate of sparingly soluble drug substances has been demonstrated in many instances by the superior dissolution rates observed after size reduction. Examples of compounds studied in such work include methylprednisolone (Higuchi et al., 1963), 1-isopropyl-7-methyl-4-phenylquinazolin-2(1H)-one (Kornblum and Hirschorn, 1970), griseofulvin (Ullah and Cadawader, 1971), monophenylbutazone (Habib and Attia, 1985), nitrofurantoin (Eyjolfsson, 1999), and piroxicam (Swanepoel et al., 2000).

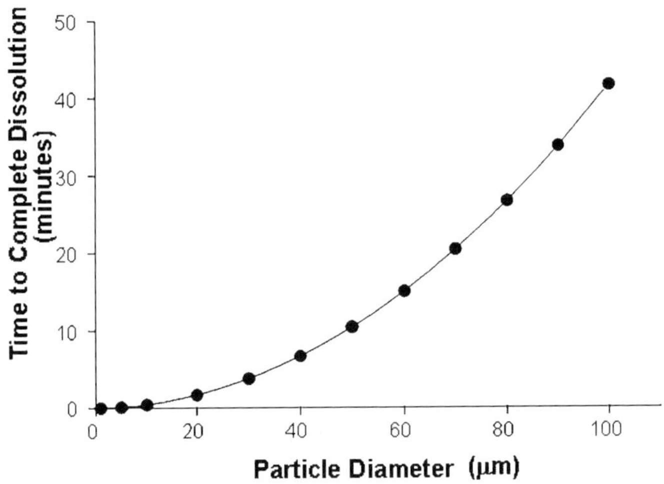

Figure 3. Relationship between particle size and the time required to completely dissolve particles of a given size.

Since the dissolution rate of a loosely suspended substance will depend on the particle size and surface area of the solid, the technique of intrinsic dissolution has been developed. In this method, the solid of interest is compressed into a die and embedded in a rotational disc where only one face of the compressed solid remains exposed to the dissolution medium. Under these circumstances, the area of the solid-liquid interface must remain constant during the dissolution process.

The dissolution rate (dm/dt) of a given solid is usually directly proportional to the wetted surface area (A) of the dissolving solid:

$$dC/dt = J \times A \tag{52}$$

where J is the mass flux, or the dissolution rate per unit surface area. J is usually termed the intrinsic dissolution rate. But since dC/dt is also defined according to the Noyes-Whitney equation (48), it follows that:

$$J = B \times (C_s - C) \tag{53}$$

where B is the mass transfer coefficient, defined as:

$$B = \frac{DS}{Ah} \tag{54}$$

At the earliest stage of an intrinsic dissolution study, $C << C_s$, so:

$$J_{t \to 0} = BC_s \tag{55}$$

It can therefore be concluded that if the surface area of the dissolving solid is kept constant, the intrinsic dissolution rate will be directly proportional to the equilibrium solubility.

Levich (1962) deduced a mathematic equation for the intrinsic dissolution rate:

$$J = 0.620 D^{\frac{2}{3}} \nu^{-\frac{1}{6}} (C_s - C) \omega^{\frac{1}{2}} \quad (56)$$

where ν is the kinematic viscosity of the dissolution medium, ω is the angular velocity of the disk in radian per second. Using this equation, one can calculate the diffusion coefficient of a dissolving solute.

Summary

Solubility phenomenon is strongly related to the intermolecular forces between solute-solute, solvent-solvent, and solute-solvent. Based on the classic thermodynamics of solubility, many models have been developed to predict solubility when a direct measurement is impossible. On the other hand, solubility, along with its related properties, frequently provides a way to look into the structure of a drug substance and interaction between the drug substances at the molecular level. Solubility phenomenon is especially important in pharmaceutical industry. Relative solubility of various forms of a drug substance indicates their relative stability. The bioavailability of a drug substance is essentially determined by the dynamics and kinetics of solubility of the drug substance. A better understanding of the solubility and its related properties is critical in the development process of a drug candidate.

List of Abbreviations

DSC............differential scanning calorimetry
MSGA............molecular and group surface area approach

List of Symbols

A............surface area
a............thermodynamic activity
B............mass transfer coefficient
Cp............heat capacity
C............concentration
D............diffusion coefficient
G............Gibbs free energy
H............enthalpy
J............intrinsic dissolution rate
K............equilibrium constant
k............Boltzmann constant, 1.3806×10^{-23} J/K
M............molarity
N............normality
R............gas constant, 8.3143 J/K · mol or 1.987 cal/K · mol
r............radius

S .... entropy
T .... absolute temperature
U .... molar internal energy
V .... molar volume
X .... mole fraction
δ .... solubility parameter
η .... dynamic viscosity
γ .... activity coefficient
μ .... chemical potential
υ .... kinematic viscocity
ω .... rate of angular rotation
Φ .... volume fraction
ρ .... density

References

Amidon GL, Yalkowsky SH, Anik ST, and Valvani SC. Solubility of Nonelectrolytes in Polar Solvents V: Estimation of the Solubility of Aliphatic Monofunctional compounds in Water using a Molecular Surface Area Approach. *J Phys Chem* 1975; 79: 2239–2246.

Amidon GL, Yalkowsky SH, and Leung S. Solubility of Nonelectrolytes in Polar Solvents II: Solubility of Aliphatic Alcohols in Water. *J Pharm Sci* 1974; 63: 1858–1866.

Barton AFM. *CRC Handbook of Solubility Parameters and Other Cohesion Parameters.* CRC Press, Boca Raton, FL, 1983, p. 594.

Bernstein J. *Polymorphism in Molecular Crystals.* Oxford University Press, New York, 2002.

Brittain HG. A Method for the Determination of Solubility of Metastable Crystal Phases Based on Total Light Scattering. *Langmuir* 1996; 12: 601–604.

Brittain HG. *Polymorphism of Pharmaceutical Solids.* Marcel Dekker, New York, 1999.

Brittain HG and Grant DJW. Effects of Polymorphism and Solid-State Solvation on Solubility and Dissolution Rate. In: Brittain HG (ed.), *Polymorphism of Pharmaceutical Solids.* Marcel Dekker, New York, 1999, pp. 279–330.

Davis SS, Higuchi T, and Rytting JH. Determination of Thermodynamics of Functional Groups in Solutions of Drug Molecules. In: Bean HS, Beckett AH, and Carless JE (eds.), *Advances in Pharmaceutical Sciences, Vol. 4.* Academic Press, New York, 1974, pp. 73–261.

Eyjolfsson R. Nitrofurantoin: Particle Size and Dissolution. *Drug Dev Ind Pharm* 1999; 25: 105–106.

Ghosh S and Grant DJW. Determination of the Solubilities of Crystalline Solids in Solvent Media that Induce Phase Changes: Solubilities of 1,2-dialkyl-3-hydroxy-4-pyridones and Their Formic Acid Solvates in Formic Acid and Water. *Int J Pharm* 1995; 114: 185–196.

Grant DJW and Brittain HG. Solubility of Pharmaceutical Solids. In: Brittain HG (ed.), *Physical Characterization of Pharmaceutical Solids.* Marcel Dekker, New York, 1995, pp. 321–386.

Grant DJW and Higuchi T. *Solubility Behavior of Organic Compounds.* John Wiley & Sons, New York, 1990, p. 656.

Grant DJW, Mehdizadeh M, Chow AHL, and Fairbrother JE. Non-linear Van't-Hoff Solubility-Temperature Plots and their Pharmaceutical Interpretation. *Int J Pharm* 1984; 18: 25–38.

Gu CH, Young V Jr, and Grant DJW. Polymorph Screening: Influence of Solvents on the Rate of Solvent-Mediated Polymorphic Transformation. *J Pharm Sci* 2001; 90: 1878–1890.

Habib FS and Attia MA. Effect of Particle Size on the Dissolution Rate of Monophenylbutazone Solid Dispersions in the Presence of Certain Additives. *Drug Dev Ind Pharm* 1985; 11: 2009–2019.

Hancock BC and Parks M. What is the True Solubility Advantage for Amorphous Pharmaceuticals. *Pharm Res* 2000; 17: 397–404.

Hancock BC and Zografi G. Characteristics and Significance of the Amorphous State in Pharmaceutical Systems. *J Pharm Sci* 1996; 86: 1–12.

Hansen C and Beerbower A. Solubility Parameters. In: Standen A (ed.), *Kirk-Othmer Encyclopedia of Chemical Technology, 2nd ed. Supplement Volume,* John Wiley & Sons, New York, 1971, pp. 889–910.

Higuchi T. Solubility. In: Lyman R (ed.), *Pharmaceutical Compounding and Dispensing,* Lippincott Philadelphia, PA, 1949, pp. 176–177.

Higuchi WI, Rowe EL, and Hiestand EN. Dissolution Rates of Finely Divided Drug Powders. II: Micronized Methylprednisolone. *J Pharm Sci* 1963; 52: 162–164.

Hilderbrand JH, Prausnitz JM, and Scott RL. *Regular and Related Solutions.* Van Nostrand Reinhold, New York, 1970, pp. 64–67.

Hilderbrand JH and Scott RL. *Solubility of Nonelectrolytes, 3rd edn.* Reinhold Pub., New York, 1950, p. 488.

Hilderbrand JH and Scott RL. *Regular Solutions.* Prentice-Hall, Englewood Cliffs, NJ, 1962, p. 180.

Hildebrand JH and Wood SE. Derivation of Equations for Regular Solutions. *J Chem Phys* 1933; 1: 817–822.

Hixson AW and Crowell JH. Dependence of Reaction Velocity upon Surface and Agitation. I: Theoretical Consideration. *Ind Eng Chem* 1931; 23: 923–931.

Kornblum SS and Hirschorn JO. Dissolution of Poorly Water-Soluble Drugs. I: Some Physical Parameters related to Method of Micronization and Tablet Manufacture of a Quinazoline Compound. *J Pharm Sci* 1970; 59: 606–609.

Krishnan CV and Friedman HL. Solvation Enthalpies of Hydrocarbons and Normal Alcohols in Highly Polar Solvents. *J Phys Chem* 1971; 75: 3598–3606.

Kumar R and Prausnitz JM. Solvents in Chemical Technology. In: Dack MRJ. *Solutions and Solubilities, Part 1, Techniques of Chemistry, Vol. VIII,* John Wiley & Sons, New York, 1975, pp. 259–326.

Levich VG. *Physicochemical Hydrodynamics.* Advance Publications, London, UK, 1977, p. 1078.

Martin A. *Physical Pharmacy, 4th edn.* Lippincott Williams & Wilkins, Philadelphia, PA, 1993, pp. 223–224.

Milosovish G. Determination of Solubility of a Metastable Polymorph. *J Pharm Sci* 1964; 53:484–487.

Mooney KG, Rodrigues-Gaxiola M, Mintun M, Himmelstein KJ, and Stella VJ. Dissolution Kinetics of Phenylbutazone. *J Pharm Sci* 1981; 70: 1358–1365.

Nernst W. Theorie der Reaktionsgeschhwindigkeit in Heterogenen Systemen. *Z Phys Chem* 1904; 47: 52–55.

Noyes AA and Whitney WR. The Rate of Dissolution of Solid Substances in Their Own Solutions. *J Am Chem Soc* 1897; 19: 930–934.

Rohrschneider L. Solvent Characterization by Gas-liquid Partition Coefficients of Selected Solutes. *Anal Chem* 1973; 45:1241–1247.

Sato T, Okada A, Sekiguchi K, and Tsuda Y. Difference in Physico-pharmaceutical Properties Between Crystalline and Non-crystalline 9,3"-diacetylmidecamycin. *Chem Pharm Bull* 1981; 29: 2675–2682.

Swanepoel E, Liebenberg W, de Villiers MM, and Dekker TG. Dissolution Properties of Piroxicam Powders and Capsules as a Function of Particle Size and the Agglomeration of Powders. *Drug Dev Ind Pharm* 2000; 26: 1067–1076.

Taft RW, Gurka D, Joris L, Schleyer P, von R, and Rakshys JW. Studies of Hydrogen-bonded complex formation with p-fluorophenol. V. Linear Free Energy Relationships with OH Reference Acids. *J Am Chem Soc* 1969; 91: 4801–4808.

Ullah I and Cadwallader DE. Dissolution of Slightly Soluble Powders under Sink Conditions. II: Griseofulvin Powders. *J Pharm Sci* 1971; 60: 230–233.

United States Pharmacopeia, 25th edition. United States Pharmacopeial Convention, Rockville, MD, 2002, p. 2363.

United States Pharmacopeia, 29th edition. United States Pharmacopeial Convention, Rockville, MD, 2006, p. 9.

Walden P. Über die Schmelzwärme, Spezifische Kohäsion und Molekulargrösse bei der Schmelztemperatur. *Z Eletrochem* 1908; 14: 713–724.

Yalkowsky SH. Estimation of Entropies of Fusion of Organic Compounds. *Ind Eng Chem Fundam* 1979; 18: 108–111.

Yalkowsky SH, Flynn GL, and Amidon GL. Solubility of nonelectrolytes in Polar Solvents. *J Pharm Sci* 1972; 61: 983–984.

Yalkowsky SH, Valvani SC, and Amidon GL. Solubility of Nonelectrolytes in Polar Solvents IV: Nonpolar Drugs in Mixed Solvents. *J Pharm Sci* 1976; 65: 1480–1494.

Zhu H and Grant DJW. Influence of Water Activity in Organic Solvent + Water Mixtures on the Nature of the Crystallizing Drug Phase. 2. Ampicillin. *Int J Pharm* 1996; 139: 33–43.

2

Ionic Equilibria and the pH Dependence of Solubility

HARRY G. BRITTAIN

Center for Pharmaceutical Physics, Milford, NJ

Introduction

Many drug substances can be classified as being either acids or bases in that they possess the ability to react with other acids or bases that are stronger than themselves. As such, they would also possess the ability to exist as ionic species under certain conditions. The state of ionization of a substance will often profoundly affect its degree of aqueous solubility, as evidenced by the high solubility of sodium benzoate as opposed to the low solubility of benzoic acid. The utility of salt forms as active pharmaceutical ingredients is well known, and represents one of the means to increase the degree of solubility of an otherwise intractable substance (Berge et al. 1977; Stahl and Wermuth 2002).

To develop an understanding as to how the relative acidity or basicity of a drug substance is able to influence its aqueous solubility, the principles of chemical equilibria as applied to weakly ionic substances will first be developed. Subsequent to this, the influence on solubility of the degree of ionization of substances will be considered. However, certain fundamental concepts must be developed first in order for the subsequent sections to be most meaningful.

Activity, Activity Coefficients, and the Equilibrium Constant

The difference in the free energy of a substance (G_i) in a non-standard state relative to its free energy in a defined standard state (G_i^0) is derived from its activity (a_i):

$$G_i = G_i^0 + RT \ln a_i \tag{1}$$

The activity of the substance is related to its concentration through the relation:

$$a_i = \gamma_i\,[C_i] \tag{2}$$

where γ_i is the activity coefficient of the substance in its non-standard state, and $[C_i]$ is its concentration.

If we consider the general reaction:

$$aA + bB \leftrightarrow cC + dD \tag{3}$$

then the thermodynamic equilibrium constant would be given by:

$$K = \frac{a_C{}^c\ a_D{}^d}{a_A{}^a\ a_B{}^b} \tag{4}$$

Substitution of equation (2) into equation (4) for the various species yields:

$$K = \frac{[C]^c\ [D]^d}{[A]^a\ [B]^b} \times \frac{\gamma_C{}^c\ \gamma_D{}^d}{\gamma_A{}^a\ \gamma_B{}^b} \tag{5}$$

It is common practice to define an equilibrium constant in terms of species concentrations:

$$K_C = \frac{[C]^c\ [D]^d}{[A]^a\ [B]^b} \tag{6}$$

The relationship between the concentration-based equilibrium constant and the thermodynamic equilibrium constant is therefore:

$$K = K_C \times \frac{\gamma_C{}^c\ \gamma_D{}^d}{\gamma_A{}^a\ \gamma_B{}^b} \tag{7}$$

It may be noted that in an ideal solution where all ions are completely non-interacting (a situation realized only under conditions of high dilution), the activity coefficient term would equal unity, and then K would be equal to K_C.

The ionic strength of a solution is defined as:

$$\mu = \tfrac{1}{2}\ \Sigma\ C_i\ z_i^2 \tag{8}$$

where C_i is the concentration (in units of moles/liter) of the ith ion, and z_i is its charge. For example, the ionic strength of a 0.1M solution of KCl is 0.1, and the ionic strength of a 0.1M solution of K_2SO_4 equals 0.3.

According to the Debye-Hückel theory, the activity coefficient of the ith ion in a solution having an ionic strength of μ is given by:

$$-\log \gamma_i = \frac{A\ z_i^2\ \mu^{1/2}}{1 + B\,d\ \mu^{1/2}} \tag{9}$$

where A is a constant equal to 0.51, B is a constant equal to 3.3×10^7, and d is the size of the hydrated ion. For many univalent ions, d is approximately 3 Å, so equation (9) simplifies to:

$$-\log \gamma_i = \frac{A\ z_i^2\ \mu^{1/2}}{1 + \mu^{1/2}} \tag{10}$$

Finally, in extremely dilute solutions, the quantity $\mu^{1/2}$ in the denominator is negligible relative to unity, and equation (10) further simplifies to the expression known as the Limiting Law:

$$-\log \gamma_i = 0.51\ z_i^2\ \mu^{1/2} \tag{11}$$

For an uncharged molecule, the activity exhibits a different dependence on ionic strength:

$$\log \gamma_j = k\ \mu \tag{12}$$

where k is a constant. For simple carboxylic acids, k will fall within the range of −0.05 to +0.20. The value of this constant depends on the characteristics of the uncharged molecule as well as the specific ions in the solution that contribute to the overall ionic strength.

Equilibria of Weak Acids and Bases

Autoionization of Water

Water, regardless of its degree of purity, contains low concentrations of ions that can be detected through the use of conductivity measurements. These ions arise from the transfer of a proton from water molecule to another, a process known as autoionization:

$$H_2O + H_2O \leftrightarrow H_3O^+ + OH^- \tag{13}$$

In equation (13), the H_3O^+ species is the hydronium ion, and the OH^- is the hydroxide ion. This reaction is reversible, and it is well established that the reactants proceed only slightly on to the products. Therefore, the approximation that the activity of the various species can be expressed by their concentrations is accurate, so the equilibrium constant for this reaction can be written as:

$$K_C = \frac{[H_3O^+]\ [OH^-]}{[H_2O]^2} \tag{14}$$

In aqueous solutions, the activity of water is a constant owing to its high concentration (55.55 M), so it is possible to simplify equation (14) to:

$$K_W = [H_3O^+]\ [OH^-] \tag{15}$$

K_W is the autoionization constant of water, and is sometimes identified as being the ion product of water. The magnitude of K_W is very small, being equal to 1.007×10^{-14} at a temperature of 25°C (Dean 1979).

For the sake of convenience, Sørensen (1909) proposed the "p" scale, where numbers such as K_W would be expressed as the negative of their base10 logarithms. The value of pK_W would then be calculated as:

$$pK_W = -\log\ (K_W) \tag{16}$$

and has a value equal to 13.997 at 25°C. Defining pH as:

$$pH = -\log [H_3O^+] \tag{17}$$

and:

$$pOH = -\log [OH^-] \tag{18}$$

then equation (15) can then be expressed as:

$$pK_W = pH + pOH \tag{19}$$

The autoionization of water is an endothermic reaction, so K_W increases as the temperature is increased (Dean 1979). This temperature dependence is plotted in Figure 1.

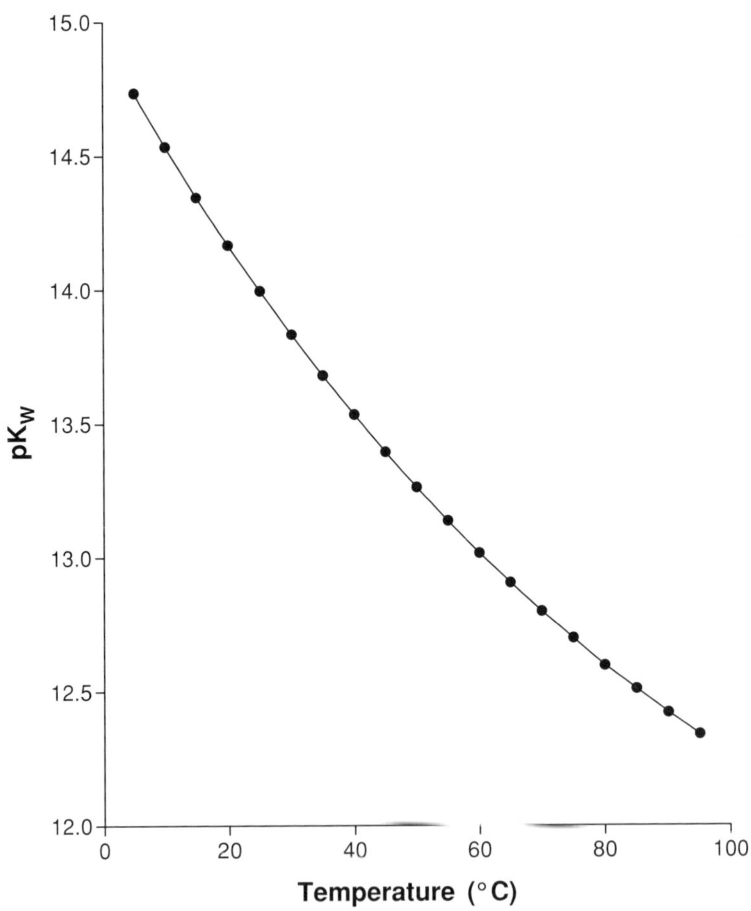

Figure 1. Temperature dependence of the autoionization constant of water. The figure was developed from data given in reference [1].

Ionic Equilibria of Acidic and Basic Substances

A large number of definitions for acids and bases have been introduced, but the 1923 definitions of J.N. Brønsted and T.M. Lowry are the most useful for discussions of ionic equilibria in aqueous systems. According to the Brønsted-Lowry model, an acid is a substance capable of donating a proton to another substance, such as water:

$$HA + H_2O \leftrightarrow H_3O^+ + A^- \tag{20}$$

The acidic substance (HA) that originally donated the proton becomes the conjugate base (A^-) of that substance, since the conjugate base could conceivably accept a proton from an even stronger acid than the original substance.

Recalling the discussion above about water and its activity, the thermodynamic equilibrium constant expression for equation (20) would be:

$$K = \frac{[H_3O^+]\ [A^-]}{[HA]} \times \frac{\gamma_H{}^+\ \ \gamma_A{}^-}{\gamma_{HA}} \tag{21}$$

For an acid capable of ionizing into a univalent anion, γ_{H+} and γ_{A-} will be approximately equal, and γ_{HA} will be approximately equal to one. Writing the concentration-based equilibrium constant expression as:

$$K_A = \frac{[H_3O^+]\ [A^-]}{[HA]} \tag{22}$$

it follows that:

$$K = K_A\ \gamma_H{}^+\ \gamma_A{}^- \tag{23}$$

For weak acids, the magnitude of K_A is very small, and as a result the resulting H_3O^+ and A^- ions will be produced in small amounts. Under those conditions, both $\gamma_H{}^+$ and $\gamma_A{}^-$ will be approximately equal to one, and then one can approximate the thermodynamic equilibrium constant, K, by the concentration-based ionization constant, K_A.

A strong acid is defined as a substance that reacts completely with water, so that the acid ionization constant defined in equation (21) or (22) is effectively infinite. This situation can only be achieved if the conjugate base of the strong acid is very weak. A weak acid will be characterized by an acid ionization constant that is considerably less than unity, so that the position of equilibrium in the reaction represented in equation (20) favors the existence of un-ionized free acid. One can define pK_A as:

$$pK_A = -\log\ (K_A) \tag{24}$$

A discussion of the ionic equilibria associated with basic substances exactly parallels that just developed for acidic substances. A base is a substance capable of accepting a proton donated by another substance, such as water:

$$B + H_2O \leftrightarrow BH^+ + OH^- \tag{25}$$

The basic substance (B) that originally accepted the proton becomes the conjugate acid (BH^+) of that substance, since the conjugate acid could conceivably donate a proton to an even stronger base than the original substance. The concentration-based ionization constant expression corresponding to equation (25) is:

$$K_B = \frac{[BH^+]\,[OH^-]}{[B]} \tag{26}$$

and pK_B is defined as:

$$pK_B = -\log\,(K_B) \tag{27}$$

A strong base is a substance that reacts completely with water, so that the base ionization constant defined in equation (26) is effectively infinite. This situation can only be realized if the conjugate acid of the strong base is very weak. A weak base will be characterized by a base ionization constant that is considerably less than unity, so that the position of equilibrium in the reaction represented in equation (25) favors the existence of un-ionized free base.

Ionic Equilibria of Conjugate Acids and Bases

Once formed, the conjugate base of an acidic substance (*i.e.*, the anion of that acid) is also capable of reacting with water:

$$A^- + H_2O \leftrightarrow HA + OH^- \tag{28}$$

Since aqueous solutions of anions are commonly prepared by the dissolution of a salt containing that anion, reactions of the type described by equation (28) are often termed hydrolysis reactions. Equation (28) is necessarily characterized by its base ionization constant expression:

$$K_B = \frac{[HA]\,[OH^-]}{[A^-]} \tag{29}$$

and a corresponding pK_B defined in the usual manner. But since:

$$[OH^-] = K_W/[H_3O^+] \tag{30}$$

it follows that:

$$K_B = \frac{[HA]\,K_W}{[A^-]\,[H_3O^+]} \tag{31}$$

Equation (31) contains the right-hand side expression of equation (22), so one deduces that:

$$K_B = K_W/K_A \tag{32}$$

or

$$K_W = K_A\,K_B \tag{33}$$

The same relation between ionization constants of a conjugate acid-base pair can be developed if one were to begin with the conjugate acid of a basic substance, so equation (33) is recognized as a general property of conjugate acid-base pairs.

Knowledge of K_A for a weak acid (or K_B for a weak base) facilitates estimation of the concentrations of the various species after equilibrium is established. When accurate solutions for equilibrium concentrations of a weak acid, for example, are required, the exact approach requires solving four simultaneous equations. Two of these have already been discussed, one being the acid ionization expression of equation (22), and the other being the water dissociation expression of equation (15). The third necessary equation is the charge balance equation:

$$[H_3O^+] = [A^-] + [OH^-] \tag{34}$$

and the fourth equation is the mass balance equation:

$$C_A = [HA] + [A^-] \tag{35}$$

where C_A is the initial concentration of weak acid in the system. It is beyond the scope of this chapter to detail the exact solutions to these equilibria calculations, but interested readers can consult chapters 5 and 6 of the text by Frieser and Fernando (1963), which contains solutions to effectively all problems of standard interest.

Ionic Equilibria of Buffer Systems

A buffer can be defined as a solution that maintains an approximately equal pH value even if small amounts of acidic or basic substances are added. To function in this manner, a buffer solution will necessarily contain either an acid and its conjugate base, or a base and its conjugate acid.

The action of a buffer system can be understood through the use of a practical example. Consider acetic acid, for which $K_A = 1.82 \times 10^{-5}$ (pK = 4.74). The following pH values can be calculated (for solutions having a total acetate content of 1.0 M) using its acid ionization constant expression:

Acetic Acid, [HA]	Acetate Ion, [A$^-$]	Calculated pH
0.4	0.6	4.92
0.5	0.5	4.74
0.6	0.4	4.56

When an acidic substance is added to a buffer system it would immediately react with the basic component, as a basic substance would react with the acidic component. One therefore concludes from the table that the addition of either 0.1 M acid or 0.1 M base to a buffer system consisting of 0.5 M acetic acid and 0.5 M acetate ion would cause the pH to change by only 0.18 pH units. This is to be contrasted with the pH changes that would result from the addition of 0.1 M acid to water (*i.e.*, 7.0 to 1.0, for a change of 6.0 pH units), or from the addition of 0.1 M base to water (*i.e.*, 7.0 to 13.0, also for a change of 6.0 pH units).

A very useful expression for describing the properties of buffer system can be derived from consideration of ionization constant expressions. For an acidic substance, equation (22) can be rearranged as

$$[H_3O^+] = \frac{K_A\,[HA]}{[A^-]} \tag{36}$$

Taking the negative of the base10 logarithms of the various quantities yields the relation known as the Henderson-Hasselbalch equation:

$$pH = pK_A + \log\{[A^-]/[HA]\} \tag{37}$$

Equation (37) indicates that when the concentration of acid and its conjugate base are equal (*i.e.*, $[HA] = [A^-]$), then the pH of the solution will equal the pK_A value. As a result, a buffer system is chosen so that the target pH is approximately equal to the pK_A value.

A buffer system can be envisioned as consisting of a partially completed neutralization reaction:

$$HA + OH^- \leftrightarrow A^- + H_2O \tag{38}$$

where comparable amounts of HA and A^- are present in the solution. The buffer region within a neutralization reaction is shown in Figure 2, where the horizontal region in the graph of anion concentration and observed pH reveals the buffer region of the system. For practical purposes, the buffer region would extend over $[HA]/[A^-]$ ratios of approximately 0.2 to 0.8.

It is clear from the preceding discussion that a buffer solution acts to reduce the change in pH when small amounts of either acid or are added to the solution. However, all buffer solutions will necessarily have a limited capacity to absorb amounts of acid or base without significant pH change, as can be illustrated by the example of a buffer solution that contains 1 mole each of a weak acid and its conjugate base (in the form of its salt). Since the buffer absorbs OH^- through reaction with its weak acid component, the capacity of the buffer to respond will be exhausted once 1 mole of base is added. Alternatively, since the buffer absorbs H_3O^+ through reaction with its conjugate base component, the capacity of the buffer to respond will be exhausted once 1 mole of acid is added. In practice, the ability of the buffer to respond would have been seriously impaired well before the equivalent amount of acid or base had been added.

The term buffer capacity refers to the amount of base (or acid) that a given buffer solution can absorb without experiencing a significant change in pH. It is clear that since the amount of acid or base that can be absorbed by a buffer solution depends on the initial concentrations of components, the buffer capacity of a system will be strongly dependent on those initial concentrations. In particular, the more dilute a buffer solution is prepared to be, the less capacity it will have.

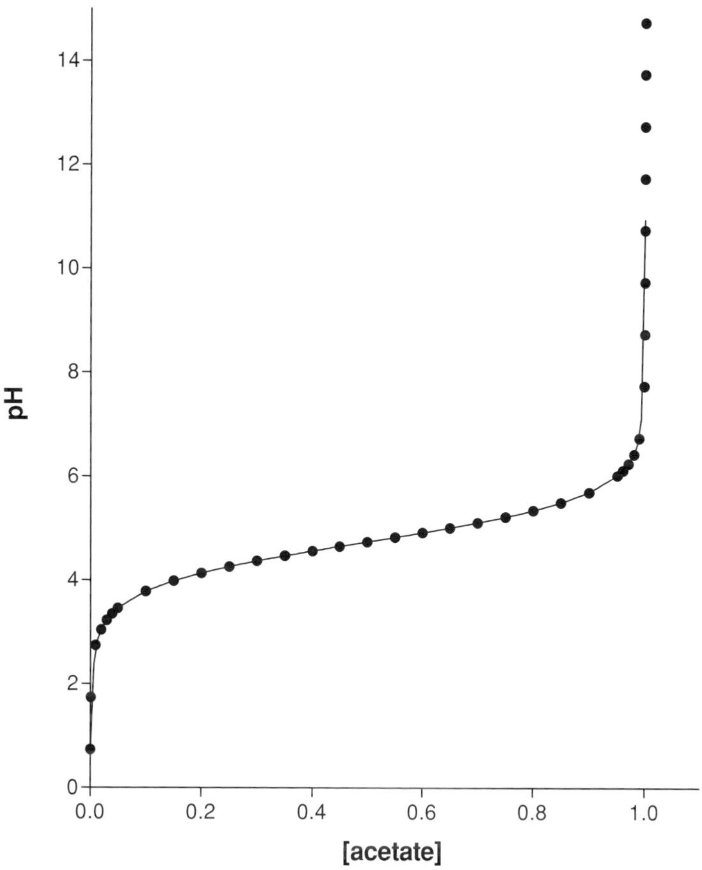

Figure 2. Neutralization curve obtained during the titration of 1.0 M acetic acid, plotted as a function of the acetate ion concentration.

The selection of a buffer system for use in a pharmaceutical dosage form is relatively straight-forward. It is evident from the preceding discussion that the most important prerequisite for a buffer is the approximate equality of the pK_A value of the buffer with the intended optimal pH value for the formulation. Knowledge of the pH stability profile of a drug substance enables one to deduce the pH range for which formulation is desirable, and the basis for the most appropriate buffer system would be the weak acid or base whose pK_A or pK_B value was numerically equal to the midpoint of the pH range of stability.

There are, of course, other considerations that need to be monitored, such as compatibility with the drug substance. Boylan (1986) has provided a summary of the selection criteria for buffering agents:

1. The buffer must have adequate capacity in the desired pH range.
2. The buffer must be biologically safe for the intended use.
3. The buffer should have little or no deleterious effect on the stability of the final product.
4. The buffer should permit acceptable flavoring and coloring of the product.

Basis for buffering system	pK_1	pK_2	pK_3	Martell and Smith reference*
Acetic acid	4.56	—	—	Vol-3; p. 3
Adipic acid	5.03	4.26	—	Vol-3; p. 118
Arginine	9.01	2.05	—	Vol-1; p. 43
Benzoic acid	4.00	—	—	Vol-3; p. 16
Boric acid	8.97	—	—	Vol-4; p. 25
Carbonic acid	10.00	6.16	—	Vol-4; p. 37
Citric acid	5.69	4.35	2.87	Vol-3; p. 161
Diethanolamine	8.90	—	—	Vol-2; p. 80
Ethanolamine	9.52	—	—	Vol-2; p. 15
Ethylenediamine	9.89	7.08	—	Vol-2; p. 36
Glutamic acid	9.59	4.20	—	Vol-1; p. 27
Glycine	9.57	2.36	—	Vol-1; p. 1
Lactic acid	3.66	—	—	Vol-3; p. 28
Lysine	10.69	9.08	2.04	Vol-1; p. 58
Maleic acid	5.83	1.75	—	Vol-3; p. 112
Phosphoric acid	11.74	6.72	2.00	Vol-4; p. 56
Tartaric acid	3.95	2.82	—	Vol-3; p. 127
Triethanolamine	7.80	—	—	Vol-2; p. 118
Tromethamine	8.09	—	—	Vol-2; p. 20

Table 1. Acids and bases suitable for use as buffer systems in pharmaceutical products.

The references are to the first 4 volumes of Critical Stability Constants, Martell, A.E.; Smith, R.M. eds.; Plenum Press, New York. These volumes are identified as:

- Volume 1 (Amino Acids): 1974 [3]
- Volume 2 (Amines): 1975 [4]
- Volume 3 (Other Organic Ligands): 1977 [5]
- Volume 4 (Inorganic Complexes): 1976 [6]

A practical consequence of equation (35) is that as long as the concentration of a buffer is not overcome by reaction demands, a buffer system will exhibit adequate capacity within ±1 pH unit with respect to its pK_A or pK_B value.

The second criterion from the preceding list restricts buffering agents to those deemed to be pharmaceutically acceptable. A list of appropriate buffer systems is provided in Table 1, along with values for their pK_A or pK_B values

(Martell and Smith 1974 and 1977; Smith and Martell 1975 and 1976). The use of buffering agents is most critical for parenteral formulations, and it has been noted over the years that phosphate, citrate, and acetate are most commonly used for such purposes (Wang and Kowal 1980; Nema et al. 1997). Ethanolamine and diethanolamine are also used to adjust pH and form their corresponding salts, while lysine and glycine are often used to buffer protein and peptide formulations. Akers (2002) has reviewed the scope of drug-excipient interactions in parenteral formulations, and has provided an overview of the effect of buffers on drug substance stability.

pH Dependence of Aqueous Solubility

It is well known that the solubility of a compound containing one or more ionizable functional groups will usually be a strong function of the pH of the dissolving aqueous medium, with the graphical representation of these qualities being known as the pH-solubility profile. Generally the solubility of a free acid is much less than the solubility of its ionized form, as has been illustrated in Figure 3

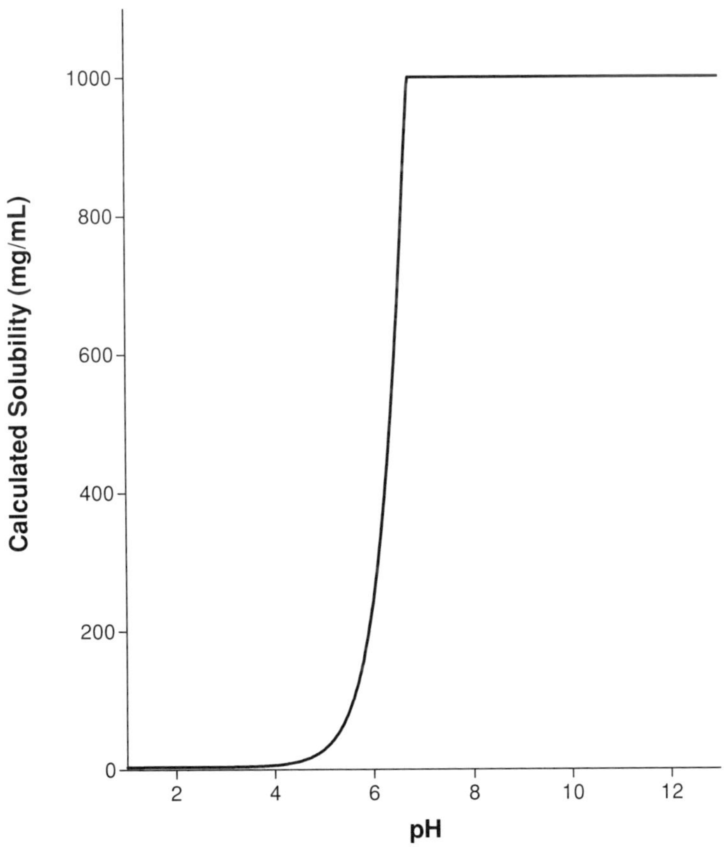

Figure 3. Calculated pH dependence of the aqueous solubility of benzoic acid, generated using the program PhysChem 7.0 (Advanced Chemistry Development, Toronto, CA).

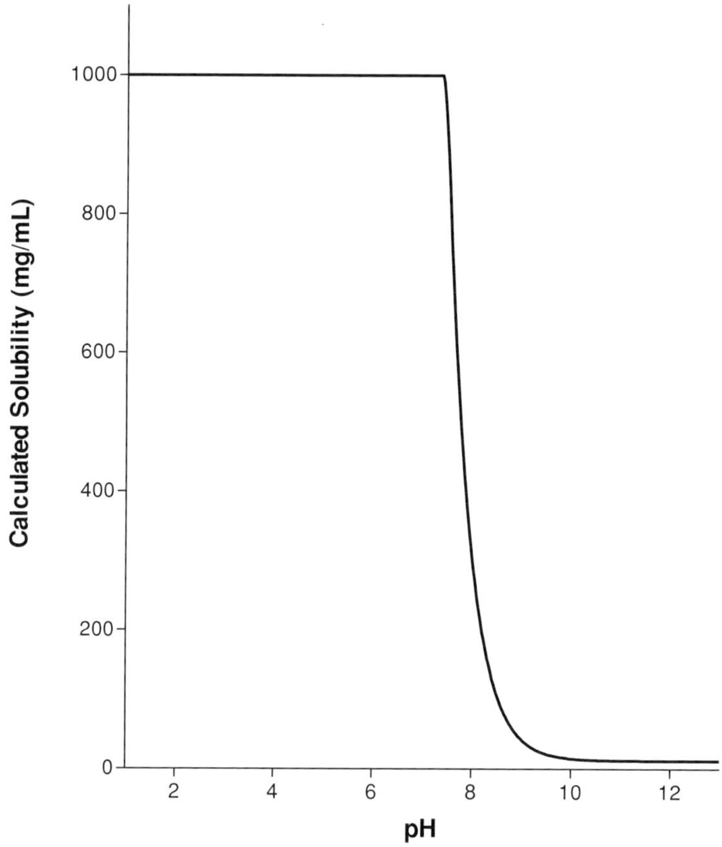

Figure 4. Calculated pH dependence of the aqueous solubility of benzylamine, generated using the program PhysChem 7.0 (Advanced Chemistry Development, Toronto, CA).

for the example of benzoic acid. Similarly, the solubility of a free base is much less than the solubility of its protonated form, as shown in Figure 4 for benzylamine. Finally, for molecules containing more than one ionizable functional group, the pH dependence of the aqueous solubility can be fairly complicated. Figure 5 gives the example for 4-(aminomethyl)benzoic acid.

The solubility of a solid can be understood using a simple model. For a solid to dissolve, the forces of attraction between solute and solvent molecules must overcome the attractive forces holding the solid intact and the solvent aggregates together. In other words, the solvation free energy released upon dissolution must exceed the lattice free energy of the solid plus the free energy of cavity formation in the solvent for the process to proceed spontaneously. The balance of the attractive and disruptive forces will determine the equilibrium solubility of the solid in question (which is an exponential function of the free energy change of the system). The enthalpy change and the increase in disorder of the system (*i.e.*, the entropy change) determine the Gibbs free energy change. Finally, the act of dissolution may be endothermic or exothermic in nature, so

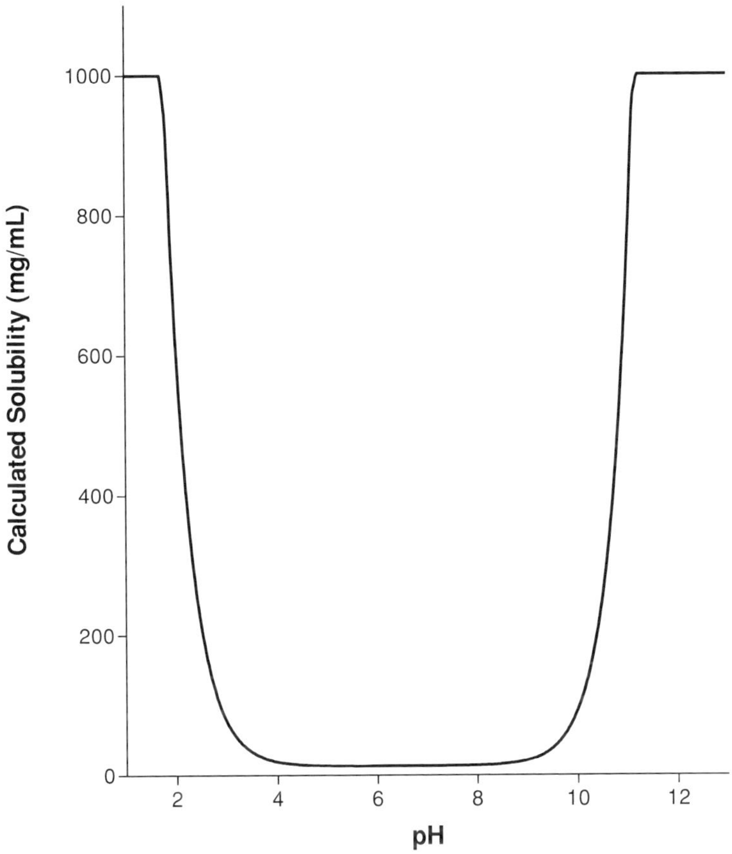

Figure 5. Calculated pH dependence of the aqueous solubility of 4-(aminomethyl) benzoic acid, generated using the program PhysChem 7.0 (Advanced Chemistry Development, Toronto, CA).

that measurements of solution calorimetry can be used to provide important information on the substance under study.

It follows that since the nature of the undissolved solid cannot be affected by the pH of an aqueous solution to which it is exposed, the nature of the interactions between the dissolved species and the solvent must be very different for the ionized (or protonated) form of the molecule relative to its neutral form in order to affect the equilibrium solubility. For ionic equilibria, it is the strong solvation energy released when the anionic or cationic form of the weak acid or base interacts with the aqueous medium that results in its superior degree of solubility over the neutral form which cannot experience such forces.

The capacity of any system to form solutions has limits imposed by the phase rule of Gibbs:

$$F + P = C + 2 \tag{39}$$

where F is the number of degrees of freedom in a system consisting of C components with P phases. For a system of two components and two phases (*e.g.*, solid and liquid) under the pressure of their own vapor and at constant temperature, F equals zero. If one of the phases consists solely of one component (a pure substance), the equilibrium solubility at constant temperature and pressure is a fixed quantity which is given as the amount of solute contained in the saturated solution in a unit amount of the solvent or solution.

For any case in which F is zero, a definite reproducible solubility equilibrium can be reached. Complete representation of the solubility relations is accomplished in the phase diagram, which gives the number, composition, and relative amounts of each phase present at any temperature in a sample containing the components in any specified proportion. Solubilities may therefore be expressed in any appropriate units of concentration, such as the quality of the solute dissolved (defined mass, number of moles) divided by the quantity either of the solvent (defined mass, volume, or number of moles) or of the solution (defined mass, volume, or number of moles). Jacques et al. (1981) have provided a compilation of the expressions for concentration and solubility.

Measurement of Solubility

Prior to entering a discussion of the pH dependence of aqueous solubility, a few remarks concerning the determination of solubilities are in order. Methods for the determination of solubility have been thoroughly reviewed (Mader et al. 1959; Grant and Higuchi 1990; Yalkowsky and Banerjee 1992), especially with respect to the determination of such data for compounds having pharmaceutical interest (Grant and Brittain 1995). Solubility is frequently highly dependent on temperature, so the temperature must be recorded for each solubility measurement in addition to the precise nature of the solvent and the solid phase at equilibrium. Plots of solubility against temperature are commonly used for characterizing pharmaceutical solids, and such graphical methods have been extensively discussed (Grant et al. 1984; Grant and Higuchi 1990). Frequently (especially over a relatively narrow temperature range), a linear plot may be given either by the van't Hoff relationship:

$$\ln X_2^{sat} = \frac{-a}{RT} + c' \tag{40}$$

or by the Hildebrand relationship:

$$\ln X_2^{sat} = \frac{b \ln T}{R} + c'' \tag{41}$$

In equations (40) and (41), X_2^{sat} is the mole fraction solubility of the solid solute at an absolute temperature T, a is the apparent molar enthalpy of solution, b is the apparent molar entropy of solution, and c′ and c″ are constants. The combined equation, attributed to Valentiner, has been used by Grant et al. (1984) in the form:

$$\ln X_2^{sat} = \frac{-a}{RT} + \frac{b \ln T}{R} + c''' \tag{42}$$

This three-parameter equation enables solubility to be simulated and correlated quite accurately over a wide range of temperatures.

Two general methods, the analytical method and the synthetic method (Grant and Brittain 1995), are available for determining solubility. In the analytical method, the temperature of equilibration is fixed, while the concentration of the solute in a saturated solution is determined at equilibrium by a suitable analytical procedure. The analytical method can be either the traditional, common batch agitation method, or the more recent flow column method. In the synthetic method, the composition of the solute-solvent system is fixed by appropriate addition and mixing of the solute and solvent, then the temperature at which the solid solute just dissolves or just crystallizes is carefully bracketed.

It is usually not difficult to determine the solubility of solids which are moderately soluble (greater than 1 mg/mL), but the direct determination of solubilities much less than 1 mg/mL is not straightforward. Problems such as slow equilibrium resulting from a low rate of dissolution, the influence of impurities, and the apparent heterogeneity in the energy content of the crystalline solid (Higuchi et al. 1979), can lead to large discrepancies in reported values.

Avdeef (1998) has reported an automated potentiometric titration method for the determination of solubilities of drug substances containing ionizable groups, where a graphical procedure is used for the estimation of solubility constants based on Bjerrum difference plots. One useful relation derived in this work was:

$$\text{Log}\,(S_O) = \log\,(C/2) - |\Delta\, pK_a| \tag{43}$$

where S_O is the intrinsic solubility of the non-ionized substance, C is the sample concentration, and Δ pK_a is the difference between the apparent pK_a determined under conditions of precipitation and the true pK_a measured in the absence of any solid phase. It should be noted that equation (43) is valid only in a solution where part of the sample precipitates at the early stages of the titration. Work conducted on flurbiprofen (as an exemplary acidic substance) and buprenorphine (as an exemplary basic substance) was used to illustrate the methodology. In subsequent works, the correlation between the acid-base titration and saturation shake-flask methods was investigated (Avdeef et al. 2000), as was a dissolution titration template method for the determination of solubility (Avdeef and Berger 2001).

Solubility and Ionic Equilibria

As indicated above, an equilibrium condition requires the presence of excess solid phase in equilibrium with the dissolved substance. For a weak acid, it is convenient to consider the various equilibria as dividing a pH-solubility profile

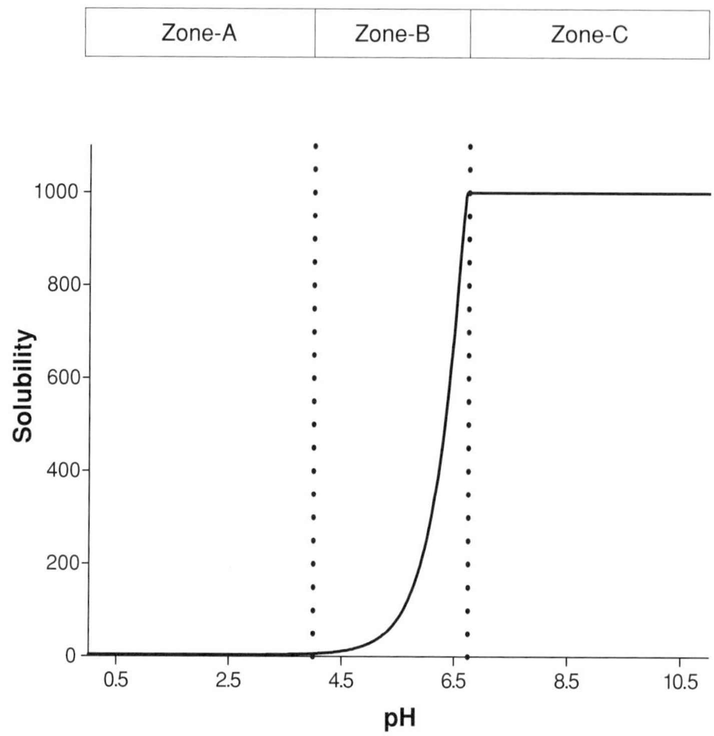

Figure 6. pH-solubility profile of a weak acid having a pK_a value of 4.20, illustrating the three zones of equilibrium interest.

into the three zones shown in Figure 6. The processes taking place in Zone-A are effectively only those consisting of solid free acid in equilibrium with dissolved free acid, while the processes taking place in Zone-C are effectively only those consisting of the solid salt of the acid in equilibrium with dissolved salt. Zone-B is effectively a buffer region, where the overall solubility consists of the sum of dissolved free acid and salt. Continuing the example of the pH-solubility profile of a weak acid, the processes taking place in each zone will each be discussed in turn.

It is to be recognized, however, that the same general discussion would apply to the solubility processes associated with a weak base, except that the characters of zones A and C would be reversed.

Zone-A: Solubility of the Free Weak Acid

The equilibria existing in Zone-A are very simple, consisting effectively of the equilibrium between solid free acid and the quantity dissolved in the aqueous phase:

$$HA_{(SOLID)} \leftrightarrow HA_{(DISSOLVED)} \tag{44}$$

The concentration of dissolved HA is the intrinsic solubility of the free acid, which can be expressed as S_{HA}. The equilibrium constant expression for equation

(44) would be:

$$K = \frac{a_{HA-DISS}}{a_{HA-SOL}} \tag{45}$$

where $a_{HA-DISS}$ is the activity of the dissolved free acid, and a_{HA-SOL} is the activity of the solid salt. Recognizing that the activity of $MA_{(SOLID)}$ is a constant, expanding out the activities of equation (45) in terms of concentrations and activity coefficients yields:

$$K = [HA]\ \gamma_{HA} \tag{46}$$

As noted above, S_{HA} values are typically fairly low in magnitude, and would be independent of pH as long as no ionization of the acid takes place. Consequently, the activity coefficient γ_{HA} would be effectively equal to one, making the value of the equilibrium constant approximately equal to the concentration of the dissolved free acid (i.e., $K = S_{HA}$). If more accuracy was required, however, the activity coefficient of the dissolved free acid could be approximated using equation (12).

Zone-C: Solubility of the Salt of the Weak Acid

In Zone-C, the pH of the aqueous medium is such that all of the free acid, HA, has been converted into its salt form, MA, where M^+ will be a unipositive cation for the purposes of this discussion. The equilibria existing in this zone will therefore consist mainly of the equilibrium between solid salt and its dissolved ions:

$$MA_{(SOLID)} \leftrightarrow M^+ + A^- \tag{47}$$

In this zone, the concentration of dissolved MA is the intrinsic solubility of the salt of the free acid, which can be expressed as S_{MA}.

The equilibrium constant expression for equation (47) would be:

$$K = \frac{a_{M^+}\ a_{A^-}}{a_{MA-SOL}} \tag{48}$$

where a_{M+} is the activity of the cation, a_{A-} is the activity of the A^- anion, and a_{MA-SOL} is the activity of the solid salt. Recognizing that the activity of $MA_{(SOLID)}$ is a constant, expanding out the activities of equation (48) in terms of concentrations and activity coefficients yields the solubility product constant:

$$K_{SP} = [M^+]\ [A^-]\ \gamma_{M+}\ \gamma_{A-} \tag{49}$$

In the usual consideration for sparingly soluble salts, the activity coefficients γ_{M+} and γ_{A-} would each equal unity, and the solubility product constant can be derived using only the concentrations of the ions.

The condition of Zone-C could be established in two different ways. The first is where a saturated solution of MA is established by equilibrating an excess amount of solid salt with water at a fixed temperature, and the second is where the weak acid has been exactly neutralized by the strong base MOH. In either case, it follows that $[M^+]$ must equal $[A^-]$, and therefore equation (49) can be rearranged to yield an expression for the concentration of dissolved salt:

$$S_{MA} = \{K_{SP}/(\gamma_{M+}\ \gamma_{A-})\}^{1/2} \tag{50}$$

Since it is the usual situation that the solubility of the salt form is quite high, inclusion of the activity coefficient terms is necessary to obtain accurate results. These can be approximated using equation (10).

Zone-B: Solubility of the Weak Acid and Its Salt Under Equilibrium Conditions
Inside of the pH region enclosed by the limits of Zone-B, one finds both the free acid and its salt form. The mass balance relationship in this zone defines the total solubility (S_T) at any particular pH value as the sum of the concentration of the free acid plus the concentration of its salt form:

$$S_T = [HA] + [A^-] \tag{51}$$

As long as the solubility of the generated salt is not exceeded, the concentration of $[A^-]$ generated by neutralization of HA by the strong base MA is calculated from knowledge of the acid ionization constant:

$$[A^-] = \frac{K_A\ [HA]}{[H_3O^+]} \tag{52}$$

or:

$$[A^-] = \frac{K_A\ S_{HA}}{[H_3O^+]} \tag{53}$$

Therefore, as long as the formed salt is completely soluble, the total solubility in Zone-B, is given by:

$$S_T = S_{HA}\ \{1 + (K_A/[H_3O^+]\} \tag{54}$$

In other words, equation (54) is only valid for those pH conditions for which the solubility product constant of the salt is not exceeded.

A different set of equations holds when the solubility of the salt is exceeded, since the amount of solubilized $[A^-]$ will be limited by the solubility product constant. Recognizing that:

$$[HA] = \frac{[H_3O^+]\ [A^-]}{K_A} \tag{55}$$

The equation (51) becomes:

$$S_T = \frac{[H_3O^+]\ [A^-]}{K_A} + [A^-] \tag{56}$$

or:

$$S_T = [A^-]\ \{1 + ([H_3O^+]/K_A\} \tag{57}$$

However, since:

$$[A^-] = \{K_{SP}/(\gamma_{M+}\ \gamma_{A-})\}^{1/2} \tag{58}$$

One obtains the equation that describes the solubility behavior for pH values where the solubility product constant of the salt is exceeded:

$$S_T = \{K_{SP}/(\gamma_{M+}\ \gamma_{A-})\}^{1/2}\ \{1 + ([H_3O^+]/K_A\} \quad (59)$$

At the exact pH for which the solubility constant is exceeded, one may equate the S_T relations of equations (54) and (59), solve for the hydronium ion concentration, and convert to the "p" scale. After solving the resulting quadratic equation, one obtains an expression for this critical pH value:

$$pH_{CRIT} = pK_A + \log\ \{S_{HA}/[K_{SP}/(\gamma_{M+}\ \gamma_{A-})]^{1/2}\} \quad (60)$$

Several papers have been published that detail the equilibria associated with pH-dependent phenomena and solubility limitations, and interested readers can consult these for additional information (Kramer and Flynn 1972; Bogardus and Blackwood 1979; Streng et al. 1984; Pudipeddi et al. 2002).

Examples of the Effects of pH on Aqueous Solubility

With the continuing development of compounds exhibiting low degrees of intrinsic aqueous solubility, the combination of pH control and complexing agents in formulations has become important. A theoretical analysis of the synergistic effect observed in the combined systems has been developed, and used to explain the solubilization noted for flavopiridol (Li et al. 1998). In this work, the total solubility was determined by the addition of the concentrations of the four components present in the solution (free non-ionized drug substance, free non-ionized drug complex, and their ionized counterparts), with the analysis indicating that the complexation constants and the ionization pH of the drug being critical parameters.

In a subsequent work, the solubilization of this substance by pH control combined with co-solvents, surfactants, or complexing agents was investigated (Li et al. 1999). Control of the solution pH was effected in combination with various surfactants (polysorbates 20 and 80) or cyclodextrin complexing agents. It was determined that the theoretical model developed by the authors could be successfully used to characterize the effects of pH, pKa, complexation constants, micelle partition coefficients, or solubilizing power of cosolvents on the solubility of ionized and non-ionized compounds.

The combined effect of pH and surfactants on the dissolution of piroxicam has been reported (Jinno et al. 2000). In this system, the dissolution rate and solubility of the drug substance could be adequately estimated by a simple additive model for the effect of pH and surfactant, where the total dissolved concentration equaled the summation of the dissolved amount of non-ionized substance, the amount of dissolved ionized substance, and the amount of substance solubilized in the surfactant micelles. It was suggested that the model developed in this work could be useful for predicting the dissolution of a water-insoluble drug containing ionizable functional groups as a function of pH and surfactant concentration, and in establishing *in vitro – in vivo* correlations that would be useful for establishing bioequivalence criteria.

An equilibrium-based model was proposed to characterize the drug-surfactant interactions observed in the system consisting of furbiprofen and polysorbate 80 in solutions having different pH conditions (Li and Zhao 2003). The model reflected both interactions and interdependence among all drug-containing species, namely non-ionized drug in water, ionized drug in water, non-ionized drug in micelles, and ionized drug in micelles. The mathematical treatment also enabled modeling of the drug solubilization in the pH-surfactant solutions without requiring the use of inappropriate approximations. The solubility data estimated by the proposed model were found to be more reliable when the surfactant concentration was high in the system. This finding confirmed that consideration of interrelations and interdependence of all drug species in the various solutions was appropriate for this model.

When the pH conditions used for a given solubility determination are set through the use of buffers, the possible solubilization of the buffering systems must be established. The desirable condition is where the buffers used to establish pH conditions have no effect on the observed solubilities, as was reported, for example, during the determination of the solubilities of trimethoprim and sulfamethoxazole at various pH values (McDonald and Faridah). On the other hand, the particular buffer system used to establish a solution pH often requires correction for such buffer effects, as noted, for example, during studies of some isoxazoyl- naphthoquinone derivatives (Granero et al. 1991).

Concluding Remarks

Since many drug substances contain acidic or basic functional groups, one of the most defining characteristic of a drug substance is its pattern of ionization constants. It is also well established that the solubility of a compound containing one or more ionizable functional groups is usually a strong function of the pH of the dissolving aqueous medium, which leads to the concept of the pH-solubility profile. Generally the solubility of a free acid is less than the solubility of its ionized form, and the solubility of a free base is usually less than the solubility of its protonated form. For molecules containing more than one ionizable functional group, the pH dependence of the aqueous solubility can be fairly complicated. Recent advances have enabled calculation of the pH dependence of aqueous solubility once knowledge of the pK_A values is in hand.

Buffers and buffering agents have been widely used for the solubilization of insoluble compounds, and for the stabilization of pharmaceutical formulations. Such phenomena also play a vitally important role during drug characterization studies, since they enter into the conduct of solubility and drug stability studies in many guises. The range of pharmaceutically acceptable buffer systems spans all useful pH values, and it can be said that there is a buffer available for every intended purpose. As long as the working scientist verifies that the buffer system itself does not have a direct effect on the observed reactions, their use in studies of solubility and stability is invaluable.

List of Abbreviations

a_i .... activity of a substance
C_i .... concentration (in units of moles/liter) of the ith ion
d .... size of a hydrated ion
G_i .... free energy of a substance in a non-standard state
G_i^0 .... free energy of a substance in a defined standard state
γ_i .... activity coefficient of the substance in its non-standard state
K .... thermodynamic equilibrium constant
K_A .... concentration-based equilibrium constant expression of an acid
K_B .... concentration-based equilibrium constant expression of a base
K_C .... equilibrium constant in terms of species concentrations
K_{SP} .... solubility product constant
K_W .... autoionization constant of water
pH .... $-\log [H_3O^+]$
pK_A .... $-\log (K_A)$
pK_B .... $-\log (K_B)$
pOH .... $-\log [OH^-]$
pK_W .... $-\log (K_W)$
R .... Universal Gas Constant
S_O .... intrinsic solubility of a non-ionized substance
T .... absolute temperature
X_2^{sat} .... mole fraction solubility of a solid solute
z_i .... charge of the ith ion

References

Akers MJ. Excipient-Drug Interactions in Parenteral Formulations. *J Pharm Sci* 2002; 91: 2283–2300.

Avdeef A. pH-metric solubility. 1. solubility-pH profiles from Bjerrum plots. Gibbs buffer and pKa in the solid state. *Pharm Pharmacol Commun* 1998; 4: 165–178.

Avdeef A and Berger CM. pH-metric solubility. 3. Dissolution titration template method for solubility determination. *Eur J Pharm Sci* 2001; 14: 281–291.

Avdeef A, Berger CM, and Brownell C. pH-metric solubility. 2. Correlation between the acid-base titration and the saturation shake-flask solubility-pH methods. *Pharm Res* 2000; 17: 85–89.

Berge SM, Bighley LD, and Monkhouse DC. Pharmaceutical Salts. *J Pharm Sci* 1977; 66: 1–19.

Bogardus JB and Blackwood RK. Solubility of doxycycline in aqueous solution. *J Pharm Sci* 1979; 68: 188–194.

Boylan JC. Liquids. In: Lachman L, Lieberman HA, and Kanig J (eds.), *The Theory and Practice of Industrial Pharmacy*. Lea & Febiger, Philadelphia, PA, 1986, pp. 459–460.

Dean JA *Lange's Handbook of Chemistry*, 12th edition. McGraw-Hill, New York, NY, 1979, pp. 5–7.

Freiser H and Fernando Q. *Ionic Equilibria in Analytical Chemistry*. John Wiley & Sons, New York, NY, 1963, p. 334.

Granero G, de Bertorello MM, and Brinon MC. Solubility profiles of some isoxazolyl-naphthoquinone derivatives. *Int J Pharm* 1991; 190: 41–47.

Grant DJW and Brittain HG. Solubility of Pharmaceutical Solids. In: Brittain HG (ed.), *Physical Characterization of Pharmaceutical Solids*. Marcel Dekker, New York, NY, 1995, pp. 321–386.

Grant DJW and Higuchi T. *Solubility Behavior of Organic Compounds*. John Wiley & Sons, New York, NY, 1990, p. 600.

Grant DJW, Mehdizadeh M, Chow AH-L, and Fairbrother JE. Non-linear van't Hoff solubility-temperature plots and their pharmaceutical interpretation. *Int J Pharm* 1984; 18: 25–38.

Higuchi T, Shih F-ML, Kimura T, and Rytting JH. Solubility Determination of Barely aqueous-soluble organic solids. *J Pharm Sci* 1979; 68: 1267–1272.

Jacques J, Collet A, and Wilen SH. *Enantiomers, Racemates and Resolutions*. John Wiley & Sons, New York, NY, 1981, p. 447.

Jinno J, Oh D-M, Crison JR, and Amidon GL. Dissolution of ionizable water-insoluble drugs: The combined effect of pH and surfactant. *J Pharm Sci* 2000; 89: 268–274.

Kramer SF and Flynn GL. Solubility of organic hydrochlorides. *J Pharm Sci* 1972; 61: 1896–1904.

Li P, Tabibi SE, and Yalkowsky SH. Combined effect of complexation and pH on solubilization. *J Pharm Sci* 1998; 87: 1535–1537.

Li P, Tabibi SE, and Yalkowsky SH. Solubilization of flavopiridol by pH control combined with cosolvents, surfactants, or complexants. *J Pharm Sci* 1999; 88: 945–947.

Li P and Zhao L. Solubilization of flurbiprofen in pH-surfactant solutions. *J Pharm Sci* 2003; 92: 951–956.

Mader WJ, Vold RD, and Vold MJ. Determination of solubility. In: Weissberger A (ed.), *Physical Methods of Organic Chemistry*, 3rd edn., Vol. 1, Part I. Interscience Publishers, New York, NY, 1959, pp. 655–688.

Martell AE and Smith RM. *Critical Stability Constants. Volume 1: Amino Acids.* Plenum Press, New York, NY, 1974, p. 469.

Martell AE and Smith RM. *Critical Stability Constants. Volume 3: Other Organic Ligands.* Plenum Press, New York, NY, 1977, p. 495.

McDonald C and Faridah H. Solubilities of trimethoprim and sulfamethoxazole at various pH values and crystallization of trimethoprim from infusion fluids. *J Parenteral Sci Tech* 1991; 45: 147–151.

Nema S, Washkuhn RJ, and Brendel RJ. Excipients and their use in injectable products. *PDA J Pharm Sci Tech* 1997; 51: 166–171.

Pudipeddi M, Serajuddin ATM, Grant DJW, and Stahl PH. Solubility and Dissolution of Weak Acids, Bases, and Salts. In Stahl PH and Wermuth CG. *Handbook of Pharmaceutical Salts.* Wiley-VCH, Weinheim, Germany, 2002, pp. 19–39.

Smith RM and Martell AE. *Critical Stability Constants. Volume 2: Amines.* Plenum Press, New York, NY, 1975, p. 415.

Smith RM and Martell AE. *Critical Stability Constants. Volume 4: Inorganic Complexes Amino Acids.* Plenum Press, New York, NY, 1976, p. 257.

Sørensen SPL. The measurement and importance of the hydrogen ion concentration in enzyme reactions. *Biochem. Z.* 1909; 21: 131–304.

Stahl PH and Wermuth CG. *Handbook of Pharmaceutical Salts.* Wiley-VCH, Weinheim, Germany, 2002, p. 374.

Streng WH, Hsi SK, Helms PE, and Tan HGH. General treatment of pH-solubility profiles of weak acids and bases and the effects of different acids on the solubility of a weak base. *J Pharm Sci* 1984; 73: 1679–1684.

Wang Y-CJ and Kowal RR. Review of excipients and pH's for parenteral products used in the United States. *J Parenteral Drug Assoc* 1980; 34: 452–462.

Yalkowsky SH and Banerjee S. *Aqueous Solubility Methods of Estimation for Organic Compounds,* Marcel Dekker, New York, 1992, pp. 149–154.

3

Solvent Systems for Crystallization and Polymorph Selection

JONATHAN M. MILLER AND NAÍR RODRÍGUEZ-HORNEDO

Department of Pharmaceutical Sciences, College of Pharmacy University of Michigan, Ann Arbor MI, USA

ANTHONY C. BLACKBURN, DAINIUS MACIKENAS, AND BENJAMIN M. COLLMAN

Materials Science, Research Formulations, Pfizer Global Research and Development, Ann Arbor Laboratories, Ann Arbor, MI, USA

Introduction

Crystallization plays an important role in the synthesis, scale-up, processing, formulation, and stability of active pharmaceutical ingredients (API) (Rodríguez-Hornedo and Murphy, 1999; Shekunov and York, 2000; Rodríguez-Hornedo and Sinclair, 2002). Crystallization from solvent is a particularly important process, as this is the primary means of purification during the intermediate and final stages of drug synthesis. Moreover, solution crystallization determines the final solid-state modification of the API namely polymorphs, solvates, and hydrates.

Polymorphs are crystalline solids that have the same chemical composition, yet adopt different molecular arrangements in the crystal lattice (Grant, 1999; Byrn et al., 1999; Vippagunta et al., 2001; Bernstein, 2002). Crystalline solids may also incorporate solvent into the lattice during crystallization to form a solvate, or a hydrate in the case of water, an occurrence that is commonly referred to as pseudopolymorphism (Byrn et al., 1999; Nangia and Desiraju, 1999). Adequate control over the crystallization of solid forms is of utmost importance, as each form can exhibit different pharmaceutically relevant properties including solubility, dissolution rate, bioavailability, physical and chemical stability, and mechanical properties (Grant, 1999; Bernstein, 2002).

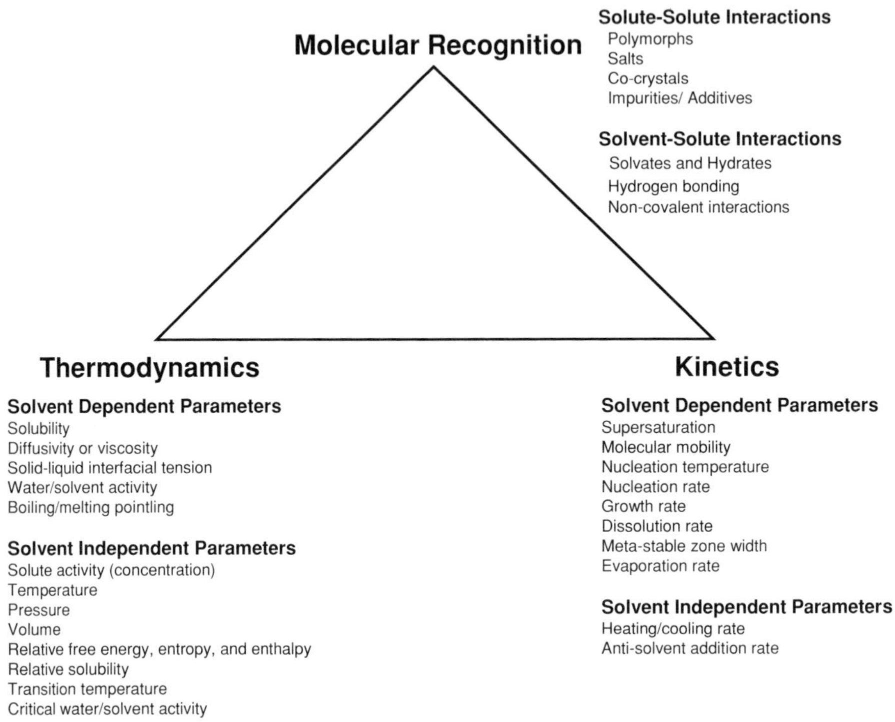

Figure 1. Schematic diagram showing the interplay between thermodynamic, kinetic, and molecular recognition phenomena that governs crystallization. After Rodríguez-Spong et al., (2004).

The rate and mechanisms by which crystallization occurs are determined by numerous thermodynamic, kinetic, and molecular recognition factors. (Nyvlt et al., 1985; Söhnel and Garside, 1992; Mersmann, 1995; Mullin, 2001; Myerson, 2002) These factors are summarized in Figure 1. The solvent plays a key role in crystallization as many of the factors depend directly on the solvent (Davey, 1982). Therefore, the intricate balance between thermodynamic, kinetic, and molecular recognition must be considered when designing experiments for polymorph screening, selection, and isolation.

In this chapter, the effects of these thermodynamic, kinetic, and molecular recognition phenomena on crystallization and the role of solvent in these processes will be described. The role of solvent on crystallization, polymorphic outcome, and phase transformations will also be discussed. Experimental approaches for polymorph screening will be presented with an emphasis on the important considerations and strategies for solvent selection.

Thermodynamics

Free Energy Relationships and Solid-State Stability

The relative thermodynamic stability of solids and the driving force for a transformation at constant temperature and pressure is determined by the difference in Gibbs free energy between forms (ΔG) and is given by:

$$\Delta G = \Delta H - T\Delta S \qquad (1)$$

The enthalpy difference between the forms, ΔH, reflects the lattice or structural energy differences and the entropy difference, ΔS, is related to the disorder and lattice vibrations. The relative stability is given by the algebraic sign of ΔG as follows:

1. ΔG is negative when the free energy decreases. The transformation can occur spontaneously and a change has the potential to continue to occur as long as the free energy of the system decreases
2. $\Delta G = 0$ when the system is at equilibrium with respect to the transformation and the free energy of the two phases is the same
3. ΔG is positive when the free energy increases and the transformation is not possible under the specific conditions.

Free Energy Relationships Between Polymorphs

Consider the Gibbs free energy curves for a hypothetical system of polymorphs A, B, and C shown in Figure 2. The free energy relationships between the forms are classified as *monotropic* or *enantiotropic*. Form C has the lowest free energy at all temperatures below the melting points, and is therefore *monotropically* related to forms A and B. Thus, form C is the most thermodynamically stable polymorph at all relevant temperatures. Forms A and B are *enantiotropically* related as there is a *transition temperature* below the melting points of the two forms. At the transition temperature, the free energies of the two forms are equal ($G_A = G_B$). Above and below the transition temperature, the relative free energy between the forms is reversed. Form A is more thermodynamically stable below the transition temperature ($G_A < G_B$) whereas form B is more thermodynamically stable above the transition temperature ($G_B < G_A$).

The difference in free energy between polymorphs is directly proportional to their relative solubilities. This is expressed for polymorphs A and B by the following equation[1]:

$$\Delta G = (G_B - G_A) = RT \ln\left(\frac{S_B}{S_A}\right) \qquad (2)$$

Where R is the universal gas constant and S_A and S_B are the solubilities of polymorph A and B, respectively. Thus, polymorph B will be more soluble (less stable) than polymorph A below the transition temperature whereas the relative solubilities will be reversed above the transition temperature.

The difference in free energy between polymorphs is also directly proportional to their relative intrinsic dissolution rates (D_B/D_A) as expressed by the

[1] Equation 2 assumes that concentrations (solubilities) can be substituted for activities if the ratio of the activity coefficients of the two polymorphs is approximately 1.

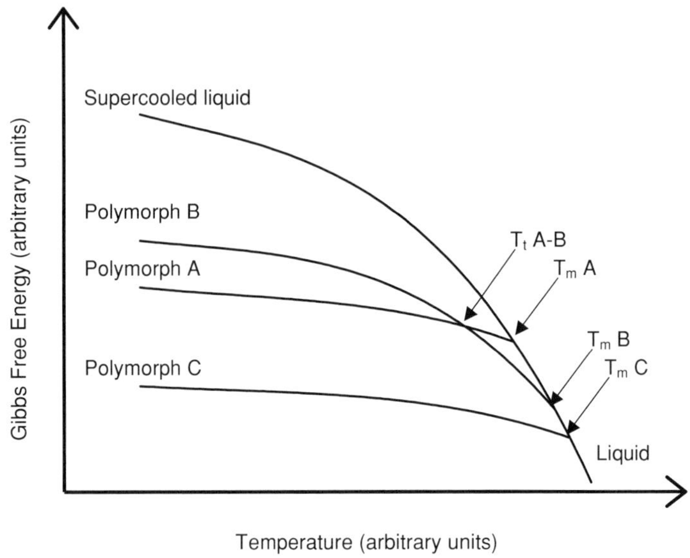

Figure 2. Gibbs free energy curves for a hypothetical system of polymorphs A, B, and C. The systems are classified as: *monotropic* (forms A and C, forms B and C) or *enantiotropic* (forms A and B) with a transition temperature, T_t. Melting points, T_m, for the polymorphs are shown by the intersection of the curves for the crystalline and liquid states. Adapted from Rodríguez-Spong et al., (2004) according to the relationships developed by Shalaev and Zografi (2002).

following equation:

$$\Delta G = (G_B - G_A) = RT \ln \left(\frac{D_B}{D_A} \right) \tag{3}$$

It is important to note that the difference in free energy between polymorphs is identical in different solvents at the same temperature, assuming that the activity coefficient in each solvent is independent of concentration. *Hence, the thermodynamic stability relationship between polymorphs depends only on the temperature at constant pressure and is completely independent of the solvent.* This is an important point to emphasize because crystallizations from different solvents frequently yield different polymorphs. In these instances, the solvent is not changing the relative thermodynamic stabilities of the polymorphs. These observations are the result of kinetic and/or molecular recognition effects on crystallization processes, which are discussed in subsequent sections of this chapter.

Free Energy Relationships Between Solvates and Non-Solvates

While the relative thermodynamic stability of polymorphs depends only on the temperature at constant pressure, the stability relationship between a solvate and a non-solvated form, or of two solvates, depends not only on the temperature but also on the activity of the solvent at constant pressure. Most of the following discussion will focus on hydrates but the concepts can also be applied to other solvates.

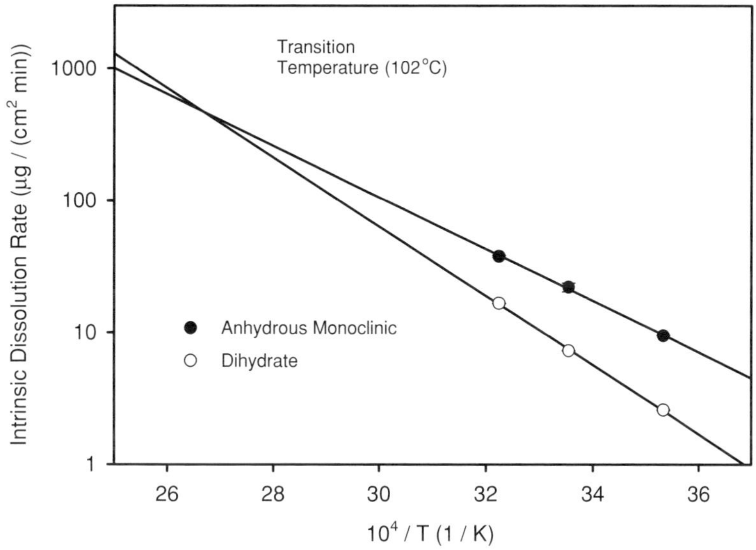

Figure 3. Van't Hoff-type plot of intrinsic dissolution rates (log scale) against the reciprocal of the absolute temperature for the anhydrous form III and dihydrated forms of carbamazepine in water. Data obtained from Murphy et al. (2002).

First, consider the free energy relationship between a solvate and a non-solvated form in the solvent forming the solvate (i.e. when the activity of the solvent is 1). As is the case for polymorphs, the free energy relationship between the solvate and non-solvated form is directly proportional to the relative solublities and the relative intrinsic dissolution rates of the two forms as expressed in Equations 2-3.

For example, Murphy et al. (2002) determined the relative intrinsic dissolution rates of carbamazapine anhydrate (form III)[2] and dihydrate forms in water at 10, 25, and 37 °C. Figure 3 shows the Van't Hoff-type plot of the intrinsic dissolution rates of the forms. A transition temperature of 102 °C was calculated by extrapolation of the intrinsic dissolution rate lines to the point of intersection. The anhydrous form is more stable above the transition temperature whereas the dihydrate is more stable below the transition temperature.

A general observation applying to solvates is that the solvated form is usually less soluble than the unsolvated form in the solvent forming the solvate at near ambient temperatures (Pudipeddi and Scrajuddin, 2005). However, this general rule is valid only at temperatures below the transition temperature. Indeed, some systems can have transition temperatures below room temperature such that the non-solvated form is less soluble and more thermodynamically stable even when the solvent activity is equal to 1.

[2] There has been some confusion in the literature on the nomenclature of carbamazepine anhydrous polymorphs. For this chapter, we will use the following nomenclature: Form I (Triclinic), Form II (Trigonal), Form III (*P*-monoclinic), and Form IV (*C*-monoclinic).

Next, consider the free energy relationship between a solvate and a non-solvated form when the activity of the solvent forming the solvate is less than 1 (i.e. solvent composition <100%). The dependence of the relative thermodynamic stability between anhydrous and hydrated crystalline forms on the water activity has been described by Grant and Higuchi (1990). The formation of hydrate crystals from an anhydrous phase is represented as:

$$A_{(S)} + mH_2O \leftrightarrow A \cdot mH_2O_{(S)}$$

$$K_h = \frac{a[A \cdot mH_2O_{(s)}]}{a[A_{(s)}]\,a[H_2O]^m} \tag{4}$$

where K_h is the equilibrium constant for the hydration/ dehydration equation, m is the number of moles of water taken up by 1 mole of the anhydrous phase, $a[A \cdot mH_2O_{(S)}]$ is the thermodynamic activity of the hydrate, $a[A_{(S)}]$ is the activity of the anhydrate, and $a[H_2O]$ is the water activity. If pure solids of the hydrate and anhydrate represent the standard state of unit activity, then the equilibrium constant can be simplified to:

$$K_h = a[H_2O]^{-m} \tag{5}$$

Therefore, the water activity, abbreviated a_w, in the crystallization solvent will determine the relative thermodynamic stability between anhydrate and hydrate, as well as the lowest energy hydration state of the hydrate in a given solvent system. At a given temperature, there is a *critical* a_w below which the anhydrous form is more stable and above which the hydrated form is more stable. Literature data on the activity coefficient of water, γ_w, as a function of mole fraction water, x_w, is available for many water-organic solvent mixtures (Gölles, 1961; Otsuka et al., 1967; Washburn, 1928; Sokolova and Morachevskii, 1967). The water activity as a function of x_w, can be determined using the following relationship:

$$a_w = \gamma_w x_w \tag{6}$$

Figure 4 contains the a_w as function of x_w plots for methanol, ethanol, and isopropanol at near ambient temperatures. Note that the relationships are not linear with a slope = 1, as predicted by Raoult's law. Rather all plots show positive deviations from Raoult's law and the magnitude of the deviation increases with increasing akyl chain length. This is an important point to consider when choosing solvent compositions in aqueous-organic mixtures, as the water activity in a given solvent may be high even at relatively low water fractions (Figure 5).

Zhu and co-workers showed that the critical a_w of a drug substance at a given temperature can be determined by suspending the hydrate and anhydrate in various organic solvent/water mixtures at controlled a_w and constant temperature (Zhu et al., 1996; Zhu and Grant, 1996). The slurries equilibrate to the lowest energy form and the relative thermodynamic stability at a given temperature can be determined as a function of a_w. This method has been used to determine the relative physical stability of many hydrate-anhydrate systems. (Khankari and Grant, 1993; Beckmann and Winter, 1999; Ticehurst et al., 2002; Sacchetti, 2004) For example, Ticehurst et al. (2002) determined the a_w of the

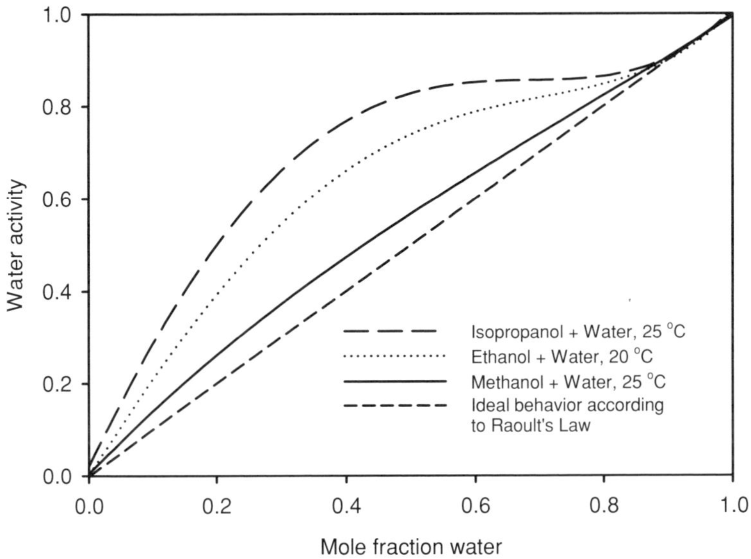

Figure 4. a_w as function of x_w plots for methanol, ethanol, and isopropanol at near ambient temperatures. Data obtained from (Gölles, 1961; Otsuka et al., 1967; Washburn, 1928).

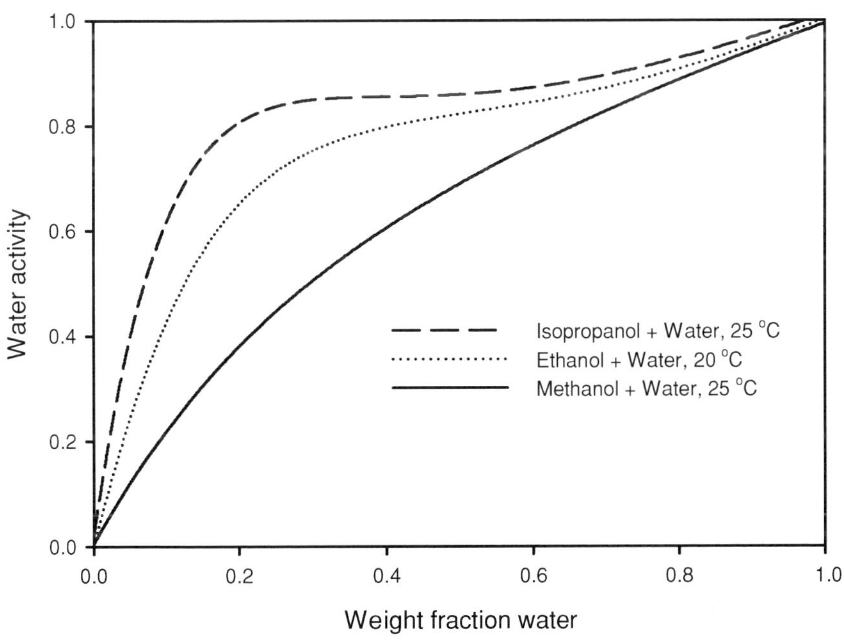

Figure 5. a_w as function of weight fraction water plots for methanol, ethanol, and isopropanol at near ambient temperatures. Data obtained from (Gölles, 1961; Otsuka et al., 1967; Washburn, 1928).

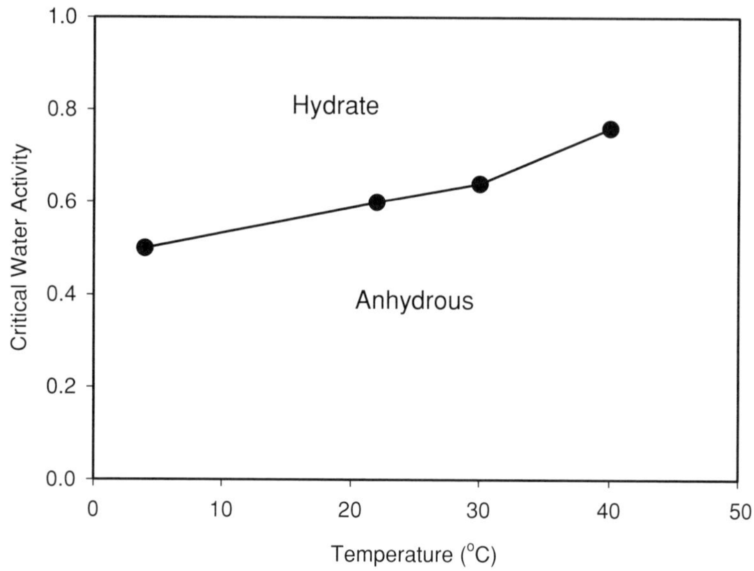

Figure 6. Theophylline hydrate/anhydrate phase diagram from 4 to 40 °C. Reproduced from (Ticehurst et al., 2002) with permission from Elsevier.

theophylline hydrate-anhydrate system from 4 to 40 °C. The data can be used to construct a hydrate/anhydrate phase diagram (Figure 6). In addition to defining the thermodynamic stability relationship between a hydrate/anhydrate system, these phase diagrams can be used to rationally design a crystallization process to selectively isolate either the anhydrate or hydrate phases of a drug substance.

It is important to note that the solubility of the drug in these organic solvent/water mixtures must be high enough to facilitate phase transformation. The factors that determine the kinetics of solvent mediated phase transformations are discussed in more detail in subsequent sections of this chapter.

Kinetics and Molecular Recognition

The kinetics of crystallization can be divided into three processes: 1) supersaturation 2) nucleation and 3) crystal growth. These processes are also significantly influenced by molecular recognition phenomena in solution. This section establishes the theoretical framework for these processes with specific focus on those parameters that are influenced by the solvent. These processes are also discussed extensively elsewhere (Rodríguez-Hornedo, 1991; Rodríguez-Hornedo and Murphy, 1999; Rodríguez-Hornedo and Sinclair, 2002).

Supersaturation

A solid phase is crystallized from solution if the chemical potential of the solid phase is less than that of the dissolved component. A solution in which the chemical potential of the solute is the same as that of the corresponding solid phase is in equilibrium with the solid phase under the given conditions (temperature, pH, and concentration) and called a *saturated solution.* In order for crystallization from solution to occur, however, this equilibrium concentration or solubility must be exceeded. This excess concentration or chemical potential, called the *supersaturation,* is the driving force for nucleation and crystal growth. The supersaturation is most commonly expressed as the concentration divided by the solubility (c/s). Thus, supersaturated states may be created by increasing the solute concentration or decreasing the solute solubility using a variety of methods including:

1) Methods that create supersaturation by increasing the solute concentration include:
   a) solvent evaporation
   b) dissolution of a metastable solid phase (transformations of amorphous to crystalline, anhydrous to hydrate, more soluble to less soluble polymorph, and salt to free acid or free base).
2) Methods that create supersaturation by decreasing the solute solubility include:
   a) temperature change (cooling)
   b) addition of solvent or additives that lower the solute solubility (i.e. anti-solvent addition)
   c) pH change
   d) addition of ions that participate in precipitation of the solute (i.e. salting out)

The concentration versus temperature curve shown in Figure 7 shows the hypothetical scenario in which supersaturation is created by cooling. When a solution represented by point A is cooled without the loss of solvent (line ABC), spontaneous nucleation will occur, when conditions corresponding to point C are attained. The boundary between the solubility line (solid line) and the line (points B to C) at which spontaneously nucleation occurs is referred to as the *metastable zone.* When the metastabe zone is exceeded, the nucleation rate increases rapidly, and the crystallization process becomes uncontrolled. Within the metastable zone, the nucleation rate is slower such that control over the crystallization process may be achieved. Nucleation may also be facilitated within the metastable zone by the addition of crystalline seeds of the solute of interest (i.e. secondary nucleation).

The width of the metastable zone is affected by the solvent as well as a number of other factors including the agitation rate, the cooling rate, the presence of soluble additives, and the thermal history of the solution (Birchall and Davey, 1981; Garti et al., 1981; Nakai et al., 1973). The solvent influences the metastable zone width primarily because the nucleation rate of a given compound will vary from solvent to solvent. This is because nucleation rate is directly affected by the supersaturation and solubility a compound may attain in a given solvent, as well as molecular recognition phenomena between solute and solvent, as discussed in the next sections.

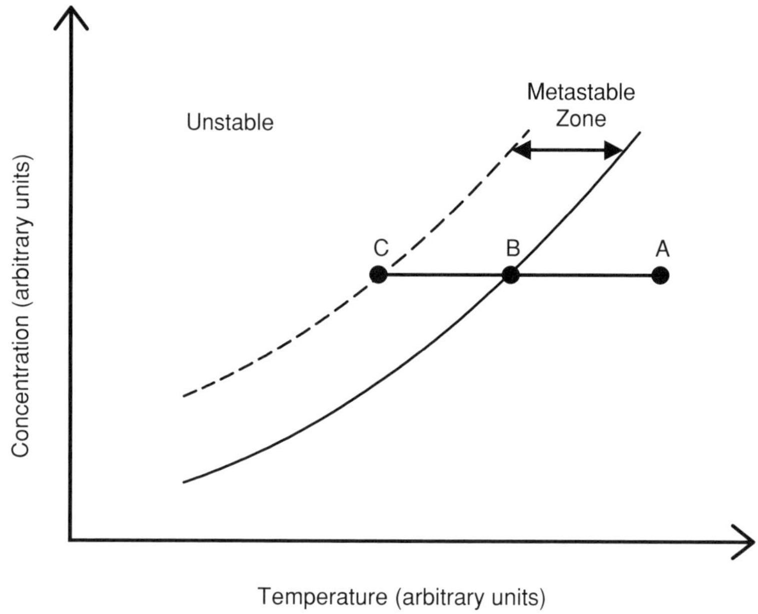

Figure 7. Concentration versus temperature curve for a hypothetical system in which supersaturation is created by cooling. The solid line represents the solubility curve, the dashed line represents the point of spontaneous nucleation. The region between these two lines is the metastable zone. Adapted from Rodríguez-Hornedo (1991).

Nucleation

Nucleation Kinetics

Nucleation may occur spontaneously, or it may be induced by a foreign surface. These two cases are frequently referred to as *homogeneous* and *heterogeneous* nucleation, respectively. Nucleation that is promoted by crystals of the crystallizing solute is known as secondary nucleation. These mechanisms are thoroughly discussed by Mullin (2001), Myerson (2002), and Zettlemoyer (1969). Of the two nucleation mechanisms, the solvent has a much larger effect on homogeneous nucleation. Thus, this section discusses homogeneous nucleation theory with focus on those parameters that are affected by the solvent.

The rate for homogeneous nucleation (J) of spherical assemblies can be expressed by the classical nucleation equation: (Gibbs, 1948; Volmer, 1939; Becker and Döring, 1935; Turnbull and Fisher, 1949):

$$J = N_0 \nu \exp\left(\frac{-16\pi \upsilon^2 \gamma_{12}^3}{3(k_B T)^3 \left(\ln\left(\frac{c}{s}\right)\right)^2}\right) \tag{7}$$

Where J is the number of nuclei formed per unit time per unit volume; N_0 is the number of molecules of the crystallizing phase in a unit volume; ν is the frequency of atomic or molecular transport at the nucleus–liquid interface; υ is the molecular volume of the crystallizing solute; γ_{12} is the interfacial energy per unit area between the crystallization solvent, 1, and the nucleating cluster, 2; k_B is the Boltzmann constant, T is temperature, c is the solute concentration, s is the solubility, and c/s is the supersaturation.

Equation 7 shows that the nucleation rate can be experimentally controlled via the following parameters: molecular or ionic transport, viscosity, supersaturation, solubility, solid–liquid interfacial tension, and temperature. Of these parameters, the solubility and solid-liquid interfacial tension are most likely to be significantly influenced by the solvent (Davey, 1982; Rodríguez-Hornedo, 1991; Rodríguez-Hornedo and Murphy, 1999; Rodríguez-Hornedo and Sinclair, 2002). This is because pharmaceutical materials generally exhibit a wide range of solubilities in different solvents. Thus, a change in crystallization solvent will most likely result in a change in solubility. From Equation 7, one can reach the conclusion that an increase in solubility, due to solvent change, will increase the nucleation rate because increased solubility will lead to an increase in the probability of molecular collisions by increasing N_0 (Rodríguez-Hornedo and Murphy, 1999; Gu et al., 2001). Moreover, the interfacial tension is inversely proportional to the logarithm of the solubility (Mersmann, 1990) such that changes in crystallization solvent will also result in changes in interfacial tensions. For example, when a change in solvent leads to an increase in solubility, the interfacial energy decreases because the affinity between crystallizing medium and crystal increases, due to the increase in adsorption of the solvent at the interface (Mersmann, 1990).

True nucleation rates are difficult to measure in practice, due to the inability to detect the formation of very small nuclei. Nucleation rates may be approximated by the measurement of induction times, which are inversely proportional to nucleation rates. Induction times are defined as the time that elapses between the achievement of supersaturation and the visual appearance of crystals in a given system (Mullin, 2001).

The dependence of nucleation rate on solubility is also consistent with Ostwald's (1897) law of stages regarding the preferential crystallization of metastable crystalline phases. It states that: "When leaving an unstable state, a system does not seek out the most stable state, rather the nearest metastable state which can be reached with minimal loss of free energy." This indicates that a metastable (more soluble) crystalline phase will generally crystallize before a more thermodynamically stable (less soluble) crystalline phase, because it will have a higher nucleation rate.

However, Ostwald's law of stages is not universally valid because the appearance and evolution of solid phases are determined by the kinetics of nucleation and growth under the specific experimental conditions (Bernstein et al., 1999; Davey et al., 1997; Threlfall, 2000) and by the link between molecular assemblies and crystal structure (Davey et al., 2002; Blagden et al., 1998; Gu et al., 2001).

Molecular Recognition Effects on Nucleation

Although solubility is an important factor in determining nucleation rates, the dependence of nucleation rate on solubility may not be followed in the presence

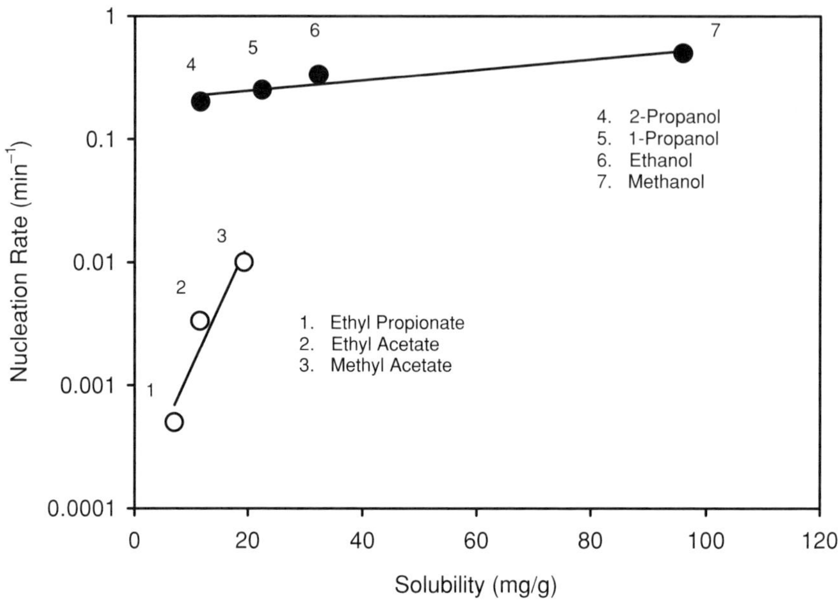

Figure 8. Relationship between nucleation rate and solubility of carbamazepine form III at 25 °C and c/s = 2.0. Data obtained from Kelly (2003).

of strong solvent–solute interactions that interfere with the formation of molecular assemblies compatible with those in the crystalline state (Gu et al., 2001; Kelly, 2003). Prior to nucleation, solute molecules in solution are closely associated with solvent molecules. These solvated solute molecules must de-solvate such that the necessary solute-solute interactions may occur to bring about the formation of supramolecular assemblies that will lead to crystallization. Thus, when strong interactions exist between solvent and solute molecules in solution, nucleation kinetics may be significantly altered. For example, Kelly (2003) has shown that the nucleation rate of the form III of carbamazepine is significantly influenced by solvent-solute interactions. Figure 8 shows the nucleation rate of carbamazepine form III as a function of solubility for two solvent classes: 1) alcohols and 2) esters. Note that within each class of solvent, the nucleation rate increases with solubility at constant supersaturation, as predicted by equation 7. However, the classical nucleation equation is not followed when comparing the two solvent classes as the nucleation rates in alcohols are 2-3 orders of magnitude greater than those in esters at similar solubilities. This observation may be explained by the effect of molecular recognition phenomena on nucleation kinetics. Inspection of the molecular structure of carbamazepine (Figure 9) shows an excess of hydrogen bond donors (1 hydrogen bond acceptor, 2 hydrogen bond donors). Thus, solvents with a strong propensity to accept hydrogen bonds (i.e. esters) interact more strongly with carbamazepine in solution as compared to solvents that possess relatively equal hydrogen bond donating and accepting propensity (i.e. alcohols). This results in slower nucleation of carbamazepine (III) in esters as compared to alcohols. This phenomenom also gives rise to solvent effects on

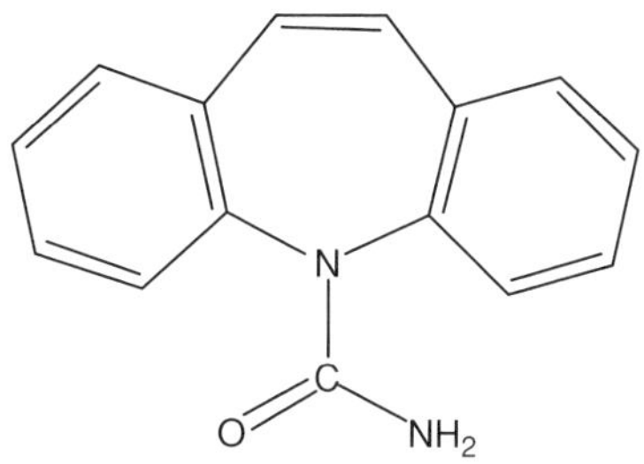

Figure 9. Chemical structure of carbamazepine.

polymorphic outcome of carbamazepine as discussed in subsequent sections of this chapter.

In addition to influencing the nucleation rate of non-solvated crystals, solvent-solute interactions may be so strong that solute molecules do not de-solvate, rather the solvent molecules are incorporated in the crystal lattice along with the solute to form a solvate. The propensity of an API molecule to form solvates has been related to molecular structures, hydrogen bond patterns, and crystal packing (Nangia and Desiraju, 1999; Görbitz and Hersleth, 2000; Bingham et al., 2001; Infantes and Motherwell, 2002; Gillon et al., 2003; Morris and Rodríguez-Hornedo, 1993; Khankari and Grant, 1995; Morris, 1999). The Cambridge Structural Database (CSD) has been useful to analyze the nature of supramolecular networks in solvated crystals (Nangia and Desiraju, 1999; Görbitz and Hersleth, 2000; Infantes and Motherwell, 2002; Gillon et al., 2003) and to identify molecular recognition events that lead to incorporation of solvents in crystal structures. Nangia and Desiraju (1999) have searched the CSD to assess the propensity of hydrogen bonding organic solvents to be incorporated into molecular crystals. The results of their CSD survey are contained in Table 1. The results are expressed as values for usage corrected occurrence (O_{corr}). Solvents with O_{corr} values greater than 1 tend to form solvates more frequently whereas those with values less than 1 tend to form solvates less frequently. They found that the propensity of organic solvents to be included in molecular crystals depends on their ability to effectively participate in hydrogen bonding, and that multi-point recognition via strong and weak hydrogen bonds between solvent and solute molecules facilitates solvate formation. Solvents such as N,N-dimethylformamide, dimethylsulfoxide, and 1,4-dioxane contain multiple sites for strong and weak hydrogen bonding and are therefore more likely to be incorporated into molecular crystals. Conversely, ethyl acetate, ethanol, and diethyl ether have a lower propensity to form solvates because these solvents do not have the multiple hydrogen bonding sites necessary to effectively participate in multi-point hydrogen bonding. Nangia and Desiraju (1999) also recognize the importance of the kinetics and energetics of solvation and desolvation events during crystallization. Because formation of an unsolvated crystal by crystallization from solvent requires that solute–solvent interactions be replaced

Solvent	N	N_{ord}	N_{Acta}	O^a_{corr}
N,N-Dimethylformamide	122	63	5	5.69
Dimethylsulfoxide	142	70	7	4.73
1,4-Dioxane	161	83	8	4.70
1,4-Xylene	49	22	3	3.81
Benzene	471	271	32	3.43
Tetrahydrofuran	188	54	14	3.13
Acetonitrile	396	181	31	2.98
Acetic acid	105	54	10	2.45
Carbon tetrachloride	59	14	7	1.97
Toluene	266	65	36	1.72
Dichloromethane	455	183	77	1.38
Chloroform	386	132	67	1.34
Methanol	795	433	158	1.17
Acetone	346	169	82	0.98
1-Propanol	65	33	16	0.95
Cyclohexane	56	9	21	0.62
Ethyl acetate	155	52	75	0.48
Ethanol	406	168	206	0.46
Diethyl ether	144	54	96	0.35
Hexane	84	21	181	0.11
Total	4851	2131	1132	

Table 1. Occurences of 20 common solvents in organic crystal structures in decreasing order of O_{corr}. Reproduced from (Nangia and Desiraju, 1999) by permission of The Royal Society of Chemistry.

[a] Calculated by the formula $(N/4851)\ (N_{Acta}/1132) = 0.233(N/N_{Acta})$

N = raw frequency of occurrence, N_{ord} = number of fully ordered crystal structures in each case N, N_{acta} = number of times solvent was used for crystallization in *Acta Crys. C* from 1986–1996, O_{corr} = usage corrected occurrence

by solute–solute interactions, strong solute–solvent interactions may lead to nucleation of solvated crystals.

In summary, the nucleation rate is governed by an intricate balance between the solubility of the compound in a given solvent and the magnitude of molecular recognition phenomena between solute and solvent (Gu et al., 2002). Thus, the nucleation rate is expected to be fastest in solvents providing an optimal compromise between the solubility of the compound and the strength of solvent-solute interactions. When solvent-solute interactions are extremely strong, the de-solvation step necessary to allow solute-solute molecular recognition may not occur and the solvent may be incorporated into the crystal lattice to form a solvate.

Crystal Growth

Crystal Growth Kinetics

Once the nucleation step has been overcome, nuclei grow into macroscopic crystals. This stage of the crystallization process is known as crystal growth. The crystal growth process consists of several stages through which the growth units (i.e., the crystallographic basis that is "tacked onto" the space lattice to form the crystal structure) pass. These include:

(1) transport of the growth unit from or through the bulk solution to an impingement site, which is not necessarily the final growth site
(2) adsorption of the growth unit at the impingement site
(3) diffusion of the growth units from the site of impingement to a growth site
(4) incorporation into the lattice

Desolvation of the growth unit may occur anywhere in steps 2–4, or the solvent may be adsorbed with the growth unit. Any of these steps can be rate-limiting in the crystal growth process and which step is rate-limiting will depend on the solvent. When the diffusion of molecules from the bulk to the crystal surface is the rate-limiting step, crystal growth is *volume-diffusion controlled* whereas if the incorporation of a growth unit into the lattice is the slowest process growth is *surface-integration controlled.* (Nyvlt et al., 1985; Mullin, 2001; Söhnel and Garside, 1992; Myerson, 2002). Thus, the growth mechanism and kinetics may vary in different solvents, since the rate-limiting step depends on the solvent. For example, if the crystal growth mechanism in a given solvent is volume-diffusion controlled, the viscosity of the solvent may significantly influence the growth rate.

For surface-integration controlled growth, the roughness of the crystal surface determines whether growth occurs by a *continuous* (relatively rough surfaces) or a *layer* (relatively smooth surfaces) mechanism. The crystal growth rate is generally faster for a continuous mechanism because a rough surface contains more "kink" sites, at which growth units may be incorporated. The degree of molecular roughness has been quantified in terms of the α factor, which defines the enthalpy changes that take place when a flat interface is roughened (Bourne and Davey, 1976). The value of α is inversely related to surface roughness, that is, as α decreases, the degree of surface roughness increases and the crystal growth rate increases. For solution growth, α is related to the solubility

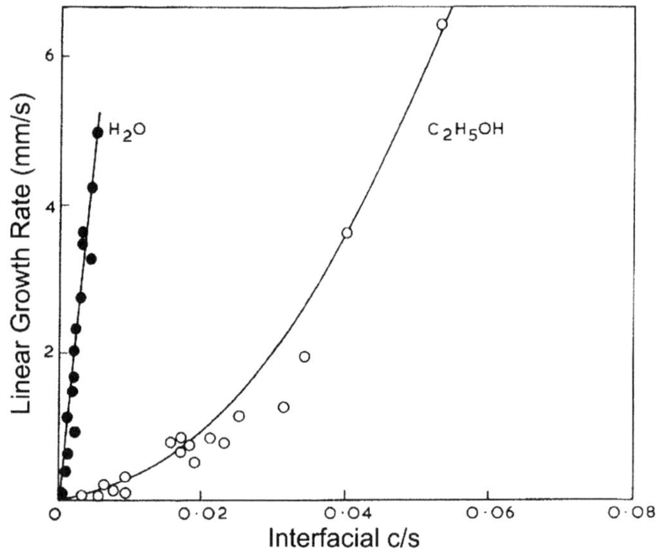

Figure 10. Growth kinetics of hexamethylene tetramine from water and ethanol. Reproduced from Davey (1979) with permission from Elsevier.

by the following equation: (Davey, 1982)

$$\alpha = \xi \left[\frac{\Delta H_f}{RT_m} - \ln X_s(T) \right] \tag{8}$$

Where ΔH_f is the heat of fusion of the solute, R is the universal gas constant, T_m is the melting point of the solute, X_s is the mole fraction solubility of the solute in the solvent, and ξ is a crystallographic factor representing the ratio of the number of first nearest neighbors in the surface to that in the bulk. The important implications of equation 8 are that as solubility increases, alpha decreases such that the surface roughness increases and the overall growth rate increases. Therefore changes in crystallization solvent will be expected to bring dramatic changes in crystal growth mechanism, kinetics, and crystal morphology. For example, Figure 10 shows the data for hexamethylene tetramine crystals growing in ethanolic and aqueous solutions (Davey, 1979). The solubility is ten times greater in water than in ethanol and leads to an order of magnitude increase in the growth rate on changing solvents. The calculated α values are 0.6 and 5.2 for growth from water and ethanol, respectively.

Molecular Recognition Effects on Crystal Growth

It is important to note that in cases where a strong interaction exists between solvent and solute, the crystal growth rate may be slowed despite high solubility (Gu et al., 2001). As discussed previously for nucleation, strong solvent-solute interactions may hinder the desolvation step necessary for crystal growth, such that the overall crystal growth rate is impeded. Moreover, strong solvent-solute interactions may lead to adsorption of the solvent at the growing surface, which may also slow the growth rate. Thus, the crystal growth rate is expected to be fastest in solvents providing an optimal compromise between the solubility of the compound and the strength of solvent-solute interactions (Gu et al., 2001).

Interplay Between Thermodynamics, Kinetics, and Molecular Recognition

When a compound exists in various solid-state forms there are two important questions to address: (1) what is their relative thermodynamic stability, or the conditions and direction in which a transformation can occur, and (2) how long will it take for the transformation to reach equilibrium? Thermodynamics provides information about the first question and kinetics about the second. While thermodynamics establishes the stability domains of the various solid states, once a metastable domain is encountered the kinetic pathways will determine which form will be created and for how long it can survive.

The balance between the kinetic and thermodynamic factors is illustrated by the free energy-reaction progress diagram (Figure 11) for a transition from

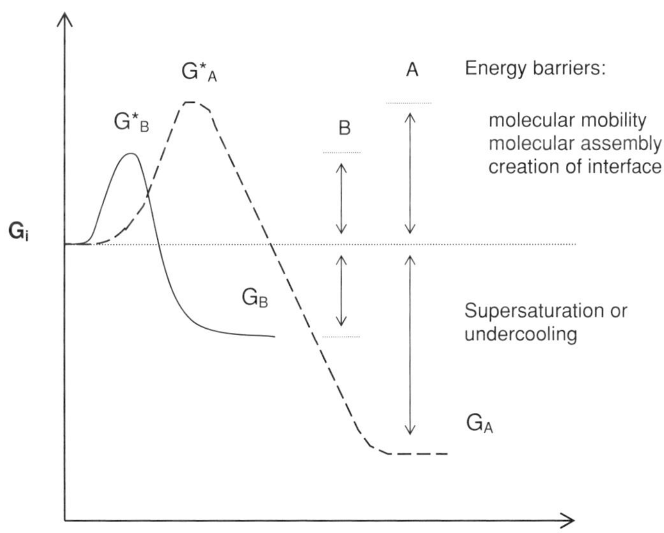

Figure 11. Schematic diagram for a hypothetical transition from the initial state, G_i, to two different solid forms A or B, with free energies G_A and G_B. Form A is more stable and less soluble than B. A transition from the initial state G_i to state A or B will depend on the energy barrier and according to this reaction pathway the height of the energy barrier for structure A, $(G^*_A - G_i)$ is greater than that for B, $(G^*_B - G_i)$. Because the rate of nucleation is related to the height of the energy barrier on the reaction path, B will nucleate at a faster rate than A even though the change in free energy is greater for A $(G_A - G_i)$ than for B $(G_B - G_i)$. From Rodríguez-Spong et al., (2004).

the initial state G_i, to two different solid forms A or B. Form A is more stable and less soluble than B ($G_A < G_B$). G_i may represent a supersaturated solution in a single or multiple component system. The reaction follows a path through an energy maximum between the initial and final states. This resistance to the transition from G_i to G_A or G_B arises because there is an energy barrier for molecular diffusion, molecular assemblies, and for the creation of an interface. For a chemical reaction in a homogeneous system, this energy maximum is the transition state and reflects elementary reactions, bimolecular or trimolecular, that yield products with new covalent bonds. In comparison, a crystallization event or phase transformation leads to heterogeneous systems in which a separate new phase is created from a supramolecular assembly by formation of non-covalent bonds.

A transition from the initial state G_i to state A or B will depend on the energy barrier and according to the reaction pathway in Figure 11, the height of the energy barrier for structure A ($G^*_A - G_i$) is greater than that for B ($G^*_B - G_i$). Because the rate of nucleation is related to the height of the energy barrier on the reaction path, B will nucleate at a faster rate than A even though the change in free energy is greater for A ($G_A - G_i$) than for B ($G_B - G_i$). This is one of the possible behaviors that could be observed in the order of appearance of polymorphs and is referred to as Ostwald's ripening (Ostwald 1897). However, Ostwald's Law is not always obeyed because the kinetics of nucleation and growth of a given phase may be significantly influenced by molecular recognition phenomena between solute and solvent, as discussed previously.

Solvent Effects on Polymorphic Outcome

Thermodynamic and Kinetic Effects on Polymorphic Outcome

Because of the interplay between thermodynamic factors (free energies, solubilities, concentrations, interfacial tensions), temperature, and molecular recognition in determining nucleation of a new phase, it is essential to consider the effects of thermodynamic and kinetic factors when using solvents to selectively nucleate polymorphs. Threlfall (2000) has thoroughly considered thermodynamic and kinetic factors and the conditions in which the solvent may or cannot affect polymorphic outcomes. The analysis is briefly summarized here.

Consider the solubility curves and metastable zone widths for a hypothetical enantiotropic system of polymorphs I (high melting) and II (low melting) with transition temperature T_t shown in Figure 12. The lines initiated at points A, B, C, D, E, F, and G represent the cooling of undersaturated solutions of various concentrations to a temperature of crystallization (A1, B1, etc.). In general, the solvent will not directly affect the polymorphic outcome when the temperature of crystallization is adequately above or below the transition temperature. These concentration and temperature regimes are represented by the shaded regions shown in Figure 12. The only role the solvent plays in these regimes is to provide the solubility needed to reach the necessary concentration at a given temperature. The polymorphic outcome will be under thermodynamic control in these

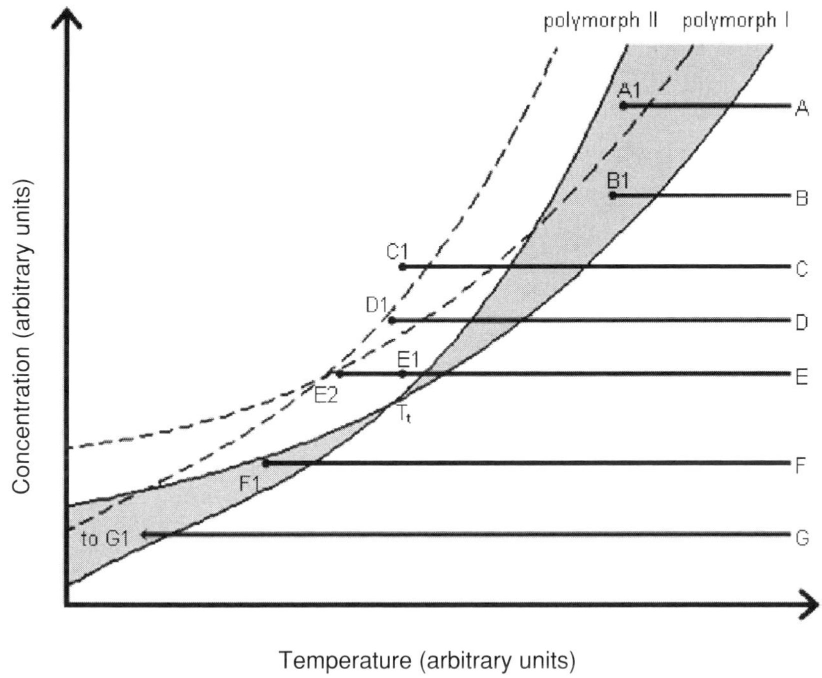

Figure 12. Polymorphic system of two enantiotropically related polymorphs I and II with transition temperature, T_t. Solubility curves, full lines; metastable zone limits, dashed lines. A-G, initial state of hot, undersaturated solutions; A1-G1 and E2, state of solution at point of initial crystallization. The shaded regions represent concentration and temperature regimes in which the the system is under thermodynamic control and the solvent will not directly affect the polymorphic outcome. Reproduced from (Threlfall, 2000) with permission. Copyright 2000 American Chemical Society.

regimes. For example, a solution of concentration A cooled to a crystallization temperature A1 will spontaneously crystallize as polymorph I, since it is outside the metastable zone of polymorph I but is still undersaturated with respect to polymorph II. Collection of the material at temperature A1 would ensure exclusive isolation of polymorph I. However, the solution may be cooled further and collected as polymorph I, provided the solvent-mediated transformation to form II does not occur within the timeframe of the experiment. The polymorphic outcome for solution B cooled to crystallization temperature B1 will be exactly the same as for A. The only difference is that seeding will be necessary to nucleate polymorph I, since B1 will have not yet crossed the metastable zone for polymorph I. Solutions G and F are directly analogous to solutions A and B, respectively, except polymorph II will be exclusively produced regardless of the solvent.

When the concentration and temperature of crystallization falls outside of the shaded regimes, the polymorphic outcome will be under kinetic control and therefore may depend on the solvent. For instance, a solution of concentration C cooled to a crystallization point C1 will be outside the metastable zones of both polymorphs I and II, such that the either or both may crystallize. The polymorphic outcome will be governed by the relative nucleation rates of the polymorphs, which depend directly on the solvent. Solution D cooled to point D1 will be outside the metastable zone of polymorph I yet still within the metastable zone of polymorph II. The expected outcome would be polymorph I in this case although the solvent may be important. Cooling a solution of concentration E to points E1 or E2 will result in solutions that are within the metastable zones of both polymorphs I and II. Here the polymorphic outcome will also depend directly on solvent and will be particularly sensitive to accidental seeding. Both polymorphs are likely to crystallize in this regime leading to so-called concomitant polymorphs (Bernstein, 1999).

Threlfall's analysis has direct applications in the isolation of the desired polymorphs on scale. For this, it is essential to know the solubility as a function of temperature and metastable zone widths of the polymorphic system in order to reach those regimes in which the desired polymorph is exclusively crystallized. The analysis also has implications for polymorph screening, where the emphasis in experimental design is commonly focused on solvent diversity. This is born from the pervasive assumption that polymorphic outcome is primarily governed by specific interactions between solute and solvent. While solvent diversity is certainly an important consideration, Threlfall's analysis shows that other factors such as supersaturation, concentration, cooling rate, nucleation temperature, and collection temperature should also receive adequate attention in experimental design for polymorph screening.

Molecular Recognition Effects on Polymorphic Outcome

When the concentration and temperature of crystallization fall outside of the shaded regimes in Figure 12 (i.e. kinetically controlled regimes), strong solvent-solute interactions may significantly influence molecular aggregation processes in solution. Consequently, the solvent may govern the polymorphic outcome in these regimes. Specific solvent-solute interactions have been shown to direct the polymorphic outcome in the crystallization of several compounds including: carbamazepine (Kelly, 2003), sulfathiazole (Koshkhoo and Anwar, 1993; Blagden et al., 1998; Blagden, 2001), sulfamerazine (Gu et al., 2001), and 2,6-dihydroxybenzoic acid (Davey et al., 2000). Carbamazepine and sulfathiazole are discussed in more detail here as examples.

Kelly (2003) investigated the role that organic solvents have on directing nucleation of carbamazepine polymorphs II and III. Form III is the thermodynamically most stable form under ambient conditions and is monotropically related to form II. Recall from Figure 8, that the nucleation rate for form III of carbamazepine is 2-3 orders of magnitude slower in hydrogen bond accepting solvents (i.e. esters) as compared to solvents that may act as hydrogen bond donors or acceptors (i.e alcohols). The nucleation rates for form II were not significantly different from those observed for form III in alcohols, which lead to

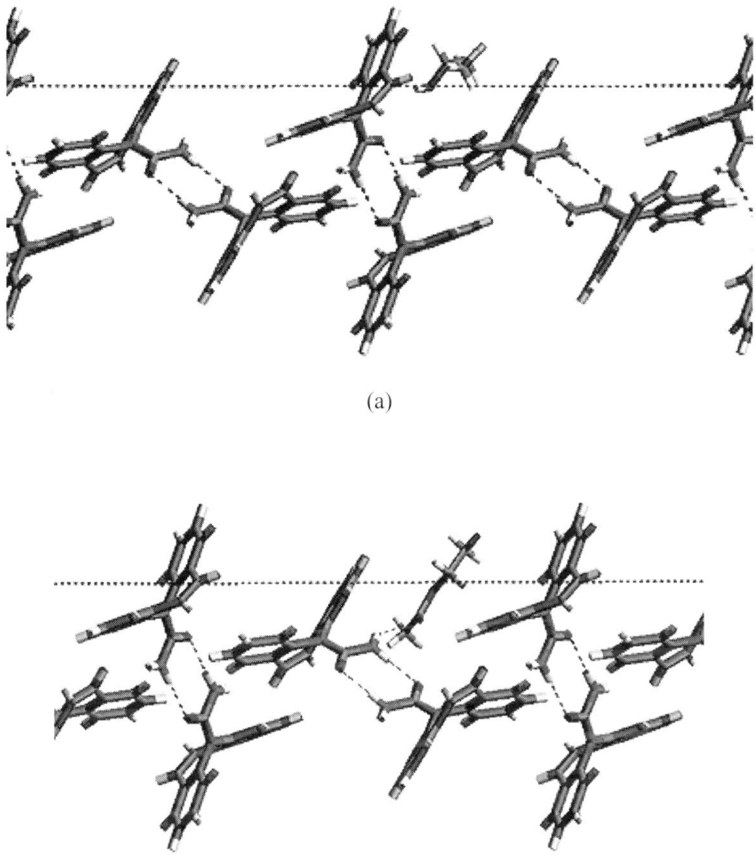

(a)

(b)

Figure 13. Solvent interaction at the (011) crystal face of carbamazepine form III. (a) 2-propanol and (b) ethyl acetate. Reprinted from Kelly, 2003 with permission. (See color insert after Index.)

concomitant crystallization of both polymorphs in alcohols. However, in esters, the nucleation rates for form II were observed to be 3-4 orders of magnitude faster than those observed for form III, which leads to selective nucleation of form II in ester solvents. Kelly (2003) then carried molecular simulation studies to evaluate the role of specific carbamazepine-solvent interactions in directing nucleation events. The molecular simulation predicts ethyl acetate to selectively hydrogen bond with the second amido hydrogen exposed on the carbamazepine form III {011} crystal face (Figure 13). This hydrogen bonding interaction is not predicted to occur in 2-propanol (Figure 13). This specific interaction of form III leads to the selective nucleation of the unaffected polymorph (form II) in hydrogen bond accepting solvents (i.e. ethyl acetate).

Sulfathiazole is another pharmaceutical molecule in which polymorphic outcome is significantly influenced by solvent-solute interactions (Koshkhoo and Anwar, 1993; Blagden *et. al.*, 1998; Blagden, 2001). Sulfathiazole has five polymorphs with known crystal structures. Forms I, II, III, and IV have been the most studied polymorphs with respect to solvent effects on polymorphic outcome. The relative thermodynamic stability of these sulfathiazole polymorphs under ambient conditions is IV > III > II > I. Koshkhoo and Anwar (1993) first reported that some solvents selectively favored the crystallization of particular sulfathiazole form(s). They observed that crystallizations from *n*-propanol exhibited extraordinary selectivity for form I, regardless of the supersaturation. The authors hypothesized that the apparent dependency of polymorphic outcome on the crystallization solvent was due to the selective absorption of the solvent to certain faces of some of the polymorphs thereby inhibiting their nucleation such that other polymorphs may nucleate. Blagden et al. (1998), also reported that form I selectively nucleated from *n*-propanol and did not convert to any of the more thermodynamically stable forms in suspension for up to 1 year at 30 °C, whereas suspensions of form I in water completely transformed to form IV (through forms II and III) within 24 hours at 30 °C. Consistent with the hypothesis of Koshkhoo and Anwar (1993), Blagden et al. (1998) proposed a mechanism in which the apparent selectivity of form I by n-propanol may be explained by solvent directed nucleation processes. A schematic of the proposed mechanism is shown in Figure 14. All four forms of sulfathiazole have been shown to crystallize as dimers (Blagden et al., 1998). The dimer in forms II, III, and IV, designated as β, is formed via hydrogen bonds between the sulfato oxygen and aniline hydrogen as well as aniline hydrogen and thiazole hydrogen. Form I has been shown to form a unique dimer, designated as α, consisting of two

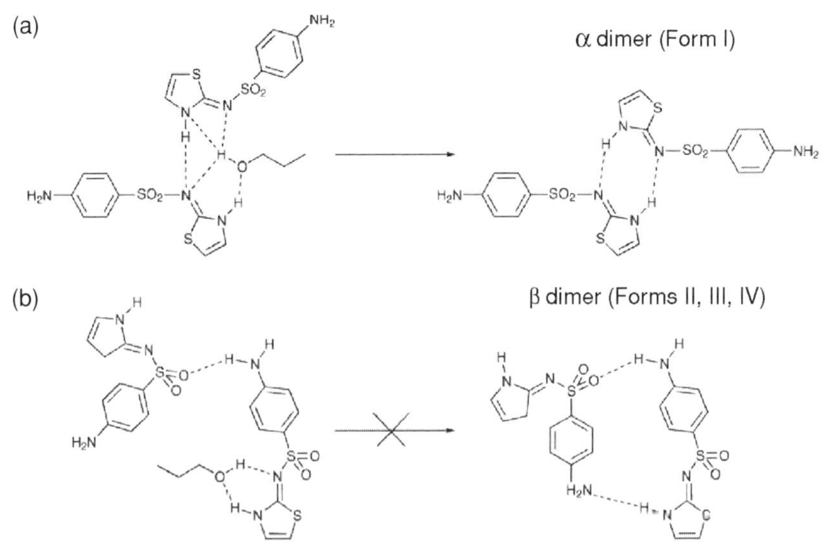

Figure 14. Proposed supramolecular assembly processes of sulfathiazole polymorphs in *n*-propanol. Reprinted from Blagden et al. (1998) with permission from The Royal Society of Chemistry.

imine nitrogens bonded to two thiazole hydrogens. The authors proposed that *n*-propanol selectively interacts with the imine nitrogens and thiazole hydrogens such that the solvated intermediate necessary to form the α dimer is stabilized while formation of the β dimer is blocked due to the formation of a steric barrier (Figure 14). This steric barrier is not as significant in other solvents (i.e. ethanol, nitromethane, water, ammonia), such that formation of the β dimer necessary to nucleate forms II, III, and IV is able to occur.

In summary, the particular polymorph that results from crystallization from a given solvent is a consequence of thermodynamic, kinetic, and molecular recognition events. Knowledge of the solubility curves and metastable zone widths is necessary to determine when the polymorphic outcome is governed by thermodynamics as opposed to kinetics and/or molecular recognition. In general, the polymorphic outcome is under thermodynamic control when the concentration and temperature of crystallization fall within the shaded regimes in Figure 12 (i.e. significantly above or below the transition temperature). Outside of these regimes, the polymorphic outcome may be significantly influenced by the relative nucleation rates of the polymorphs and/or specific interactions between a given polymorph and the solvent. Therefore, the intricate balance between thermodynamic, kinetic, and molecular recognition must be considered when designing experiments for polymorph screening, selection, and isolation.

Solvent-Mediated Phase Transformations

Consider a solid that may exist as multiple crystalline phases, polymorphs and solvates included. Recall from the Thermodynamics section of this chapter that at a given temperature and solvent activity (excluding transition temperatures and critical solvent activities) only one solid phase has the lowest free energy and hence is the most thermodynamically stable of all the phases in the system. All other phases in the system have higher free energies and are therefore metastable with respect to the most thermodynamically stable form. Given sufficient time, these metastable phases will eventually transform to the most stable phase of the system, because of the thermodynamic drive toward minimizing the free energy of the system. On the other hand, these metastable phases may persist for relatively long periods of time, despite their higher free energy as compared to the thermodynamically most stable form. Transformation to the stable phase can occur in the solid-state or may be mediated by a solution or vapor phase (Cardew and Davey, 1985). In this section, we are concerned with systems in which the phase transformation is mediated by a solvent.

As discussed earlier in the Thermodynamics section of this chapter, the difference in free energy between solid phases (i.e. polymorphs and solvates) is directly proportional to their relative solubilities (Equation 2). Therefore, a saturated solution of a less stable (more soluble) phase is super-saturated with respect to a more stable (less soluble) phase. Since super-saturated solutions are metastable, a more stable phase will eventually crystallize in order to establish equilibrium and remove supersaturation. According to Ostwald's (1897) Law of Stages, a supersaturated state does not spontaneously transform directly into that phase that is the most stable of the possible states, rather into the phase that is next more stable than itself. Nevertheless, a saturated solution of any metastable phase is supersaturated with respect to the most stable phase of the system. Thermodynamics guarantees that the most stable phase will eventually crystallize, whereas the path and the time-scale for the transformation to the most stable phase are governed by the kinetics of solvent-mediated conversion.

The process of solvent-mediated phase transformation consists of three steps (Cardew and Davey, 1985; Gu et al., 2001): 1) dissolution of the metastable phase to form a solution which is supersaturated with respect to the more stable phase 2) nucleation of the more stable phase and 3) growth of the more stable phase. Cardew and Davey (1985) developed a theoretical frame-work to investigate solvent-mediated transformations in terms of dissolution kinetics of one phase and nucleation and growth of a second phase. The model represents the time development of the supersaturation with respect to the stable phase, or solute concentration in solution, and to the solid phase composition during the transformation. This experimental approach involves saturating the solution with respect to the metastable phase under consideration and to monitor both solution concentration and solid phase composition in the presence of the metastable phase, under constant external conditions. More useful information is obtained from the concentration or de-supersaturation profiles than from the solid phase composition profiles with time because the former is related to the driving forces that regulate the transformation rate and can be used to identify the rate-controlling process: dissolution or growth. Experimental studies of the phase transitions of organic crystals have shown this model to be applicable in explaining the solution-mediated transformation kinetics of polymorphs (Cardew and Davey, 1985; Davey et al., 1986; Nass, 1991; Davey et al., 2002; Ferrari et al., 2003; Kelly, 2003) and solvates (Rodríguez-Hornedo et al., 1992; Murphy et al., 2002; Rodríguez-Hornedo and Murphy, 2004) For example, Figure 15 shows the de-supersaturation profile for the solvent-mediated polymorphic transformation of carbamazepine in ethyl acetate at 25 °C and initial $c/s_{CBZ(III)} = 2.0$ (Kelly, 2003). Initially, form II nucleates as this polymorph is favored due to inhibition of form III nucleation as a result of specific solvent-solute interactions, as discussed earlier. This is followed by dissolution/equilibration of form II to form a solution that is supersaturated with respect to form III. This leads to subsequent nucleation and growth of form III until the entire suspension consists of the thermodynamically most stable form, carbamazepine form III.

The kinetics for a solvent mediated phase transformation will be governed by the kinetics of dissolution, nucleation, and crystal growth. These rates will depend directly on the solvent and any step may be rate limiting. As discussed in earlier sections of this chapter, the solvent influences the nucleation rate and crystal growth rate via two factors: 1) solute solubility and 2) specific solvent-solute interactions. The dissolution rate will also be solvent dependent as pharmaceutical materials generally exhibit a wide range of dissolution rates in different solvents.

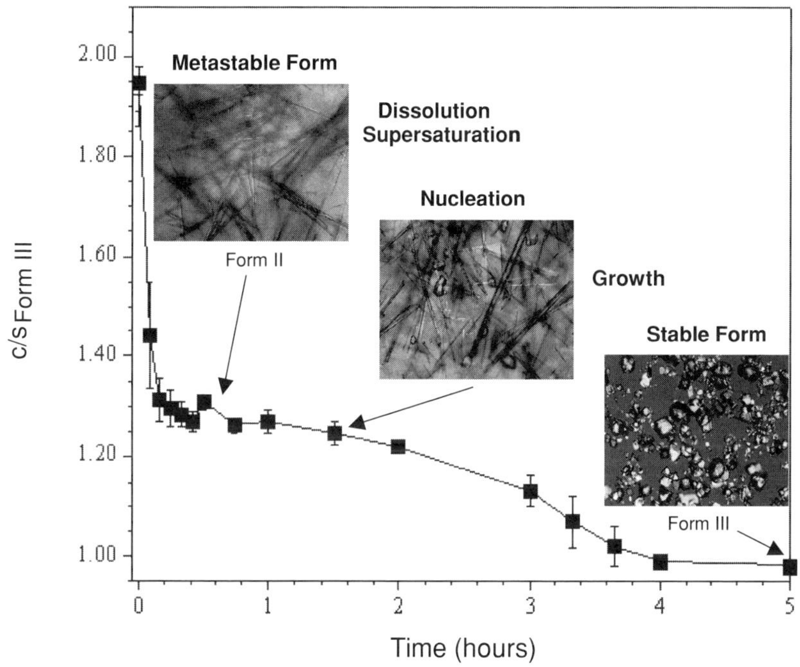

Figure 15. Solvent-mediated transformation of carbamazepine polymorphs in ethyl acetate at 25 °C and initial $c/s_{FormIII} = 2.0$. Reprinted from Kelly, 2003 with permission. (See color insert after Index.)

Gu et al., (2001) studied the influence of solvent on the rate of solvent-mediated polymorphic transformation using sulfamerazine as a model compound. The authors concluded that the overall rate of polymorphic transformation is governed by the nucleation rate of the stable form. They showed that the nucleation rate is a function of the balance between solubility and strength of solvent–solute interactions, and that the fastest transformation rates will be observed in solvents with a relatively high solubility but moderate solvent-solute interactions. For sulfamerazine, a solubility of at least 8 mM was needed in order to ensure transformation to the stable phase. Data obtained by Miller et al. (2005) on two other model drugs, including the HIV protease inhibitor ritonavir, support this finding. In solvents in which the solubility is low (less than 8 mM), because of a high interfacial energy, the metastable zone may be wider than the solubility difference between two polymorphs, such that the critical free energy barrier for nucleation cannot be overcome (Gu et al., 2001). Therefore, nucleation of the stable form is less probable in those solvents giving lower solubilities, such that the metastable form is kinetically stable in those solvents. The authors point out that solvent-mediated polymorphic transformation is an effective technique for obtaining the most stable polymorph of a drug. Likewise, solvents in which the transformation is slow may be utilized for isolation of a metastable polymorph. For instance, Okamato et al. (2004) showed that solvent-mediated polymorphic transformation may be utilized for exclusive production of a metastable form of a drug substance.

Polymorph Screening Approaches

Due to the important role that crystal form plays in the Active Pharmaceutical Ingredient (API) of a drug product, screening for polymorphs of APIs has become a common practice. The extent and type of polymorph screening performed depends on the stage of development and the business strategy of the screening company. Since exhaustive polymorph screening is resource intensive and attrition of early drug candidates is very high, exhaustive screening would not be practical from a business perspective for early drug candidates. On the other hand, an absolute minimalist approach of not performing any type of polymorph screening, while it would save resources early in development, would lead to a host of potential problems including: uncertainty around the robustness of the process to make a reproducible API form; uncertainty about the physical stability of the API form; lack of intellectual property on solid forms; costly need to switch crystal forms well into drug development.

Polymorph screening approaches may be categorized as: 1) rational design for resource saving and 2) comprehensive design to discover all possible solid forms of a drug candidate. The goal of a rationally designed polymorph screen is to discover all relevant forms of a drug candidate that may be encountered during development, particularly the thermodynamically most stable form, as early as possible while expending minimal resources. In a comprehensive polymorph screen, the objective is to gain further confidence that all relevant forms have been identified and to secure freedom to operate with, or exclusive rights to, all possible solid forms of a drug candidate.

Table 2 summarizes the polymorph screening approaches that are rationally designed for resource saving and a comprehensive design. Table 3 lists many commonly used solvents for rational and comprehensive polymorph screening along with those properties that are relevant for solvent selection. The solvents in bold are those that are commonly used for scale-up and processing. These solvents are used in rational screening approaches because those screens are often used early in the drug-development process, and thus the methods for crystallizing any solid forms that are identified may be readily transferred to the development chemist. In this section, each approach will be discussed and the rationale for solvent selection will be highlighted. It should be noted that combinations or hybrids of these approaches are often used.

Prediction-Based Screening

In-Silico *Polymorph Screening*

Structure prediction and subsequent calculation of relative energies of polymorphs and their physical properties without having to synthesize and characterize them would revolutionize polymorph screening, as well as drug discovery.

Polymorph screening approach		Focus of approach	Rationale for approach Solvent selection	Important solvent dependant parameters
Rational designs to save resources	Prediction-based screening (*in-silico* or combination of *in-silico* and experimental techniques)	Molecular Recognition	Reduce the number of crystallization solvents, find all predicted polymorphs 5–10 solvents chosen to interact with specific functional groups	HBD/HBA propensity Solvent class (H-bonding functionality)
	Stable Form Screening (slurries to identify stable polymorphs and solvates)	Thermodynamics	Targeted to find the most stable polymorph and stable solvates 10–20 solvents (neat or mixed) with variety in their properties and focus on those that provide high solubility	Solubility, HBD/HBA propensity, polarity, dipole moment, dielectric constant
	Single Solvent Screening (rapid cooling of solutions to multiple crystallization temperatures)	Thermodynamics, Kinetics	Reduce the number of crystallization trials, explore many temperatures and varying degrees of supersaturation 1–5 solvents chosen to provide a wide operating temperature range in liquid state and moderate solubility	Solubility, boiling point, freezing point
Comprehensive design	Comprehensive Screening (multi-solvent recrystallizations by evaporation, cooling, antisolvent addition, slurries)	Thermodynamics, Kinetics, Molecular Recognition	Comprehensive to explore as much phase-space as possible and identify all possible polymorphs 20–100 neat solvents and additional mixed solvents covering a wide range of properties	Solubility, boiling point, freezing point, HBD/HBA propensity, polarity, dipole moment, dielectric constant, cohesive energy density, viscosity

Table 2. Selected polymorph screening approaches.

Solvent	Molecular weight (g/mol)	Melting[a] point (°C)	Boiling[b] point (°C)	Density[c] (g/cm³)	Σ[d] HBA	Σ[e] HBD	Dielectric[f] constant	Cohesive[g] energy density (J/cm³)	Viscosity[h] (mPa s)
Acetic acid	60.052	16.64	117.9	1.0446[1]	0.61	0.44	6.2528[2]	363.22	1.056
Acetic anhydride	102.089	−74.1	139.5	1.08712			22.45[20]		0.843
Acetone	58.079	−94.7	56.05	0.7844	0.04	0.49	20.493[2]	385.06	0.3060
Acetonitrile	41.052	−48.82	81.65	0.7766	0.07	0.32	35.688[2]	576.25	0.369
Acetophenone	120.149	20.5	202	1.02382	0	0.48	17.44	450.95	1.681
Aniline	93.127	−6.02	184.17	1.0175	0.26	0.41	6.8882[2]	582.91	3.85
Anisole	108.138	−37.13	153.7	0.98932	0	0.29	4.2247[2]	406.39	1.056
Benzene	78.112	5.49	80.09	0.8736	0	0.14	2.2706[2]	350.63	0.604
Benzonitrile	103.122	−13.99	191.1	1.0006	0	0.33	25.592[2]	514.27[3]	1.267
Benzyl alcohol	108.138	−15.4	205.31	1.04127	0.33	0.56	12.457[2]		5.47
Bromobenzene	157.008	−30.72	156.06	1.4882	0	0.09	5.3954[2]	398.68	1.074
Butanenitrile	69.106	−111.9	117.6	0.7865	0	0.36	24.291[2]	419.4	0.553
1-Butanol	74.121	−88.6	117.73	0.80575	0.37	0.48	17.332[2]	542.13	2.54
2-Butanol	74.121	−88.5	99.51	0.80241	0.33	0.56	15.944[2]	511.42	3.10
2-Butanone	72.106	−86.64	79.59	0.7997	0	0.51	18.246[2]	358.35	0.405
Butyl acetate	116.158	−78	126.1	0.87636	0	0.45	4.9941[2]	278.02	0.685

Table 3. Solvents used for polymorph screening and their important properties.

Bold Solvents = Solvents especially relevant for pharmaceutical development.

Butylamine	73.137	−49.1	77	0.73683	0.16	0.61	4.6178[2]	334.89	0.574
Carbon disulfide	76.141	−112.1	46	1.2555	0	0.07	2.6105[2]	412.74	0.352
Chlorobenzene	112.557	−45.31	131.72	1.1009	0	0.07	5.6968[2]	376.47	0.753
m-Cresol	108.138	12.24	202.27	1.03019	0.57	0.34	12.44	564.27	12.91
Cumene	120.191	−96.02	152.41	0.8640[1]	0	0.16	2.3712[2]	306.6	0.737
Cyclohexane	84.159	6.59	80.73	0.77389	0	0	2.0165[2]	280.75	0.894
Cyclohexanone	98.142	−27.9	155.43	0.9478^{20}[1]	0	0.56	15.619[2]	411.63	2.02
Cyclopentanone	84.117	−51.9	130.57	0.9487^{20}[1]	0	0.52	13.58	454.32	2.25^{15}[4]
n-Decane	142.282	−29.6	174.15	0.72635	0	0	2.006[2]	249.84	0.838
cis-Decalin	138.25	−42.9	195.8	0.89288	0	0	2.2139[2]	315.58[3]	3.04
1,2-Dibromoethane	187.861	9.84	131.6	2.1687[1]	0.1	0.17	4.9313[2]	453.12	1.595
Dibromomethane	173.835	−52.5	97	2.4969^{20}[1]	0.1	0.1	7.2273[2]	496.01	0.980
Dibutylether	130.228	−95.2	140.28	0.7641	0	0.45	3.0473[2]	249.31	0.637
o-Dichlorobenzene	147.002	−17	180	1.3033	0	0.04	9.9949[2]	423.18	1.324
1,1-Dichloroethane	98.959	−96.9	57.3	1.1679	0.1	0.1	10.1	332.12	0.464
1,2-Dichloroethane	98.959	−35.7	83.5	1.2454[1]	0.1	0.11	10.125[2]	411.29	0.779
2,2-Dichlorethanol	114.958	−35.7	83.5	1.2454[1]					

(*continued*)

Solvent	Molecular weight (g/mol)	Melting[a] point (°C)	Boiling[b] point (°C)	Density[c] (g/cm^3)	Σ[d] HBA	Σ[e] HBD	Dielectric[f] constant	Cohesive[g] energy density (J/cm^3)	Viscosity[h] (mPa s)
1,1-Dichloroethene	96.943	−122.56	31.6	1.1747	0.08	0.03	4.60^{20}	290.83	0.358^{20}[4]
Dichloromethane	84.933	−97.2	40	1.31678	0.1	0.05	8.93	408.38	0.413
1,2-Diethoxyethane	118.174	−74	121.2	0.8351			3.866[2]	287.76	0.7^{20}[4]
Diethylamine	73.137	−49.8	55.5	0.7016	0.8	0.69	3.5766[2]	276.57	0.319
Diethylene glycol	106.12	−10.4	245.8	1.1164^{20}[1]			30.925[2]		30.2
Diethyl ether	74.121	−116.2	34.5	0.70782	0	0.41	4.24[2]	235.12	0.224
Diethylene glycol dimethyl ether	134.173	−68	162	0.9384			7.23[2]	295.22	0.989[4]
Diethyl sulfide	90.187	−103.91	92.1	0.8362^{20}[1]	0	0.32	5.72	309.33	0.422
Diiodomethane	267.836	6.1	182	3.3079	0.05	0.23	5.32	579.25[3]	3.4^{15}[4]
Diisopropylamine	101.19	−61	83.9	0.71	0.08	0.73		225.45	0.393
Diisopropyl ether	102.174	−85.4	68.4	0.71854	0	0.41	3.81^{30}	208.45	0.379[4]
1,2-Dimethoxyethane	90.121	−69.2	84.5	0.8637[1]	0.3	0.83	7.185[2]	325	0.455[4]
Dimethoxymethane	76.095	−105.1	42	0.8593^{20}[1]			2.652[2]	298.72[3]	0.340^{15}[4]
N, *N*-Dimethylacetamide	87.12	−18.59	165	0.9372[1]	0	0.78	37.781[2]	513.79	0.927
Dimethyl disulfide	94.199	−84.67	109.74	1.0625^{20}[1]	0	0.28	9.6	399.54	0.585[4]

N, N-Dimethylformamide	73.094	−60.48	153	0.9445[1]	0	0.74	37.219[2]	573.87	0.794
2,4-Dimethylpyridine	107.153	−64	158.38	0.9271	0	0.63	9.4176[2]	389.44	0.887^{20}[4]
2,6-Dimethylpyridine	107.153	−6.1	144.01	0.91806	0	0.63	7.1735[2]	367.22	0.869^{20}[4]
Dimethyl sulfoxide	78.133	17.89	189	1.1010[1]	0	0.88	46.826[2]	710.22[3]	1.987
1,4-Dioxane	88.106	11.85	101.5	1.02797	0	0.64	2.2099[2]	421.44	1.177
Dipropyl Ether	102.174	−114.8	90.08	0.7419	0	0.45	3.371[2]	241.15	0.396
n-Dodecane	170.334	−9.57	216.32	0.74518	0	0	2.006[2]	258.29	1.383
Ethanol	46.068	−114.14	78.29	0.78493	0.37	0.48	24.852[2]	678.83	1.074
2-Ethoxyethanol	90.121	−70	135	0.9253[1]			13.38	469.53	1.85[4]
Ethyl acetate	88.106	−83.8	77.11	0.89455	0	0.45	5.9867[2]	336.28	0.423
Ethylbenzene	106.165	−94.96	136.19	0.8626[1]	0	0.15	2.4339[2]	323.06	0.631
Ethylene diamine	60.098	11.14	117	0.8931			13.519[2]	631.6	1.54[4]
Ethylene glycol	62.068	−16.69	197.3	1.11	0.9	0.52	40.245[2]	1168.18[3]	16.06
Ethyl formate	74.079	−79.6	54.4	0.9153	0	0.38	8.331[2]	364.26	0.380
N-Ethyl-N-isopropyl-2-propanamine	129.244		126.5	0.742[1]					
Ethyl phenyl ether	122.164	−29.43	169.81	0.96049	0	0.32	4.1797[2]	381.8	1.197

(continued)

Solvent	Molecular weight (g/mol)	Melting[a] point (°C)	Boiling[b] point (°C)	Density[c] (g/cm^3)	Σ[d] HBA	Σ[e] HBD	Dielectric[f] constant	Cohesive[g] energy density (J/cm^3)	Viscosity[f] (mPa s)
Fluorobenzene	96.102	−42.18	84.73	1.0225^{20}[1]	0	0.1	5.47^{20}	341.99	0.550
Formamide	45.041	2.49	220	1.12915	0.62	0.6	108.94[2]	1445.78	3.34
Formic acid	46.023	8.3	101	1.21405	0.75	0.38	51.1	464.83	1.607
Glycerol	92.094	18.1	290	1.2559	1.21	0.51	45.637[2]	1167.9[3]	934
***n*-Heptane**	100.202	−90.55	98.4	0.67946	0	0	1.9113[2]	231.17	0.387
2-Heptanone	114.185	−35	151.05	0.81123	0	0.51	11.658[2]	318.01	0.714
Hexafluorobenzene	186.054	5.03	80.26	1.60732	0	0	2.03	287.08	2.79
n-Hexane	86.175	−95.35	68.73	0.65484	0	0	1.8819[2]	220.99	0.300
2-Hexanone	100.16	−55.5	127.65	0.8067	0	0.51	14.136[2]	327.49	0.583
Iodobenzene	204.008	−31.3	188.4	1.8229	0	0.12	4.547[2]	420.87[3]	1.554
1-Iodobutane	184.018	−103	130.5	1.6072	0	0.15	6.173[2]	333.21	0.826[4]
Isobutyl acetate	116.158	−98.8	116.5	0.8702			4.975[2]	275.1[3]	0.676
Isopropyl acetate	102.132	−73.4	88.6	0.8702	0	0.47		295.84	0.569^{20}[4]
Methanol	32.042	−97.53	64.46	0.78637	0.43	0.47	32.613[2]	857.76	0.544
Mesitylene	120.191	−44.72	164.74	0.86111	0	0.19	2.265[2]	322.55	1.154^{20}[4]

2-Methoxyethanol	76.095	−85.1	124.1	0.96024	0.3	0.84	17.2	538.72	1.60[4]
Methyl acetate	74.079	−98.25	56.87	0.9279	0	0.45	6.8615[2]	373.41	0.364
Methyl benzoate	136.149	−12.4	199	1.0837[1]	0	0.46	6.7367[2]	422.59	1.857
2-Methyl-2-butanol	88.148	−9.1	102.4	0.805	0.3	0.6	5.78	434.89	3.55[4]
3-Methyl-1-butanol	88.148	−117.2	131.1	0.8071	0.37	0.48	14.954[2]	486.48	3.69
Methylcyclohexane	98.186	−126.6	100.93	0.76506	0	0	2.02[20]	256.21	0.679
Methyl formate	60.052	−99	31.7	0.9664	0	0.38	8.8377[2]	411.19	0.325
Methyl isopropyl ketone	86.132	−93.1	94.33	0.8051			10.288[2]	320.62	
4-Methyl-2-pentanone	100.158	−84	116.5	0.7963	0	0.51	12.887[2]	303.16	0.545
2-Methylpropanoic acid	88.106	−46	154.45	0.94288	0.6	0.49	2.58[20]	351.24	1.226
2-Methyl-1-propanol	74.121	−101.9	107.89	0.7978	0.37	0.48	16.777[2]	520.32	3.33[4]
2-Methyl-2-propanol	74.121	25.69	82.4	0.7812	0.31	0.6	12.47	465.96	4.31
4-Methylpyridine	93.127	3.67	145.36	0.9502	0	0.54	11.957[2]	429.36	
N-Methyl-2-pyrrolidone	99.131	−23.09	202	1.0230[1]	0	0.77	32.2		1.67[4]
Methyl *tert*-butyl ether	88.148	−108.6	55	0.7353[1]	0	0.4	4.5	228.07	
2-Methyltetrahydrofuran	86.132	−137.15	78	0.8540[20][1]	0	0.53	6.97	293.01	
Morpholine	87.12	−4.8	128	0.99547	0.29	0.7	7.42	474.04[3]	2.02

(*continued*)

Solvent	Molecular weight (g/mol)	Melting[a] point (°C)	Boiling[b] point (°C)	Density[c] (g/cm^3)	Σ[d] HBA	Σ[e] HBD	Dielectric[f] constant	Cohesive[g] energy density (J/cm^3)	Viscosity[f] (mPa s)
Nitrobenzene	123.11	5.7	210.8	1.19835	0	0.28	34.809[2]	511.34	1.863
Nitroethane	75.067	−89.5	114	1.0448[1]	0.02	0.33	28.29[2]	544.36[3]	0.688
Nitromethane	61.041	−28.38	101.19	1.13128	0.06	0.31	36.562[2]	663.32	0.630
1-Nitropropane	89.094	−108	131.1	0.9961[1]	0	0.31	23.73[2]	457.4[3]	0.798
2-Nitropropane	89.094	−91.3	120.2	0.9821[1]	0	0.33	25.654[2]	428.37[3]	0.721[4]
n-Octane	114.229	−56.82	125.67	0.69862	0	0	1.9406[2]	238.59	0.508
1-Octanol	130.228	−14.8	195.16	0.82157	0.37	0.48	9.8629[2]	432.15	7.29
n-Pentane	72.149	−129.67	36.06	0.62139	0	0	1.84[20]	206.28	0.224
Pentanoic acid	102.132	−33.6	186.1	0.9339[1]	0.6	0.45	2.6924[2]	610.37[3]	1.97[4]
1-Pentanol	88.148	−77.6	137.98	0.8108	0.37	0.48	15.13	501.68	3.62
2-Pentanone	86.132	−76.8	102.26	0.8015	0	0.51	15.2[2]	334.26	0.470
3-Pentanone	86.132	−39	101.7	0.80945	0	0.51	16.78[2]	338.71	0.444
Perfluoroheptane	388.049	−51.2	82.5	1.7333[1]			1.847[16]		
Perfluorohexane	338.042	−88.2	56.6	1.6995[1]			1.76		
Propanenitrile	55.079	−92.78	97.14	0.77649	0.02	0.36	29.324[2]	472.99	0.294

Propanoic acid	74.079	−20.5	141.15	0.9882[1]	0.6	0.45	3.44	395.67	1.030
Propanoic anhydride	130.141	−45	170	1.0057			18.30[20]		1.06[4]
1-Propanol	60.095	−124.39	97.2	0.7996	0.37	0.48	20.524[2]	598.37	1.945
2-Propanol	60.095	−87.9	82.3	0.78126	0.33	0.56	29.264[2]	557.86	2.04
Propyl acetate	102.132	−93	101.54	0.88303	0	0.45	5.5205[2]	321.98	0.544
Propylene carbonate	102.089	−48.8	242	1.1951			64.906[2]		2.53[4]
1,2-Propylene glycol	76.095	−60	187.6	1.0328			28.378[2]	840.42[3]	40.4
Pyridine	79.101	−41.7	115.23	0.97824	0	0.52	12.978[2]	466.62	0.879
Quinoline	129.159	−14.78	237.16	1.0977[15][1]	0	0.54	9.003[2]	483.62	3.34
1,1,2,2-Tetrachloroethane	167.849	−42.4	145.2	1.58666	0.16	0.12	8.50[20]	408.66	1.84[15][4]
Tetrachloroethene	165.833	−22.3	121.3	1.61432	0	0	2.27[30]	362.14	0.844
Tetrachloromethane	153.823	−22.62	76.8	1.58436	0	0	2.228[2]	308.49	0.908
Tetrahydrofuran	72.106	−108.44	65	0.8833[1]	0	0.48	7.4257[2]	361.51	0.456
Tetralin	132.202	−35.7	207.6	0.9645[1]	0		2.77	384.85[3]	2.14[4]
Toluene	92.139	−94.95	110.63	0.86219	0	0.14	2.3741[2]	332.48	0.560
Tribromomethane	252.731	8.69	149.1	2.8788[1]	0.15	0.06	4.2488[2]	496.31	1.857
1,1,1-Trichloroethane	133.4	−30.01	74.09	1.3299	0	0.09	7.0826[2]	299.29	0.793

(continued)

Solvent	Molecular weight (g/mol)	Melting^{a} point (°C)	Boiling^{b} point (°C)	Density^{c} (g/cm^{3})	Σ^{d} HBA	Σ^{e} HBD	Dielectric^{f} constant	Cohesive^{g} energy density (J/cm^{3})	Viscosity (mPa s)
1,1,2-Trichloroethane	133.404	−36.3	113.8	1.43213	0.13	0.13	7.19	405.38	1.19^{20}[4]
Trichloroethene	131.388	−84.7	87.21	1.4642^{20}[1]	0.08	0.03	3.422[2]	357.75	0.545
Trichloromethane	119.378	−63.41	61.17	1.4788[1]	0.15	0.02	4.7113[2]	356.77	0.537
Triethylamine	101.19	−114.7	89	0.72305	0	0.79	2.3832[2]	231.23	0.347
Trifluoroacetic acid	114.023	−15.2	73	1.5351[1]			8.319[2]		0.808
2,2,2-Trifluoroethanol	100.039	−43.5	74	1.3842^{20}[1]	0.57	0.25	26.726[2]	574.67[3]	1.99^{20}[4]
Trimethoxymethane	106.12	15	104	0.9676[1]					
(Trifluoromethyl)benzene	146.11	−28.95	102.1	1.18129	0	0.1	9.22	283.95	0.574^{20}[4]
2,2,4-Trimethylpentane	114.229	−107.3	99.22	0.68781	0	0	1.9358[2]	196.66	0.504^{20}[4]
Water	18.015	0	100	0.9970[1]	1.17	0.47	78.36	2296.78	0.890
m-Xylene	106.165	−47.8	139.12	0.86009[1]	0	0.16	2.3478[2]	325.44	0.581
o-Xylene	106.165	−25.2	144.5	0.87594	0	0.16	2.5454[2]	337.88	0.760
p-Xylene	106.165	13.25	138.37	0.85661	0	0.16	2.2705[2]	322.11	0.603

a = From Lide, 2005.

b = From Lide, 2005.

c = Values at 25°C unless noted by superscript (Riddick, 1986; Lide 2005). Superscript '[1]' denotes values obtained from the latter reference.

d = Summation of hydrogen bond donor propensities of the solvent (Abraham 1993 a,b; Abraham, 1994).

e = Summation of the hydrogen bond acceptor propensities of the solvent (Abraham 1993 a,b; Abraham, 1994).

f = Values at 25°C unless noted otherwise by superscript (Lide, 2005). Superscript '[2]' indicates value was computed from equation for temperature dependence of the dielectric constant, $\varepsilon(T) = a + bT + cT^2 + dT^3$. T = 298.15 Kelvin, and a, b, c, and d are coefficients obtained from Lide, 2005.

g = Calculated from $\Delta H_{vap} - RT/V$ where ΔH_{vap} is the enthalpy of vaporization at 25°C, R is the universal gas constant, T is temperature (Kelvin) and V is the molar volume (cm^3/mole). Temperature value used was the same as the temperature at which the density is reported. The molar volume was calculated via V = MW/Density. ΔH_{vap} values were collected from the literature (Lide, 1995; Lide, 2005). Values denoted with the superscript '[3]' used ΔH_{vap} values from the former reference.

h = Values at 25°C unless otherwise noted by superscript (Lide, 1995; Lide, 2005). Values obtained from the former reference are denoted with the superscript '[4]'.

Besides the obvious human resource savings, the rationale for this approach, in theory, is to learn very early about the likelihood of polymorphism and predict the most stable polymorph, which could then be the solid form targeted for crystallization and development.

Since many critical physical and mechanical properties of pharmaceutical compounds are in large part dependent on crystal form, accurate prediction of crystal structure would be a highly valued tool in the pharmaceutical industry. Unfortunately, to date, reliable crystal structure prediction is only feasible for rigid, low molecular weight molecules, which do not represent the size and flexibility of pharmaceutical molecules.

Significant progress has been made to improve crystal structure predictions with better definitions of charges and use of clever algorithms (Gavezzotti, 1994; Desiraju 1997; Beyer et al., 2001; Motherwell et al., 2002; Day et al., 2004; Day et al., 2005). Coupled with rapidly increasing computational power, these improvements will enable more reliable crystal structure predictions in the future.

Crystal Structure Prediction and Graph Set Analysis

It is possible to combine *in-silico* predictions and experimentation. This is done by analyzing the hydrogen-bonding modes of a molecule in its predicted crystal structures and then strategically selecting solvents with functional groups that promote or block specific hydrogen-bonding modes. This approach has been shown to identify most or all known polymorphs for model compounds (Blagden *et. al.*, 2001; Blagden and Davey, 2003; Cross et al., 2003). A schematic of this approach is illustrated in Figure 16 (Blagden and Davey, 2003). In summary, crystal structures are predicted and ranked according to calculated lattice energy or density by a computer program. Then a graph-set (hydrogen-bond motif) (Etter, 1990; Etter et al., 1990) analysis is carried out on the resulting crystal structures to identify common packing motifs; while as many as 100 predicted structures may be ranked, there are often only a few unique graph-sets. The number of unique graph-sets represents a realistic number of polymorphs to be expected. Crystallization solvents are then selected to manipulate the crystallization outcome to favor or inhibit the specific packing motifs identified. Thus, the important properties for solvent selection for this approach are primarily hydrogen bond donor/acceptor propensity as well as the specific hydrogen bonding functionalities of the solvent (e.g., ester vs. alcohol).

Solvent selection for this approach places a strong emphasis on molecular recognition. It should be noted that although solvents can be chosen to favor a particular polymorph through molecular recognition, the free energy relationship between polymorphs can make the choice of solvent immaterial under a given crystallization condition (Threlfall, 2000), as discussed earlier in this chapter. In order to fully realize the directing power of a given solvent, varying degrees of supersaturation at multiple temperatures of crystallization must be explored. This would increase the odds for each solvent to have at least one condition under which molecular recognition by solvent-solute interaction can direct nucleation and growth of a specific polymorph.

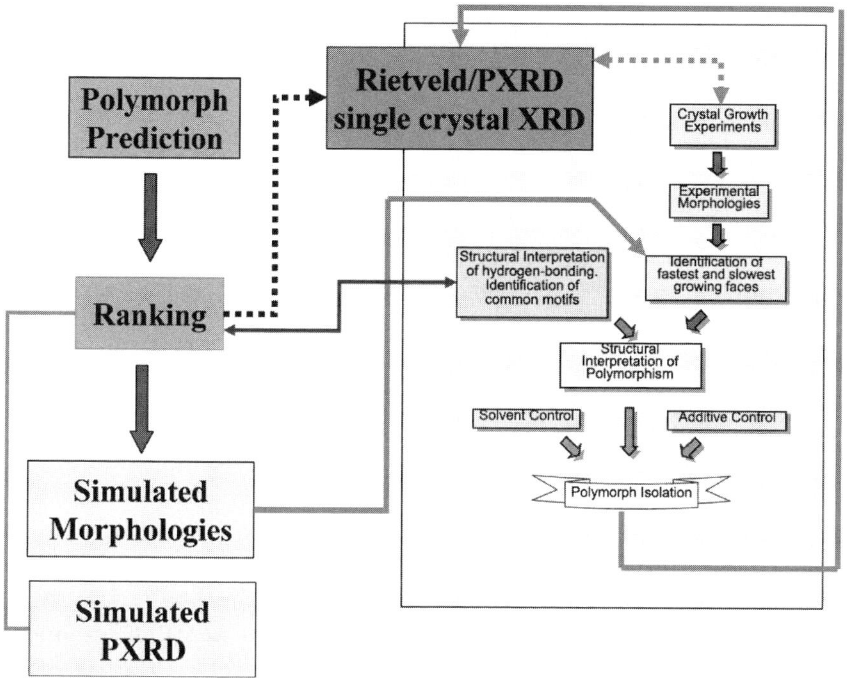

Figure 16. Polymorph selection strategy. Reproduced from Blagden and Davey (2003) with permission. Copyright (2003) American Chemical Society.

Stable Form Screening

Stable Polymorph Screening

Since the thermodynamically most stable polymorph at room temperature is usually the desired crystal form for drug development, there is good reason to design a polymorph screen specifically to identify this form. A screening method for identifying the stable polymorph that utilizes solvent-mediated phase transformation by aging the API in slurries has been described in the literature (Gu et al., 2001; Miller et al., 2005). Possible outcomes from such a screen, if performed properly, are limited to polymorphs or solvates that are thermodynamically more stable than the starting crystal form(s), putting a premium on thermodynamics. The method for crystallization does not include evaporation, cooling, or anti-solvent addition. Rather, crystallization is limited to solvent-mediated phase transformation (i.e. slurries). For this reason, there are fewer conditions to set-up and analyze, allowing for resource savings. The technique can also be compound-sparing as slurry volumes can be made very small (<0.5 mL). Since this type of screen is resource-saving and targets the most important crystal phase of the API, it is ideal for early development.

To perform this type of screen, excess starting API is suspended in a diverse set of solvents for a designated period of time. For practical purposes, slurry times for this type of screen tend to be 1–3 weeks, but may depend on how much time is afforded. Given chemical stability, longer slurry times provide more confidence that the most stable form will be identified from this type of screen.

The best chance of successfully nucleating the thermodynamically most stable polymorph is afforded if solvents are selected with an emphasis on achieving high solubility. Thus, a solvent set for stable polymorph screening is initially selected with a wide range of polarity (i.e. dielectric constants, cohesive energy densities) to give the best chance of achieving high solubility in some of the solvents. Examples have been shown that suggest solubility greater than approximately 8 mM may be necessary to ensure conversion of a metastable polymorph to the most stable form in a reasonable amount of time (Gu et al., 2001; Miller et al., 2005). Since high solubility is such an important parameter in this approach, measurement of solubility is essential for proper experimental design and interpretation of results. Measurement of solubility in an initial list of solvents also gives the scientist the ability to customize the final slurry solvents. Solubility measurement can be done *in-situ* since slurries eventually become saturated solutions with respect to the excess solid form. There are many methods that can be used to determine solubility. High-performance liquid chromatography (HPLC) with UV detection is one option, provided the time and personnel to develop the HPLC method are available. Generic gradient HPLC methods may be a satisfactory option. A semi-automated gravimetric technique to determine solubility quickly with little sample consumption has been described (Greene et al., 2005). This technique requires no method development or preparation of standard solutions and provides adequate precision and accuracy for stable polymorph screening. After considering the solubility results, mixed solvents are often utilized to achieve multiple slurries with appropriate solubility. To reduce sample consumption it may be necessary on occasion to lower solubility by adding anti-solvent to those solvents that provide extremely high solubility. Table 4 shows the utility of stable polymorph screening for identifying the thermodynamically most stable polymorph of ritonavir (Miller et al., 2005). The transformation to the stable polymorph was slow or did not occur within 2 weeks in solvents that gave low solubility, demonstrating the practical influence of solubility on the solvent-mediated transformation.

The greatest concern with this approach is lack of nucleation of the most stable form. If the starting API does not contain seeds of the stable form, a successful screen will produce nucleation of the stable form within the duration of the slurry experiments in at least one of the solvents. Lack of adequate solubility, or inhibition of nucleation due to solvent-solute interactions, may preclude transformation to the stable polymorph (Gu et al., 2001). Moreover, impurities or additives, even in trace amounts, can dramatically affect the rate of solvent-mediated polymorphic transformation (Gu et al., 2002; Okamata et al., 2004; Mukuta et al., 2005). Thus, it is important to have starting API with the highest possible purity. Recrystallization prior to starting this or any type of polymorph screen can be used in an attempt to purify the API. While this could reduce an impurity that might stifle solvent-mediated transformation to the most stable polymorph, it should be noted that recrystallization from solution does not

Solvent	Solid form 2 days	Solid form 2 weeks	Solubility (mg/mL) 2 weeks
Water	I	II	<0.10 +/− 0.00
Hexanes	I	I, II mixture	<0.10 +/− 0.00
Methyl *t*-butyl ether	I	I	0.30 +/− 0.05
1,2-Xylene	I	I	1.03 +/− 0.01
Toluene	I	I	2.26 +/− 0.02
Nitromethane	II	II	5.21 +/− 0.03
Ethyl acetate	II	II	5.24 +/− 0.11
Acetonitrile	II	II	6.97 +/− 0.07
2-Propanol	II	II	14.1 +/− 0.1
2-Butanone	II	II	18.3 +/− 0.4
Acetone	II	II	22.4 +/− 0.3
1,2-Dimethoxyethane	II	II	28.1 +/− 0.5
Ethanol	II	II	56.6 +/− 1.2
Tetrahydrofuran	–	–	>200
1,4-Dioxane	–	–	>200
Methanol	–	–	>200
N,N-Dimethylformamide	–	–	>200
Chloroform	–	–	>200

Table 4. Results of stable polymorph screen of ritonavir. Data from Miller et al. (2005).

guarantee purification beyond the starting API. Nor does multiple samples recrystallized from solution guarantee observation, or even seeds, of the most stable form. Thus, the greatest chance of successfully nucleating the thermodynamically most stable polymorph is afforded by using API with the highest possible purity and using multiple solvents that provide high solubility.

Stable Hydrate Screening

One solvent that is essential to include in the list of slurry solvents is water, to evaluate the potential existence of hydrates. Missing a hydrated crystal form of the API in the early stages of development can be critical. This is because hydrates often have significantly reduced aqueous solubility compared to differences in solubility of anhydrous polymorphs (Pudipeddi and Serajuddin, 2005). Thus, *in-vivo* absorption of the compound may suffer more when a stable hydrate is discovered versus a more stable anhydrous polymorph. Moreover, crystallization of a more stable hydrate is of particular concern for parenteral formulations, which typically contain a high percentage of water.

As discussed earlier in this chapter, hydrates may be more or less stable than their anhydrous counterparts, depending on the a_w and temperature of the solvent. The slurry method described in the previous section can also be used to target thermodynamically more stable hydrate crystal forms, if they exist (Zhu et al., 1996; Zhu and Grant, 1996; Ticehurst et al., 2002; Sacchetti, 2004). Just as with a stable polymorph screen, the starting API is slurried in water or water-organic solvent mixtures with high water activity for a defined period of time. Again, high solubility is critical for nucleation of a more stable hydrate if it is not present in the starting API. Many organic molecules, particularly free-acids and free-bases, are often not soluble in water. Therefore, transformation to an undiscovered more stable hydrate may be slow or may not occur in water, even though the water activity is maximized. To address this common problem, water can be combined with organic solvents. Figure 5 contains the a_w versus weight fraction water plots for alcohol-water mixtures, that may be employed for this purpose. The resulting solvent mixture will ideally have high water activity and the API will have high solubility, to give the best chance of nucleating a more stable hydrate. Care must also be taken to estimate the water activity when using solvent mixtures in which the solubility of the compound is extremely high (>1 M), because the a_w of the solvent may be altered by the solute (Sacchetti, 2004).

It should be noted that it is possible to crystallize metastable hydrates from aqueous solution. A metastable hydrate is one that when saturated in water, where the water activity is maximal, there exists an anhydrous crystal form that is thermodynamically more stable. For a stable hydrate screen, solvent mediated conversion will occur in time, converting the metastable hydrate to the stable anhydrous form. Thus, a metastable hydrate should not be observed in a stable hydrate screen, provided the slurries have reached equilibrium.

Single-Solvent Polymorph Screening

It has been pointed out that under some conditions, crystallization of a given polymorph will be under thermodynamic control, for which the nature of the solvent will be immaterial (Threlfall 2000). This point was illustrated theoretically earlier in the chapter using Figure 8. It follows that by systematically exploring a wide range of temperature-concentration regimes, it is possible to produce the relevant stable and metastable forms of a drug candidate using a single solvent.

In this single-solvent polymorph screening approach, the solvent is added to the API and heated to an appropriate temperature until the solid is completely dissolved. The resulting solution is quickly cooled to a specific crystallization temperature chosen to create the desired degree of supersaturation. Rapid cooling

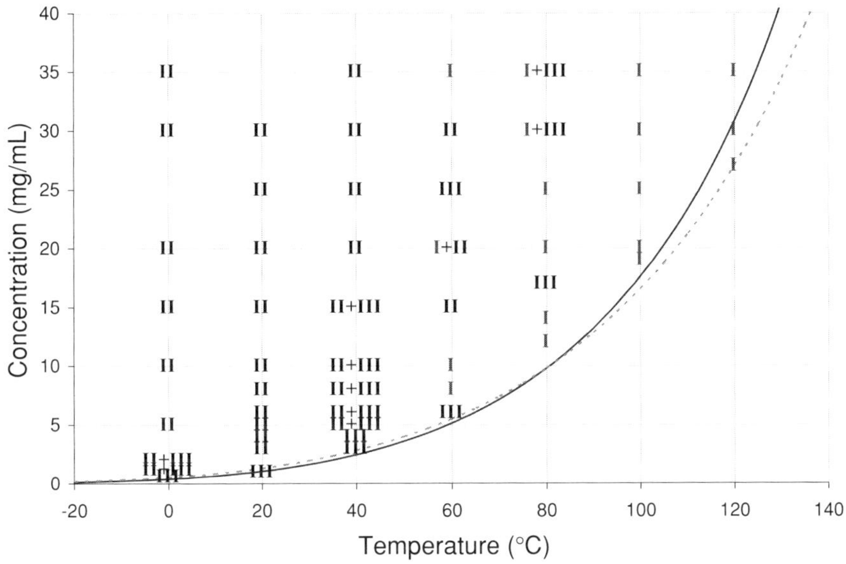

Figure 17. Polymorph screen for carbamazepine using one solvent, cumene; temperature-solubility curves (forms I and III) and solid forms initially crystallized at various temperature-concentration conditions. The blue (solid) line and the green (dotted) line represent the temperature-solubility curves of form III and form I respectively. From Getsoian et al., 2006.

is utilized in an attempt to ensure that the material does not crystallize until the desired isothermal temperature is reached. The resulting solid is collected immediately to capture the initial polymorph that crystallized. This approach is illustrated in Figure 17 via a single-solvent polymorph screen of carbamazepine in the solvent cumene (Getsoian et al., 2005). A phase diagram indicating the metastable and stable forms isolated can be constructed from the series of recrystallizations at various degrees of supersaturations and multiple temperatures (Figure 17). The most relevant forms (I, II, III) of carbamazepine were identified using this single-solvent approach, using minimal resources. Another advantage of this approach is that the phase diagram may be used to guide efforts to isolate the desired polymorph on a large scale.

For this approach, it is important to select a solvent with a relatively low melting point (T_m) but high boiling point (T_b) such that a wide range of temperatures may be achieved. Another important parameter for solvent selection for this approach is the solubility of the solute as a function of temperature. Ideally, the solubility should be high enough so that the material may dissolve and precipitate upon cooling, but not too high such that supersaturation is not achieved and precipitation does not occur. From room-temperature solubility and a reasonable number of solute and solvent parameters, it is possible to calculate semi-empirical temperature-solubility curves (Frank et al., 1999). Figure 18 compares the semi-empirical and experimental solubility-temperature profiles for carbamazepine in 2-propanol, for example. Using the calculated curve as

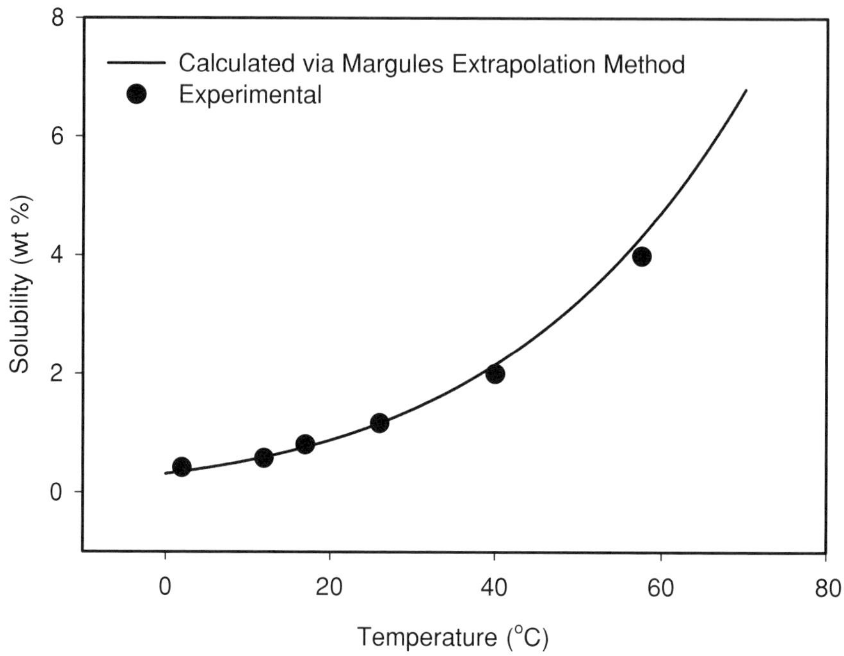

Figure 18. Solubility as a function of temperature for carbamazepine (III) in 2-propanol. Experimental data obtained from Behme and Brooke (1991). Calculated curve via Margules Extrapolation Method as described by Frank et al. (1999) using the following input data: solute-molecular weight, melting point, heat of fusion, and solubility in 2-propanol at 26 °C; solvent-molecular weight and density.

a guide, samples may be prepared to target specific levels of supersaturation for each temperature of interest. For the single-solvent polymorph screen of carbamazepine, the solvent cumene (T_m −96 °C, T_b 153 °C, carbamazepine solubility 1.1 mg/mL at 20 °C) was selected using the above criteria to accommodate sample handling, compound consumption, and maximize the range of concentration-temperature regimes over which the crystallizations can occur. Note that this approach could obviously be expanded to more than one solvent. However, multiple solvents providing the solution-state properties desired are sometimes difficult to find, and a large number of solvents would take this approach out of the resource-saving category.

Comprehensive Polymorph Screening

The approaches detailed in the previous sections can be described as resource saving when compared to comprehensive polymorph screening. Comprehensive polymorph screening requires more resources because the number of crystallization conditions can range in the thousands. A comprehensive polymorph screen implies that an exhaustive set of crystallization conditions has uncovered all possible polymorphs of the API. This is typically carried out during the later stages of development, when confidence in the compound making it to the market place is high and resources spent on the screening are easily justified. The rationale for doing such a screen is to establish exclusivity or freedom to operate with all the relevant forms that could be used in, or on the way to, a product. In theory, the number of experimental conditions to be explored is virtually infinite, so a rational design is needed to cover as much phase-space as possible with the available resources.

Although a comprehensive polymorph screen provides a final opportunity to find a potentially undiscovered most stable form, it is a reasonable assumption that most relatively stable polymorphs and hydrates are known from earlier stages of development, which likely included stable form screening. Therefore a large fraction of the crystallization conditions for a comprehensive polymorph screen are executed in such a way as to find, isolate, and characterize metastable forms and solvates that may only exist under specific conditions and may readily convert to more stable forms after isolation (Dunitz and Bernstein, 1995; Blagden and Davey, 2003). In general, samples should be analyzed as close to *in-situ* conditions as practical without unnecessary delays to maintain the integrity of the form that initially crystallized. Since much of the screen targets crystallization of metastable forms, the ability to achieve high supersaturation before nucleation followed by fast isolation and characterization is required. In addition to crystallizations from solution, a variety of specialized techniques are commonly being applied including: crystallization from melts, crystallization from supercritical fluids (Velaga et al., 2002; Gosselin et al., 2003), laser induced nucleation (Garetz et al., 1996; Zaccaro et al., 2001), crystallization in capillaries (Chyall et al., 2002), templated crystallization (Mitchell et al., 2001), and polymer heteronucleation (Lang et al., 2002; Price et al., 2005). Some of these techniques have been reviewed elsewhere (Rodríguez-Spong et al., 2004). For the purposes of this chapter we will focus on polymorph screening approaches that involve crystallization from solvent systems.

Several hundred solvents may be utilized, either pure or in combination. Each solvent system is used to explore conditions designed to achieve varying degrees of supersaturation over a wide range of temperatures. Advances in high-throughput (HT) sample preparation have enabled thousands of crystallization experiments to be carried out in short periods of time (van Langevelde and Blomsma, 2002; Carlson et al., 2003; Hilfiker et al., 2003). Commercial characterization tools such as powder x-ray diffraction, Raman spectroscopy, and polarized light microscopy, have been designed for HT data collection. These solid form characterization tools have been reviewed elsewhere (Rodríguez-Spong et al., 2004). These recent advances in HT sample preparation and characterization have shifted the focus to information technology for analysis of large data sets resulting from comprehensive polymorph screening.

Although large numbers of conditions generated by a HT screen appear impressive, without proper experimental design, the results may be unimpressive due to lack of crystallization or near redundant crystallization conditions. As with other types of polymorph screening, the most useful data to help guide the design of a comprehensive polymorph screen is the solubility of the drug candidate in the solvents to be used. As discussed previously, the solubility of the API governs the level of supersaturation that can be achieved, and potentially the ability of the solvent to direct nucleation and crystal growth. Moreover, it is essential to have knowledge of the solubility in order to select those solvents that will lead to a meaningful crystallization result. Methods for determining solubility are reviewed elsewhere in this book. Simple methods for estimating solubility have also been described (Frank et al., 1999). The solubility as a function of temperature is useful to know for the design of cooling crystallization, but these profiles are difficult to determine for every solvent. As described earlier for single solvent polymorph screening, semi-empirical temperature-solubility curves can serve as useful surrogates (Frank et al., 1999).

The solvents for a comprehensive polymorph screen are typically selected from a large set containing a breadth of properties, as well as variety in functionality (Kolá et al., 2002; Carlson, 2003; Gu et al., 2004). Table 3 contains a list of solvents with a wide range of properties that may be employed for comprehensive polymorph screening. There are some practical limitations that should be considered. Acidity/basicity can be problematic, since acetic acid or pyridine may form salts with an API in question. Lack of solubility or chemical instability might also place limitations on the solvents that can be used for screening a particular API. Beyond these practical considerations, there are effective statistical tools that enable scientists to narrow down the list of solvents while preserving ample diversity (Katritzky et al., 1999; McKay et al., 2003; Gu et al., 2004), if experimentation in all possible solvents and their mixtures is not desired or cannot be afforded.

Using a diverse solvent set with a wide range of properties coupled with a variety of methods for generating supersaturation creates the best opportunities for discovering new polymorphs through various thermodynamic, kinetic, and molecular recognition phenomena. As discussed earlier in the chapter, supersaturation may be achieved using a variety of methods. The most frequently used methods for creating supersaturation for polymorph screening and corresponding considerations for solvent selection will be discussed next.

Solvent Evaporation

Solvent evaporation is perhaps the easiest way to generate supersaturation. The method is guaranteed to create supersaturation in the solution at some point during solvent evaporation. In order to provide the best chance of achieving high supersaturation, and to crystallize forms different from the starting material, it is best to start solvent evaporation from solutions devoid of the initial solid phase. Saturated solutions are generally prepared by filtering suspensions, or heating suspensions until all solids have dissolved. Alternatively, sub-saturated solutions may be prepared initially, as supersaturation will occur as the solvent evaporates. For HT preparation, parallel filtration (e.g., by 96-well filtration plate) of all samples, solutions or suspensions, is an easy way to ensure solvent evaporation starts with solutions.

Different solvents produce different evaporation rates depending on their volatilities, making it convenient to explore different supersaturation conditions. However, this also makes it difficult to control HT crystallizations by solvent evaporation. HT crystallizations normally take place in a 96-well plate or similarly formatted crystallizer (Carlson et al.; 2003). While some solutions require elevated temperature and/or reduced pressure to evaporate, others may have already evaporated to dryness leaving their samples vulnerable to desolvation or phase transformation. Therefore it is necessary to group solvents with an appropriately small range of volatilities (boiling points) to optimize a solvent evaporation experiment in a HT format. This is generally carried out by grouping the range of boiling points of solvents in such a crystallization plate. Some solvents with extremely high boiling points (i.e. dimethyl sulfoxide), are difficult to evaporate, therefore reducing the likelihood of crystallization, particularly of a metastable polymorph. Such high-boiling solvents require elevated temperature for the evaporation experiments to complete on a reasonable timescale. Elevated temperature, as well as reduced pressure, can be used to create faster evaporation rates for any solvent. Note that fast evaporation will create significant temperature lowering at the surface of the sample solution. Technologies, such as temperature controlled vacuum evaporators, are commercially available to minimize this issue.

Knowledge of solubility is important for the design of solvent evaporation experiments and may help with interpretation of results. For samples with very high solubility, reducing solvent volume often leads to a highly viscous medium in which an oily residue is all that is left after solvent evaporation (i.e. oiling out). Likewise, solvent evaporation experiments in solvents that provide minimal solubility are meaningless.

Cooling

Cooling of solutions is another effective means of creating supersaturation. As discussed earlier for the single-solvent polymorph screen, temperature-solubility profiles are very useful in selecting solvents for a successful experiment. These profiles may be readily estimated from an experimental solubility at one temperature (Frank et al. 1999). The calculated temperature-solubility profile of carbamazepine (III) in 2-propanol is shown in Figure 18, for example. From the temperature-solubility curve of the starting crystal form in a given solvent, solutions are made by preparing suspensions with known excess solid and heating to an appropriate temperature to ensure complete solubilization. Cooling can then be done very slowly, to favor crystallization of stable forms, or very rapidly to provide a better opportunity for high supersaturation which favors crystallization of metastable forms. A similar type of solvent grouping as mentioned above for evaporation must be done if a HT crystallizer is being used, to avoid boiling away solvents with high volatility.

There are some problems that may be encountered when using cooling as a crystallization technique. While well-designed temperature control units ensure sufficient temperature control under many conditions, it is difficult to achieve rapid cooling throughout a 96-well crystallizer. This is more easily done with individual vials, for example, by placing a sample vial into dry-ice and acetone. Moreover, cooling crystallizations are often designed to hold at temperatures well above ambient, and therefore require filtration of saturated solutions at elevated temperatures. Hot filtration is difficult to accomplish without crystallization during the filtration process, especially when using a parallel HT crystallizer.

Anti-Solvent Addition

The use of anti-solvents is another way to supersaturate solutions by lowering the solubilization ability of the media. An advantage of this method is that microenvironments of extremely high supersaturation can be created, making it possible to observe metastable solid forms that might not be observed otherwise. Samples are prepared by creating solutions that are saturated with API, but devoid of excess solids. Each saturated solution is then mixed with an anti-solvent. The order of mixing of anti-solvent with saturated solution is important. Adding the anti-solvent to the saturated solution of the API is the standard approach. Alternatively, relatively small volumes of a saturated solution may be delivered into a larger volume of anti-solvent (i.e. reverse anti-solvent addition). This approach provides the best route to microenvironments with extremely high supersaturation. Temperature, anti-solvent addition rate, and mixing are all parameters that may be varied to affect the crystallization outcome. Anti-solvent vapor diffusion is also commonly employed to achieve very slow rates of anti-solvent addition.

Competition Slurries

If multiple forms have been observed in the comprehensive polymorph screen, competition (i.e. bridging) slurries may be set up to determine the relative thermodynamic stabilities of the forms and to verify that the most stable form has been identified. Here, the scientist slurries together all the forms observed in the comprehensive screen and determines their relative thermodynamic stabilities via solvent-mediated phase transformation. Given sufficient time, this technique will convert metastable forms to more stable forms until ultimately the most stable form of the system is reached. As pointed out earlier for stable form screening, special attention should be paid to the solubility in the slurry solvents. Selection of solvents providing high solubility will give faster rates of solvent mediated phase transformation (Gu et al., 2001). For hydrates, bridging slurries should be carried out in solvents with a wide range a_w, in order to determine the critical a_w of the system under study. Finally, prior to adding all the solids into a chosen solvent, it should be saturated with the predominant form to avoid dissolving away polymorphs that may be present in trace amounts.

Summary and Conclusions

In this chapter, we have discussed the significance of thermodynamic, kinetic, and molecular recognition phenomena in directing crystallization outcomes and the role of solvent in these processes. Both rational and comprehensive approaches for polymorph screening have been presented including important considerations for the selection of solvents and crystallization conditions. It is recognized that the interplay between molecular recognition, thermodynamics, and kinetics can be carefully controlled by the selection of the appropriate solvents and crystallization conditions to design successful experiments for polymorph screening, selection, and isolation.

List of Abbreviations

API............Active Pharmaceutical Ingredient
a............Activity
γ_w............Activity coefficient of water
A............Anhydrous phase
T_b............Boiling Point
k_B............Boltzmann constant
CSD............Cambridge Structural Database
cm............Centimeter
O_{corr}............Corrected occurrence
ξ............Crystallographic factor representing ratio of number of nearest neighbors in surface to that in bulk
α............Degree of molecular roughness
°C............Degrees celsius
ε............Dielectric Constant
ΔH............Enthalpy difference
ΔH_{vap}............Enthalpy of vaporization
ΔS............Entropy difference
K_H............Equilibrium constant for a hydrate and its corresponding anhydrous form
ν............Frequency of atomic or molecular transport at the nucleus-liquid interface
G............Gibbs free energy
ΔG............Gibbs free energy difference
G_i............Gibbs free energy of the initial state
G^*............Gibbs free energy of the transition state
g............Gram
ΔH_f............Heat of fusion
HT............High Throughput
HPLC............High-Performance Liquid Chromotography
J............Homogeneous nucleation rate
HBA............Hydrogen Bond Acceptor
HBD............Hydrogen Bond Donor
γ_{12}............Interfacial energy per unit area between the crystallization solvent and the nucleating cluster
D............Intrinsic dissolution rate

J .... Joules
mm/s .... Linear growth rate
T_m .... Melting point
mPas .... Millipascals per second
min .... Minutes
M .... Molar
υ .... Molar Volume
MW .... Molecular weight
x_w .... Mole fraction of water
X_S .... Mole fraction solubility of solute in the solvent
m .... Moles
N_{ord} .... Number of fully ordered crystal structures
N_0 .... Number of molecules of the crystallizing phase in a unit volume
N_{Acta} .... Number of times solvent was used for crystallization in *Acta Crys. C* from 1986–1996
PXRD .... Powder X-Ray Diffraction
N .... Raw Frequency of Ocurrence
S .... Solubility
Σ .... Summation
c/s .... Supersaturation
T .... Temperature
T_t .... Transition temperature
UV .... Ultra Violet
R .... Universal gas constant
a_w .... Water activity
wt% .... Weight percent
XRD .... X-Ray Diffraction

References

Abraham MH. Hydrogen Bonding. 31. Construction of A Scale of Solute Effective or Summation Hydrogen-Bond Basicity. *J Phys Org Chem* 1993, 6: 660–684.

Abraham MH. Scales of Solute Hydrogen-bonding: Their Construction and Application to Physicochemical and Biochemical Processes. *Chem Soc Rev* 1993, 73–83.

Abraham MH, Chadha, HS, Whiting GS, and Mitchell RC. Hydrogen Bonding. 32. An Analysis of Water-Octanol and Water-Alkane Partitioning and the ΔLog P Parameter of Seiler. *J Pharm Sci* 1994, 1085–1100.

Becker R and Döring W. The Kinetic Treatment of Nuclear Formation in Supersaturated Vapors. *Ann Phys* 1935; 24: 719–752.

Beckmann W and Winter G. Stability of the Hydrate and Anhydrate of Pyrazzocarnil Hydrochloride in Water-Ethanol Mixtures and in Moist Air. *Int Symp Ind Cryst* 1999; 14: 236–245.

Behme RJ and Brooke D. Heat of Fusion Measurement of a Low Melting Polymorph of Carbamazepine that Undergoes Multiple-Phase Changes During Differential Scanning Calorimetry Analysis. *J Pharm Sci* 1991; 80: 986–990.

Bernstein J. *Polymorphism in Molecular Crystals.* Oxford University Press, Oxford, 2002.

Bernstein J, Davey RJ, and Henck JO. Concomitant Polymorphs. *Angew Chem Int Ed* 1999; 38:3441–3461.

Beyer T, Day GM, and Price SL. The Prediction, Morphology, and Mechanical Properties of the Polymorphs of Paracetamol. *J Am Chem Soc* 2001; 123:5086–5094.

Bingham AL, Hughes DS, Hursthouse MB, Lancaster RW, Tavener S, and Threlfall TL. Over One Hundred Solvates of Sulfathiazole. *Chem. Commun* 2001; 7: 603–604.

Birchall JD and Davey RJ. The Crystallization of Sodium Chloride from Aqueous Solution in the Presence of Polysaccharides. *J Cryst Growth* 1981; 54: 323–329.

Blagden N. Crystal Engineering of Polymorph Appearance: The Case of Sulphathiazole *Powder Tech* 2001; 121:46–52.

Blagden N, Cross WI, Davey RJ, Broderick M, Pritchard RG, Roberts RJ, and Rowe RC. Can Crystal Structure Prediction be used as Part of an Integrated Strategy for Ensuring Maximum Diversity of Isolated Crystal Forms? The Case of 2-amino-4-nitrophenol. *Phys Chem Chem Phys* 2001; 3: 3819–3825.

Blagden N and Davey RJ. Polymorph Selection: Challenges for the Future? *Cryst Growth Des* 2003; 3:873–885.

Blagden N, Davey RJ, Lieberman HF, Williams L, Payne R, Roberts R, Rowe R, and Docherty R. Crystal Chemistry and Solvent Effects in Polymorphic Systems-Sulfathiazole *J Chem Soc Faraday Trans* 1998; 94: 1035–1044.

Blagden N, Davey RJ, Rowe R, and Roberts R. Disappearing Polymorphs and the Role of Reaction By-Products: The Case of Sulphathiazole. *Int J Pharm* 1998; 172: 169–177.

Bourne JR and Davey RJ. The Role of Solvent-Solute Interactions in Determining Crystal Growth Mechanisms from Solution I. The Surface Entropy Factor. *J Cryst Growth* 1976; 36: 278–286.

Byrn SR, Pfeiffer RR, and Stowell JG. *Solid-State Chemistry of Drugs.* SSCI, West Lafayette, IN, 1999.

Cardew PT and Davey RJ. The Kinetics of Solvent-Mediated Phase Transformations. *Proc R Soc Lond* 1985; A398: 415–428.

Carlson ED, Cong P, Chandler Jr. WH, Chau HK, Crevier T, Desrosiers PJ, Doolen RD, Freitag C, Hall LA, Kudla T, Luo R, Masui C, Rogers J, Song L, Tangkilisan

A, Ung KQ, and Wu L. An Integrated High Throughput Workflow for Pre-Formulations: Polymorph and Salt Selection Studies. *Pharma Chem* 2003; 2:10–15.

Chyall LJ, Tower JM, Coates DA, Houston TL, and Childs SL. Polymorph Generation in Capillary Spaces: The Preparation and Structural Analysis of a Metastable Polymorph of Nabumetone. *Cryst Growth Des* 2002; 2: 505–510.

Cross WI, Blagden N, Davey RJ, Pritchard RG, Neumann MA, Roberts RJ, and Rowe RC. A Whole Output Strategy for Polymorph Screening: Combining Crystal Structure Prediction, Graph Set Analysis, and Targeted Crystallization Experiments in the Case of Diflunisal. *Cryst Growth Des* 2003; 3: 151–158.

Davey RJ. The Control of Crystal Habit. In: de Jong EJ and Jancic SJ. *Industrial Crystallization 78*. North-Holland, New York, NY, 1979, pp. 169–183.

Davey RJ. Solvent Effects in Crystallization Processes. In: Kaldis E (ed.). *Current Topics in Materials Science*. North-Holland, Amsterdam, 1982, pp. 429–479.

Davey RJ, Allen K, Blagden N, Cross WI, Lieberman HF, Quayle MJ, Righini S, Seton L, and Tiddy GJT. Crystal Engineering- Nucleation, the Key Step. *Cryst Eng Commun* 2002; 4: 257–264.

Davey RJ, Blagden N, Potts GD, and Docherty R. Polymorphism in Molecular Crystals: Stabilization of a Metastable Form by Conformational Mimicry. *J Am Chem Soc* 1997; 119: 1767–1772.

Davey RJ, Blagden N, Righini S, Alison H, and Ferrari ES. Nucleation Control in Solution Mediated Polymorphic Phase Transformations: The Case of 2,6-Dihydroxybenzoic Acid. *J Phys Chem B* 2002; 106: 1954–1959.

Davey RJ, Blagden N, Righini S, Alison H, Quayle MJ, and Fuller S. Crystal Polymorphism as a Probe for Molecular Self-Assembly during Nucleation from Solutions: The Case of 2,6-Dihydroxybenzoic Acid *Cryst Growth Des* 2001; 1: 59–65.

Davey RJ, Cardew PT, Mcewan D, and Sadler DE. Rate Controlling Processes in Solvent-Mediated Phase Transformations. *J Cryst Growth* 1986; 79: 648–653.

Day GM, Chisholm J, Shan N, Motherwell WDS, and Jones W. An Assessment of Lattice Energy Minimization for the Prediction of Molecular Organic Crystal Structures. *Cryst Growth Des* 2004; 4: 1327–1340.

Day GM, Motherwell WDS, and Jones W. Beyond the Isotropic Atom Model in Crystal Structure Prediction of Rigid Molecules: Atomic Multipoles versus Point Charges. *Cryst Growth Des* 2005; 5: 1023–1033.

Desiraju GR. Crystal Gazing: Structure Prediction and Polymorphism. *Science* 1997; 278: 404–405.

Dunitz JD and Bernstein J. Disappearing Polymorphs. *Acc Chem Res* 1995; 28: 193–200.

Etter MC. Encoding and Decoding Hydrogen-Bond Patterns of Organic Compounds. *Acc Chem Res* 1990; 23: 120–126.

Etter MC, MacDonald JC, and Bernstein J. Graph-Set Analysis of Hydrogen-Bond Patterns in Organic Crystals. *Acta Cryst B Struct Sci* 1990; B46: 256–262.

Ferrari ES, Davey RJ, Cross WI, Gillon AL, and Towler CS. Crystallization in Polymorphic Systems: The Solution-Mediated Transformation of β to α Glycine. *Cryst Growth Des* 2003; 3: 53–60.

Frank TC, Downey JR, and Gupta SK. Quickly Screen Solvents for Organic Solids. *Chem Eng Progress* 1999; 95: 41–61.

Garetz BA, Aber JE, Goddard NL, Young RG, and Myerson AS. Nonphotochemical, Polarization-Dependent, Laser Induced Nucleation in Supersaturated Aqueous Urea Solutions. *Phys Rev Lett* 1996; 77: 3475–3476.

Garti N, Leci CL, and Sarig S. The Effect of Solvents on Crystal Habit of 1,4-Di-Tert-Butylbenzene (DTBB). *J Cryst Growth* 1981; 54: 227–231.

Gavezzotti A. Are Crystal Structures Predictable? *Acc Chem. Res* 1994; 27: 309–314.

Getsoian A, Lodaya R, and Blackburn AC. Single-Solvent Polymorph Screening. *Int J Pharm* 2006; In Preparation.

Gibbs JW. *Collected Works.* Yale University Press, New Haven, 1948.

Gillon AL, Feeder N, Davey RJ, and Storey R, Hydration in Molecular Crystals-A Cambridge Structural Database Analysis. *Cryst Growth Des* 2003; 3: 663–673.

Gölles F. Examination and Calculation of Thermodynamic Data from Experimental Measurements I. The Numerical Integration of the Vapour-Pressure Curves of the System Methanol-Water. *Montash Chem* 1961; 92: 981–991.

Görbitz CH and Hersleth HP. On the Inclusion of Solvent Molecules in the Crystal Structures of Organic Compounds. *Acta Crys B* 2000; B56: 526–534.

Gosselin PM, Thibert R, Preda M, and McMullen JN. Polymorphic Properties of Micronized Carbamazepine Produced by RESS. *Int J Pharm* 2003; 252: 225–233.

Grant DJW. Theory and Origin of Polymorphism. In: Brittain HG. *Polymorphism in Pharmaceutical Solids.* Marcel Dekker, New York, 1999, pp. 1–33.

Grant DJW and Higuchi T. *Solubility Behavior of Organic Compounds.* Wiley, New York, 1990.

Greene LR, Miller JM, and Blackburn AC. Rapid, Small-Scale Determination of Organic Solvent Solubility Using a Thermogravimetric Analyzer. *J Pharm Biomed Anal* 2005; 35: 344–347.

Gu C, Chatterjee K, Young Jr. V, Grant DJW. Stabilization of a Metastable Polymorph of Sulfamerazine by Structurally Related Additives. *J Cryst Grow* 2002; 235: 471–481.

Gu CH, Li H, Gandhi RB, and Raghavan K. Grouping Solvents by Statistical Analysis of Solvent Property Parameters: Implication to Polymorph Screening. *Int J Pharm* 2004; 283: 117–125.

Gu CH, Young Jr. V, and Grant DJW. Polymorph Screening: Influence of Solvents on the Rate of Solvent-mediated Polymorphic Transformation. *J Pharm Sci* 2001; 90: 1878–1890.

Hilfiker R, Berghausen J, Blatter F, Burkhard A, De Paul SM, Freiermuth B, Geoffroy A, Hofmeier U, Marcolli C, Siebenhaar B, Szelagiewicz M, Vit A, and von Raumer M. Polymorphism - Integrated Approach from High-Throughput Screening to Crystallization Optimization. *J Them Anal Calor* 2003; 73: 429–440.

Infantes L and Motherwell S. Water Clusters in Organic Molecular Crystals. *Cryst Eng Commun* 2002; 4: 454–461.

Katritzky AR, Tamm T, Wang Y, and Karelson M. A Unified Treatment of Solvent Properties. *J Chem Inf Comp Sci* 1999; 39: 692–698.

Kelly RC. A Molecular Approach to Understanding the Directed Nucleation and Phase Transformation of Carbamazepine and Nitrofurantoin in Aqueous and Organic Solutions. *PhD Thesis University of Michigan* 2003.

Khankari RK and Grant DJW. Prediction of Nedocromil Sodium Hydrate Phases in Isopropanol-Water Mixtures from the Relative Humidity Phase Diagram. *Pharm Res* 1993; 10:S-153 (PT 6106).

Khankari RK and Grant DJW, Pharmaceutical Hydrates. *Thermochim Acta* 1995; 248: 61–79.

Kolář P, Shen J, Tsuboi A, and Ishikawa T. Solvent Selection for Pharmaceuticals. *Fluid Phase Equil* 2002; 194–197: 771–782.

Koshkhoo S and Anwar J. Crystallization of Polymorphs: The Effect of Solvent *J Phys D: Appl Phys* 1993; 26: B90–B93.

Lang MD, Grzesiak AL, and Matzger AJ. The Use of Polymer Heteronuclei for Crystalline Polymorph Selection. *J Am Chem Soc* 2002; 124: 14834–14835.

Lide DR. *Handbook of Organic Solvents.* CRC Press, Boca Raton, FL, 1995.

Lide DR, Database. CRC Handbook of Chemistry and Physics, 85th Edition, 2004–2005. http://www.hbcpnetbase.com/ (accessed October 2005).

McKay B, Hoogenraad M, Damen EWP, and Smith AA. Advances in Multivariate Analysis in Pharmaceutical Process Development. *Curr Opin Drug Disc Dev* 2003; 6: 966–977.

Mersmann A. Calculation of Interfacial Tensions. *J Cryst Growth* 1990; 102: 841–847.

Mersmann A. *Crystallization Technology Handbook.* Marcel Dekker, Inc., New York, 1995.

Miller JM, Collman BM, Greene LR, Grant DJW, and Blackburn AC. Identifying the Stable Polymorph Early in the Drug Discovery-Development Process. *Pharm Dev Tech* 2005; 10: 275–281.

Mitchell CA, Yu L, and Ward MD. Selective Nucleation and Discovery of Organic Polymorphs Through Epitaxy with Single Crystal Substrates. *J Am Chem Soc* 2001; 123: 10830–10839.

Morris KR. Structural Aspects of Hydrates and Solvates. In: Brittain HG. *Polymorphism in Pharmaceutical Solids.* Marcel Dekker, New York, 1999, pp. 125–181.

Morris KR and Rodríguez-Hornedo N. Hydrates. In: *Encyclopedia of Pharmaceutical Technology.* Marcel Dekker, New York, 1993, pp. 393–440.

Motherwell WDS, Ammon HL, Dunitz JD, Dzyabchenko A, Erk P, Gavezzotti A, Hofmann DWM, Leusen FJJ, Lommerse JPM, Mooij WTM, Price SL, Scheraga H, Schweizer B, Schmidt MU, van Eijck BP, Verwern P, and Williams DE. Crystal Structure Prediction of Small Organic Molecules: A Second Blind Test. *Acta Cryst B* 2002; B58: 647–661.

Mukuta T, Lee AY, Kawakami T, and Myerson AS. Influence of Impurities on the Solution-Mediated Phase Transformation of an Active Pharmaceutical Ingredient. *Cryst Growth Des* 2005; 5: 1429–1436.

Mullin JW. *Crystallization.* Butterworth-Heinemann Ltd, Oxford, 2001.

Murphy D, Rodríguez-Cintrón F, Langevin B, Kelly RC, and Rodríguez Hornedo N. Solution-Mediated Phase Transformation of Anhydrous to Dihydrate Carbamazepine and the Effect of Lattice Disorder. *Int J Pharm* 2002; 246: 121–134.

Myerson AS. *Handbook of Industrial Crystallization.* Butterworth-Heinemann Ltd, Oxford, 2002.

Nakai T, Aoyama Y, and Miyake K. Application of the Thermal History of Copper Sulfate Solution to Industrial Crystallization I. Crystallization with Pilot Plant. *Krist. u. Technik* 1973; 8: 1313–1324.

Nangia A and Desiraju GR. Pseudopolymorphism: Occurrences of Hydrogen Bonding Organic Solvents in Molecular Crystals. *Chem Commun* 1999; 7: 605–606.

Nass KK. Process Implications of Polymorphism in Organic Compounds. *AIChE* 1991; 284: 72–81.

Nyvlt J, Söhnel O, Matuchova M, and Broul M. *The Kinetics of Industrial Crystallization.* Elsevier, New York, 1985.

Okamoto M, Hamano M, Igarashi K, and Ooshima H. The Effects of Impurities on Crystallization of Polymorphs of a Drug Substance AE1-923. *J Chem. Eng Jap* 2004; 37: 1224–1231.

Okamoto M, Hamano M, and Ooshima H. Active Utilization of Solvent-Mediated Transformation for Exclusive Production of Metastable Polymorph Crystals of AE1-923. *J Chem Eng Jap* 2004; 37: 95–101.

Ostwald W. Studien Uber Die Bildung und Umwandlung Fester Korper. *Z Physik Chem* 1897; 22: 289–302.

Otsuka M, Kaneniwa N, Kawakami K, Umezawa O, Udovenko VV, and Mazanko TF. Liquid-Vapour Equilibrium in the 2-Propanol-Water and 2-Propanol-Benzene System. *Rus J Phys Chem* 1967; 41: 863–866.

Price CP, Grzesiak AL, and Matzger AJ. Crystalline Polymorph Selection and Discovery with Polymer Heteronuclei. *J Am Chem Soc* 2005; 127: 5512–5517.

Pudipeddi M and Serajuddin ATM. Trends in Solubility of Polymorphs. *J Pharm Sci* 2005; 94: 929–939.

Riddick JA, Bunger WB, and Sakano TK. *Organic Solvents.* Wiley, New York, NY, 1986.

Rodríguez-Hornedo N, Lechuga-Ballesteros D, and Wu HJ. Phase Transition and Heterogeneous/epitaxial Nucleation of Hydrated and Anhydrous Theophylline Crystals. *Int J Pharm* 1992; 85: 149–162.

Rodríguez-Hornedo N and Murphy D. Significance of Controlling Crystallization Mechanisms and Kinetics in Pharmaceutical Systems. *J Pharm Sci* 1999; 88: 651–660.

Rodríguez-Hornedo N and Murphy D. Surfactant-Facilitated Crystallization of Dihydrate Carbamazepine During Dissolution of Anhydrous Polymorph. *J Pharm Sci* 2004; 93: 449–460.

Rodríguez-Hornedo N and Sinclair BD. Crystallization: Significance in Product Development, Processing, and Performance. In: Boylan JC, and Swarbrick J (eds.), *Encyclopedia of Pharmaceutical Technology.* Marcel Dekker, New York, NY, 2002, pp. 671–690.

Rodríguez-Hornedo N. Crystallization and the Properties of Crystals. In: Boylan JC, and Swarbrick J (eds.), *Encyclopedia of Pharmaceutical Technology.* Marcel Dekker, New York, NY, 1991, pp. 399–434.

Rodríguez-Spong B, Price CP, Jayasankar A, Matzger AJ, and Rodríguez-Hornedo N. General Principles of Pharmaceutical Solid Polymorphism. A Supramolecular Perspective. *Adv Drug Deliv Rev* 2004; 56: 241–274. .

Sacchetti M. Determining the Relative Physical Stability of Anhydrous and Hydrous Crystal Forms of GW2016. *Int J Pharm* 2004; 273: 195–202.

Shalaev E and Zografi G. The Concept Of 'Structure' in Amorphous Solids From the Perspective of the Pharmaceutical Sciences. In: *Amorphous Food and Pharmaceutical Systems*. Cambridge: The Royal Society of Chemistry; 2002. 11–30.

Shekunov BY and York P. Crystallization Processes in Pharmaceutical Technology and Drug Delivery Design. *J Cryst Growth* 2000; 211: 122–136.

Söhnel O and Garside J. *Precipitation: Basic Principles and Industrial Applications*. Butterworth-Heinemann Ltd, Oxford, 1992.

Sokolova EP and Morachevskii AG. Thermodynamic Properties of the Acetone-Water System. *Vestn Leningr Univ Fiz Khim* 1967; 3: 110–115.

Threlfall T. Crystallization of Polymorphs: Thermodynamic Insight into the Role of Solvent. *Org Process Res Dev* 2000; 4: 384–390.

Ticehurst MD, Storey RA, and Watt C. Application of Slurry Bridging Experiments at Controlled Water Activities to Predict the Solid-State Conversion Between Anhydrous and Hydrated Forms Using Theophylline as a Model Drug. *Int J Pharm* 2002; 247: 1–10.

Turnbull D and Fisher J.C. Rate of Nucleation in Condensed Systems. *J Chem Phys* 1949; 17: 71–73.

van Langevelde A and Blomsma E. High-Throughput Screening in Solid Form Section. *Acta Cryst* 2002; A58: C9.

Velaga SP, Berger R, and Carlfors J. Supercritical Fluids Crystallization of Budesonide and Flunisolide. *Pharm Res* 2002; 19: 1564–1571.

Vippagunta SR, Brittain HG, and Grant DJW. Crystalline Solids. *Adv Drug Deliv Rev* 2001; 48: 3–26.

Volmer M. *Kinetic der Phasenbildung*. Steinkopff, Leipzig, 1939.

Washburn, EW. *International Critical Tables of Numerical Data, Physics, Chemistry, and Technology*. McGraw Hill, New York, 1928. p. 290.

Zaccaro J, Matic J, Myerson AS, and Garetz BA. Nonphotochemical, Laser-Induced Nucleation of Supersaturated Aqueous Glycine Produces Unexpected Gamma-Polymorph. *Cryst Growth Des* 2001; 1: 5–8.

Zettlemoyer AC. *Nucleation*. Marcel Dekker, New York, 1969.

Zhu HJ and Grant DJW. Influence of Water Activity in Organic Solvent Plus Water Mixtures on the Nature of the Crystallizing Drug Phase 2. Ampicillin. *Int J Pharm* 1996; 139: 33–43.

Zhu HJ, Yuen CM, and Grant DJW. Influence of Water Activity in Organic Solvent Plus Water Mixtures on the Nature of the Crystallizing Drug Phase 1. Theophylline. *Int J Pharm* 1996; 135: 151–160.

4

Solubility Issues in Early Discovery and HTS

LI DI AND EDWARD H. KERNS

Wyeth Research, Princeton, NJ

Introduction

Drug discovery programs begin with target identification and validation for diseases with unmet medical needs (Figure 1). Drug discovery targets that are currently available mostly fall into two categories, with 45% being receptors and 28% being enzymes (Table 1) (Drews, 2000). There are only 483 targets addressed by all drugs in the pharmaceutical industry, which is a relatively small number compared to the estimated 3000–10,000 disease relevant genes (Meisner et al., 2004). It is evident that the drug target universe is far from being fully exploited.

Hits for a specific drug target can be identified through high throughput screening (HTS), computational approaches and information from literature and patents (Figure 2). After hits are confirmed, they are characterized and optimized to generate lead series. Leads are optimized and generate development candidates. Through clinical development, the clinical candidates optimally reach the market. The process takes an average of 12 years and costs more than $800 million per new chemical entity (NCE) with a success rate of 10% in human clinical trials (Dickson and Gagnon, 2004; Lombardino and Lowe, 2004). HTS is one of the first steps towards this long, high risk and costly drug discovery and development processes. The decisions taken at this stage have far-reaching consequences for success later in lead optimization and in clinical development (Bleicher et al., 2003). Therefore, the quality of HTS is of critical importance for drug discovery programs.

The success of hit identification through HTS relies not only on the robustness of biological assays, but also on the quality and the diversity of compound libraries (Gribbon and Andreas, 2005). In addition to the purity and identity of the compound library (Kerns et al., 2005), drug-like properties, such as solubility in the HTS assay media and in DMSO stock, significantly impact HTS and biological screening results (Lipinski et al., 1997). Insoluble compounds have unknown concentrations in screening assays. They have either the incorrect starting

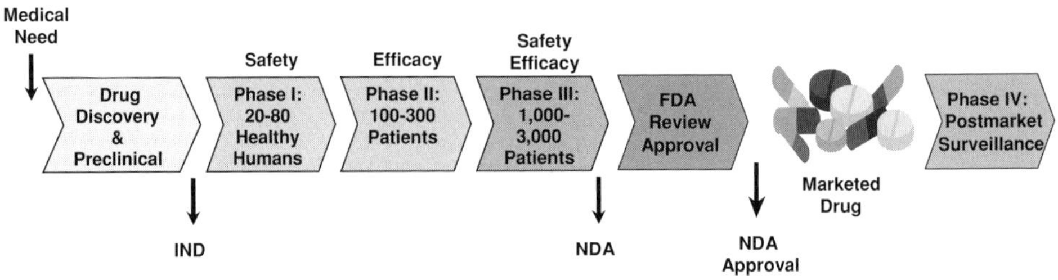

Figure 1. Drug discovery and development processes.

concentration, due to low solubility in DMSO stock solution, or incorrect final concentration, due to precipitation in aqueous assay media. Poor solubility in DMSO can give both false positive and false negative assay results, depending on whether any solid material is transferred and re-dissolves in aqueous buffer. Precipitation in assay media tends to give false negative results in HTS. It can also miss important pharmacophores and generate inaccurate SAR.

Numerous examples have been reported in which assays could not be conducted at the desired concentration or false negatives were generated, due to the limit of solubility (Andreani et al., 2001; Gezginci et al., 2001; Blackman et al., 2002; Dalvit et al., 2002; Davidson et al., 2004; Faria et al., 2004; Roehrl et al., 2004; Miller et al., 2005). One common example of this is as follows: when a compound was initially tested in a receptor binding assay, it was not very active, with an IC_{50} of 1 μM. However, the chemist had a lot of faith in the compound and it was retested. During the retest, the biologist found that it was not soluble and used special conditions to dissolve the compound. Under these conditions, the IC_{50} was 1 nM. By completely dissolving the compound, the IC_{50} improved by 1000 fold.

In a second example, a compound was initially tested in a CYP450 2D6 inhibition assay and the IC_{50} was greater than 10 μM. However, when the compound was

Target classes	Percentage
Receptors	45%
Enzymes	28%
Homones and factors	11%
Ion Channels	5%
Nuclear receptors	2%
Nuclear acids	2%
Unknown	7%

Table 1. Therapeutic target classes of current drugs (Drews, 2000).

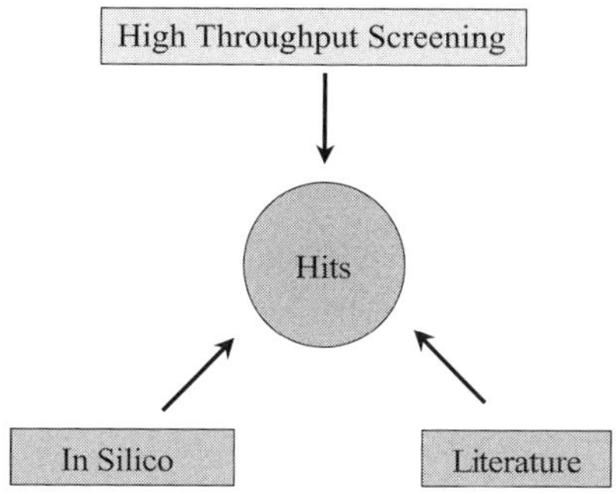

Figure 2. Hit generation.

retested. It was found to be insoluble in the DMSO stock solution. It was, therefore, retested using a lower concentration DMSO stock solution with ethanol as a cosolvent. The IC_{50} in this retest was less than 1 μM. The compound was, thus, a potent 2D6 inhibitor. Because of the poor solubility of the compound, the initial assay underestimated the potential toxicity due to drug-drug interaction.

In a third example, when a compound was initially tested in a microsomal stability assay, no metabolism was observed, with 100% of the compound remaining after a 15-min incubation. When it was retested under the same conditions, only 50% remained after 15 minutes. The results were irreproducible and erratic. The reason was that the compound had poor aqueous solubility. It precipitated during the first dilution into the aqueous media (Dilution is essential to reduce the DMSO content, because it inhibits Cytochrome P450 activity). The solid precipitate was not metabolized, but it was quantitated in the assay when acetonitrile was added to the aqueous reaction mixture. The compound, thus, appeared artificially to be more stable. Precipitation is quite a variable process and depends on many different factors, such as temperature, incubation time, seeding conditions and agitation. The completeness of precipitation affects the final results, thus, assay results tend to be more variable for insoluble compounds.

Solubility issues are universal for activity and property screening, both HTS and traditional bench-top assays. It significantly impairs the quality of biological assays, ADME/TOX screens and *in vivo* activity measurements. This chapter focuses on solubility issues and potential solutions for screening assays in early drug discovery. The effects of solubility on absorption and oral bioavailability will be addressed elsewhere in this book.

Solubility in DMSO

Solubility Issues in DMSO

Compounds that are used in HTS, bioassays and ADME screening are typically dissolved and stored in DMSO at a concentration of 10–30 mM. It has been estimated that 10–20% of compounds used in these activities are insoluble in DMSO at these concentrations at room temperature (Balakin, 2003; Oldenburg et al.,

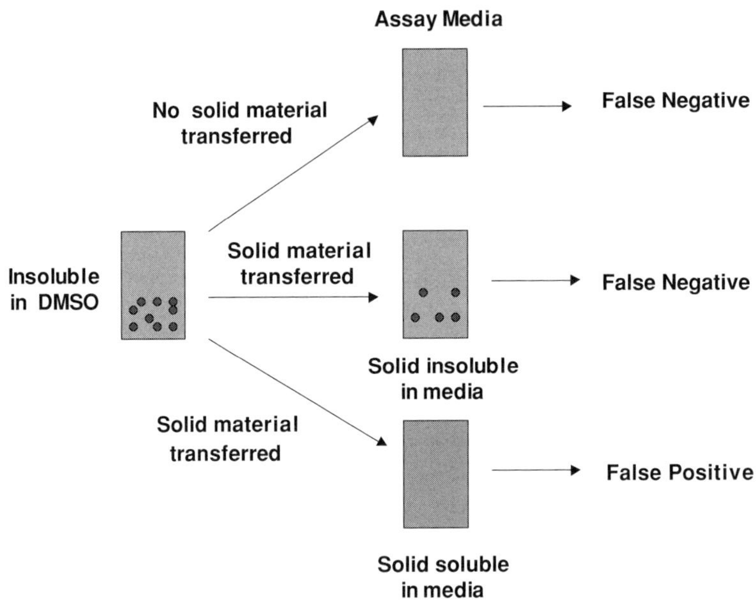

Figure 3. Effects of DMSO-insoluble compounds on bioassay results. Both false positive and false negative results can be generated.

2005). Low solubility in DMSO significantly affects the screening concentrations and data quality. The starting concentrations for these compounds are essentially unknown due to incomplete dissolution and erroneous data can be generated. Furthermore, particulates of undissolved compounds can block pipeters and cause an incorrect volume to be delivered to the assay. This can be an especially large problem for low volume robotics and microfluidic systems, where clogging can cause major downtime for a screening campaign (Hertzberg and Pope, 2000). Clearly, assay quality and reliability suffer if compounds are insoluble in DMSO. Some companies routinely identify compounds that are insoluble in DMSO by visual inspection and remove them from the main screening collection. This improves data quality, minimizes the time and resources expended on insoluble compounds and generates more reliable data (Walters and Namchuk, 2003).

If compounds are insoluble in DMSO, they can give either false negative or false positive assay results (Figure 3). If the pipetting process does not transfer any solid material to the assay media or if the transferred solid material is insoluble in the assay media, the actual assay media concentration will be lower than the target concentration. The subsequent dilution of this assay media will give lower concentration of test solutions than anticipated and lead to false negative results. False negatives are active compounds that go undetected in the screening process. For example, if a screen is performed at 10 μM and the requirement to make the primary hit list is 80% inhibition, then a compound with 50% inhibition and an actual concentration of 1 μM (due to low solubility in

DMSO) will be missed, even though it is actually a 1 μM inhibitor and it was mistakenly thought to have an IC_{50} of 10 μM. This compound will not make it to the hit list due to the incorrect starting concentration. Screening at this much lower concentration than expected could potentially miss important hits and key pharmacophores.

If solid material was transferred to the assay media during pipetting and the solid was soluble in aqueous buffer, this would give a higher concentration than expected and lead to false positives. False positives are compounds identified as hits in the HTS that are inherently inactive against the target. For example, if the screen that was expected to be at 10 μM gave 80% inhibition for a compound with an actual concentration of 100 μM, due to solid material being transferred and re-dissolved in the aqueous buffer, then this is a false positive. The compound would not have made it to the hit list if the actual concentration had been known. The quality of HTS and bioassays are compromised if compounds are insoluble in the DMSO stock solution, because both false negative and false positive hits can be generated.

Effect of Concentration on Solubility Issues in DMSO

When compound stock solutions are stored at lower concentrations, solubility in DMSO is less of a problem. The optimal storage concentrations were found to be 2–5 mM, which maximizes solubility and minimizes precipitation in DMSO (Popa-Burke et al., 2004; Schopfer et al., 2005). Storing compounds at lower concentrations greatly reduces DMSO solubility problems, because it reduces the solubility requirement. For example, compounds might only need to have a solubility in DMSO of 2–5 mM, rather than the 10-30 mM used in many screening labs. The drawback of storing compounds at a low concentration is that it can limit the test concentrations of compounds in assays that can tolerate only a small percentage of DMSO in the assay buffer, such as cell-based assays and certain enzyme assays. For example, if a cell-based serial-dilution assay can only tolerate 1% DMSO and it is designed for a maximum concentration of 100 μM, the starting DMSO stock solution needs to be at least 10 mM. A lower DMSO stock solution concentration will necessitate either a lower maximum screening concentration (less than 100 μM), or a higher DMSO content. There is a fixed ratio between sample concentration and percent DMSO. If the DMSO stock concentration is only 5 mM, dilution to 100 μM will give a final DMSO concentration of 2%. If it is necessary to keep the % DMSO at 1%, the screen would have a maximum concentration of 50 μM. A higher % DMSO could compromise the cell integrity. However, having a maximum concentration of 50 μM could reduce the assay signal, or generate an incomplete IC_{50} curve, due to the lower starting concentration. In general, DMSO can interfere with most cell-based assays at a concentration of greater than 1% (Johnston and Johnston, 2002). DMSO tolerance is assay-dependent and it can vary greatly from assay to assay. For example, some enzyme assays can tolerate 10% DMSO, but in others 0.2% DMSO can inhibit enzyme activity. Hence, percent DMSO in a specific assay needs to be optimized based on individual assay conditions. Storage at lower concentrations can alleviate solubility issues, however, it is not always feasible for every screening program and it can restrict the number of compounds that can be tested at higher concentrations without increasing the DMSO content.

Effects of Freeze-Thaw Cycles, Water Uptake and Crystallization

DMSO stock plates are usually sealed and stored below ambient temperature (−20 to 4°C). For each assay, plates are thawed and specific volumes are dispensed. Repeated use of such plates requires several freeze-thaw cycles, which can lead to precipitation and decomposition (Cheng et al., 2003; Kozikowski et al., 2003b; Kozikowski et al., 2003a; Lipinski and Hoffer, 2003; Lipinski, 2004a; Lipinski, 2004b; Oldenburg et al., 2005). While compound stability during storage has been an issue of concern in the past, solubility in DMSO is at least as serious of a problem. It was found that precipitation is the major reason for loss of compound during the first six months of frequent plate usage. After 6 months of usage, compound decomposition becomes the major cause of compound loss (Hoever and Zbinden, 2004).

Freeze thaw cycles increase precipitation by two factors: absorption of water into DMSO solutions and crystallization of compounds from DMSO. DMSO is very hygroscopic and will absorb up to 10% water in as little as 6 hours under laboratory conditions (Fillers, 2004). In fact, it is technically difficult to limit uptake of water into DMSO. A few companies deliberately introduce about 10% water into the DMSO and store stock solutions in a cool but nonsolid matrix (Schopfer et al., 2005). At 10% water/DMSO, the freezing point of DMSO drops from 18°C to below 4°C, so stock solutions can be stored at 4°C without causing freeze-thaw cycles. There are two major advantages of this approach: (1) it allows fast access to the samples, since no melting is required, and (2) the stock concentration is less variable by avoiding volume increases through water absorption. The ambiguity of the final assay concentration is probably the single most influential parameter in the validation rate of primary screening data.

Repeated dispensing from the same stock plate and freeze-thaw cycles increase the water content of the DMSO, and result in decreased solubility and precipitation. Solubility of compounds in wet-DMSO is much lower than in dry DMSO. A small amount of water in the DMSO can decrease solubility dramatically, due to the non-ideal properties of DMSO-water mixtures. Cavity formation during dissolution is more difficult in wet DMSO than in dry DMSO or in pure water (Oldenburg et al., 2005). The impact of water absorption during freeze-thaw cycles led several groups to build large storage facilities for single-use mini tubes.

Freeze-thaw cycles also create an environment of slow cooling which enhances crystallization from DMSO. The amorphous material, that initially dissolved in DMSO, can crystallize at low temperature in the refrigerator and become more difficult to re-dissolve. This is because the high-energy amorphous material has higher solubility in DMSO than the low-energy crystalline material. As a result, compounds become less and less soluble with successive freeze-thaw cycles. Compounds are frequently more active when freshly prepared in DMSO from powder, than after they have undergone several freeze-thaw-cycles, due to the higher solubility of the initial amorphous compounds than of the crystalline precipitate.

Increasing water content and formation of crystalline material are two synergistic causes for decreased solubility in DMSO after freeze-thaw cycles. Storage of DMSO stocks at low temperature is a good strategy if stability is an issue, but it is not a wise practice if solubility is an issue. Lipinski recommends storing DMSO stock solutions at room temperature for short term use to minimize solubility effects (Lipinski, 2004b).

Classes of Compounds That Are Insoluble in DMSO

Two types of compounds tend to have low DMSO solubility (Balakin et al., 2004). Type A compounds have a strong molecular lattice for crystal packing. A typical example of Type A compounds is an organic salt. Salts tend to have lower solubility in DMSO than free bases and acids. Amines are frequently prepared as hydrochloride salts to increase chemical stability upon storage. They usually have poor solubility in DMSO due to strong crystal packing, but are more soluble in DMSO-water mixed solvent. A mixed solvent of 1:1 DMSO: water is commonly used to dissolve salts that are not soluble in DMSO. Type A compounds tend to have good aqueous solubility, which is a desirable pharmaceutical property. Removal of Type A compounds from the screening library due to low DMSO solubility could potentially throw out highly valuable pharmacophores and drug-like compounds. Dissolving Type A compounds in DMSO/water mixed solvent can avoid the mistake of rejecting valuable compounds. Type B compounds have low DMSO solubility due to poor solvation. They tend to have higher molecular weight, higher Log P, more rotatable bonds and greater solvent-accessible surface area. An example of this class of compound is itraconazole. It has a MW of 705, cLogP of 6 and 9 rotatable bonds. Type B compounds have poor solubility in both DMSO and aqueous media.

Prediction of Solubility in DMSO

Solubility in DMSO is less intuitive than aqueous solubility, based on examination of the chemical structure. Chemists can usually differentiate between compounds that are soluble or insoluble in water, but it is much harder to predict compounds that are soluble or insoluble in DMSO (Balakin, 2003). Solubility in DMSO is determined by a subtle balance of oppositely-directed inter- and intramolecular forces. Computational models have been developed to predict DMSO solubility with greater than a 90% success rate (Balakin, 2003; Balakin et al., 2004; Japertas et al., 2004; Lu and Bakken, 2004; Delaney, 2005). Software can be used to provide an alert to a compound with a low DMSO solubility and care should be taken in biological screening of these compounds and in prioritizing the hits.

Strategies to Overcome Solubility Issues in DMSO

Several strategies can be applied to overcome solubility issues in DMSO. These are listed in Table 2 and are discussed below.

In-Well Sonication

High throughput in-well sonication has been applied to dissolve compounds that are insoluble in DMSO in 96, 384, and 1536 well formats (Oldenburg et al., 2005). Compounds that precipitated from DMSO stocks, due to either water uptake that reduced solubility or low instrinsic solubility that promoted crystallization, can be re-dissolved by low energy sonication. Sonication can accelerate compound dissolution in seconds and, in some cases, drive the solution to supersaturation, due to energy input and elevated temperature. This process can bring many precipitates back into solution and has no effect on compound stability. Sonication of DMSO stocks or concentrated aqueous solutions can improve HTS hit rates and enhance biological assay results.

Strategies	Advantages	Disadvantages
In-well Sonication	Increase solubility and dissolution rate	Cross contamination
Use 90% DMSO/Water	Consistent water content, volume and concentration Fast access as liquid Favor drug-like properties	Instability for certain compounds
Lower Stock Concentration	Reduce precipitation	Might limit test concentration
Use DMSO/Water for Salts	Enhance salt solubility	Cherry picking salts
Determine Exact Concentration	Increase accuracy Known concentration	Time and resource consuming, costly
Non-DMSO Storage Approach	Overcome stability and solubility issues	High initial investment Not suitable for all assays

Table 2. Strategies to enhance DMSO solubility.

Store at Lower Concentrations

Dissolving and storing compounds at low stock concentrations in DMSO can reduce precipitation. A lower stock concentration reduces the solubility requirements. More compounds will be soluble at lower concentrations. A DMSO stock concentration of 2–5 mM is recommended (Popa-Burke et al., 2004; Schopfer et al., 2005).

Use DMSO/Water Cosolvent for Salts

Use of water/DMSO mixed solvent can enhance solubility of salts in the stock solution and avoid missing valuable compounds with good aqueous solubility. Typically 1:1 water/DMSO is used to dissolve salts. Besides DMSO, water mixable solvents can sometimes be used to dissolve certain classes of compounds, such as methanol, ethanol, acetonitrile, THF, pyridine and DMF (Buchli et al., 2005). Tolerance for different organic solvents needs to be carefully evaluated for each assay during assay method development.

Determine Exact Stock Concentration

Another common practice in rescuing DMSO-insoluble compounds is to determine the exact concentration after removal of insoluble particulates by centrifugation or filtration. The concentration of the solution can be determined using chemiluminescent nitrogen detectors (CLND) or evaporative light scattering detectors (ELSD) with LC-UV-MS. CLND is particularly useful for quatitation because its response is proportional to the molar amount of compound. No standard of the compound is necessary and CLND tends to have good linearity at high concentrations. Several high throughput methods have been developed to purify compound libraries and determine the exact concentration of DMSO stock solutions before screening (Yurek et al., 2002; Yan et al., 2003; Popa-Burke et al., 2004; Yan et al., 2004; Kerns et al., 2005). The actual concentrations determined experimentally can be used to calculate the IC_{50} to determine the activity of the compound.

Non-DMSO Storage Options

Storage of compounds in DMSO and having them readily available greatly facilitates HTS and different screening assays. However, because of solubility and stability issues associated with compounds stored as DMSO solution, several non-DMSO storage concepts have been developed.

NanoStore technology uses 1536 well plates on a NanoCarrier™ to store compounds after the DMSO is evaporated (Benson et al., 2005). Compounds in a dry state avoid issues associated with DMSO storage. The evaporation procedure eliminates DMSO, thus, it can be used in assays that are sensitive to DMSO.

The DotFoil™ concept was developed to store compounds in a dry state as films (Topp et al., 2005). Compounds are more stable as dry films in DotFoils than as DMSO solutions. Compounds stored in DotFoils can be redissolved in a fast, reliable, and easy manner and then directly used in 96- or 384-well plates.

ChemCards™ have been developed to store compounds for microarray compound screening (μARCS). With ChemCards™ technology, the compounds are dried immediately after spotting onto the cards, and then sealed in gas-tight, light-proof pouches under an inert atmosphere. No plate-preparation steps are required before running the assays, since cards are taken from the storage device and used directly in the assay. The associated savings in time and resources are significant (Hoever and Zbinden, 2004).

Solubility in Biological Assay Buffers

Solubility Issues in Aqueous Buffers

It has been estimated that about 30% of discovery compounds have aqueous solubility less than 10 μM (Lipinski, 2001). Ten μM is a typical concentration used in primary biological assays and HTS. Screening at this concentration can potentially underestimate the biological activity for 30% of the compounds and lead to false negative results. It has been found that increasing compound concentration in screening can enhance the statistical power of detecting hits of a given potency. For example, screening at higher concentrations, such as 10–30 μM, generates more hits and leads to greater structural diversity than screening at below 10 μM (Walters and Namchuk, 2003). Screening at higher concentrations, however, puts more demand on compound solubility. More compounds will precipitate during screening and the assay quality is compromised. At high screening concentrations, compounds have a higher propensity to form aggregates and generate false positives (see Section 3.4). Such compounds interact with targets in a nonspecific manner and are resistant to structural optimization. They have limited positive impact on lead discovery. A typical primary screening concentration is 3–10 μM.

Effect of Solubility on HTS Hit Rate

Compound libraries with poor solubility tend to have a lower HTS hit rate than soluble libraries, because the actual screening concentration is much lower than anticipated. The HTS results for a set of 2797 compounds screened against 52 enzyme targets showed that the soluble compounds have a hit rate of 32%, but the insoluble compounds only have a hit rate of 4% (Figure 4) (Popa-Burke et al., 2004). Precipitation in assay buffer reduces compound concentration and leads to a lower success rate in identifying valuable hits and pharmacophores.

Solubility Causes Discrepancies between Isolated Target and Cell-Based Assays

Screening assays are typically performed using two formats: (1) isolated targets, such as enzymes or receptors; and (2) cell-based assays. Isolated target assays

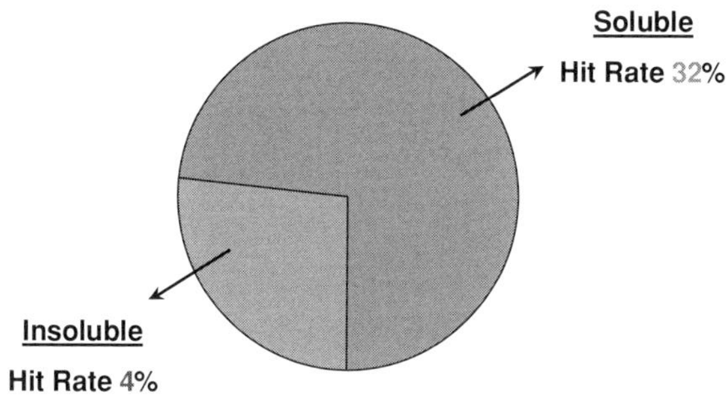

Figure 4. Effect of compound solubility on HTS hit rates.

offer the ideal assay format for screening, because they measure a functional consequence of ligand activity at the target protein within an isolated target environment (Moore and Rees, 2001). There is no permeability limitation for compounds caused by cell membrane penetration and no interference due to cytotoxicity. Furthermore, isolated targets tend to have higher tolerance for organic solvents (e.g., DMSO), serum protein and various buffer components than cell-based assays. These additives help increase solubility of test compounds in the assay media and generate more consistent and reliable data for development of SAR.

Cell-based assays are usually considered more "physiological" and have higher "information content" than isolated targets. However, the complexity of the cell environment and its many mechanisms all combine to confound the observed SAR. These complexities include biochemical mechanisms such as: multiplicity of targets within the cell environment, non-target-oriented mechanisms of action, and variable coupling between binding and response. In addition, physicochemical, metabolic and toxicity mechanisms impact cellular penetration, efflux, metabolic stability, and cytotoxicity. SAR is typically less straight forward using cell-based assays as compared to isolated targets. Cell-based assays are most useful where the whole cell mechanism is needed to capture a specific biological activity, such as cell-surface receptors or ion channel targets. Certain complex targets cannot be adequately configured or reconstituted in an isolated-target assay. Such targets could involve complex interactions between receptors, co-activators, co-repressors, response elements and other cellular factors that cannot be adequately reproduced outside the cell. Cell-based functional assays can therefore provide information on the nature of the pharmacological activity of a compound that cannot be obtained from a biochemical assay. Cellular formats can generate data on compound cytotoxicity and cellular membrane permeability that could serve to provide an alert for problems of chemical series early in the lead generation process (Johnston and Johnston, 2002).

In drug discovery programs, it is common to use isolated target assays as primary screen and cell-based functional assays as secondary screens to verify the observed activity in a more physiological cell environment. Discrepancies in biological activity between the two assays are often observed. Besides the complex mechanisms in cell-based assays that cause the difference between the two assays, solubility differences in the different assay media contributes to the difference in activity between the two assays. Isolated targets usually can tolerate a higher amount of DMSO than cell-based assays. DMSO can help solubilize the compounds. The amount of DMSO in cell assays is typically less than 1%. The low amount of DMSO can reduce solubility of compounds in the assay media and cause a discrepancy between primary enzyme or receptor screening and functional cell-based assays.

Inert carrier proteins, such as bovine serum albumin, are commonly added to the assay media at high concentrations for screening of isolated targets in order to saturate the potential protein binding surfaces and reduce the loss of target proteins due to adsorption onto the surfaces of the containers, such as plates and vials. The added benefit of inert proteins is that they can bind to lipophilic compounds and increase the solubility of insoluble compounds in the assay buffer. The presence and absence of serum protein can cause difference in solubility and cause a discrepancies between the two assays.

Table 3 shows an example of three compounds from a discovery program at Wyeth Research. The project team found there were large discrepancies between the receptor binding assay and the cell-based assay. The cell-based assay had much weaker activity than the receptor binding assay. Both assays had the same target screening concentration of 10 μM. For the receptor binding assay, all three compounds were soluble in the assay buffer. The actual concentrations were very close to the targeted concentration of 10 μM. For the cell-based assay, however, all the compounds precipitated during the assay. The actual concentration in the media was much lower than the target concentration of 10 μM. They varied from 1.4 to 4.8 μM. The difference in solubility was mainly due to the difference in buffer composition of the two assays. The receptor binding assay buffer had 25-fold higher amount of DMSO than the cell-based assay (2.5% vs 0.1% DMSO), and it also contained serum protein to keep the compounds in solution (5% vs. 0% BSA). The lower solubility in the cell-based assay contributed to the weaker biological activity. Differences in buffer composition led to different solubility and caused a difference in activity.

Compound	Solubility in receptor binding assay buffer (μM)	Solubility in cell-based assay buffer (μM)
1	11	2.4
2	10	4.8
3	10	1.4
Assay Buffer	5% BSA, 2.5% DMSO	0.1% DMSO

Table 3. Example of the comparative effect of buffer composition on solubility and activity in receptor binding vs. cell-based assays.

Aggregates as Promiscuous False Positive HTS Hits

It has been observed that certain compounds appear as frequent hitters in HTS regardless of the biological target, especially when screened at higher than 10 μM concentration. Figure 5 shows examples of promiscuous HTS hits (McGovern et al., 2002; McGovern et al., 2003; McGovern and Shoichet, 2003). These spurious hits have low potency at 1–10 μM, noncompetitive and reversible inhibition, flat SAR and poor selectivity. The potency increases with incubation time, but decreases with higher enzyme concentration, temperature, ionic strength and with the addition of urea or BSA. The inhibition is not reversible in a dialysis experiment, suggesting formation of aggregates (Feng et al., 2005).

The aggregates of frequent hitters are in the size range of 30–400 nm and they are not micelles or vesicles. Some of the aggregates will pass through a 0.2 μM filter (200 nm), which is commonly used in solubility assays to separate the insoluble solid from the soluble compounds in solution. It has been speculated that the aggregation is a consequence of compounds being tested at supersaturated concentrations. This can occur when poorly soluble compounds are introduced into aqueous media from DMSO stock solutions. The aggregation state may be a kinetically transient state that precedes crystallization (Lipinski, 2004b). The exact mechanism by which aggregates appear as "actives" in HTS is unclear. They seem to inhibit enzyme activity through adsorption or absorption to the target proteins. Aggregate formation is concentration dependent. The higher the screening concentration, the more likely that compounds will show promiscuous inhibition due to aggregate formation. In a screening study of 1,030 compounds, promiscuous inhibitors dropped from 19% to 1.4%, when the screening concentration was reduced from 30 μM to 5 μM. Structurally, the

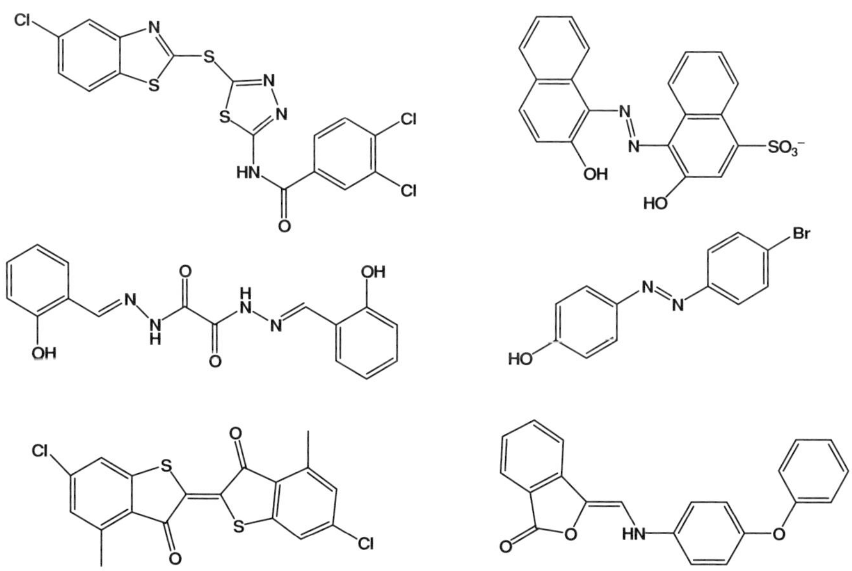

Figure 5. Examples of promiscuous false positive HTS hits.

phony hittters can be drug-like or non-drug-like. Even some marketed drugs form aggregates at high concentrations.

Spurious HTS hits can drain medicinal chemistry resources and downstream biological efforts. Screening at high concentration can be counterproductive in lead discovery. Early elimination of these hits from the screening process, or removal from the screening library can increase discovery productivity. There are several approaches that can be used to recognize and overcome false positive hits due to aggregate formation.

(1) Re-screen hits in the presence and absence of the detergent Triton X-100 (0.1%) (Ryan et al., 2003; Feng et al., 2005). This is the most effective way to break up the aggregates. It has been found that detergent-dependent enzyme assays give the fastest and most reliable single indication of aggregate-based inhibition. Molecules that inhibit only in the absence of detergent are considered likely promiscuous aggregators. The potency of promiscuous inhibitors is reduced in the presence of 0.1% Triton X-100.

(2) Use a dynamic light scattering (DLS) plate reader to measure aggregate formation. The identification of aggregate particle formation using a DLS plate reader seems reliable. However, other aggregation phenomena, such as precipitation, can also lead to light scattering and be mis-identified as aggregation. Precipitation and aggregation are two distinct phenomena. Signals from aggregates can be weak, which can lead to ambiguous results. Molecules with optical properties can interfere with observation in DLS, making the results un-reliable. For example, Congo Red absorbs light at 514.4 nm and interferes with DLS measurement. Congo Red forms aggregates, as observed by transmission electron microscopy (TEM), and shows promiscuous inhibition. However, it does not form particles detectable by DLS due to interference (McGovern et al., 2002; Feng et al., 2005).

(3) Examine the IC_{50} curves to see if they are very steep compared to normal hits (Figure 6) (McGovern and Shoichet, 2003). Careful examination of IC_{50} curves for abnormality can help diagnose aggregate formation. Aggregate-based target protein inhibition

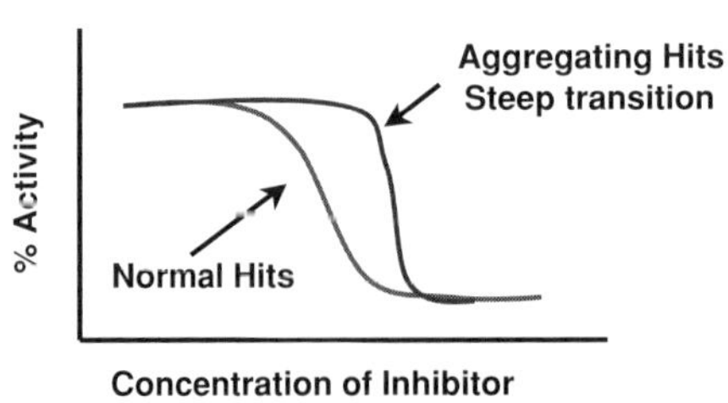

Figure 6. Steep concentration dependence of aggregate-based inhibition.

has a steep dependence on concentration. Screening at lower concentrations, below this steep rise, can minimize false hits from aggregates.

(4) Use computational methods that have been developed to rapidly and automatically identify potential frequent hitters (Roche et al., 2002; Seidler et al., 2003; Feng et al., 2005). Predictive models show some potential for predicting aggregation-based promiscuity in large libraries. A caveat of computational models is that they remain too crude to capture the concentration dependence of aggregate formation.

Solubility Screening Methodologies

Many methods are available in the industry for solubility measurement (Kerns, 2001; Kerns and Di, 2004; Kerns and Di, 2005). Both kinetic solubility and thermodynamic solubility methods are commonly used in drug discovery. In kinetic solubility, equilibrium is not established between solid and solution. The solid material can be in meta-stable crystal form, such as amorphous and unstable polymorphous. In thermodynamic solubility, the solution and the solid material is in equilibrium. The crystal form is the most stable polymorph. Kinetic solubility is more applicable in early drug discovery, because the assay conditions closely mimic biological assays in terms of: initial dissolution of compound in DMSO, the DMSO content of the assay, incubation time and meta-stable crystal forms. It is typically performed in a high throughput plate-based format. Thermodynamic solubility is more relevant for late stage drug discovery and development to aid formulation development and regulatory filing. The most commonly used methods are briefly described below.

Direct UV Method

In the direct UV method, compounds are dissolved in DMSO stock solution at 10 mg/mL. A small volume is added to an aqueous buffer and mixed. If the target concentration exceeds the solubility of the compound, the insoluble material will precipitate. The solution is allowed to settle for certain period of time (e.g. overnight) and is then filtered to remove the precipitate. The concentration of the supernatant is determined by using a UV plate reader and the solubility is derived against a single point standard (Avdeef, 2001).

Nephelometric Method

The nephelometric method is performed by serial dilution in aqueous buffer (Bevan and Lloyd, 2000; Goodwin, 2003). At high concentration, the compound will precipitate and scatter light, resulting in low light transmission. At low concentration, the solution is clear and will have high light transmission. A plot of the intensity of light transmission vs. compound concentration indicates the concentration at which the compound started to precipitate, which is taken as the solubility.

Turbidimetric Method

In the turbidimetric method, a DMSO stock solution is added consecutively in small microliter increments to an aqueous buffer in a cuvette (Lipinski et al., 1997). The solution is mixed by a stir bar and the light transmission is determined with a UV-Vis spectrometer. The solubility is the concentration when the compound starts to precipitate.

Equilibrium Shake Flask Method

The equilibrium shake flask method is the gold standard for thermodynamic solubility determination (Glomme et al., 2005). The assay is performed by adding a solvent to solid material, and mixing for 24–48 hrs. The solution is filtered and the supernatant is diluted and assayed using LC-UV-MS. This method is low throughput compared to the needs of HTS and early discovery.

Prediction of Aqueous Solubility

Numerous computational methods have been developed to predict aqueous solubility from molecular structure (Jorgensen and Duffy, 2002; Delaney, 2005). Many of them have an accuracy of 1 log unit. Commercial solubility software is widely available and some of these products estimate the solubility-pH profile for ionizable compounds. Software is most useful for virtual screening of large libraries, prioritization of compounds prior to synthesis and scoring of HTS hits (Oprea et al., 2005). Calculated values can be used to alert teams to potential solubility issues.

Strategies to Overcome Solubility Issues in Aqueous Media

Several strategies have been developed to overcome solubility issues in assay media (Table 4). These are discussed below.

Optimization of Dilution Protocols

Dilution protocols can affect the actual assay concentration. Even if the final target concentration and buffer composition are the same, different dilution procedures can generate different assay concentrations. Table 5 shows an example with the same final target concentration, but with different dilution protocols. In the first protocol, because of the high concentration of the first dilution, most compounds from the project will precipitate after the first dilution. The concentrations from the subsequent dilutions are quite variable, depending on whether any solid material is transferred. The actual concentration tends to be lower than the target concentration. In the second protocol, because of the low first dilution concentration, the compounds stay in solution after the first dilution and no precipitation is observed. The assay results are more accurate and consistent. The compound concentration and percent DMSO can be balanced to maximize solubility during dilution steps.

Strategies	Advantages	Disadvantages
Optimize Assay Protocol	Conditions with maximal solubility and minimal precipitation	Early investment of assay development
Determine Solubility in Assay Buffer	Know exact assay concentration	Low throughput
Correction of Bioactivity using Solubility Data	Rescue insoluble series	Can over- or under-correct bioactivity
Avoid Dilution	Minimize precipitation and maximize bioactivity	Require low volume accuracy from robotics
In Well Sonication	Increase solubility	Potential cross contamination and degradation
Screening at Low Concentration	Reduce precipitation Reduce aggregate formation and false positives	Lower signal-to-noise ratio

Table 4. Strategies to reduce aqueous solubility issues in discovery assays.

There are two dilution protocols that are commonly used in the pharmaceutical industry for IC_{50} determination (Di and Kerns, 2005). These are: 1) serial dilution in aqueous solution and 2) serial dilution in DMSO. The first dilution protocol is to dilute directly into an aqueous buffer from a

Protocols	Protocol # 1	Protocol # 2
DMSO Stock 1st Dilution	2 mM 100 μM, 5% DMSO Precipitated	2 mM 5.3 μM, 0.27% DMSO No Precipitation
Target Final Concentration	2 μM, 0.1% DMSO	2 μM, 0.1% DMSO
Actual Final Concentration	<2 μM due to precipitation in 1st dilution	2 μM due to lack of precipitation in 1st dilution

Table 5. Example of the effects of dilution protocols on final concentration.

concentrated DMSO stock solution to give a high aqueous concentration (e.g. 100 μM with 1% DMSO), followed by serial dilution from the high concentration in aqueous solution to the various lower concentrations. In this protocol, the first dilution, at high concentration and low DMSO percentage, can form a precipitate if the compound has low solubility. Depending on whether the solid precipitates were transferred during the subsequent dilution, the concentrations of the serial dilution can be quite variable. Frequently, biologists find that the assay results are more reproducible where they use rapid up and down mixing with the pipet before transferring an aliquot of solution for dilution. This is because mixing can break the large precipitated particles to finer particles and re-suspend them into the solution to form a more homogeneous suspension, so that the subsequent transfer is more consistent. While this is better than a static approach, it is still inherently more variable than if the compound is completely dissolved.

In the second dilution protocol, the serial dilution is performed in DMSO from the high concentration to lower concentrations. The diluted DMSO stock solutions are then added to the aqueous buffer to give the final target concentration. In this protocol, if the resulting high aqueous buffer concentrations precipitate, they will not affect the lower concentrations. We recommend using the second protocol for serial dilution to minimize the effect of precipitation on IC_{50} determination.

Determination of Solubility in Bioassay Buffers

Solubility screening data is very useful to provide an early alert to potential issues related to biological assays. However, solubility is very sensitive to its environment. It changes with buffer composition, percent DMSO, dilution procedure, incubation time and temperature. Therefore, determination of compound concentrations under the conditions of the biological assay protocol can help diagnose potential solubility issues in screening and help optimize the assay protocol. Similar to "formulation development", which is used to maximize solubility and dissolution rate of insoluble compounds for *in vivo* dosing, "screening solubility development" for biological assays can be used to optimize solvent systems and dilution protocols to maximize solubility and dissolution rate in bioassays in order to generate high quality screening data. Water miscible cosolvents, such as DMSO, methanol, ethanol, acetonitrile, DMF, dioxane, and excipients (e.g. cyclodextrin) can be added to improve solubility in screening assays (Dean et al., 2003; Hoever and Zbinden, 2004; Schmidt and Bornscheuer, 2005).

Determination of solubility under assay conditions can help optimize assay protocols for biological screening and HTS. The samples for solubility determination in bioassay buffers are typically prepared step-by-step according to bioassay protocols. The solutions are then centrifuged and the supernatants are assayed using LC-MS against a fully soluble standard in solvent. The data can be used iteratively on a set of structurally diverse compounds to guide assay development and maximize solubility.

Correction of Biological Activity Using Solubility Data

Efforts have been made to quantitate the actual concentration of the compound in buffer solution and correct the dose-response curve for the biological activity (Popa-Burke et al., 2004). A generic protocol is often used to determine solubility in aqueous buffer. Compounds are added to a generic phosphate buffer solution from a 3-mM DMSO stock to give a final concentration of 150 μM with 5% DMSO. The actual concentrations of the test compounds in the buffer are determined using CLND, which requires no standards for calibration, as long as the compounds contain nitrogen. This greatly increases throughput and efficiency. The linearity of CLND over a wide range of concentrations is particularly useful for concentrated samples. The limitation of current CLND technology is low sensitivity, with a detection limit of ~5 μM. This is too high to detect compounds in the actual screening setting with low μM concentration. Because the detection is based on nitrogen, the actual assay buffer may not be used if it contains components with nitrogen, such as HEPES. As discussed earlier, solubility is very sensitive to its environment, especially buffer composition, % DMSO, dilution protocol, incubation time and temperature and crystal forms. Screening using a generic protocol will give a general idea about aqueous solubility of the compounds. However, this data might not be relevant to the specific assay conditions used for the screening assay, which can vary greatly from the phosphate buffer. While correction of biological activity based on generic solubility is better than assuming that all compounds are fully soluble, it can over- or under- correct activity and introduce additional errors compared to solubility determination in the actual bioassay buffer.

Table 6 shows examples of assay buffers used for biological screening of ten different discovery programs. The buffer components and percent organic solvents are very different for each project. A generic buffer system would not be able to closely mimic the different conditions used in the assays. The solubility of the compounds will vary in the different buffers containing drastically different amounts of organic solvents, serum proteins, pHs, chelating agents, surfactants and salts. Correction of activity using solubility data measured in generic buffer can introduce additional uncertainty into the activity data.

Even if the solubility is measured using a specific bioassay protocol, we do not recommend correction of bioactivity using solubility data. Undoubtedly, correction of biological activity using solubility data can lead to identification of new chemical series. However, this approach favors insoluble compounds. Assuming everything being equal, the least soluble compound may become the most apparently potent compound after solubility correction. This apparent "activity" might never be achievable due to the limited solubility of the compound. This approach can lead to pursuing very insoluble non-drug-like compounds, which are difficult and costly to develop. The recommended approach is to optimize the assay protocol so that it can dissolve most of the compounds. In this way, activities can be measured accurately from the beginning. This also greatly reduces the number of analyses that would otherwise be performed for every sample that is screened.

Projects	% Organic	Buffer components
Project 1	10% DMSO	50 mM HEPES, pH 7.5 100 mM, NaCl, 5 mM $CaCl_2$ 0.1% CHAPS, 5% glycerol
Project 2	0.5% DMSO	50 mM Tris, pH 7.5 150 mM NaCl 10 μg/mL BSA 0.01% Tween 80
Project 3	0.5% DMSO	50 mM Na_2HPO_4,pH 6.6 200 mM NaCl 1 mM EDTA
Project 4	1% DMSO 10% Ethanol	50 mM Tris, pH 7.5 10 mM $MgCl_2$ 0.5 mM EDTA
Project 5	0.1% DMSO	RPMI
Project 6	1% DMSO	RPMI 10% FSB
Project 7	1.4% DMSO	F12 HAM Media
Project 8	5% DMSO	DMEM
Project 9	1% DMSO	20 mM PIPES pH 7.2 100 mM NaCl, 1 mM EDTA 0.1% CHAPS, 10% Sucrose
Project 10	2% DMSO	50 mM HEPES, pH 7.4 100 mM NaCl, 5 mM $CaCl_2$ 0.005% Triton X-100

Table 6. Examples of buffer components used by different discovery teams.

Eliminate the Dilution Step into Aqueous Media

A dilution step before adding compound solutions to the assay mix is sometimes essential in order to achieve the target compound concentration, amount of DMSO and pipetting accuracy. Insoluble compounds tend to precipitate in the dilution step and cause a right shift of the IC_{50}, resulting in lower potency. Final assay media typically contain various protein components, cell membrane, or lipids in microsomes, which can help dissolve insoluble compounds. Furthermore, adding concentrated DMSO stock solution directly into the assay can cause super-saturation and slow precipitation. This will provide a time window for compounds to interact with biological targets in solution. When dilution steps are added, however, compounds need to be added to the dilution buffers, mixed and added to the final assay wells, which takes a longer time before the compounds are in contact with the target proteins. Under these conditions, they have a higher probability of precipitating from supersaturated solution and missing the window for activity. Elimination of the aqueous dilution step, by adding sample stock solution from DMSO or organic solvents directly into the assay mix, can minimize precipitation and maximize the interaction with biological targets to demonstrate activity.

In order to achieve this, the robotic pipetting system must be able to accurately and precisely deliver very low volumes of DMSO. This is necessary to keep the DMSO concentration in the assay as low as possible. DMSO concentration can significantly affect assay results. Current state-of- the- art robotic systems can deliver small volumes (0.5–1.0 μL) with high accuracy. This allows addition of concentrated DMSO stock solution directly into final assay mix and minimize precipitation. One microliter of DMSO added to a 99 μL assay results in a 1% DMSO concentration. The precision and accuracy of delivery of such low volumes should be studied during the screening solubility development stage.

Other Approaches to Increase Aqueous Solubility

Strategies used to improve DMSO solubility can also be applied to increase aqueous solubility. In-well sonication and screening at a lower concentration (e.g., 3 μM vs. 10 μM), help to minimize precipitation.

If a dilution step is required, dilution into a pH 7.4 buffer is better than directly into water, because buffer can help ionize basic or acidic compounds and improve solubility. Water has no buffer capacity, and a compound added from DMSO will be in its neutral state and have only limited intrinsic solubility. Salt forms tend to be soluble in water. Dissolving in buffer will ensure no conversion back to the free acid or base and the solution will have a consistent pH.

Conclusions

Solubility significantly impacts HTS, bioassays and ADME/TOX screens in early drug discovery. Strategies and new technologies are available to overcome solubility issues in both DMSO and water. Careful optimization of the assay protocol to maximize solubility, during a screening solubility development stage, is the key to achieving high quality screening data.

Acknowledgments. The authors would like to thank Guy Carter and Magid Abou-Gharbia for their support, encouragement, and leadership, and the Pharmaceutical Profiling team for their contributions, especially Susan Petusky, Susan Li, Zhen Lin, Ian Bezar and Hong Jin. We would also like to thank Janet Weiss and Phyllis Zeban in the library for providing useful references.

References

Andreani A, Cavalli A, Granaiola M, Guardigli M, Leoni A, Locatelli A, Morigi R, Rambaldi M, Recanatini M, and Roda A. Synthesis and Screening for Antiacetylcholinesterase Activity of (1-Benzyl-4-oxopiperidin-3-ylidene)methylindoles and -pyrroles Related to Donepezil. *J Med Chem.* 2001; 44: 4011–4014.

Avdeef A. Physicochemical Profiling (Solubility, Permeability and Charge State). *Curr Top Med Chem* 2001; 1: 277–351.

Balakin KV. DMSO Solubility and Bioscreening. *Curr Drug Discov* 2003; 27–30.

Balakin KV, Ivanenkov YA, Skorenko AV, Nikolsky YV, Savchuk NP, and Ivashchenko AA. In Silico Estimation of DMSO Solubility of Organic Compounds for Bioscreening. *J Biomol Screen* 2004; 9: 22–31.

Benson N, Boyd HF, Everett JR, Fries J, Gribbon P, Haque N, Henco K, Jessen T, Martin WH, Mathewson TJ, Sharp RE, Spencer RW, Stuhmeier F, Wallace MS, and Winkler D. NanoStore: A Concept for Logistical Improvements of Compound Handling in High-Throughput Screening. *J Biomol Screen* 2005; 10: 573–580.

Bevan CD and Lloyd RS. A High-Throughput Screening Method for the Determination of Aqueous Drug Solubility Using Laser Nephelometry in Microtiter Plates. *Anal Chem* 2000; 72: 1781–1787.

Blackman MJ, Corrie JET, Croney JC, Kelly G, Eccleston JF, and Jameson DM. Structural and Biochemical Characterization of a Fluorogenic Rhodamine-Labeled Malarial Protease Substrate. *Biochemistry* 2002; 41: 12244–12252.

Bleicher KH, Bohm H-J, Muller K, and Alanine AI. Hit and Lead Generation: Beyond High-Throughput Screening. *Nat Rev Drug Discov* 2003; 2: 369–378.

Buchli R, VanGundy RS, Hickman-Miller HD, Giberson CF, Bardet W, and Hildebrand WH, Development and Validation of a Fluorescence Polarization-Based Competitive Peptide-Binding Assay for HLA-A*0201-A New Tool for Epitope Discovery. *Biochemistry* 2005; 44: 12491–12507.

Cheng X, Hochlowski J, Tang H, Hepp D, Beckner C, Kantor S, and Schmitt R. Studies on Repository Compound Stability in DMSO under Various Conditions. *J Biomol Screen* 2003; 8: 292–304.

Dalvit C, Flocco M, Knapp S, Mostardini M, Perego R, Stockman BJ, Veronesi M, and Varasi M. High-Throughput NMR-Based Screening with Competition Binding Experiments. *J Am Chem Soc* 2002; 124: 7702–7709.

Davidson W, Frego L, Peet GW, Kroe RR, Labadia ME, Lukas SM, Snow RJ, Jakes S, Grygon CA, Pargellis C, and Werneburg BG. Discovery and Characterization of a Substrate Selective p38 Inhibitor. *Biochemistry* 2004; 43: 11658–11671.

Dean KES, Klein G, Renaudet O, and Reymond J-L. A Green Fluorescent Chemosensor for Amino Acids Provides a Versatile High-throughput Screening (HTS) Assay for Proteases. *Bioorg Med Chem Lett* 2003; 13: 1653–1656.

Delaney JS. Predicting Aqueous Solubility from Structure. *Drug Discov Today* 2005; 10: 289–295.

Di L and Kerns EH. Application of Pharmaceutical Profiling Assays for Optimization of Drug-like Properties. *Curr Opin Drug Discov & Devel* 2005; 8: 495–504.

Dickson M and Gagnon JP. Key Factors in the Rising Cost of New Drug Discovery and Development. *Nat Rev Drug Discov* 2004; 3: 417–429.

Drews J. Drug Discovery: A Historical Perspective. *Science* 2000; 287: 1960–1964.

Faria TN, Timoszyk JK, Stouch TR, Vig BS, Landowski CP, Amidon GL, Weaver CD, Wall DA, and Smith RL. A Novel High-Throughput PepT1 Transporter Assay Differentiates between Substrates and Antagonists. *Mol Pharm* 2004; 1: 67–76.

Feng BY, Shelat A, Doman TN, Guy RK, and Shoichet BK. High-throughput assays for promiscuous inhibitors. *Nat Chem Biol* 2005; 1: 146–148.

Fillers WS. Compound Libraries—Cost of Ownership. *Drug Discov World* 2004; 5: 86–90.

Gezginci MH, Martin AR, and Franzblau SG. Antimycobacterial Activity of Substituted Isosteres of Pyridine- and Pyrazinecarboxylic Acids. *J Med Chem* 2001; 44: 1560–1563.

Glomme A, März J, and Dressman JB. Comparison of a Miniaturized Shake-Flask Solubility Method with Automated Potentiometric Acid/Base Titrations and Calculated Solubilities. *J Pharm Sci* 2005; 94: 1–16.

Goodwin JJ. Flow Cell System for Solubility Testing, in p 12 pp, (Becton, Dickinson and Company, USA). Application: EP, 2003.

Gribbon P and Andreas S. High-throughput Drug Discovery: What Can We Expect from HTS? *Drug Discov Today* 2005; 10: 17–22.

Hertzberg RP and Pope AJ. High-throughput Screening: New Technology for the 21st Century. *Curr Opin Chem Biol* 2000; 4: 445–451.

Hoever M and Zbinden P. The Evolution of Microarrayed Compound Screening. *Drug Discov Today* 2004; 9: 358–365.

Japertas P, Verheij H, and Petrauskas A. DMSO Solubility Prediction. LogP 2004, Zurich, Switzerland, 2004.

Johnston PA and Johnston PA. Cellular Platforms for HTS: Three Case Studies. *Drug Discov Today* 2002; 7: 353–363.

Jorgensen WL and Duffy EM. Prediction of Drug Solubility from Structure. *Adv Drug Deliv Rev* 2002; 54: 355–366.

Kerns EH. High Throughput Physicochemical Profiling for Drug Discovery. *J Pharm Sci* 2001; 90: 1838–1858.

Kerns EH and Di L. Physicochemical Profiling: Overview of the Screens. *Drug Discovery Today: Technologies* 2004; 1: 343–348.

Kerns EH and Di L. Automation in Pharmaceutical Profiling. *J Association for Laboratory Automation* 2005; 10: 114–123.

Kerns EH, Di L, Bourassa J, Gross J, Huang N, Liu H, Kleintop T, Nogle L, Mallis L, Petucci C, Petusky S, Tischler M, Sabus C, Sarkahian A, Young M, Zhang M-y, Huryn D, McConnell O, and Carter G. Integrity Profiling of High Throughput Screening Hits Using LC-MS and Related Techniques. *Comb Chem High Throughput Screen* 2005; 8: 459–466.

Kozikowski BA, Burt TM, Tirey DA, Williams LE, Kuzmak BR, Stanton DT, Morand KL, and Nelson SL. The Effect of Freeze/Thaw Cycles on the Stability of Compounds in DMSO. *J Biomol Screen* 2003a; 8: 210–215.

Kozikowski BA, Burt TM, Tirey DA, Williams LE, Kuzmak BR, Stanton DT, Morand KL, and Nelson SL. The Effect of Room-Temperature Storage on the Stability of Compounds in DMSO. *J Biomol Screen* 2003b; 8: 205–209.

Lipinski C. Solubility in the Design of Combinatorial Libraries. *Chem Anal, New York* 2004a; 163: 407–434.

Lipinski CA. Avoiding Investment in Doomed Drugs. *Curr Drug Discov* 2001; 17–19.

Lipinski CA. Solubility in Water and DMSO: Issues and Potential Solutions. *Biotechnology: Pharmaceutical Aspects* 2004b; 1: 93–125.

Lipinski CA and Hoffer E. Compound Properties and Drug Quality. *Practice of Medicinal Chemistry (2nd Edition)* 2003, pp. 341–349.

Lipinski CA, Lombardo F, Dominy BW and Feeney PJ. Experimental and Computational Approaches to Estimate Solubility and Permeability in Drug Discovery and Development Settings. *Advanced Drug Delivery Reviews* 1997; **23**:3–25.

Lombardino JG and Lowe JA. The Role of the Medicinal Chemist in Drug Discovery—Then and Now. *Nat Rev Drug Discov* 2004; 3: 853–862.

Lu J and Bakken GA. Building Classification Models for DMSO Solubility: Comparison of Five Methods. *Abstracts of Papers, 228th ACS National Meeting*, Philadelphia, PA, United States, August 22–26, 2004, CINF-045.

McGovern SL, Caselli E, Grigorieff N, and Shoichet BK. A Common Mechanism Underlying Promiscuous Inhibitors from Virtual and High-Throughput Screening. *J Med Chem* 2002; 45: 1712–1722.

McGovern SL, Helfand BT, Feng B, and Shoichet BK. A Specific Mechanism of Nonspecific Inhibition. *J Med Chem* 2003; 46: 4265–4272.

McGovern SL and Shoichet BK. Kinase Inhibitors: Not Just for Kinases Anymore. *J Med Chem* 2003; 46: 1478–1483.

Meisner N-C, Hintersteiner M, Uhl V, Weidemann T, Schmied M, Gstach H, and Auer M. The Chemical Hunt for the Identification of Drugable Targets. *Curr Opin Chem Biol* 2004; 8: 424–431.

Miller KA, SureshKumar EVK, Wood SJ, Cromer JR, Datta A, and David SA. Lipopolysaccharide Sequestrants: Structural Correlates of Activity and Toxicity in Novel Acylhomospermines. *J Med Chem* 2005; 48: 2589–2599.

Moore K and Rees S. Cell-Based Versus Isolated Target Screening: How Lucky Do You Feel? *J Biomol Screen* 2001; 6: 69–74.

Oldenburg K, Pooler D, Scudder K, Lipinski C, and Kelly M. High Throughput Sonication: Evaluation for Compound Solubilization. *Comb Chem High Throughput Screen* 2005; 8: 499–512.

Oprea TI, Bologa CG, Edwards BS, Prossnitz ER, and Sklar LA. Post-High-Throughput Screening Analysis: An Empirical Compound Prioritization Scheme. *J Biomol Screen* 2005; 10: 419–426.

Popa-Burke IG, Issakova O, Arroway JD, Bernasconi P, Chen M, Coudurier L, Galasinski S, Jadhav AP, Janzen WP, Lagasca D, Liu D, Lewis RS, Mohney RP, Sepetov N, Sparkman DA, and Hodge CN. Streamlined System for Purifying and Quantifying a Diverse Library of Compounds and the Effect of Compound Concentration Measurements on the Accurate Interpretation of Biological Assay Results. *Anal Chem* 2004; 76: 7278–7287.

Roche O, Schneider P, Zuegge J, Guba W, Kansy M, Alanine A, Bleicher K, Danel F, Gutknecht E-M, Rogers-Evans M, Neidhart W, Stalder H, Dillon M, Sjogren E, Fotouhi N, Gillespie P, Goodnow R, Harris W, Jones P, Taniguchi M, Tsujii S, von der Saal W, Zimmermann G, and Schneider G. Development of a Virtual Screening Method for Identification of "Frequent Hitters" in Compound Libraries. *J Med Chem* 2002; 45: 137–142.

Roehrl MHA, Wang JY, and Wagner G. Discovery of Small-Molecule Inhibitors of the NFAT-Calcineurin Interaction by Competitive High-Throughput Fluorescence Polarization Screening. *Biochemistry* 2004; 43: 16067–16075.

Ryan AJ, Gray NM, Lowe PN, and Chung C-w. Effect of Detergent on "Promiscuous" Inhibitors. *J Med Chem* 2003; 46: 3448–3451.

Schmidt M and Bornscheuer UT. High-throughput Assays for Lipases and Esterases. *Biomol Eng* 2005; 22: 51–56.

Schopfer U, Engeloch C, Stanek J, Girod M, Schuffenhauer A, Jacoby E, and Acklin P. The Novartis Compound Archive—From Concept to Reality. *Comb Chem High Throughput Screen* 2005; 8: 513–519.

Seidler J, McGovern SL, Doman TN, and Shoichet BK. Identification and Prediction of Promiscuous Aggregating Inhibitors among Known Drugs. *J Med Chem* 2003; 46: 4477–4486.

Topp A, Zbinden P, Wehner HU, and Regenass U. A Novel Storage and Retrieval Concept for Compound Collections on Dry Film. *J Association for Laboratory Automation* 2005; 10: 88–97.

Walters WP and Namchuk M. Designing Screens: How To Make Your Hits a Hit. *Nat Rev Drug Discov* 2003; 2: 259–266.

Yan B, Collins N, Wheatley J, Irving M, Leopold K, Chan C, Shornikov A, Fang L, Lee A, Stock M, and Zhao J. High-Throughput Purification of Combinatorial Libraries I: A High-Throughput Purification System Using an Accelerated Retention Window Approach. *J Comb Chem* 2004; 6: 255–261.

Yan B, Fang L, Irving M, Zhang S, Boldi AM, Woolard F, Johnson CR, Kshirsagar T, Figliozzi GM, Krueger CA, and Collins N. Quality Control in Combinatorial Chemistry: Determination of the Quantity, Purity, and Quantitative Purity of Compounds in Combinatorial Libraries. *J Comb Chem* 2003; 5: 547–559.

Yurek DA, Branch DL, and Kuo M-S. Development of a System To Evaluate Compound Identity, Purity, and Concentration in a Single Experiment and Its Application in Quality Assessment of Combinatorial Libraries and Screening Hits. *J Comb Chem* 2002; 4: 138–148.

5

Practical Aspects of Solubility Determination in Pharmaceutical Preformulation

WEI-QIN (TONY) TONG

Novartis Pharmaceuticals Corporation, Pharmaceutical and Analytical Development, One Health Plaza, East Hanover, NJ

Introduction

Solubility is one of the most important physicochemical properties studied during pharmaceutical preformulation. For liquid dosage form development, accurate solubility data are essential to ensure the robustness of the finished product. For solid dosage forms, solubility data are important in determining if an adequate amount of drug is available for absorption *in vivo.* If a compound has a low aqueous solubility, it may be subject to dissolution rate-limited or solubility-limited absorption within the gastrointestinal (GI) residence time (Lobenberg et al., 2000). The importance of solubility, in biopharmaceutical terms, is highlighted by its use in the biopharmaceutics classification system (BCS) described by Amidon et al. (1995). This system defines low solubility compounds as those whose aqueous solubility in 250 mL of pH 1-7.5 aqueous solution is less than the total dose. Solubility data are also used to estimate the maximum absorbable dose (MAD) (Johnson and Swindell, 1996). MAD is a conceptual tool that relates the solubility requirement for oral absorption to the dose, permeability and GI volume and transit time. It is defined as:

$$\text{MAD (mg)} = \text{S (mg/mL)} \times K_a \text{ (1/min)} \times \text{SIWV (mL)} \times \text{SITT (min)}$$

where S is solubility at pH 6.5 reflecting typical small intestine condition; K_a is the trans-intestinal absorption rate constant determined by a rat intestinal perfusion experiment; SIWV is the small intestine water volume, generally considered to be 250 mL; and SITT is the small intestine transit time, typically about 270 min.

Solubility is influenced by many variables including temperature, pH (for ionizable compounds), solvents used for the solubility determination, state of the solid, common ions in the medium, and so on. For poorly soluble compounds, determining solubility in the presence of various solubilizing agents presents a special set of challenges.

The aims of this chapter are to summarize solubility determination methods commonly used in pharmaceutical preformulation and to discuss various factors to be considered in designing and carrying out these solubility studies.

Experimental Methods

Saturation Shake-Flask Method

The shake-flask method is based on the phase solubility technique that was developed 40 years ago and is still the most reliable and widely used method for solubility measurement today (Higuchi and Connors, 1965). The method can be divided into five steps: sample preparation, equilibration, separation of phases, analysis of the saturated solution and residual solid, and data analysis and interpretation (Yalkowsky and Banerjee, 1992, Winnike, 2005).

Sample Preparation

A solubility sample is typically prepared by adding an excess amount of solid to the solubility medium in a stoppered flask or vial. The amount added does not need to be accurately measured. While it is important to ensure that enough material is added so the sample is a suspension, it is also important not to add too much material to significantly alter the properties of the solubility medium including its pH.

Equilibration

Depending on the type of agitation used, the drug substance properties, amount of material used, and the equilibration method used, the time to reach equilibrium varies. With good agitation, samples generally reach equilibrium reasonably quickly, often within 24 hours. However, for poorly soluble compounds, the equilibration time may be unrealistically long due to the poor dissolution rate that is further depressed as the equilibrium process advances and the concentration in solution gets closer to the solubility. One way to speed up the process is to increase the effective surface area for dissolution. This can be achieved by either vortexing or sonicating samples prior to equilibration. Creating a supersaturated solution may also be helpful in overcoming the problem of a slow dissolution rate. This can be achieved by adding a certain amount of amorphous material to the samples, or by cycling the sample temperature to higher and lower temperatures during the equilibration process.

Another challenge for determining solubility of poorly soluble compounds is their poor wettability and their tendency to float. Ways to get around this problem include using small glass microspheres (Glassperlen) to de-aggregate the particles with agitation or sonication, and adding an amount of sodium dodecyl sulfate below the critical micelle concentration to serve as a wetting agent (Lötter et al., 1983).

There is no better way to accurately determine the end point for equilibration than by performing an actual analysis. Saturation or equilibrium is considered to be achieved when multiple samples assayed after different equilibration time periods give the same apparent solubility. If solid-state form transitions occur during equilibration, the equilibration time may be longer, especially if the solubility differences between various forms are small. To ensure that equilibrium is indeed reached, it is a good idea to demonstrate that the same equilibrium state (solubility) can be reached from different directions; for example, from undersaturation and supersaturation as well as from constant temperature or from temperature variation by means of temperature cycling.

Separation of Phases

Filtration and centrifugation both have been commonly employed to separate the saturated solution from the solute phase. Filtration is easily accomplished, but filter sorption can be a significant source of error. Generally, filter sorption is more significant for hydrophobic and poorly soluble compounds, and obviously it is directly proportional to the filter surface area. Typically, pre-rinsing the filter with a few milliliters of the saturated solution can remedy the problem. However, in some extreme cases where the solubility of the compound is very low, a much larger volume may be needed to saturate the filter adsorption sites.

Centrifugation or ultracentrifugation may be preferable for certain samples that are difficult to filter. Solubility samples in co-solvent systems with high viscosity are such examples. If the solute is less dense than the solubility medium, it will float on the surface, making it difficult to sample the solution. This may be particularly problematic for compounds with low solubility where a single particle carried over to the solution may cause significant overestimation of the true solubility.

Theoretically, the solid should always be separated from the saturated solution at the equilibrium temperature. Obviously this is more important when equilibrium is reached quickly. For poorly soluble compounds, equilibrium is typically reached slowly, thus filtration at ambient temperature may not introduce a significant error.

Centrifugal filter devices such as the UniPrep® filter have become commercially available in recent years, making it possible to combine both approaches (Glomme et al., 2004; Winnike, 2005).

Analysis of the Saturated Solution and Residual Solid

High performance liquid chromatography (HPLC) is the most commonly used analytical tool for the analysis of saturated solutions. Its advantage over the ultraviolet method is that it can detect impurities and any instability. A generic gradient method can be made readily available that is stability-indicating enough for multiple compound analyses without the need to make major adjustments in the column or mobile phase.

When determining the solubility of meta-stable forms, the application of a fiber optic probe, which permits the detection of the drug concentration every few seconds, may prove to be very useful.

Examination of the residual solid from solubility samples is one of the most important but often overlooked steps in solubility determinations. Powder X-ray diffraction (PXRD) is the most reliable method to determine whether any solid state form change has occurred during equilibration. The sample should be studied both wet and dry to determine if any hydrate or solvate exists. Thermal analysis techniques such as differential scanning calorimetry (DSC) can also be used to identify any solid-state transformations, although they may not provide as definitive an answer as the PXRD method. Other methods useful in identifying any solid-state changes include microscopy, Raman and infrared spectroscopy, and solid-state NMR (Brittain, 1999). When changes in solid-state properties are identified in solubility studies, it is important to link the new properties to the properties of known crystal forms so the solubility result can be associated with the appropriate crystal form.

Data Analysis and Interpretation

Solubility theory based on pH-solubility profiles for weak acids and bases is well established (Grant and Higuchi, 1990, Butler, 1998). From a knowledge of the intrinsic solubility of the unionized form, the dissociation constant (pK_a') and the solubility of a salt, one should be able to construct the pH-solubility profile. If multiple solubility data are available, data can be analyzed through the use of a non-linear regression model to calculate pK_a'. If the solubilities of the various salts are also determined, the complete pH-solubility profile can be constructed.

Deviations from the theoretical pH-solubility profiles may be an indication of experimental error. They may also suggest other interactions not predicted by the solubility theory. Examples for the causes of such deviations include changes in solid-state properties, self-association and micelle formation of the drug in solution. Figure 1 shows an example of a compound that forms micelles at a pH above 9 (Winnike, 2005). Further addition of sodium hydroxide does not increase the pH; rather it enhances solubility through micelle formation. In any of these cases, it is important to identify the causes of the deviation so that appropriate formulation decisions can be made based on the solubility data.

Non-Equilibrium Methods

Any methods that do not contain steps to ensure the establishment of equilibrium can be considered non-equilibrium methods. In the last few years, several methods commonly used for solubility measurements in the early discovery setting have been reported (Lipinski et al., 1997; Pan et al., 2001). These methods typically begin with dimethylsulfoxide (DMSO) solutions or with amorphous material. Turbidity and ultraviolet detection are commonly used because they

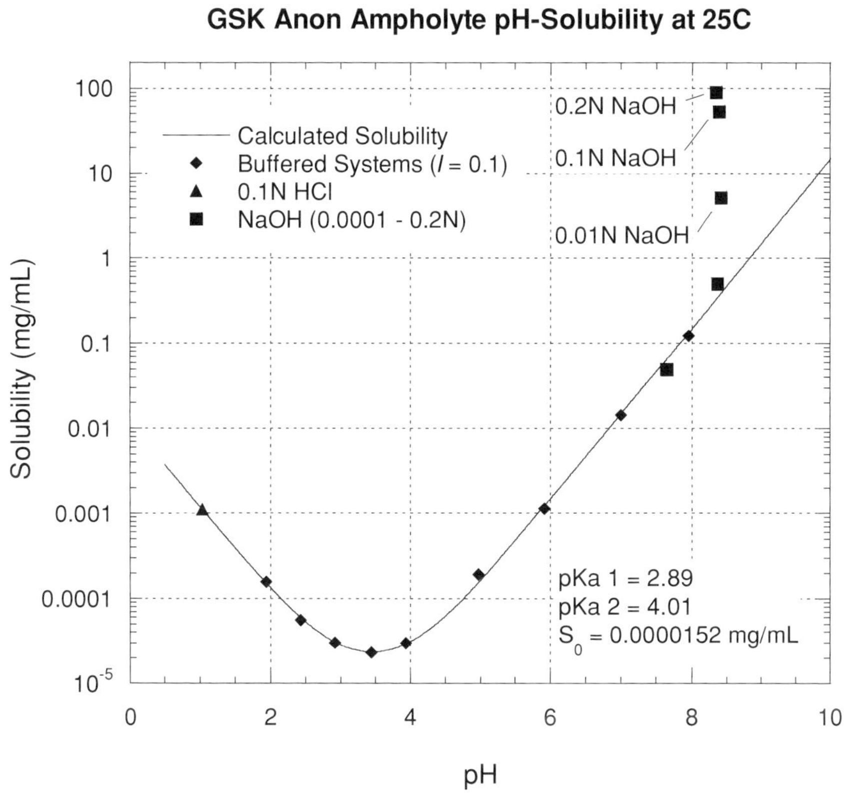

Figure 1. pH-solubility profile of a compound that forms micelles at high pH values.

easily can be designed into high-throughput instrumentation. A potentiometric method has also been reported (Avdeef, 2003).

The usefulness of the solubility data from these non-equilibrium methods often is questionable. Some pharmaceutical companies use these data as a first criterion to eliminate poorly soluble compounds. However, because the contribution of crystallinity to solubility is not controlled in non-equilibrium methods, the reliability of the data cannot be guaranteed. If experimental error is minimized, it is generally safe to assume that solubility can only be less when solid crystalline material is later used to determine equilibrium solubility. Therefore, the use of these solubility data as a gatekeeper seems to be justified. However, it is questionable whether data generated by these methods are any better for this purpose than those generated by computational methods. In addition, since for highly potent drug candidates the solubility requirement is dose-dependent, compounds, whose solubility is in the microgram range, may still be developable. Therefore, setting the right criteria to eliminate poorly soluble compounds may be challenging. It is the author's opinion that in order to make informed decisions, one must understand why these data are needed, and how they will be used, prior to initiating solubility studies.

Attention Points in Solubility Determination

pH-Solubility Profile

For drugs with ionizable functional groups, determining solubility as a function of pH is an important preformulation task. pH-solubility profiles define the range of opportunities for liquid formulation development, and they provide baseline guidance to solubilization strategies for poorly soluble compounds.

Typically, there are two ways to control pH. One approach is to use buffers. Since multiple buffer systems are needed to control the entire pH range, the solubility results may be complicated by salt formation with the buffer species (Tong and Whitesell, 1998). This can be detected by examining the residual solid from solubility determinations.

Another way to control pH is through the use of a pH-stat, where pH is controlled by titrating with acidic and/or basic solutions (Todd and Winnike, 1994). Ionic equilibrium can be monitored continuously by measuring the solution pH. Equilibrium can be considered to have been reached when the pH no longer changes over a period of time.

For poorly soluble compounds, depending on what material is used at the starting point for a solubility determination, e.g. the unionized form vs. its salt, a different pH-solubility profile may be observed. This may be due to the slower dissolution rate of the unionized form, which can cause a delay in reaching true equilibrium. Or it may be due to supersaturation of the salt solution because of a delay in nucleation of the ionized form. The methods previously described for increasing the powder dissolution rate should permit the system to reach equilibrium more quickly, thus reducing or eliminating the difference.

In certain instances, if the ionic form of a drug candidate is used as the starting material, the apparent solubility can be different when varying amounts of salt are added to the solution. For example, when the excess amount of the di-hydrochloride (2·HCl) salt of an experimental compound, E2050, is used to determine the pH-solubility profile, the solubility in the pH region where the mono-hydrochloride salt controls the solubility is suppressed by the excess chloride ion resulting from the conversion of the di-HCl salt to the mono-HCl salt (Wang et al., 2002). Figure 2 shows the three pH solubility curves determined by using different amounts of di-HCl salt. The difference in salt solubility also causes the pH_{max} to vary.

Solubility of Salts

Solubility determination for pharmaceutical salts using the equilibrium method may be challenging for certain compounds such as those with poor intrinsic solubility. Theoretically, after an excess amount of solid salt is equilibrated in water, the solution concentration at equilibrium should represent the solubility of the salt. However, this is only true if the pH of the saturated solution is below pH_{max}. For compounds with low intrinsic solubilities and weak basicity or acidity, their salts may convert to the unionized form in the solubility medium. In such

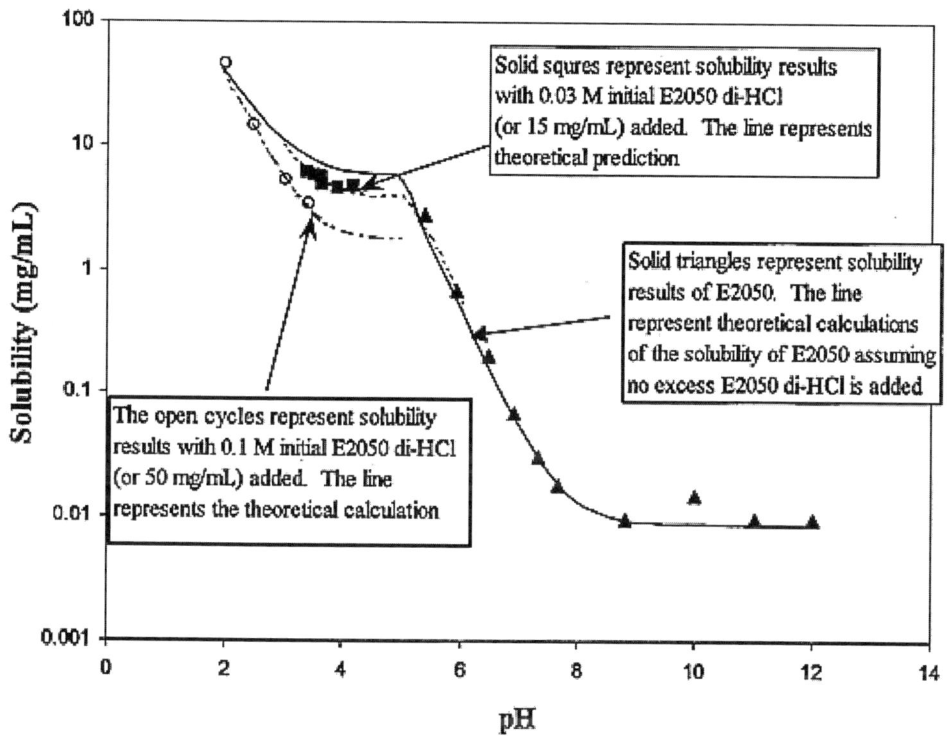

Figure 2. pH-solubility profiles of a new chemical entity E2050 constructed with different amounts of starting material as the Di-HCl salt.

cases the measured solubility is only the solubility of the unionized form at those particular pH values. For example, the solubility of the phosphate salt of the developmental candidate, GW1818X, was found to be 6.8 mg/mL when the solution pH was 5.0 (Tong and Whitesell, 1998). The pH_{max} is approximately 4 in this case. Analysis of the residual solid showed that the solution was in equilibrium with the free base, indicating that the solubility determined did not adequately represent the solubility of the salt. An additional complication is that the pH of the solubility sample may vary depending on the lots of drug substance used. This is because different lots of material may contain different amounts of residual acid, base, or solvent.

There are several ways to overcome this type of problem. One approach is to determine the solubility in a diluted acidic solution using the same acid that formed the salt with the base. The concentration of the acid solution needs to be such that the solution pH is lower than the pH_{max}. The solubility can then be estimated by correcting for the common ion effect from the acid used in the solubility study. A second approach to ensure a lower solution pH than pH_{max} is to use a high ratio of drug to solvent (Pudipeddi, 2002). However, this may not be possible for every compound.

When determining the solubility of salts in simulated gastric fluid, or in pH 1 or pH 2 hydrochloric acid solutions, the salt may convert to the hydrochloride salt depending on the relative solubility of the salts. If the simulated gastric fluid contains sodium chloride, the common ion effect of the chloride ion may significantly depress the solubility of the hydrochloride salt. Therefore, examination of the residual solid from these experiments is even more important. A different PXRD pattern may be indicative of a different crystal form of the same salt or the hydrochloride salt.

In-Situ Salt Screening

For ionizable compounds, pH adjustment is often one of the most important ways to improve solubility. Sometimes, solubility data for salts may be needed. For example, when developing a solution formulation, the buffer selected should not form a less soluble salt with the drug substance. However, the actual salt may not be readily available. In these cases, the in-situ salt screening method may be useful in estimating the solubility of various salts (Tong and Whitesell, 1998).

In this method, an accurately known amount of free base is added to a known concentration of acid. The acid concentration is chosen so that there is an excess amount of acid in the solution to ensure that the pH of the suspension is lower than the pH_{max}. The solubility of the compound is measured by the equilibrium method described previously. After correcting for the common ion effect, the K_{sp} and the solubility of the salt formed in-situ can then be calculated.

Examination of the residual solid is critically important in this case. Sometimes, the residual solid may not be a perfectly crystalline solid salt. If this is the case, it is obvious that the solubility determined only represents the solubility of the particular form that is in equilibrium with the saturated solution.

Solubility Determination in Non-Aqueous Solutions

Special precautions are required when determining solubility in non-aqueous solvents. Since many non-aqueous systems are viscous, it may be more practical to use weight (W/W) instead of volume (W/V) to represent solubility. Since not all filters are compatible with non-aqueous solvents, it is essential to choose the correct type of filter. Upon dilution of the saturated solution for analysis, it is important to ensure that the compound does not precipitate. Precipitation may occur in many co-solvent systems because the solubility changes that accompany dilution are log-linear.

Solubility dependence on temperature may be different for different solvent compositions. Therefore, it is important to use a statistical factorial design to study the effect of composition and temperature simultaneously.

Solubility Determination for Polymorphs and Solvates

The solubility of a solid substance, by definition, is the concentration at which the solution phase is in equilibrium with a given solid phase at a stated temperature and pressure (Butler, 1998). When a substance exists in more than one crystal form, only the least soluble form at a given temperature is considered to be the most physically stable form, all others are considered to be metastable.

The thermodynamic activity of each crystalline form, represented by its solubility, may change quite differently as a function of temperature. Monotropic systems are defined as systems where a single form is always more stable regardless of the temperature. Enantiotropic systems are defined as systems where the relative stability of the two forms inverts at some transition temperature (Byrn et al., 1999).

Determining the transition temperature of polymorphs is a necessary step in understanding the relationship of various forms. The transition temperature can be determined by plotting log solubility as a function of inverse temperature (Brittain and Grant, 1999; Byrn et al., 1999).

To determine the solubility of each form, one needs to monitor the solution concentration as a function of time more frequently. Enough data points need to be collected so the equilibrium concentration of each form can be assessed. Theoretically, a single experiment starting with the least stable form should generate solubility data for all the other forms. However, since the transition temperature is typically unknown initially, it is best to conduct the solubility experiment with each form.

The solubility difference between different polymorphs is independent of the solvent used provided the solvent used does not form a solvate with the drug substance (Brittain and Grant, 1999). The solvent selected for the solubility study should afford reasonable solubility. It should be high enough so the solubility can be measured accurately, but low enough so the amount of drug substance consumed is minimized. Sometimes, it may be a good idea to use two different solvent systems to determine the same transition temperature in order to increase one's confidence in the results.

If the drug substance can form a hydrate, all polymorphs will eventually convert to the hydrate(s), since hydrates are typically less soluble in aqueous media than anhydrous forms (Grant and Higuchi, 1990). Hydrate formation should be detected when the residual solid is characterized as part of the solubility study. Care must be taken to make sure that samples are examined both wet and dry since some hydrates may readily convert to the anhydrate form upon drying.

Miniaturization, High-Throughput, and Automation in Solubility Measurement

Solubility is not only important in preformulation studies. It is also important in lead selection and optimization during drug discovery. As discussed earlier, the usefulness of the data generated by non-equilibrium methods can be questionable, thus, it is desirable to have methods that can determine equilibrium solubility with as little compound as possible and with a high enough throughput to support the need for lead optimization. A miniaturized shake-flask method was reported recently (Glomme et al., 2004) that can provide data on up to 20 compounds a week with a single set-up. All the steps of the equilibration method are included but on a much smaller scale.

Systems that automate all the steps to measure equilibrium solubility have also been developed and are commercial available. The Symyx Discovery Tools Solubility and Liquid Formulations Workflow is a good example of such a system (www.symyx.com). This integrated software and instrumentation workflow is equipped with automated sample preparation, pH measurement, filtration and sample analysis. The manufacturer claims an annual throughput of more than 60,000 experiments.

Analytical techniques such as Raman spectroscopy have been developed for examining residual solids from solubility samples. Since the largest difference in solubility is observed between crystalline and amorphous materials, it may be sufficient to know if the material in equilibrium with the saturated solution is crystalline or amorphous.

Typical Solubility Studies to Support Formulation Development

Table 1 summarizes the commonly studied solvents for solubility and the solubilization strategies these solubility results support. A combination of various

	Solubilization strategies	
Solvents	**Solution formulation**	**Solid dosage form**
Aqueous solutions of various pHs (buffered or unbuffered)	pH-adjustment and salt formation	Salts
Non-aqueous solvents and their mixtures	Co-solvents	Co-solvents, lipid based systems in soft or hard gelatin capsules
Surfactants and phospholipid in aqueous and non-aqueous media	Micellar solubilization Liposome	Lipid based systems in capsules
Aqueous solutions containing complexing agents such as cyclodextrins with or without pH adjustment	Complexation	Complexation
Oil, Intralipid® or pre-made liposomes	Emulsion Liposome	Emulsion, micro-emulsion, solid micro-emulsion in capsules

Table 1. Solubility studies to support various solubilization strategies.

techniques is often required to maximize the solubilization potential. Chemical stability studies are typically done in parallel with solubility measurements. Solutions from solubility measurements can be used as solution stability samples.

Other solubility studies often required to support various development activities include:

- solubility studies in physiologically relevant media such as simulated intestinal fluids for dissolution method development, and
- solubility in organic solvents for analytical method development, crystal form screening, and cleaning verification.

Summary

Solubility determination is an important step in pharmaceutical preformulation. Although the experiment seems simple and well known, depending on the properties of the drug substance, special care must be taken to ensure the reliability of the results. Establishment of equilibrium and identifying what solid material is in equilibrium are two of the most important considerations in any solubility experiment.

Acknowledgements

I would like to thank my professor Dr. Keith Guillory of the University of Iowa for reviewing and editing this manuscript. He remains to be my endless source of advice, suggestions and inspiration.

List of Abbreviations

MAD .... maximum absorbable dose
SIWV .... small intestine water volume
SITT .... small intestine transit time
HPLC .... high performance liquid chromatography
PXRD .... powder x-ray diffraction
DSC .... differential scanning calorimetry
DMSO .... dimethylsulfoxide

References

Amidon GL, Lennernas H, Shah VP, and Crison JR. A Theoretical Basis for a Biopharmaceutic Drug Classification: the Correlation of in vitro Drug Product Dissolution and in vivo Bioavailability. *Pharm Res* 1995; 12: 413–420.

Avdeef A. *Absorption and Drug Development, Solubility, Permeability, and Charge State.* John Wiley & Son, Inc., Hoboken, NJ, 2003.

Brittain HG. Methods for the Characterization of Polymorphs and Solvates. In: Brittain HG (ed.), *Polymorphism in Pharmaceutical Solids.* Marcel Dekker, Inc., New York, NY, 1999.

Brittain HG and Grant DJW. Effects of Polymorphism and Solid-State Solvation on Solubility and Dissolution Rate. In: Brittain HG (ed.), *Polymorphism in Pharmaceutical Solids.* Marcel Dekker, Inc., New York, NY, 1999.

Butler JN. *Ionic Equilibrium, Solubility and pH Calculations.* John Wiley & Sons, Inc., New York, NY, 1998.

Byrn SR, Pfeiffer RR, and Stowell JG. *Solid State Chemistry of Drugs.* Second Edition, SSCI, Inc., West Lafayette IN, 1999.

Glomme A, Marz J, and Dressman JB. Comparison of a Miniaturized Shake-Flask Solubility Method with Automated Potentiometric Acid/Base Titrations and Calculated Solubilities. *J Pharm Sci* 2004; 94: 1–16.

Grant DJW and Higuchi T. *Solubility Behavior of Organic Compounds.* Wiley, New York, NY, 1990.

Higuchi T and Connors KA. Phase-Solubility Techniques. *Advan Anal Chem Instr* 1965; 4: 117–212.

Huang LF and Tong WQ. Impact of Solid State Properties on Developability Assessment of Drug Candidates. *Adv Drug Del Rev* 2004; 56: 321–334.

Johnson KC and Swindell AC. Guidance in the Setting of Drug Particle Size Specifications to Minimize Variability in Absorption. *Pharm Res* 1996; 13: 1795–1798.

Lipinski, CA, Lombardo F, Dominy, BW, and Feeney PJ. Experimental and Computational Approaches to Estimate Solubility and Permeability in Drug Discovery and Development Settings. *Adv Drug Del Rev* 1997; 23: 3–25.

Lobenberg R, Amidon GL, and Vieira M. Solubility as a Limiting Factor to Drug Absorption. In: Dressman JB and Lennernas H (eds.), *Oral Drug Absorption, Prediction and Assessment.* Marcel Dekker, Inc., New York, NY, 2000, pp. 137–153.

Lötter AP, Flanagan, Jr. DR, Palepu NR, and Guillory JK. A Simple Reproducible Method for Determining Dissolution Rates of Hydrophobic Powders. *Pharm Tech.* 1983; 4: 56–66.

Pan L, Ho Q, Tsutsui K, and Takahashi L. Comparison of Chromatographic and Spectroscopic Methods used to Rank Compounds for Aqueous Solubility, *J Pharm Sci* 2001; 90: 521–529.

Pudipeddi, M, Serajuddin ATM, Grant DJW, and Stahl PH. Solubility and Dissolution of Weak Acids, Bases and Salts. In: Stahl PH, and Wermuth CG (eds.), *Handbook of Pharmaceutical Salts, Properties, Selection, and Use.* Zurich, Switzerland: Verlag Helvetica Chimica Acta and Weinheim, Wiley-VCH, Weinheim, Federal Republic of Germany: Wiley-VCH, 2002, pp. 19–39.

Todd D and Winnike RA. A Rapid Method for Generating pH-Solubility Profiles for New Chemical Entities. *Pharm Res* 1994; 11(10): S-271.

Tong WQ. Preformulation Aspects of Insoluble Compounds. In: Liu R (ed.), *Water-Insoluble Drug Formulation.* Interpharm Press, Denver, Colorado, 2000.

Wang ZR, Burrell LS, and Lambert WJ. Solubility of E2050 at Various pH: A Case in Which Apparent Solubility is Affected by the Amount of Excess Solid. *J Pharm Sci* 2002; 91: 1445–1455.

Winnike R. Solubility Assessment in Pharmaceutical Development, Practical Considerations for Solubility Profiling and Solubilization Techniques. AAPS Short Course on Fundamentals of Preformulation in Pharmaceutical Product Development, Nashville, TN, Nov. 5, 2005.

Yalkowsky SH and Banerjee S. *Aqueous Solubility, Methods of Estimation for Organic Compounds.* Marcel Dekker, New York, NY, 1992.

6

Biorelevant Dissolution Media

ANETTE MÜLLERTZ

Department of Pharmaceutics and Analytical Chemistry,
The Faculty of Pharmaceutical Sciences, University of Copenhagen

Introduction

Oral administration is the most convenient way to deliver drugs, and therefore the most preferred. However, the oral route is very complex based on the physiological conditions encountered by the drug as it passes from the mouth to the absorptive sites in the intestine. When moving from the stomach through the pylorus into the small intestine, the drug will meet a rapidly changing environment including bile and pancreatic secretions which will introduce different enzymes and surface active bile components, and increase in pH from acidic to neutral. Physiological factors such as the rate of gastric emptying, intestinal motility, blood flow, as well as volume, composition and pH of alimentary secretions are known to impact the rate and/or extent of drug absorption. The basic parameters determining the absorption of a drug compound after oral administration are its solubility and permeability in the conditions associated with the gastrointestinal (GI) tract.

An increasing problem for the pharmaceutical industry, partly arising from the introduction of high throughput screening, is the discovery of highly hydrophobic active pharmaceutical candidates with low water solubility (Lipinski, 2000). Many of these compounds have a high permeability and are consequently classified as Class 2 compounds according to the Biopharmaceutical Classification System (BCS). The solubilization and/or dissolution in the gastrointestinal tract are the rate-limiting steps for the absorption of Class 2 compounds; as a result, classic solid formulations of Class 2 compounds often exhibit low and/or variable oral bioavailability which is highly influenced by the dietary state (fasted/post-prandial).

For the *in vitro* assessment of *in vivo* behavior of a drug formulation, the dissolution test is most often used, both for development purposes and for quality control. For quality control, simple dissolution media like aqueous buffers are preferred, due to cost and reproducibility. When using dissolution for development purposes, the use of more complex biorelevant media, simulating the composition of the GI fluids, is relevant. These media will subject the drug formulation to conditions that are closer to those encountered in the intestinal tract, which can increase the possibility of predicting differences between formulations.

The relationship between *in vitro* dissolution and *in vivo* bioavailability/absorption is far from being fully explored. It is important to keep in mind the limitations of the dissolution test when it is used as a predictor of *in vivo* performance. By gaining further knowledge of the intraluminal environment, *in vitro* dissolution tests will be able to more accurately simulate the conditions in the GI tract, leading to more powerful prediction of *in vivo* performance.

This chapter will describe the current knowledge of the physiological conditions encountered by a drug formulation during transit through the GI tract, and based on these conditions give examples of suitable biorelevant dissolution media simulating GI fluids. For information on conventional dissolution media, the reader is referred to the Guidelines issued by FDA and other world-wide boards of health.

Conditions in the Gastrointestinal Tract

Biorelevant Lipids and Surfactants

Amphiphiles (surface-active compounds or surfactants) are characterized by a molecular structure with both hydrophilic and hydrophobic domains. They tend to adsorb to surfaces and interfaces, with a concomitant lowering of surface tension. At the critical micellar concentration (CMC), the limit of surface tension reduction is reached and a spontaneous self-assembly takes place with the formation of aggregates (micelles). The size and structure of the micelles depend on the type and concentration of surfactant(s) present.

In the small intestine, the main relevant amphiphiles are bile acids/salts, lyso-phospholipids and cholesterol, secreted from the bile, and fatty acids and monoglycerides from lipid digestion. Based on their ability to interact with water, the amphiphiles are characterized as *soluble*, *insoluble swelling*, or *insoluble non*-swelling (Carey and Small, 1970). Bile salts and lyso-phospholipids are characterized as *soluble* amphiphiles, and they possess an enormous capacity to solubilize *insoluble swelling* amphiphiles. Only soluble amphiphiles are able to form micelles on their own. Other lipids can be solubilized and transported by these micelle-forming amphiphiles. Phospholipids and monoglycerides belong to the class of *insoluble swelling* amphiphiles. However, above their phase transition temperature, water can be incorporated between the polar groups, which create a swollen lipid structure. Triglycerides, diglycerides, cholesterol and protonated long-chain fatty acids are insoluble as well, but not able to swell in water; they belong to the class of *insoluble non-swelling* amphiphiles (Figure 1).

Bile salts are planar molecules, with all the hydrophilic hydroxyl groups placed on one side, and the hydrophobic groups on the other. The molecular structure of bile salts is shown in Figure 2. Bile salt molecules tend to orient

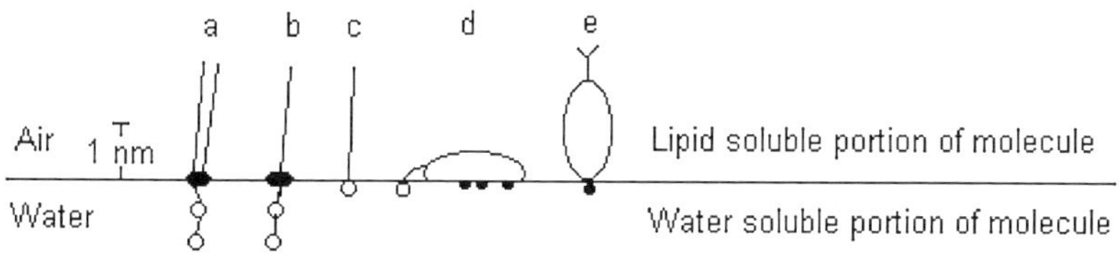

Figure 1. Schematic presentation of some polar lipids and their orientation at an air-water or oil-water interface, modified from Carey and Small (1970). (a) Lecithin, (b) lysolecithin, (c) ionized fatty acid, (d) bile salt, (e) cholesterol.

themselves in micelle-like structures with the hydrophobic side projecting towards the interior of the micelles, and the hydrophilic side directed outwards, in order to maximize the hydration of the hydrophilic part of the molecule. Recent molecular dynamic simulations have shown that bile salt micelles are very dynamic structures, taking shapes ranging from prolate, roughly spherical to oblate. An aggregation number between 8 and 16 has been observed (Warren et al., 2006). As compared to that of typical ionic surfactants, bile salt self-association occurs over a broader concentration range (Kratohvil et al., 1983). The CMC of bile salts is characterized by the following rank order: unconjugated > glycine conjugates > taurine conjugates, with the CMC for taurocholic acid being around 5 mM (Roda et al., 1983; Carey, 1984). The bile salt micelles (primary micelles) can aggregate by hydrogen bonding to form secondary micelles. Aggregation does not necessarily create a larger hydrophobic domain, but rather a distinct interfacial region (Wiedmann and Kamel, 2002).

In the presence of counter ions, lyso-phospholipid, phospholipids or other lipophilic molecules, the CMC of mixed bile salt micelles is reduced (Carey,

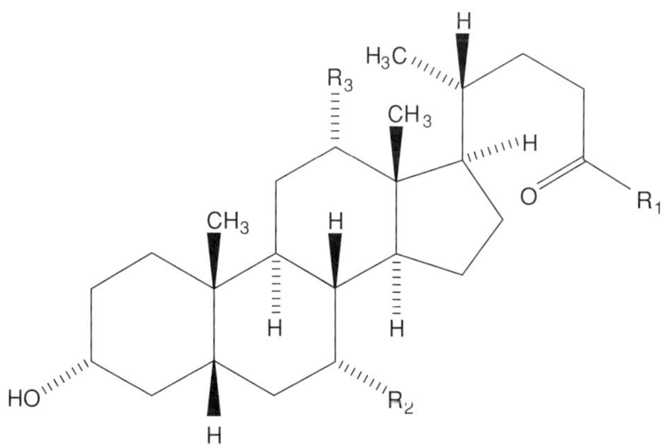

R_1 = OH — Unconjugated
R_1 = $NHCH_2COO$- — Glycine conjugated $pK_a \cong 3.7$
R_1 = $NHCH_2CH_2SO_2O$- — Taurine conjugated $pK_a \cong 1.5$

R_2 = OH & R_3 = OH — Cholic acid (primary)
R_2 = OH & R_3 = H — Chenodeoxycholic acid (primary)
R_2 = H & R_3 = OH — Deoxycholic acid (secondary)
R_2 = H & R_3 = H — Lithocholic acid (secondary)

Figure 2. Molecular structure of bile acids.

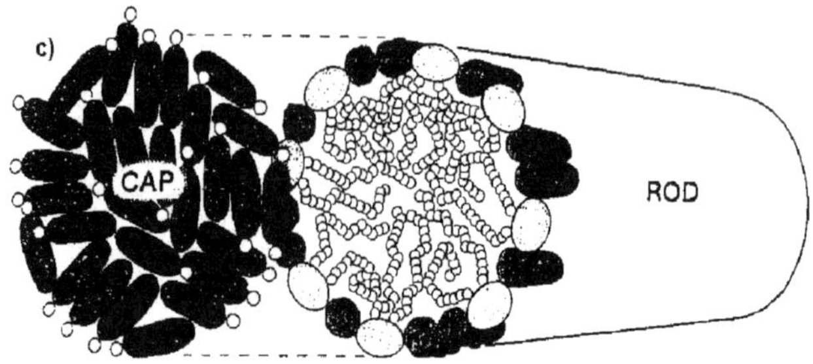

Figure 3. Rod shaped mixed micelles, containing bile salt and phospholipids (Nichols and Ozarowski, 1990) reproduced with permission.

1984; Hofmann and Mysels, 1992). The CMC for taurocholate is decreased from 4.7 mM to approximately 0.25 mM when phospholipids are present (Naylor et al., 1993a).

Bile salts can solubilize a certain amount of phospholipids by incorporating them in micelles with an approximate radius of 3 nm (Hjelm et al., 1992; Marrink and Mark, 2002). The kinetics of solubilization of phospholipids by bile salts is strongly dependent on the type of bile salt, being most rapid for glycochenodeoxycholate (Lindenbaum and Rajagopalan, 1984). When the bile salt : phospholipids ratio exceeds 1:1, the micelles transform into a rod-like shape (Nichols and Ozarowski, 1990). In the rod-like micelles, the phospholipids are oriented radially with respect to the long axis (Figure 3). Bile salts sit at the surface and act as wedges, filling in the space between the polar phospholipid headgroups; they also form the caps at the end of the rod-like micelle. Hjelm et al. (2000) showed that rod-like micelles are also formed when fatty acids or monoglycerides interact with bile salts at ratios above 1:1. With further addition of fatty lipids, hexagonal, lamellar and other liquid crystalline phases are formed (Hjelm et al., 2000).

Human Bile

Bile acids are synthesized by hepatocytes in the liver and stored in the gall bladder. In the gall bladder, water and salts are extracted from the bile thereby concentrating the bile salts 5- to 20-fold (Berne and Levy, 1996). Human gall bladder bile consists of approximately 84% water, 11.5% bile salts, 3% phospholipids, 0.5% cholesterol and 1% other components (bile pigments, proteins and inorganic ions and cations) (Carey and Small, 1970). After intake of a meal, hormonal responses facilitate secretion of bile from the gall bladder into the duodenum. After release from the gall bladder, the bile fluid is diluted in the hepatic duct to a total concentration of bile salt of 10–20 mM (Kararli, 1989). The concentration of bile salts is further diluted when entering the intestine, but will still be above the CMC, which is approximately 0.35 mM in a taurocholate:lysophospholipid (5:1 on a molar base) system (unpublished data).

Bile acids are secreted from the gall bladder as sodium and potassium salts conjugated by peptide linkages to either taurine or glycine (Figure 2). Reported compositions of human bile in the upper small intestine are shown in Table 1. As can be seen, there is a large variation in the composition, probably

	Hofmann et al., 1976	Alvaro et al., (n = 3) 1986	Persson et al., 2005 Fasted state	Persson et al., 2005 (n = 3) Fed state	Brouwers et al., 2006 (n = 4) Fasted state	Perez de la cruz Moreno et al., 2006 (n = 13) Fasted state
Taurine-conjugated						
Taurocholic acid	12	9.4 ± 0.8	23	18	30.1 ± 13.4	44.95 ± 18.17
Taurochenodeoxycholic acid	12	11.6 ± 2.2	21	17	8.9 ± 2.7	7.89 ± 2.16
Taurodeoxycholic acid	8	5.0 ± 0.4	3	8	3.8 ± 2.9	4.71 ± 4.47
Total %	**32**	**26**	**47**	**43**	**43**	**58**
Glycine-conjugated						
Glycocholic acid	24	26.1 ± 3.2	37	26	20.0 ± 4.8	17.21 ± 6.37
Glycochenodeoxycholic acid	24	31.9 ± 3.4	14	16	20.3 ± 4.7	13.77 ± 9.59
Glycodeoxycholic acid	16	16.0 ± 2.2	2	15	15.9 ± 9.0	10.40 ± 7.32
Total %	**64**	**74**	**53**	**57**	**55**	**41**

Table 1. Bile acid composition in human intestinal fluids, compiled from the literature.

due to a combination of individual variations and analytical limitations. However, it is clear that cholic acid (a trihydroxy bile acid) and chenodeoxycholic acid (a dihydroxy bile acid) are the major bile acids in human bile, being conjugated to either taurine or glycine. They are *primary bile acids* because they are synthesized from cholesterol in the liver (Hofmann, 1984). Deoxycholic acid is a *secondary bile acid* formed by bacterial dehydroxylation of cholic acid in the digestive tract.

Conjugated bile acids are more completely ionized and thus more water soluble than unconjugated bile acids (Berne and Levy, 1996). In the pH range encountered in the small intestine (pH 5 – 6.8), taurine conjugates (pK_a ~ 1.5) are ionised, whereas glycine conjugates (pK_a ~3.7) exist as a mixture of ionised and unionised molecules (Davenport, 1977). The water solubility of fully dissociated conjugated bile salts is as high as 1–2 M (Carey, 1984). Human bile fluid contains less than 1% unconjugated bile acids, with a pKa value of 5.0. At physiological pH, the unconjugated bile salts are poorly soluble (Hay and Carey, 1990).

A small fraction of unionized bile salts are absorbed by passive diffusion in the duodenum and jejunum. Ionized bile salts, being large and negatively charged ions at the pH prevailing in the intestinal lumen, need an active transporter, which is situated in the distal ileum. Consequently, the concentration of the majority of bile salts remains high, until fat digestion and absorption is complete. After absorption into the portal blood, bile salts return to the liver, in what is called enterohepatic circulation. The hepatocytes in the liver re-conjugate the de-conjugated bile acids and hydroxylate some of the secondary bile acids (Berne and Levy, 1996) after which they are secreted into the gall bladder.

The Composition of Human Intestinal Fluids

In the following sections, literature on the composition of the gastric and intestinal fluids during fasted and fed state is reviewed.

Fasted State

The Stomach

The volume of gastric fluid in the fasted state is in the range of 15–50 ml (Delgado-Aros et al., 2004; Jellish et al., 2005). Average gastric pH ranges between 1.5 and 2.9 (Dressman et al., 1990; Lindahl et al., 1997; Maltby et al., 2004; Kalantzi et al., 2006). However, large inter-individual variations in pH values have been reported, ranging from 1.2 to 7.5. Gastric fluid is hypo-osmotic ($\leq$200 mOsm/kg) (Mahe et al., 1992; Lindahl et al., 1997; Pedersen et al., 2000b), the main cation being Na^+ and the main anion Cl^-. Surface tension in fasted gastric fluids has been reported in the range of 30 to 50 Nm/m, which is lower than water (approximately 70 Nm/m) (Finholt and Solvang, 1968; Efentakis and Dressman, 1998; Kalantzi et al., 2006). Pepsin can only partly be responsible of this reduced surface tension, as pepsin is not able to reduce the surface tension to less than 55 Nm/m (Vertzoni et al., 2005). The additional presence of low amounts of bile salts, refluxed from the duodenum, or lyso-phospholipids, from breakdown of gastric mucosal cells, might explain the low surface tension in the stomach.

The Upper Small Intestine

Moving from the stomach to the duodenum, the pH of the GI fluids changes from acidic to neutral within a distance of a few centimeters. Average pH values between 6 and 7.1 have been observed for the fasted duodenum and upper jejunum, bicarbonates being the principal species responsible of the buffer capacity (Fallingborg et al., 1989; Dressman et al., 1990; Lindahl et al., 1997; Kalantzi et al., 2006). Osmolality is increased compared to the gastric fluids (270 mOsm/kg), with Na^+ and Cl^- still being the dominant ions (Lindahl et al., 1997d). Bile is supposedly only secreted as a response to food intake, but the fasted small intestine also contains low amounts of bile, which are partly responsible for the low surface tension (between 30 and 34 mN/m), that has been observed (Pedersen et al., 2000b; Kalantzi et al., 2006). In human small intestinal fluids, mean values between 1.5 and 3 mM bile salt have been reported under fasted conditions (Lindahl et al., 1997; Pedersen et al., 2000a; Persson et al., 2005; Brouwers et al., 2006). As mentioned above, bile salts are secreted together with phospholipid, primarily phosphatidylcholine (Schersten, 1973). In the fasted state, the ratio between bile salts and phospholipid has been reported to be between 2:1 and 10:1 in the intestinal fluids (Alvaro et al., 1986; Persson et al., 2005). In this range, bile salts and phospholipid will form spherical mixed micelles.

Fed State

The Stomach

The composition of the fluids in the fed stomach will be highly dependent on the nature of the ingested meal and the time after intake. Initially, the composition of the gastric fluids will be close to the composition of the meal with regard to pH, osmolality and surface tension. Over time, with secretion of gastric juices and following gastric emptying, values will return to those of the fasted state. The time to return to fasted state values has been reported to be between 1 and 4 hours, depending on the composition and volume of the ingested meal (Malagelada et al., 1976; Dressman et al., 1990; Kalantzi et al., 2006). The capacity of the human stomach is 1–1.6 L (Kararli, 1995).

The rate of gastric emptying is dependent on a complex interplay between physiological conditions and the composition of the ingested meal. Liquids empty faster than solids, mainly because the pylorus holds back larger food particles until further breakdown has taken place (Moore et al., 1983). Gastric emptying is also dependent on the caloric content of the meal and there seems to be a dynamic feedback mechanism between the stomach and the duodenum to control the caloric delivery (Brener et al., 1983).

The Upper Small Intestine

The compositions of the fluids in the upper small intestine during the fed state will also be dependent on the type of food ingested, however, to a lower extent than in the stomach. Upon arrival in the duodenum, the chyme is mixed with bile and pancreatic secretions, both being slightly alkaline, which results in a pH in the duodenum in the range of 5.5 to 6.5; this is slightly lower than in the fasted state (Fallingborg et al., 1989; Persson et al., 2005; Kalantzi et al., 2006). Large fluctuations in pH are observed in the proximal duodenum, but the fluctuations diminish when the chyme moves along the small intestine. In addition, there is a pH gradient which increases between the duodenum and the ileum (Dressman et al., 1990). As a consequence of chyme and intestinal secretions, both buffer capacity and osmolality in the intestinal fluids have been reported to increase after meal intake. Values of approximately 18–30 mmol/l/pH and 300–400 mOsm/kg during the first 2 hours after a meal have been reported for the buffer capacity and the osmolality, respectively (Dressman et al., 1998; Kalantzi et al., 2006). In contrast, surface tension does not change significantly between fasted and fed state (Persson et al., 2005; Kalantzi et al., 2006).

Food-induced bile secretions result in bile salt levels in the small intestine between 8 and 20 mM with single measurements up to 40 mM (Ladas et al., 1981; Tangerman et al., 1986; Armand et al., 1996; Persson et al., 2005). The molar ratio between bile salts and phospholipids in the fed state will be dependent on the phospholipid level in the food and has been reported to be in the range of 2:1 to 5:1 (Schersten, 1973, Persson et al., 2006).

Lipid Digestion

Lipid digestion begins in the stomach by emulsification and initial hydrolysis of dietary lipids (Phan and Tso, 2001). In the fed stomach, the main amphiphiles are dietary phospholipids and some hydrophobic proteins. Gastric lipase, secreted by the chief cells of the gastric mucosa, removes one fatty acid from a triacylglyceride molecule, thus forming surface active diacylglycerides and free fatty acids. Gastric lipase is responsible for approximately 10–20% of dietary triacylglyceride hydrolysis (Carriere et al., 1993), which promotes emulsification. Armand et al. showed that the main emulsification of dietary lipids takes place in the stomach, whereas hydrolysis of lipids primarily takes place in the upper part of the small intestine (Armand et al., 1996).

When the lipid emulsion is transferred from the acidic environment in the stomach to the higher pH conditions in the duodenum, the fatty acids become partly ionized and migrate to the interface of the emulsion particle. Triacylglycerides and diacylglycerides will segregate into the core of the emulsion particles, that will also be covered by bile salts and lyso-phospholipids secreted from the bile.

Enzyme-rich pancreatic juice which contains several enzymes capable of hydrolysing triacylglycerides and other lipids is secreted from the pancreas and mixed with the chyme. Pancreatic phospholipase A_2 hydrolyzes phospholipid, thus producing fatty acids and lyso-phospholipids. Pancreatic triacylglyceride lipase is by far the most important lipase in triacylglyceride digestion. It is a sn1,3 specific lipase, meaning that it removes the fatty acids in the 1 and 3 position, thus releasing 2-monoacylglyceride and two free fatty acid molecules. The challenge of the pancreatic lipase is to reach the interface between the triacylglyceride core of the emulsion particle, and the surface covered with a mono- or multilamellar layer of amphiphiles (Lowe, 2002). In this case, pancreatic co-lipase plays an important role. Co-lipase is believed to clear the surface of the emulsion particles and anchor the lipase to the triacylglycerides interface where the lipase will excel its action (Larsson and Erlansonalbertsson, 1991; Patton and Carey, 1981).

The lamellar phases on the surface of the emulsion particles are mainly composed of monoacylglycerides, lyso-phospholipids and ionized fatty acids. When the phases have reached a certain size, they will desorb from the emulsion surface and form multi-lamellar vesicles, which are transformed into uni-lamellar vesicles upon increased incorporation of bile salts (Rigler et al., 1986). Upon further incorporation of bile salts, the ratio of lipid amphiphiles to bile salts will decrease to 1 or lower, whereby the uni-lamellar vesicles are transformed to mixed micelles (Staggers et al., 1990). These events are presented in Figure 4.

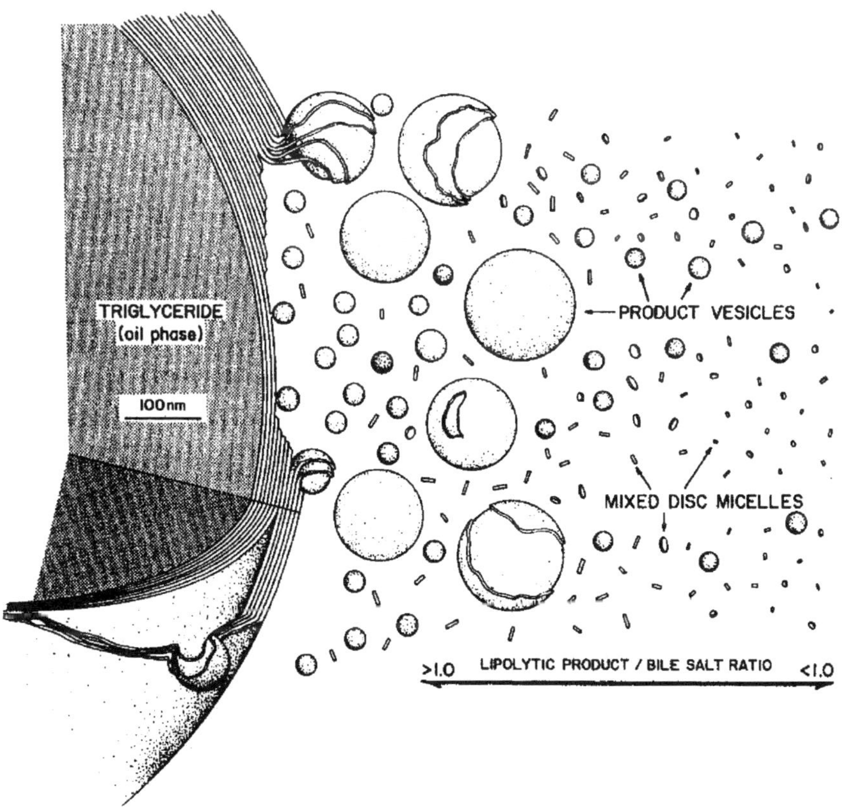

Figure 4. Emulsification and digestion of lipids, leading to accumulation of micelles and vesicles (Rigler et al., 1986) reproduced with permission.

The formed mixed micelles can diffuse to the unstirred water layer that lines the epithelium, where the micelles disintegrate and lipid amphiphiles are absorbed. Bile salts are recycled back into the lumen and continue to interact with lipid digestion. Thus at any given time during lipid digestion, a complex mixture of different colloid phases is present in the intestinal lumen (Rigler et al., 1986).

During lipid digestion in the small intestine, a steady state seems to be reached approximately 1 hour after food intake and lasts for up to 3 hours (Armand et al., 1996; Persson et al., 2006). In this period, the triglyceride content is stable, probably due to a constant delivery of chyme from the stomach. In most studies of intestinal contents after food intake, the fatty acid to monoglyceride ratio is around 4–6:1, which is above the theoretical ratio of 2:1, that would be the result of triacylglyceride hydrolysis (Ricour and Rey, 1970; Hernell et al., 1990; Persson et al., 2006). Rautureau et al. suggest that the high absorption rate of monoglycerides as compared to fatty acids can explain the high ratio of fatty acids to monoglycerides in intestinal fluids (Rautureau et al., 1981).

Biorelevant Dissolution Media

Traditional dissolution tests, while appropriate for quality control, are often not predictive of the more complex *in vivo* dissolution behavior of poorly-soluble drugs. In order to gain knowledge on the behavior of a poorly-soluble drug compound during transit of the GI tract, the use of biorelevant media, simulating the actual *in vivo* situation, is important. Biorelevant media simulating the fasted or fed stomach and upper small intestine have been described. Parameters like pH, osmolarity and surfactant type and level determine what the biorelevant media is aiming to emulate. All biorelevant media have reduced surface tension compared to water, due to the presence of different surfactants in the fluids of the GI tract. The role of the amphiphiles is to increase the wetting behavior towards solids, thereby affecting the effective surface area available for dissolution and increasing the solubilising capacity of poorly soluble drug compounds. The presence of amphiphiles improves the solubility of most poorly soluble drugs, thereby increasing the chance of having sink conditions during dissolution. By determining dissolution in media simulating both fasted and fed state, potential food effects might be elucidated.

Which media, or combination of media, to choose, will depend on the physicochemical characteristics of the compound, and, when relevant, also on the type of formulations involved.

Strict adherence to the GI conditions is not always needed when developing biorelevant media, but should depend on the purpose of the test. It may be relevant to study only the effect of a single parameter, e.g. bile salt level or pH, keeping all other parameters constant. In the following sections, examples of biorelevant dissolution media used for simulating the gastric and small intestinal environment, in both fasted and fed state, are described. The choice of biorelevant media, depending on the specific physicochemical characteristics for a drug, will also be reviewed.

Fasted State

The Stomach

For simulations of dissolution in the stomach, the creation of sink conditions is usually not required as absorption across the gastric mucosa is usually negligible. The majority of media presently used to assess dissolution under fasted gastric conditions do not take the actual physiological conditions into account. The simplest dissolution media simulating gastric fluids is the USP test fluid without pepsin, consisting essentially of 0.1 N HCl, having a surface tension basically equal to water (68 Nm/m). In order to reduce the surface tension to a value that is more biorelevant, several attempts have been made by including surfactants that are not physiological relevant [e.g. SLS (Dressman et al., 1998) or TritonX100 (Galia et al., 1999)]. Since the low surface tension in gastric fluids is most likely due to various factors (see above), Vertzoni et al. (Vertzoni et al., 2005) recently proposed a fasting gastric medium (Fasted State Simulated Gastric Fluid; FaSSGF), containing pepsin, and a low level of taurocholate and phospholipids (80 μM and 20 μM, respectively), in order to obtain a surface tension close to that found *in vivo*. For the same reasons, pH was adjusted to 1.6, instead of the 1.2 usually employed. FaSSGF was found to give a better prediction of the intraluminal dissolution of a lipophilic weak base (GR253035X), whereas the use of FaSSGF had no impact on the dissolution of atovaquone (log P 5.1; non-ionizable), probably because gastric dissolution does not contribute significantly to the overall dissolution of this type of drug (Vertzoni et al., 2005).

The Upper Small Intestine

In the late 90's, Dressmann and co-workers (Dressman et al., 1998; Galia et al., 1998) suggested a biorelevant medium, simulating the fluids in the fasted small intestine (FaSSIF, Table 2). Since then, FaSSIF has become widely used for the determination of biorelevant solubility and dissolution studies, both in industry and academia. The composition of FaSSIF was developed so as to take pH, osmolarity, buffer capacity and solubilizing capacity of the fluids in the fasted upper small intestine into account. Initially, KCl was used to increase the osmolarity, but was later replaced by NaCl (Vertzoni, 2004), which gives a better representation of the ions present in the jejunum (Lindahl et al., 1997). Recently, the buffer capacity of FaSSIF has been shown to be significantly higher than the buffer capacity in human intestinal fluids (Cruz Moreno et al., 2006). Interestingly, this issue has not yet been addressed in the development of biorelevant media. To represent the bile salts, sodium taurocholate was chosen based on its importance in human bile (Table 1) and because it has a very low pKa value, reducing the risk of precipitation and changes in micellar size due to small changes in pH.

Phospholipid is added in a 1:4 molar ratio to taurocholate, which is within the physiological relevant range. Egg phospholipid is recommended, as its fatty acid composition is close to the one in human bile phospholipids. Studies by Naylor et al. (Naylor et al., 1993; Naylor et al., 1995) demonstrated the importance of adding phospholipids, especially when dealing with hydrophobic drug

FaSSIF		FeSSIF	
pH	6.5	pH	5.0
Osmolality	270 ± 10 mOsmol	Osmolality	635 ± 10 mOsmol
Buffer Capacity	12 mmol/l pH	Buffer Capacity	76 mmol/l pH
Sodium taurocholate	3 mM	Sodium taurocholate	15 mM
Lecithin	0.75 mM	Lecithin	3.75 mM
NaH_2PO_4	28.66 mM	Acetic acid	144 mM
NaCl	106 mM	NaCl	173 mM
NaOH, qs	pH 6.5	NaOH, qs	pH 5.0

Table 2. Composition of fasted state simulated intestinal fluid (FaSSIF) and fed state simulated intestinal fluid (FeSSIF) (Galia et al., 1998, Vertzoni et al., 2004).

compounds. For danazol (log P 4.5), significant increases were found for both contact angle, solubility and dissolution rate when phospholipids were added to the taurocholate solution, whereas differences were modest for betamethasone (log P 1.9). Another option is to replace phospholipids with lyso-phospholipids, which is the phospholipid species that is actually present in the intestine *in vivo*; however, this has not yet been explored.

Due to the presence of surfactants, FaSSIF generally gives a better dissolution of most BCS class II drugs (Galia et al., 1998; Nicolaides et al., 1999) compared to compendial media.

In order to simulate the actual composition of the small intestine more closely, crude bile extract has been used. Crude extracts from both ox and porcine gall bladder are available and have been used for dissolution studies (Vertzoni et al., 2004; Sunesen et al., 2005). The composition of the bile salts in the crude bile extract reflects the composition present in the gall bladder of the animal. As an example, the composition of porcine bile extract is shown in Table 3. However, bile salt extracts often contain only approximately 50–85% bile salts, with a very large batch-to-batch variation (Zangenberg et al., 2001a; Christensen et al., 2004; Vertzoni et al., 2004). Therefore each batch of crude bile extract has to be carefully analysed with regard to bile salt concentration and composition. The phospholipid level has also shown large variation, and should therefore also be determined in each batch (Zangenberg et al., 2001b; Christensen et al., 2004) and corrected by addition of phospholipid to assure a reproducible phospholipids level. In short, the use of crude bile extract can be problematic in terms of batch analysis and batch-to-batch reproducibility; however, it will better simulate the content of the small intestine, and will also include a significant reduction in cost. The price of crude bile extract is less than

	Porcine bile extract[a]
Primary conjugates	
Glycocholate	25.2
Taurocholate	2.8
Glycochenodeoxycholate	42.6
Taurochenodeoxycholate	5.0
Glycohyocholate	14.7
Secondary conjugates	
Glycodeoxycholate	n.d.
Taurodeoxycholate	n.d.
Glycohyodeoxycholate	2.9
Glycine conjugates	85
Taurine conjugates	8
Unconjugated	7

Table 3. Bile salt composition porcine bile extract.

[a] Determined by HPLC analysis

n.d., not detected

15% on a weight to weight basis than the price of pure taurocholate. When large volumes are used in dissolution experiments, this is a significant advantage.

Fed State

The Stomach

The composition of the stomach contents in the fed state is highly dependent on the meal ingested, and thereby difficult to standardize. Milk (3.5% fat) has been used to simulate fed conditions in the stomach because it contains nutrients comparable to a typical western diet (Macheras et al., 1986; Macheras et al., 1987; Galia et al., 1998; Nicolaides et al., 1999; Kalantzi et al., 2005). However, with regard to osmolality and buffer capacity, milk has much lower values than a homogenised meal, making it less suitable as a fed state simulated gastric fluid (Klein et al., 2004). Milk has a pH of about 6.5, which might be too high even for the initial post-prandial state, and definitely too high for the later stages of gastric digestion. Under *in vivo* conditions, lipids and casein micclles in the milk are partly digested during gastric residence, so addition of enzymes might further increase its biorelevance.

Another possibility for simulating the fed stomach is to use a nutritional drink, like Ensure Plus® (Ensure) or Nutrison® (Nutricia Nordica), which has a composition that resembles a regular meal, but lacks solids. This approach has been used by Klein et al. (Klein et al., 2004), who suggested the use of Ensure Plus containing 0.45% pectin, as a viscosity enhancer, based on comparison with physicochemical parameters of a homogenized standard breakfast.

Abrahamsson and coworkers (Abrahamsson et al., 2004) also used a medium based on a nutritional drink in order to simulate the fed state in the stomach. Nutrison® was mixed 6:4 with hydrochloric acid, sodium and potassium chloride and sucrose to reach a pH of 5.4, an osmolality of 420 mOsm/kg and an ionic strength of 0.15. The medium was shown to simulate tablet disintegration in the stomach of dogs fed Nutrison®, but not a standard fatty meal. This confirms the difficulty in simulating the fed stomach, since it will be very dependent on the type and amount of food ingested.

The Upper Small Intestine

Together with FaSSIF mentioned above, a fed state simulated intestinal media (FeSSIF) was developed (Dressman et al., 1998; Galia et al., 1998). FeSSIF simulates the fluids in the fed upper small intestine in terms of bile salt and phospholipids levels, pH, osmolarity and buffer capacity. The pH was set at 5.0, which is in the lower end of what has lately been reported in the literature (see above). By addition of NaCl, the osmolality is adjusted to 635 ± 10 mOsm/kg, and acetic acid is added to maintain a higher buffer capacity of 76 mmol/l/pH, both values aiming at simulating the fed state. It should be noted that recent buffer capacity values from collected intestinal juices after feeding nutritional drinks, are lower (see section above on the composition of the intestinal fluids in the fed state). *In vivo*, the main buffer capacity probably originates from the food components, like generated peptides and amino acids.

As for FaSSIF, taurocholate is used as bile salt and egg phospholipids are recommended as phospholipids. The ratio between bile salt and phospholipids is kept at 4:1, even though ingested food often contains phospholipids that will decrease this ratio.

FeSSIF was found to be suitable for a qualitative prediction of formulation and food effects for a series of drugs tested by Nicolaides et al. (Nicolaides et al., 1999), and Löbenberg et al. (Lobenberg et al., 2000). The increased level of bile salts and phospholipids in FeSSIF often results in a better dissolution for BCS class II drugs compared to FaSSIF, suggesting possible food effects (Galia et al., 1998; Nicolaides et al., 1999). However, attention should be paid to the lower pH (pH 5) and higher ionic strength in FeSSIF, compared to FaSSIF. Both these factors can also impact drug solubility and dissolution. Mefenamic acid, an NSAID with a log P of 5.3 and a pKa of 4.2, has a better dissolution in FaSSIF as compared to FeSSIF since the drug is more soluble at pH 6.5 than 5.0 (Galia et al., 1998). On the other hand, troglitazone, an antidiabetic drug, which is also a weak acid with pKa values of 6.1 and 12.0, shows a significant better dissolution in FeSSIF as compared to FaSSIF (Nicolaides et al., 1999). Whether the bioavailability of troglitazone is dependent on food intake, has not yet been published, but bioavailability of mefenamic acid has been shown to be independent of food intake (Hamaguchi et al., 1986), indicating that a comparison between dissolution curves in FaSSIF and FeSSIF might be predicative for *in vivo* behavior.

The three-fold increase in danazol bioavailability observed upon food intake (Charman et al., 1993) was closely simulated by using FaSSIF and FeSSIF as dissolution media, whereas compendial media incorrectly suggested the absence of a food effect (Galia et al., 1998).

FeSSIF is considered a reasonable starting point for assessment of food effects on drug dissolution in the small intestine. However, it does not account for the presence of lipolytic products (free fatty acids and monoglycerides) from dietary or formulation triglycerides in the intestine and may underestimate dissolution of lipophilic drugs (Nicolaides et al., 1999).

For biorelevant dissolution media containing only bile salt and PL, dynamic light scattering determinations reveal that only one particle population in the micellar size range is present (Sunesen et al., 2005; Ilardia-Arana et al., 2006). Upon addition of free fatty acids and monoglycerides, the structure and size of the colloid phases in these media depend on absolute concentrations and the ratios between the amphiphiles. In general, a micellar fraction is still present but co-exists with particles of a vesicular size. This was also observed by Kossena and coworkers by the use of size exclusion chromatography (Kossena et al., 2003).

It is well known that the solubility of most poorly soluble drugs increases linearly with increasing bile salt concentration. However, only a few studies have investigated the impact on solubility associated with the addition of lipolysis products to the media. Studies by Sunesen et al., and Zangenberg et al., showed that solubility of danazol in biorelevant media containing crude bile extract correlates very well with the total concentration of bile salt, phospholipids, free fatty acids and monoglycerides (Zangenberg et al., 2001; Sunesen et al., 2005), as illustrated in Figure 5. Furthermore, Illardia-Arana et al. (Ilardia-Arana et al., 2006), using pure sodium glycocholate as a bile salt, showed that the solubility of estradiol in biorelevant media was also linearly dependent on the total surfactant concentration. Thus, it seems that the solubility enhancement of poorly soluble drugs in biorelevant media containing bile salts, is roughly independent of the

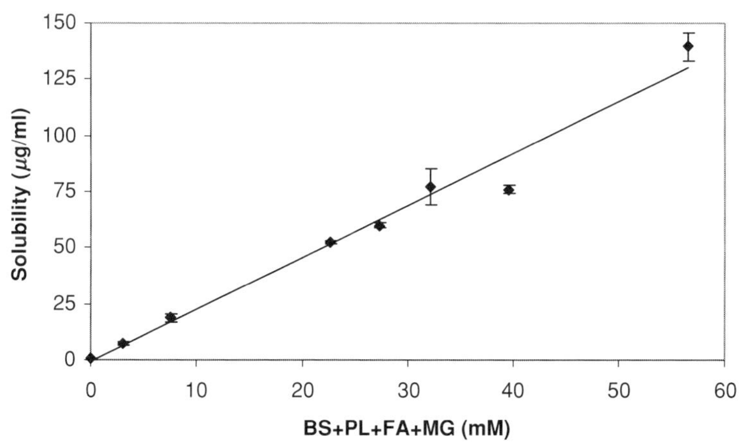

Figure 5. The solubility of danazol as a function of the total surfactant concentration ($R^2 = 0.98$) in the media listed in Table 4 (Sunesen et al., 2003).

type of biorelevant surfactant added to the system, and thus also of the colloid structures formed. This, however, requires further investigation before a final suggestion can be made.

Taking Drug Characteristics into Considerations

The choice of biorelevant dissolution media in any given situation will be dependent on the poorly water-soluble drug in question. In general, when trying to elucidate the impact of the fluids in the GI tract on a drug molecule, events occurring in both the gastric and intestinal fluids are of interest. However, for some type of drugs the gastric fluid has minor impact. This will be described below.

Lipophilic drugs, being either weak acids, weak bases or non-ionized compounds, are dissolved in the lipid fraction of the food when the macroscopic structure of the food is broken down into microscopic particles during the formation of chyme. Consequently, lipophilic molecules are predissolved in triglyceride droplets when they enter the small intestine. For these compounds, it is especially relevant to include lipolysis products in the biorelevant media simulating the intestinal fluids. Sunesen et al. (2005) could only obtain IVIVC in the fed state for a non-ionized compound (danazol), when including lipolysis products in the media.

For *non-ionized compounds,* the concentration of solubilizers in the medium seems to be the major determinant for dissolution rate. In the fasted state, the level of surfactants in the gastric fluids is fairly low and the gastric dissolution often only plays a minor role (Vertzoni et al., 2005); therefore the use of only fasted state intestinal media would be a proper choice in this case (Galia et al., 1998). Considering the fed state, the dissolution in the fed stomach may be important and should be considered. In any case, the use of a fed state dissolution medium will be necessary. In the case of a lipophilic drug, lipolysis products should be included in the media.

Weak acids tend to dissolve very slowly at low pH values which occur in gastric conditions. Nicolaides et al. (1999) found that the dissolution of atovaquone and troglitazone in the fasted stomach was reasonably simulated with water. The dissolution of weak acids in the small intestine may be modeled using FaSSIF (Galia et al., 1998). A stronger IVIVC and higher discriminatory power for glibenclamide formulations were obtained by using FaSSIF rather than media without amphiphilies (Lobenberg et al., 2000). In the fed state, the dissolution in the stomach probably has a bigger impact, since the pH will be higher; therefore the inclusion of a gastric dissolution step should be considered, in addition to a fed state dissolution media.

Weak bases have higher solubility at low pH values, and a combination of media simulating the fasted gastric fluids and media simulating the fasted intestinal fluids will be suitable to simulate their dissolution behaviour in the GI tract. The model used should be able to simulate possible precipitation of a weak base when transferred from an acidic to a neutral medium. In the fed state, the impact of dissolution in the stomach could be lower, since the pH will be higher; however, since the transit time is increased along with the level of surfactants, the inclusion of a fed state gastric dissolution step should be considered. For simulating intestinal dissolution, fed state media containing biorelevant amphiphiles should be used. The enhanced dissolution in FeSSIF as compared to FaSSIF of three weak bases studied by Kostewicz et al. (Kostewicz et al., 2002), indicated that the elevated bile salt and lecithin level combined with a lower pH value had an important influence on the dissolution rate.

Prediction of *In Vivo* Formulation Behavior Based on Biorelevant Dissolution Testing: *In Vitro In Vivo* Correlation

Although it is the ultimate goal to predict *in vivo* formulation behavior based on *in vitro* dissolution testing using biorelevant media, the number of studies in which *in vitro–in vivo* correlations (IVIVC) have been described, is limited. Finding a correlation between *in vitro* dissolution testing and *in vivo* performance will be challenging due to the multiple factors which affect the blood concentration–time profile.

Using a mathematical simulation model based on the Noyes-Whitney equation, Nicolaides and coworkers tested the predictability of two 2-step biorelevant dissolution models (fasted and fed state), compared to a compendial model using water and SIF_s media (USP). Solid formulations of four BCS Class II drug compounds [troglitazone (log P 2.7; pKa 6.1 and 12.0), atovaquone (log P 5.1; non-ionizable), sanfetrinem cilexetil (Log P 3.0 and 3.1 (two diastereoisomers; non-ionizable) and GV150013X (log P 5.4; non-ionizable)] were used. *In vitro* dissolution data were compared to clinical studies, using the respective formulation.

Fasted conditions were simulated using water for gastric dissolution and FaSSIF for small intestinal dissolution. The difference between the observed plasma profile and the *in vitro* dissolution profile, only showed a satisfactory correlation for sanfetrinem; however, the superiority of FaSSIF to compendial media was not evident, most probably due to the relatively high solubility of this drug. For atovaquone, the correlation was better for the biorelevant model than the compendial model, but still not acceptable. GV150013X may be a class IV drug (low solubility–low permeability) and this might explain that an IVIVC was not possible.

The fed state was simulated using milk for gastric dissolution and FeSSIF for dissolution in the small intestine. A satisfactory correlation was found for troglitazone and atovaquone, while the profiles of sanfetrinem dissolution in FeSSIF did not correlate with the obtained plasma curve. No difference was seen between the simulated profiles with compendial media and FeSSIF, indicating that sanfetrinem is a border-line low solubility drug and suggesting that gastric emptying might be limiting the intestinal absorption.

This study indicates that dissolution-limited absorption is a prerequisite for IVIVC using dissolution, but one should also realize that modification of the biorelevant media, in terms of composition, but also dissolution conditions, might result in development of a model that does provide for an IVIVC.

Using crude porcine bile extract, Sunesen et al., (2005) identified biorelevant dissolution conditions simulating human bioavailability of danazol (log P 4.6, non-ionizable) in the fasted and fed state. The compositions of the employed media are shown in Table 4. The flow through dissolution apparatus (USP apparatus 4) was used for the *in vitro* studies in order to include the assessment of hydrodynamics as a parameter. A clinical study was designed to focus on the impact of food intake on danazol bioavailability. A level A IVIVC was obtained by nonlinear mathematical modelling describing the relationship between the plasma concentration of danazol and the *in vitro* dissolution under fasted conditions, using the software PDx-IVIVC™ (GloboMax, USA). By simulation of the fasted state, two conditions gave an acceptable correlation for C_{max} and AUC: Fa(low) containing 2.5 mM bile salt and 0.5 mM phospholipids at a flow rate of 32 ml/min and Fa(high) containing 6.3 mM bile salt and 1.25 mM phospholipids, at a flow rate of 8 ml/min. The correlation of observed and predicted plasma concentrations are shown in Figure 6. Both media that obtained predictions errors of less than 10% had a surface tension very close to the one determined in fasted state human intestinal fluids. Data imply that a 4-fold increase in flow rate has the same

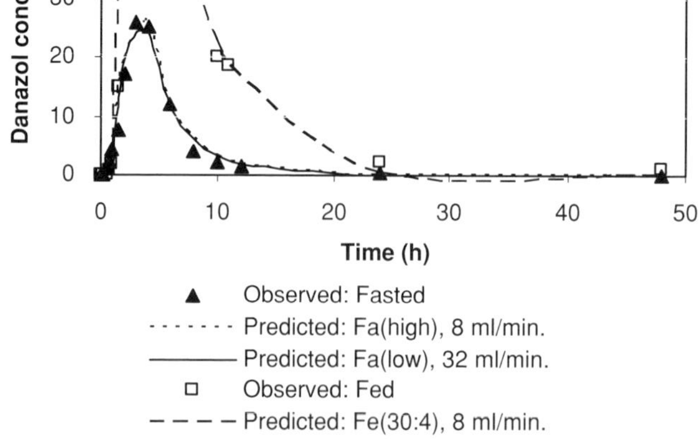

Figure 6. Observed and predicted plasma concentration profiles of danazol in the fasted and the fed state (Sunesen et al., 2005) reprinted with permission from Elsevier.

	Fa(low)	Fa(high)	Fe	Fe(3.75:1)	Fe(7.5:2)	Fe(15:2)	Fe(30:4)
pH	6.8	6.8	5.6 ± 0.1	5.6 ± 0.1	5.4	5.6 ± 0.1	5.5 ± 0.4
KH_2PO_4 (mM)	29	29					
TRIZMA® maleate (mM)			12	12	12	12	12
Bile salts (mM)	2.5	6.3	18.8	18.8	18.8	18.8	18.8
Phospholipids (mM)	0.5	1.25	3.75	3.75	3.75	3.75	3.75
Fatty acids (mM)				3.75	7.5	15	30
Monoglycerides (mM)				1	1	2	4

Table 4. Composition of biorelevant media. Fa and Fe denotes fasted and fed state media, respectively. (Sunesen et al., 2005).

effect on the dissolution profile as a 2.5-times increase in bile salt:phospholipids level.

In order to obtain fed state conditions, the authors increased bile salt and phospholipids levels, and added lipolysis products (fatty acids and monoglycerides) in varying concentrations and ratios (Table 4). The media containing only fed state levels of bile salts and phospholipids did not meet the criteria of max. 10% prediction error. Addition of lipolysis products to the dissolution media introduced a slight bend in the dissolution curve after approximately 30 min., and dissolution levelled off more rapidly than in media without lipolysis products. The change in dissolution rate was particularly distinct for media containing high concentrations of lipolysis products, and the curvature of the dissolution profile was apparently important for the possibility of yielding an IVIVC. The prediction was only satisfactory for the Fe(30:4) media at 8 ml/min. giving a prediction error of 11% for AUC and 7.3% for C_{max}. Figure 6 shows the observed and predicted plasma curves.

Apparently, it is possible to modify the biorelevant media and also the involved hydrodynamics, flow rate so as to achieve IVIVC in this particular situation. However, more studies with different BCS class II compounds are needed.

Conclusion

More and more information on the composition of the human gastric and intestinal fluids in fasted and fed state is being compiled. Based on this data, the possibilities of developing media that are closer to the *in vivo* composition are becoming a reality. However, there is still a long way to go before the impact of all the factors present in the GI tract are elucidated.

Literature describing simulated media for the fasted stomach and small intestine are available. In the fed state, the simulation of the fed stomach seems to be problematic because the composition is highly dependent on the associated meal. Therefore only a few publications of fed state gastric media are available. With regard to media simulating intestinal fluids in the fed state, several attempts have been made to develop suitable media, both with and without the addition of lipolysis product.

It has to be pointed out, however, that, although many media have been developed, only a few publications actually attempt to identify an IVIVC, thus leaving open the question of whether these media actually simulate the events taking place in the GI tract during absorption of a poorly soluble drug compound.

List of Abbreviations

BCS .... Biopharmaceutics Classification Scheme
CMC .... Critical Micellar Concentration
FaSSGF .... Fasted State Simulated Gastric Fluid
FaSSIF .... Fasted State Simulated Intestinal Fluid

FeSSIF..Fested State Simulated Intestinal Fluid
GI..Gastrointestinal
IVIVC..*In vitro in vivo* correlation

References

Abrahamsson B, Albery T, Eriksson A, Gustafsson I, and Sjoberg M. Food Effects on Tablet Disintegration. *Eur J Pharm Sci* 2004; 22: 165–172.

Alvaro D, Angelico M, Cantafora A, Dibiase A, Desantis A, Bracci F, Minervini G, Corradini SG, Attili AF and Capocaccia L. Biliary-Secretion of Phosphatidylcholine and its Molecular-Species in Cholecystectomized T-Tube Patients—Effects of Bile-Acid Hydrophilicity. *Biochem Med Metab Biol* 1986; 36: 125–135.

Armand M, Borel P, Pasquier B, Dubois C, Senft M, Andre M, Peyrot J, Salducci J, and Lairon D. Physicochemical Characteristics of Emulsions During Fat Digestion in Human Stomach and Duodenum. *Am J Physiol-Gastroint Liver Physiol* 1996; 34: G172–G183.

Berne RM and Levy MN. *Principles of Physiology*. Mosby-Year Book, Inc., St Louis, 1996.

Brener W, Hendrix TR, and Mchugh PR. Regulation of the Gastric-Emptying of Glucose. *Gastroenterology* 1983; 85:76–82.

Brouwers J, Tack J, Lammert F, and Augustijns P. Intraluminal Drug and Formulation Behavior and Integration in In Vitro Permeability Estimation: A Case Study with Amprenavir. *J Pharm Sci* 2006; 95: 372–383.

Carey MC. Bile-Salt Structure and Phase-Equilibria in Aqueous Bile-Salt and Bile Salt-Lecithin Systems. *Hepatology* 1984; 4: S138–S150.

Carey MC and Small DM. The Characteristics of Mixed Micellar Solutions with Particular Reference to Bile. *Am J Med* 1970; 49:590–608.

Carriere F, Barrowman JA, Verger R, and Laugier R. Secretion and Contribution to Lipolysis of Gastric and Pancreatic Lipases During a Test Meal in Humans. *Gastroenterology* 1993; 105: 876–888.

Charman WN, Rogge MC, Boddy AW, and Berger BM. Effect of Food and a Monoglyceride Emulsion Formulation on Danazol Bioavailability. *J Clin Pharmacol* 1993; 33: 381–386.

Christensen JO, Schultz K, Mollgaard B, Kristensen HG and Mullertz A. Solubilisation of Poorly Water-Soluble Drugs during In Vitro Lipolysis of Medium- and Long-Chain Triacylglycerols. *Eur J Pharm Sci* 2004; 23: 287–296.

Cruz Moreno MP, Oth M, Deferme S, Lammert F, Tack J, Dressman JB, and Augustijns P. Characterization of Fasted-State Human Intestinal Fluids Collected from Duodenum and Jejunum. *J Pharm Pharmacol* 2006; 58: 1079–1089.

Delgado-Aros S, Cremonini F, Castillo JE, Chial HJ, Burton DD, Ferber I, and Camilleri M. Independent Influences of Body Mass and Gastric Volumes on Satiation in Humans. *Gastroenterology* 2004; 126: 432–440.

Dressman JB, Amidon GL, Reppas C, and Shah VP. Dissolution Testing as a Prognostic Tool for Oral Drug Absorption: Immediate Release Dosage Forms. *Pharm Res* 1998; 15: 11–22.

Dressman JB, Berardi RR, Dermentzoglou LC, Russell TL, Schmaltz SP, Barnett JL, and Jarvenpaa K M. Upper Gastrointestinal (GI) Ph in Young, Healthy-Men and Women. *Pharm Res* 1990; 7: 756–761.

Efentakis M and Dressman JB. Gastric Juice as a Dissolution Medium: Surface Tension and PH. *Eur J Drug Metab Pharmacokin* 1998; 23: 97–102.

Fallingborg J, Christensen LA, Ingemannielsen M, Jacobsen BA, Abildgaard K, and Rasmussen HH. Ph-Profile and Regional Transit Times of the Normal Gut Measured by a Radiotelemetry Device. *Aliment Pharmacol Ther* 1989; 3: 605–613.

Finholt P and Solvang S. Dissolution Kinetics of Drugs in Human Gastric Juice-Role of Surface Tension. *J Pharm Sc* 1968; 57: 1322–1331.

Galia E, Horton J, and Dressman JB. Albendazole Generics—A Comparative In Vitro Study. *Pharm Res* 1999; 16: 1871–1875.

Galia E, Nicolaides E, Horter D, Lobenberg R, Reppas C, and Dressman JB. Evaluation of Various Dissolution Media for Predicting In Vivo Performance of Class I and II Drugs. *Pharm Res* 1998; 15: 698–705.

Hamaguchi T, Shinkuma D, Yamanaka Y, and Mizuno N. Bioavailability of Mefenamic-Acid—Influence of Food and Water-Intake. *J Pharm Sci* 1986; 75: 891–893.

Hay DW and Carey MC. Chemical Species of Lipids in Bile. *Hepatology* 1990; 12: 6S–14S.

Hernell O, Staggers JE, and Carey MC. Physical-Chemical Behavior of Dietary and Biliary Lipids During Intestinal Digestion and Absorption. 2. Phase-Analysis and Aggregation States of Luminal Lipids During Duodenal Fat Digestion in Healthy Adult Human-Beings. *Biochemistry* 1990; 29: 2041–2056.

Hjelm RP, Schteingart CD, Hofmann AF, and Thiyagarajan P. Structure of Conjugated Bile Salt-Fatty Acid-Monoglyceride Mixed Colloids: Studies by Small-Angle Neutron Scattering. *J Phys Chem B* 2000; 104: 197–211.

Hjelm RP, Thiyagarajan P, and Alkanonyuksel H. Organization of Phosphatidylcholine and Bile-Salt in Rodlike Mixed Micelles. *J Phys Chem* 1992; 96: 8653–8661.

Hofmann AF. Chemistry and Enterohepatic Circulation of Bile-Acids. *Hepatology* 1984; 4: S4–S14.

Hofmann AF. The Enterohepatic Circulation of Bile Acids in Man. *Adv Intern Med* 1976; 21: 501–534.

Hofmann AF and Mysels KJ. Bile-Acid Solubility and Precipitation In Vitro and In Vivo—the Role of Conjugation, Ph, and Ca2+ Ions. *J Lipid Res* 1992; 33: 617–626.

Ilardia-Arana D, Kristensen HG, and Mullertz A. Biorelevant Dissolution Media: Aggregation of Amphiphiles and Solubility of Estradiol. *J Pharm Sci* 2006; 95: 248–255.

Jellish WS, Kartha V, Fluder E, and Slogoff S. Effect of Metoclopramide on Gastric Fluid Volumes in Diabetic Patients Who Have Fasted Before Elective Surgery. *Anesthesiology* 2005; 102: 904–909.

Kalantzi L, Goumas K, Kalioras V, Abrahamsson B, Dressman JB, and Reppas C. Characterization of the Human Upper Gastrointestinal Contents Under Conditions Simulating Bioavailability/Bioequivalence Studies. *Pharm Res* 2006; 23: 165–176.

Kalantzi L, Polentarutti B, Albery T, Laitmer DS, Abrahamsson B, Dressman J, and Reppas C. The Delayed Dissolution of Paracetamol Products in the Canine Fed Stomach can be Predicted In Vitro but it does not Affect the Onset of Plasma Levels. *Int J Pharm* 2005; 296: 87–93.

Kararli TT. Gastrointestinal Absorption of Drugs. *Crit Rev Ther Drug Carrier Syst* 1989; 6: 39–86.

Kararli TT. Comparison of the Gastrointestinal Anatomy, Physiology, and Biochemistry of Humans and Commonly used Laboratory-Animals. *Biopharm Drug Disp* 1995; 16: 351–380.

Klein S, Butler J, Hempenstall JM, Reppas C, and Dressman JB. Media to Simulate the Postprandial Stomach I. Matching the Physicochemical Characteristics of Standard Breakfasts. *J Pharm Pharmacol* 2004; 56: 605–610.

Kossena GA, Boyd BJ, Porter CJH, and Charman WN. Separation and Characterization of the Colloidal Phases Produced on Digestion of Common Formulation Lipids and Assessment of their Impact on the Apparent Solubility of Selected Poorly Water-Soluble Drugs. *J Pharm Sci* 2003; 92: 634–648.

Kostewicz ES, Brauns U, Becker R, and Dressman JB. Forecasting the Oral Absorption Behavior of Poorly Soluble Weak Bases using Solubility and Dissolution Studies in Biorelevant Media. *Pharm Res* 2002; 19: 345–349.

Kratohvil JP, Hsu WP, Jacobs MA, Aminabhavi TM, and Mukunoki Y. Concentration-Dependent Aggregation Patterns of Conjugated Bile-Salts in Aqueous Sodium-Chloride Solutions—a Comparison between Sodium Taurodeoxycholate and Sodium Taurocholate. *Colloid Polymer Sci* 1983; 261: 781–785.

Ladas S, Isaacs PET, Qureshi Y, Murphy G, and Sladen G. Effect of Long and Medium Chain Triglyceride Meals on Human Ileal Flow-Rate and Luminal Bile-Acid Concentrations. *Clinical Science* 1981; 61: 39.

Larsson A and Erlansonalbertsson C. The Effect of Pancreatic Procolipase and Colipase on Pancreatic Lipase Activation. *Biochim Biophys Acta* 1991; 1083: 283–288.

Lindahl A, Ungell AL, Knutson L, and Lennernas H. Characterization of Fluids From the Stomach and Proximal Jejunum in Men and Women. *Pharm Res* 1997; 14: 497–502.

Lindenbaum S and Rajagopalan N. Kinetics and Thermodynamics of Dissolution of Lecithin by Bile-Salts. *Hepatology* 1984; 4: S124–S128.

Lipinski CA. Drug-Like Properties and the Causes of Poor Solubility and Poor Permeability. *J Pharmacol Toxicol Met* 2000; 44: 235–249.

Lobenberg R, Kramer J, Shah VP, Amidon GL, and Dressman JB. Dissolution Testing as a Prognostic Tool for Oral Drug Absorption: Dissolution Behavior of Glibenclamide. *Pharm Res* 2000; 17: 439–444.

Lowe ME. The Triglyceride Lipases of the Pancreas. *J Lipid Res* 2002; 43: 2007–2016.

Macheras P, Koupparis M, and Apostolelli E. Dissolution of 4 Controlled-Release Theophylline Formulations in Milk. *Int J Pharm* 1987; 36: 73–79.

Macheras P, Koupparis M, and Tsaprounis C. Drug Dissolution Studies in Milk using the Automated Flow-Injection Serial Dynamic Dialysis Technique. *Int J Pharm* 1986; 33: 125–136.

Mahe S, Huneau JF, Marteau P, Thuillier F, and Tome D. Gastroileal Nitrogen and Electrolyte Movements after Bovine-Milk Ingestion in Humans. *Am J Clin Nut* 1992; 56: 410–416.

Malagelada JR, Longstreth GF, Summerskill WHJ, and Go VLW. Measurement of Gastric Functions During Digestion of Ordinary Solid Meals in Man. *Gastroenterol* 1976; 70: 203–210.

Maltby JR, Pytka S, Watson NC, McTaggart RA, and Fick GH. Drinking 300 ML of Clear Fluid, Two Hours before Surgery has no Effect on Gastric Fluid Volume and PH in Fasting and Non-Fasting Obese Patients. *Can J Anaesthesia* 2004; 51: 111–115.

Marrink SJ and Mark AE. Molecular Dynamics Simulations of Mixed Micelles Modeling Human Bile. *Biochemistry* 2002; 41: 5375–5382.

Moore JG, Christian PE, Brown JA, Brophy CM, Datz F, Taylor A, and Alazraki N. Influence of Meal Weight, Caloric Content and Composition on Gastric-Emptying of Meals in Man. *Clin Res* 1983; 31: A33.

Naylor LJ, Bakatselou V, and Dressman JB. Comparison of the Mechanism of Dissolution of Hydrocortisone in Simple and Mixed Micelle Systems. *Pharmaceutical Res* 1993; 10: 865–870.

Naylor LJ, Bakatselou V, RodriguezHornedo N, Weiner ND, and Dressman JB. Dissolution of Steroids in Bile Salt Solutions is Modified by the Presence of Lecithin. *Eur J Pharm Biopharm* 1995; 41: 346–353.

Nichols JW and Ozarowski J. Sizing of Lecithin-Bile Salt Mixed Micelles by Size-Exclusion High-Performance Liquid-Chromatography. *Biochemistry* 1990; 29: 4600–4606.

Nicolaides E, Galia E, Efthymiopoulos C, Dressman JB, and Reppas C. Forecasting the In Vivo Performance of Four Low Solubility Drugs from their In Vitro Dissolution Data. *Pharm Res* 1999; 16: 1876–1882.

Nicolaides E, Symillides M, Dressman JB, and Reppas C. Biorelevant Dissolution Testing to Predict the Plasma Profile of Lipophilic Drugs after Oral Administration. *Pharm Res* 2001; 18:3 80–388.

Patton JS and Carey MC. Inhibition of Human Pancreatic Lipase-Colipase Activity by Mixed Bile Salt-Phospholipid Micelles. *Am J Physiol* 1981; 241: G328–G336.

Pedersen BL, Brondsted H, Lennernas H, Christensen FN, Mullertz A, and Kristensen HG. Dissolution of Hydrocortisone in Human and Simulated Intestinal Fluids. *Pharm Res* 2000a; 17: 183–189.

Pedersen BL, Mullertz A, Brondsted H, and Kristensen HG. A Comparison of the Solubility of Danazol in Human and Simulated Gastrointestinal Fluids. *Pharm Res* 2000b; 17: 891–894.

Persson EM, Gustafsson AS, Carlsson AS, Nilsson RG, Knutson L, Forsell P, Hanisch G, Lennernas H, and Abrahamsson B. The Effects of Food on the Dissolution of Poorly Soluble Drugs in Human and in Model Small Intestinal Fluids. *Pharm Res* 2005; 22: 2141–2151.

Persson EM, Nilsson RG, Hansson GI, Lofgren LJ, Liback F, Knutson L, Abrahamsson B, and Lennernas H. A Clinical Single-Pass Perfusion Investigation of the Dynamic in Vivo Secretory Response to a Dietary Meal in Human Proximal Small Intestine. *Pharm Res* 2006; 23: 742–751.

Phan CT and Tso P. Intestinal Lipid Absorption and Transport. *Front Biosci* 2001; 6: D299–D319.

Rautureau M, Bisalli A, and Rambaud JC. Bile-Salts and Lipids in Aqueous Intraluminal Phase During the Digestion of A Standard Meal in Normal Man. *Gastroent Clin Biol* 1981; 5: 417–425.

Ricour C and Rey J. Study of Oil and Micellar Phases During Fat Digestion in Normal Child. *Eur J Clin Biol Res* 1970; 15: 287–292.

Rigler MW, Honkanen RE, and Patton JS. Visualization by Freeze Fracture, in Vitro and in Vivo, of the Products of Fat Digestion. *J Lipid Res* 1986; 27: 836–857.

Roda A, Hofmann AF, and Mysels KJ. The Influence of Bile-Salt Structure on Self-Association in Aqueous-Solutions. *J Biol Chem* 1983; 258: 6362–6370.

Schersten T. Formation of Lithogenic Bile in Man. *Digestion* 1973; 9: 540–553.

Staggers JE, Hernell O, Stafford RJ, and Carey MC. Physical-Chemical Behavior of Dietary and Biliary Lipids During Intestinal Digestion and Absorption .1. Phase-Behavior and Aggregation States of Model Lipid Systems Patterned After Aqueous Duodenal Contents of Healthy Adult Human-Beings. *Biochemistry* 1990; 29: 2028–2040.

Sunesen VH, Pedersen BL, Kristensen HG, and Mullertz A. In Vivo in Vitro Correlations for a Poorly Soluble Drug, Danazol, Using the Flow-Through Dissolution Method With Biorelevant Dissolution Media. *Eur J Pharm Sci* 2005; 24: 305–313.

Sunesen VH. *In vitro – in vivo correlations.* PhD thesis, The Danish University of Pharmaceutical Sciences, Copenhagen; 2003, p. 185.

Tangerman A, vanSchaik A, and Vanderhoek EW. Analysis of Conjugated and Unconjugated Bile-Acids in Serum and Jejunal Fluid of Normal Subjects. *Clin Chim Acta* 1986; 159: 123–132.

Vertzoni M, Dressman JB, Butler J, Hempenstall J, and Reppas C. Simulation of Fasting Gastric Conditions and Its Importance for the in Vivo Dissolution of Lipophilic Compounds. *Eur J Pharm Biopharm* 2005; 60: 413–417.

Vertzoni M, Fotaki N, Kostewicz E, Stippler E, Leuner C, Nicolaides E, Dressman JB, and Reppas C. Dissolution Media Simulating the Intralumenal Composition of the Small Intestine: Physiological Issues and Practical Aspects. *J Pharm Pharmacol* 2004; 56: 453–462.

Warren, DB, Chalmers, DK, Hutchison, K, Dang, W, and Pouton, CW. Molecular dynamics simulations of spontaneous bile salt aggregation. *Colloids Surfaces A: Physiochem Eng Aspects* 2006; 280:182–193.

Wiedmann TS and Kamel L. Examination of the Solubilization of Drugs by Bile Salt Micelles. *J Pharm Sci* 2002; 91: 1743–1764.

Zangenberg NH, Mullertz A, Kristensen HG, and Hovgaard L. A Dynamic in Vitro Lipolysis Model I. Controlling the Rate of Lipolysis by Continuous Addition of Calcium. *Eur J Pharm Sci* 2001a; 14:115–122.

Zangenberg NH, Mullertz A, Kristensen HG, and Hovgaard L. A Dynamic in Vitro Lipolysis Model II: Evaluation of the Model. *Eur J Pharm Sci* 2001b; 14:237–244.

7

Selection of Solvent Systems for Membrane-, Cell- and Tissue-Based Permeability Assessment

FRANÇOISE INGELS

Pharmaceutical Sciences R&D, Lilly Development Centre, Mont-Saint-Guibert, Belgium

ANNA-LENA UNGELL

DMPK and Bioanalytical Chemistry, AstraZeneca R&D, Mölndal, Sweden

PATRICK AUGUSTIJNS

Laboratory for Pharmacotechnology and Biopharmacy, Katholieke Universiteit Leuven, Leuven, Belgium

Introduction

In the past, drug discovery was mainly linearly oriented: discovery compounds were primarily synthesized in view of optimizing their pharmacological activity, often resulting in poorly water-soluble and highly lipophilic compounds. In contrast to the traditional discovery paradigm, drug discovery today is more and more practiced in parallel design, where the pharmacological potency is screened concurrently with the initial ADMET profiling of compounds. For a compound to be a successful drug, it should not only have pharmacological activity, but also adequate biopharmaceutical properties enabling it to reach the site of action. This approach permits one to improve the quality of generated drug candidates, to have greater probability of success in the clinic and to guide the chemists in the selection of compounds with better biopharmaceutical properties. In the overall optimization exercise, one of the requirements is the availability of accurate, predictive, relative low-cost and high-throughput techniques for the screening of the biopharmaceutical properties (such as solubility, permeability and stability) of compounds at early developmental phases. One aspect of particular importance is the ability of the drug candidate to cross biological membranes, since permeability for a drug will impact its absorption, distribution and elimination. Several methods are known to be useful in the determination of the transport of drug candidates across a membrane. The different methodologies to assess drug permeation can be sorted into three categories: (1) computational techniques, (2) experimental methodologies using physicochemical surrogates, and (3) experimental methodologies using biological surrogates (Ungell et al., 1997; Balimane et al., 2000; Ungell 2004; Miret et al., 2004). The computational and physicochemical approaches have been shown to be useful for the prediction of passive diffusion of drugs across the intestinal mucosa as well as for the ranking of structurally related compounds. Several molecular descriptors of drug compounds such as the octanol/water partitioning coefficient (Martin, 1981; Palm et al., 1996, 1998), hydrogen bonding capacity (Conradi et al., 1991; Stenberg et al., 1999a), polar surface area (Stenberg et al., 1999b), desolvatation energy (Wright and Painter, 1992) or molecular flexibility [as measured by the number of rotatable bonds (Veber et al., 2002)] have been correlated with intestinal absorption as a single parameter or a combination of parameters collected into one equation (Egan et al. 2000; Lipinski et al., 2001). Some molecular descriptors can be computed by *in silico* approaches or experimentally determined e.g., log P and log D. Measurement of the transport rate through or affinity for artificial membranes [parallel artificial membrane permeability assay (PAMPA), immobilized artificial membrane (IAM) or capillary electrophoresis (Lundahl and Beigi, 1997; Yang et al., 1997; Kansy et al., 1998; Avdeef et al., 2001; Sugano et al. 2003; Örnskov et al., 2005)] can be predictive for the potential of compounds to pass physiological membranes. Finally, many pharmaceutical scientists use biological surrogates to evaluate the oral absorption potential of drug candidates. Different biological models have therefore been proposed, including cultured human intestinal epithelial cells (e.g., Caco-2), dog kidney cells [e.g., Madin-Darby Canine Kidney (MDCK)], *ex vivo* intestinal mucosal tissue in Ussing chambers (Ussing and Zerahn, 1951; Fix, 1996; Ungell et al. 1997; Ungell, 2002), *in situ* perfused intestinal segment (Kim et al., 1993; Annaert et al., 2000) and cannulated animal models (Griffiths et al., 1996; Stewart et al., 1997). Currently, these models are routinely used for permeability assessment in drug discovery to give insight into the absorption process in man (Artursson and Karlsson, 1991; Yee, 1997; Bohets et al., 2001). However, since the models represent different aspects of the barrier properties of the intestine (passive or active transport, enzymes) or may be based on different animal species, they may not be consistent in showing the same transport behavior for compounds.

This chapter focuses on the membrane-, cell- and tissue-based systems used for the physicochemical and biological evaluation of the permeation of drug compounds. It briefly reviews the different models with their intrinsic and practical limitations. Different options to overcome some of these limitations are discussed and decision criteria, which may be helpful for the design of experiments, are proposed.

Permeability Screening in Drug Discovery

Membrane-Based Systems

Simple models assessing the passive transcellular diffusion of drugs often consider the cell membrane as an homogeneous barrier and describe drug permeation as a process of partitioning into the lipid bilayer. This is followed by diffusion into the more hydrophilic cell interior, interstitial fluid and blood compartment. Importantly, because of the presence of proteins in the lipid bilayer, the cell membrane cannot be simply considered as an isotropic system since drug molecules entering the cell membrane will show different diffusion rates in different parts of the lipid bilayer membrane (Bassolino et al., 1996). Drug permeability can be described by a multiple-step model, whereby the drug is (1) first incorporated into the polar head regions of the phospholipids, (2) transported to the interior of the lipid bilayer, (3) passes the other polar surface of the phospholipds and (4) enters the cell. This process is repeated at the abluminal side of the cells (Marrink and Berendsen, 1994; Goodwin et al., 1999). Therefore, two-phase partition studies of drugs, where drug partitioning between water and a lipidic phase is quantified, can provide valuable information on the parameters influencing the mechanism of permeation. Different methodologies to assess drug permeation using membranes have been proposed (Mälkiä et al., 2004), and include the use of liposomes, immobilised artificial membrane (IAM) and parallel artificial membrane permeation assay (PAMPA) (Pidgeon and Venkataram, 1989; Kansy et al., 1998; Balon et al. 1999).

Liposomes were introduced as models for cellular membranes in 1965 (Bangham, 1993) and have been proposed as an alternative for octanol in drug partitioning studies (Balon et al., 1999). It has been pointed that partitioning measured in a liposome-water system may not always reflect transmembrane permeation (Palm et al., 1998) as solutes may associate with the membrane without entering the bilayer (Jacobs and White, 1989). While information of such interactions might be predictive for drug-induced membrane effects (Grinius et al., 2002), the development of a high-throughput method does not seem technically feasible.

Immobilised artificial membrane (IAM) surfaces are constructed of phospholipid analogs which are covalently bound by an alkyl chain to silica particles at high molecular surface densities, mimicking fluid phase phospholipid bilayers (Pidgeon and Venkataram, 1989; Taillardat-Bertschinger et al., 2003). Even given the advantages of IAM, including the ease of automation, the speed of screening, and the possibility to study pH-dependent partitioning, the data obtained with IAM chromatography may not always reflect biological permeation but only drug interactions with the membrane surface. Other shortcomings associated with the method include column variability and instability.

Kansy et al. (1998) proposed the use a parallel artificial membrane permeation assay (PAMPA) as a high-throughput alternative to Caco-2 monolayers for the prediction of passive drug permeation. In the PAMPA approach, aqueous donor and receiver compartments are separated by a filter, which has been impregnated with an oily solution containing lipids and surfactants designed to emulate the cell membrane (n-hexadecane; n-dodecane and lecithin). Since its introduction, researchers have developed a number of variants of the assay. Various PAMPA models have been described, differing mainly in the composition of the lipid constituents and the presence or absence of additives in the experimental aqueous buffers (Kansy et al., 2004). Sugano and coworkers emphasized that an adequate composition of lipids for the membrane formation increased the biorelevance of the model and introduced the bio-mimetic artificial membrane permeability assay or BAMPA (Sugano et al., 2001). In their study, the lipid system is composed of 0.8% L-α-phosphatidylcholine, 0.8% L-α-phosphatidylethanolamine, 0.2% L-α-phosphatidylserine, 0.2% L-α-phosphatidylinositol and 1% cholesterol. The PAMPA assay has been widely promoted as a passive permeability screen focusing on the simulation of transcellular processes; it appears to be a good alternative for cell-based models in absorption screening of research compounds. PAMPA is also a fast and relatively low-cost tool to assess passive permeation. However, the method has some drawbacks, such as the lack of paracellular and active transport mechanisms (Table 1) and remains a simplistic version of the complex *in vivo* situation (Figure 1).

In addition to the intrinsic limitations of these membrane-based systems, the experimental problems (i.e. solubility, adsorption to plastic surfaces, analytical constraints) are similar as those encountered for cell-based systems.

Cell-Based Systems

Since the early nineties, the use of cell cultures to predict drug permeation across the intestinal barrier has gained popularity (Artursson et al., 2001; Ungell and Karlsson, 2003; Balimane and Chong, 2005). From the different adenocarcinoma cell lines [e.g. Caco-2, HT-29 goblet cells, T84 crypt cells, LS180, and 2/4/A1 (rat fetal duodenal cells)], the Caco-2 cell culture model has become a widely used tool for the determination of the intestinal transport characteristics of drug candidates (Hidalgo et al., 1989; Hilgers et al., 1990; Gan and Thakker, 1997; Balimane et al., 2000; Ungell and Karlsson, 2003). Caco-2 cells differentiate spontaneously under standard culturing conditions to form confluent monolayers; they acquire many features of absorptive intestinal cells during culture (Artursson and Karlsson, 1991; Ungell and Karlsson, 2003). Another cell-based model that is frequently used for permeability assessment is the MDCK (Madin-Darby Canine Kidney) cell culture model. MDCK cells differentiate into columnar epithelial cells and form tight junctions when cultured on microporous membranes. A good correlation between permeation of passively absorbed drugs across MDCK cells and Caco-2 cells has been shown (Irvine et al., 1999). However, the potential difference in expression level of transporters between cells of different origin (canine versus human, kidney versus intestine)

Advantages	Disadvantages
Membrane-based systems	
easy	only passive diffusion
high- or medium throughput	variability/stability of columns (IAM)
low cost	
Cell-based systems	
good screening tool	no mucus
high- or medium throughput	no cellular heterogeneity
widely recognized	no natural tissue environment (nerves, blood vessels, regulation by hormones)
cells may have receptors	culture conditions can influence presence of receptors
good correlation with human fraction absorbed	
evaluation of transport mechanism	
evaluation of absorption enhancing strategies	
evaluation of toxicity of compounds	
Tissue-based systems	
evaluation of transport mechanism	limited database
segmental absorption evaluation	viability of tissue
evaluation of absorption enhancing strategies	low throughput
other tissues (nasal, buccal)	limited access to (human) tissue
comparison between species possible	
presence of nerves (e.g., local control, crypt-villus axis)	
transporters and enzymes as *in vivo*	

Table 1. Advantages and disadvantages of various absorption models.

and the lack of knowledge regarding the correlation between the transport of carrier-mediated compounds in the MDCK model and human absorption values, makes this model somewhat less attractive as an early screening tool compared to the Caco-2 model. On the other hand, MDCK cells can easily be transfected to express particular transporter proteins (e.g., P-gp, MDR1 gene product), making the model attractive for mechanistic studies.

In vitro permeation using a monolayer of cultured epithelial cells is currently recognized as a valuable screening tool and has become indispensable in drug discovery and development to estimate intestinal permeability and the absorption potential of drug candidates. Cell culture models can easily be automated and may offer a good compromise between biological relevance and rapidity of screening. In addition, these models can allow evaluation of the 'mechanism' of drug permeation by performing bi-directional transport studies in the presence

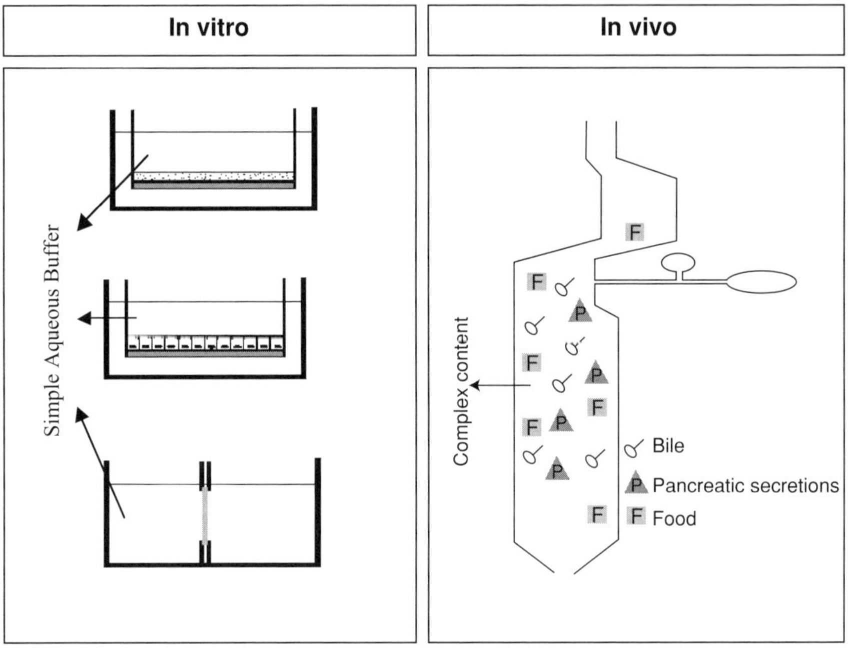

Figure 1. Schematic representation of the complexity of the *in vivo* model versus the simplistic approach followed with *in vitro* models (membrane-, cell- and tissue-based systems).

or absence of specific transporter inhibitors. Finally, cell models can be used to evaluate drug transport in order to obtain a waiver for *in vivo* bio-equivalence studies (Yu et al., 2002). Despite the proven usefulness of cell models as screening and mechanistic tools, they also have their intrinsic limitations precluding the direct translation of *in vitro* data to the *in vivo* situation (Figure 1). These include the fact that cell models are static, are often composed of only one type of cells (no mucus-producing cells, no cellular heterogeneity), suffer from variable expression of enzymes and transporter proteins under different culturing conditions and do not allow to study regional intestinal differences in oral absorption (Table 1).

Recently, Matsson et al. compared artificial membranes (hexadecane-membranes; HDM) with 2 different cell monolayer models [i.e., the rat fetal duodenal cell line (2/4/A1) and the commonly used Caco-2 cell line] with respect to their potential for predicting the fraction absorbed in man; they also successfully built a computer-aided prediction model of drug permeability using the same compound data set (Matsson et al., 2005). The three methods describe the importance of different pathways:

- HDM: passive transcellular transport
- 2/4/A1 cells: paracellular and both active and passive transcellular transport
- Caco-2 cells: active and passive transcellular transport.

This comparison revealed that the transport characteristics of most compounds were well-predicted in all three models. However, HDM failed, as expected, to predict the absorption characteristics of both carrier-mediated and paracellularly transported compounds; the two cell lines predicted most compounds equally well, but 2/4/A1 cells showed slightly better prediction, suggesting that the paracellular component of drug transport is more important than expected (Matsson et al. 2005).

Tissue-Based Systems

Tissue-based systems have been introduced in the late 1940s and early 1950s for the measurement of ion fluxes in isolated frog skin (Ussing and Zerahn, 1951). Since then, these techniques have been extended and applied to a variety of epithelia including intestine, nasal mucosa, buccal mucosa and cultured epithelial cells (Grass and Sweetana, 1988; Hidalgo et al., 1991; Quadros et al., 1996; Reardon, 1996; Ungell et al., 1997; Polentarutti et al., 1999; Annaert et al., 2000). Modified diffusion cells, derived from the original Ussing chamber have been developed for the measurement of compound permeation across intestinal tissue (Grass and Sweetana, 1988; Ungell 1997; Polentarutti et al., 1999). In the modified Ussing chambers, a segment of the small or large intestine is mounted between two half-chambers creating a bicompartmental set-up with tissue viability and integrity monitored during the experiment by the use of two electrode pairs [for current (Isc) and voltage (PD)] (Polentarutti et al., 1999). A continuous gas flow provides the buffer solution with oxygen and carbon dioxide, necessary for maintaining the tissue viability and pH, respectively. Stirring of the solutions can be provided either via a gas lift system (Grass and Sweetana, 1988) or via rotors (Polentarutti et al., 1999). The ability to mount different parts of the intestine in the chambers makes it possible to study regional differences in oral absorption (Grass and Sweetana, 1988; Narawane et al. 1993; Ungell, 1997; Annaert et al., 2000). This technique also permits the study of site-dependent carrier-mediated absorption and/or metabolism of drug compounds (Narawane et al., 1993; Makhey et al., 1998). Additionally, intestinal tissue from both animals and humans can be used, enabling the evaluation of species differences in drug transport (Davies et al., 1982; Rubas et al., 1993; Söderholm et al., 1998; Sjöström et al., 2000). The presence of a crypt-villus axis, a variety of different cell types as well as the presence of enteric nerves makes this model a tool which arguably takes into account many *in vivo* parameters and complexities. However, the capacity is low and dissection of the epithelial tissue and the muscle layers of the intestine from smaller animals like the rat or mouse can be technically difficult (Bohets et al., 2001). However, for dissection of human intestinal tissue both circular and longitudinal muscle layers are more easily removed (Sjöström et al. 2000). The careful excision and dissection of the intestinal tissue is an important part of this method and handling differences can result in large variability between experiments (Ungell 2002).

Experimental Limitations of Absorption Models

Experimental limitations of the absorption models include poor solubility of the drug candidate in the aqueous buffers used, non-specific binding, and the lack of physiological relevance of the commonly used buffers. To guarantee high-throughput, the solvent systems used should also not add challenges to the analysis of the samples.

Poor Aqueous Solubility

Common for all the membrane-, cell- and tissue-based systems used to determine drug permeation, is the need for an incubation solution preserving the viability, functionality and/or integrity of the membranes, cells or tissue. It is also important that the incubation solution does not influence the permeability characteristics of the epithelial membrane or lipid bilayer and that the drug compound is chemically and physically stable and sufficiently soluble in the buffer. In all models, aqueous incubation solutions are used, often with slight modifications depending on the purpose of the study.

Permeability studies in the PAMPA model are often performed by filling the receiver compartment of the 96-well filter plate with a 50 mM sodium phosphate buffer pH 6.0 containing up to 5% DMSO. In some cases surfactants are used in the receiver compartment to ensure that a lipophilic sink is created (Avdeef, 2003).

Transport experiments using the Caco-2 cell culture model are usually performed with, for instance, Hanks' Balanced Salt Solution (HBSS) supplemented with glucose (final concentration of 25 mM) and buffered with 10 mM HEPES (pH 7.4) or 10 mM MES (pH 6.0). In side-by-side diffusion chambers, drug transport studies are initiated by the addition of drug compound to the donor chamber containing oxygenated (O_2/CO_2 95/5) Krebs Bicarbonate Ringer's solution (KBR) (Table 2). Similar buffers slightly differing in the concentrations of Na^+, Ca^{2+}, Mg^{2+}, Cl^- or phosphates have been proposed (Zakelj et al., 2004).

One of the shortcomings in the experimental set-up of these *in vitro* or *ex vivo* models is the limited solubility of drugs in the aqueous buffers used (Figure 2). Aqueous buffers without any additive are usually preferred in screening studies in order to ensure reasonably high throughput, biocompatibility and ease of analysis; however, when screening new drug compounds, poor solubility reduces the possibility to correctly evaluate these compounds experimentally under standard conditions (Neuhoff, 2005).

Non-Specific Binding

A second drawback associated with the classical set-up used in permeation models is the potential for non-specific binding to device surfaces and/or (in)to the membrane, cells and/or tissue (Figure 2). The adsorption to the culture device and/or to the intestinal cell monolayer can lead to an erroneous estimation

Composition (g/L)	HBSS	TM	KBR
$CaCl_2.2H_2O$	0.19	0.19	0.18
KCl	0.40	0.40	0.37
KH_2PO_4	0.06	0.06	
$NaH_2PO_4.H_2O$			0.05
$MgCl_2.6H_2O$	0.10	0.10	0.24
$MgSO_4.7H_2O$	0.10	0.10	
NaCl	8.00	8.00	6.54
$NaHCO_3$	0.35	0.35	2.10
Na_2HPO_4	0.05	0.05	
$Na_2HPO_4.2H_2O$			0.28
D-Glucose	1.00	ad 25 mM	2.07
HEPES/MES		10 mM	
NaOH		ad pH 7.4/6.5	ad pH 7.4
NaGlutamate			0.83
NaPyruvate			0.54
Na_2Fumarate			0.86

Table 2. Composition of Hanks' Balanced Salt Solution (HBSS), HBSS-like transport medium (TM) and modified Krebs Bicarbonate Ringer's Solution (KBR).

(mostly underestimation) of drug transport; binding will also result in poor recovery of the compound. The need to minimize non-specific binding to the cell monolayer or to cell tissue during *in vitro* tests remains controversial, as this binding may also be present in physiological conditions. On the other hand,

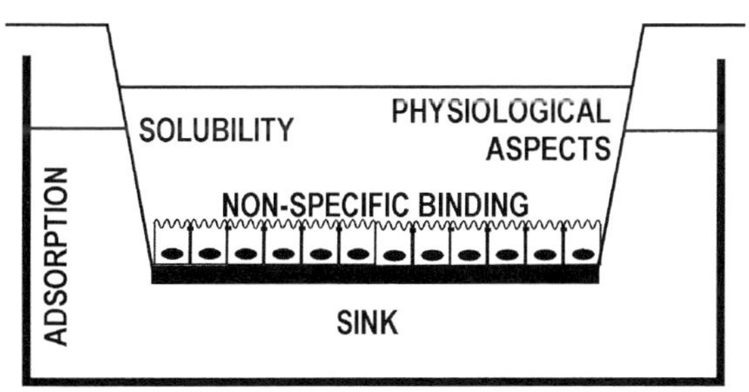

Figure 2. Experimental shortcomings of cell-based assays (may similarly be adapted to membrane-based and tissue based systems).

adsorption to the (plastic) surfaces of the experimental device is clearly an artifact of the *in vitro* system.

For cell-based systems, it has been suggested to correct or compensate for adsorption to the plastic device by:

(1) determining the adsorption to blank inserts, wells and microporous filters (Chan et al., 1996);
(2) pre-treating the device with albumin (Chan et al., 1996);
(3) include a post-experimental wash step with for instance acetonitrile, DMSO or methanol (Augustijns et al., 1993; Chan et al., 1996; Krishna et al., 2001);
(4) calculating the apparent permeability coefficient based on the disappearance of the compound from the donor compartment (Chan et al., 1996).

All these approaches are, however, not without controversy (Ingels and Augustijns, 2003). The inclusion of additives such as surfactants and/or proteins in experimental buffers in donor and/or receiver compartment will be discussed later in this chapter.

Physiological Relevance

The ultimate goal of screening drug candidates for transport characteristics during lead selection and lead optimization is to predict in the most accurate way possible intestinal absorption observed after the drug is orally ingested. While the permeability/absorption models were originally created to mimic intestinal absorption, one can certainly question the physiological relevance of the commonly used buffer solution. The absence of bile salts and the selection of pH will have a major impact on transepithelial drug transport, and thus on the prediction of the oral absorption of the compounds (Figure 3, Figure 4). A lot of

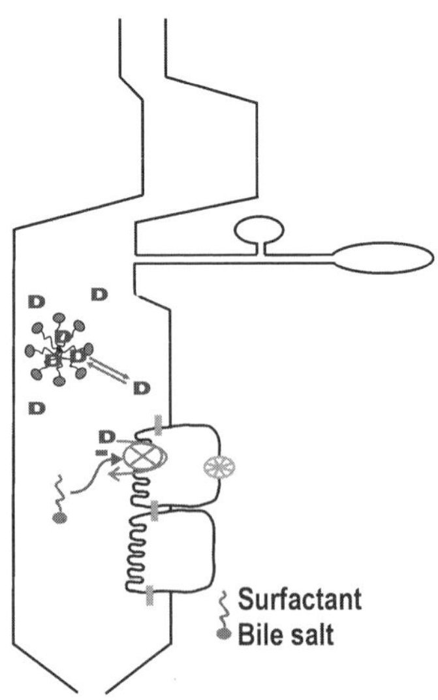

Figure 3. Potential *in vivo* effect of surfactants and/or bile salts on drug transporters.

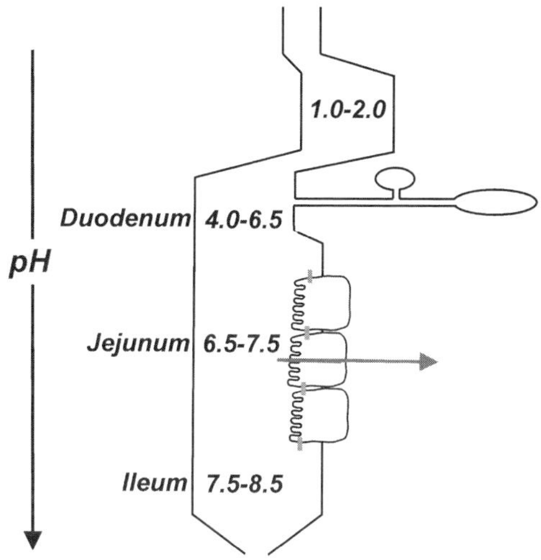

Figure 4. pH-gradient in intestinal lumen.

efforts have been made to 'adapt' the experimental set-up to better mimic the physiological conditions, with respect to pH, the composition of the media and the creation of sink conditions (Figure 2).

The media used for transport experiments in cell-based systems, and more specifically in the Caco-2 cell culture model, were historically buffered at pH 7.4 on both sides of the monolayer. However, it has been reported that the pH in the lumen in the upper gastrointestinal tract ranges from 5.0 to 6.5 under fasted conditions (Gray and Dressman, 1996). This bulk pH increases towards 7.4 in the ileum region (Fallingborg et al. 1989) (Figure 4). The upper small intestine is also characterized by the existence of an acidic microclimate operating at the surface of the epithelial cell layer. This microclimate has been estimated to be up to one pH unit more acidic than the lumenal pH (Lucas, 1983). Excised intestinal segments in Ussing chambers and intestinal tissue in *in situ* perfusions can produce an acidic microclimate close to the membrane and are therefore less prone to be affected by the pH of the bulk solution (Hanisch et al., 1994). In contrast, an acidic microclimate is not present at the surface of cultivated monolayers.

The pH in the cellular interstitial space and blood compartment is known to be about 7.4. The pH present in the GI tract will not only have an impact on the ionization of drugs, and thus on their partitioning capacity, but it will also affect the pH-dependent functionality of various carriers located in the intestinal mucosa (Tsuji and Tamai, 1996). Many of these carriers are also expressed and functionally active in the Caco-2 cell culture model (Hidalgo and Li, 1996; Ogihara et al., 1999; Friedrichsen et al., 2002; Putman et al., 2002a,b). Therefore, the pH of the apical medium can have a critical effect on ionization and transport of drugs. It has for instance been shown that the absorption of weak bases, like beta-blockers, are absorbed better in the ileum region, where the pH is around

8, than in proximal jejunum, where the pH is more acidic (Taylor et al., 1985). Thus, selection of a more physiological pH condition, for instance 6 to mimic proximal jejunum and 8 to mimic the lower parts of the small intestine, would be appropriate for *in vitro* techniques. Suitable universal conditions for screening would include a pH of 6-7 on the donor side and a pH of 7.4 on the receiver side (Yamashita et al., 2000; Krishna et al., 2001). It is obvious that the same reasoning can be applied to the tissue-based systems and the artificial membrane-based systems, with the exception that tissue based system can maintain an acidic microclimate regardless of the pH in the bulk solution (Hanisch et al., 1994).

It is clear that the presence of bile salts in the GI tract can alter drug absorption in many different ways. Bile salts may increase the total solubility of the drug compound in the GI lumen, possibly decrease the free concentration of the drug by micellar encapsulation or modulate the transport characteristics of compounds that are actively transported by the ileal bile acid transporter or by other transporters, such as P-gp (Ingels et al., 2002).

Finally, *in vivo,* drugs absorbed across the intestinal epithelium are immediately carried away by the portal blood, preserving the concentration gradient as the driving force for drug transport, i.e. sink conditions are maintained. The absence of sink conditions during *in vitro* transport experiments will negatively influence the permeability, especially for poorly soluble and/or highly permeable drugs that rapidly reach a saturation concentration in the receiver compartment. The maintenance of sink conditions during the transport experiment will have a major impact, especially when studying the active (efflux) transport mechanisms. Conditions that create high cellular accumulation of drug may result in an overestimation of the secretory transport process (Aungst et al., 2000). On the other hand, higher cellular concentrations could saturate the efflux mechanism involved and so lead to less (concentration-normalized) efflux. Available options to maintain sink conditions include (1) a frequent change of the buffer in the receiver compartment or (2) the inclusion of additives which increase the apparent solubility of the drug (e.g. albumin or surfactants).

Analytical Limitations

When screening for drug permeability in early discovery, processing the large amount of samples requires sensitive, simple and rapid analytical methods. In order to reduce the analytical workload so that no bottleneck is created, different options have been proposed, including the use of radiolabelled compounds (if they are available) or the implementation of generic LC-MS methods. The use of different additives to the media to overcome previously mentioned limitations, should not compromise the analytical method and should not require additional manipulations for sample preparation. Therefore, efforts have been made to propose and use additives that are compatible with the analytical method (discussed in section "Proteins or Micellar Excipients for Sink Conditions"). The use of analysis-friendly additives can result in a significant reduction of cycle time, workload and cost of *in vitro* screening tests; it is therefore important to carefully evaluate this option before initiating a large screening program.

Solvent Systems for Permeability Screening

To improve the performance of *in vitro* absorption models, a multitude of approaches have been proposed to overcome the limitations associated with classical experimental conditions. Unfortunately, the fact that different approaches are being used jeopardizes the ability to compare data generated in different laboratories. For early screening, it is essential to have rapid, widely applicable conditions that do not influence the quality of the data. Major efforts have been directed at establishing widely acceptable conditions. Most of the studies set up to establish new conditions have made use of the Caco-2 cell culture model (Anderberg and Artursson, 1994; Rege et al., 2001). However, as PAMPA gains currency in early drug discovery programs, various studies in which different solvent systems are tested in membrane-based assays, have appeared in recent literature (Bohets et al., 2001). The same approaches proposed for cell-based and membrane-based systems might potentially be extended to the tissue-based systems, although, at this stage, very few systematic studies are available in these systems; a limited set of excipients [e.g., DMSO (Watanabe et al., 2004), PEG (Watanabe et al., 2000), ethanol and solutol (Ungell 2005)] have been demonstrated to be acceptable as solubility enhancers in tissue-based systems.

Table 3 summarizes different additives that have been proposed (1) as solubility enhancers, (2) to reduce adsorption or (3) to increase the biorelevance (pH, bile salts, sink conditions) in *in vitro* absorption models. The maximal tolerable concentration of each was determined by assessing the flux of a hydrophilic marker, the transepithelial electrical resistance (TEER) and/or the release of enzymes or the enzymatic status of the cell monolayers.

Solubility Enhancers

Most of the approaches proposed for solubility enhancement are based on cosolvents, complexation and micellar encapsulation (e.g., DMSO, dimethyl acetamide, albumin, cyclodextrins, TPGS). As shown in Table 3, all these additives have been shown to be compatible with the Caco-2 monolayers up to a well-defined concentration. Co-solvents, such as DMSO (5%), dimethyl acetamide (DMA, 10%), polyethylene glycol (PEG 400, 20%), and 1-methylpyrrolidone (2.5%) were demonstrated to be suitable solubility enhancers for use in the Caco-2 cell culture model (Aungst et al., 2000; Ginski et al., 2000; Hugger et al., 2002a,b; Demirbas and Stavchansky, 2003). Other solubilizers proposed to increase solubility for permeation assays, include vitamin E TPGS [up to 1% (Deferme et al., 2002)], gelucire 44/14 [up to 1%, (Saha and Kou, 2000; Deferme et al., 2002)] and sodium lauryl sulfate [up to 0.004%, (Rege et al., 2001)]. These additives are included in the apical and/or basolateral compartment.

Excipients	Model			Concentration (Maximum tolerable, %)[1]		Purpose			Tested compounds	Reference
	Membrane	Cell	Tissue	Donor	Receptor	Adsorption	Solubility	Biorelevance		
Acetonitrile	×			20			×			Ruell et al., 2004
Albumin		×			4	×		×	HIV protease inhibitors, atenolol, cimetidine, chlorothiazide, chlorpromazine, losartan, phenytoin, tamoxifen, warfarin	Aungst et al., 2000
Albumin, bovine serum		×			1	×		×	Dexamethasone	Yamashita et al., 2000
Albumin, bovine serum		×		4		×	×		SCH-A, SCH-B, SCH-E, mannitol, propranolol, progesterone	Krishna et al., 2001
Albumin, bovine serum		×			4	×		×	NCE Sch 56592, Sch X, Sch Y, PEG 4000, Mannitol, Cimetidine, Phenytoin, Diltiazem, Propranolol	Saha and Kou, 2002
Albumin, bovine serum		×		4	4	×		×	Mannitol, Metoprolol, Propranolol, Inogatran, Digoxin, Quinidine, Doxorubicin	Neuhoff et al. 2005
Albumin, bovine serum		×		4	4	×		×	Felodipin, asimadoline	Neuhoff, 2005
Albumin, human serum		×			4	×		×	Mannitol, Diltiazem	Demirbas and Stavchansky, 2003
Brij 30		×		200 μM			×		Epirubicin	Lo, 2003

CD, alpha-	×	5			×		Cosalane and congeners	Udata et al., 2003
CD, beta-	×	1.8			×		Cosalane and congeners	Udata et al., 2003
CD, dimethyl-beta-	×	5			×		Cosalane and congeners	Udata et al., 2003
CD, gamma-	×	5			×		Cosalane and congeners	Udata et al., 2003
CD, hydroxypropyl-beta-	x[2]	4			×		Mefenamic acid, danazol, phenytoin	Taub et al., 2002
CD, hydroxypropyl-beta-	×	5			×		Nadolol, propranolol, FITC-dextran, lucifer yellow	Takahashi et al., 2002
CD, hydroxypropyl-beta-	×	5			×		Cosalane and congeners	Udata et al., 2003
CD, hydroxypropyl-beta-	×	5			×		Digoxin	Ingels, 2004
Cholic acid	×	10 mM		×		×	Dexamethasone	Yamashita et al., 2000
Cremophor EL	x[2]	1		×			Taxol	Hugger et al., 2002b
Cremophor EL	×	<1		×			Cyclosporin A	Seelballuck et al., 2003
Cremophor EL	×		0.001	×			Cosalane and congeners	Udata et al., 2003
Cremophor EL	×	0.1		×	×	×	Felodipin, asimadoline	Neuhoff, 2005
Culture Medium	×	100	100	×	×	×	Bepiridil	Mathieu et al., 1999
Culture Medium	×		100	×	×	×	UC-781	Deferme et al., 2002
Culture Medium	×	100		×	×	×	UC-781	Deferme et al., 2002
Dimethylacetamide	×	4 (5)	4 (2)		×		HIV protease inhibitors, amiloride, atenolo, losartan, lucifer yellow, phenol red, rhodamine 123, theophylline, warfarin	Aungst et al., 2000

(*continued*)

Excipients	Model			Concentration (Maximum tolerable, %)[1]		Purpose			Tested compounds	Reference
	Membrane	Cell	Tissue	Donor	Receptor	Adsorption	Solubility	Biorelevance		
Dimethylacetamide		×		5			×		Hydrochlorothiazide	Ginski et al., 2000
DMSO	×			2			×		23 compounds	Avdeefet al., 2001
DMSO	×			1			×		Fluoroquinolones	Bermejo et al., 2004
DMSO	×			30			×		Hydrocortisone, propranolol, ketoprofen	Sugano et al., 2001
DMSO		×		0.5			×		Quercetin, taxol, oestradiol 17-beta-D-glucuronide, propranolol	Walgren and Walle, 1999
DMSO		×		0.1			×		Hydrochlorothiazide	Ginski et al., 2000
DMSO		×		2			×		Dexamethasone	Yamashita et al., 2000
DMSO		×		1	1		×		SCH-A, SCH-B, SCH-E, mannitol, propranolol, progesterone	Krishna et al., 2001
DMSO		x[2]		1			×		Mefenamic acid, danazol, phenytoin	Taub et al., 2002
DMSO		×		10			×		Mannitol, Diltiazem	Demirbas and Stavchansky, 2003
DMSO			×	5			×		Sulfasalazine, furosemide, atenolcl, acetaminophen, propranolol, theophylline, diclofenac, metoprolol, antipyrine, naproxen	Watanabe et al., 2004

Docusate sodium		×		0.002	×		Furosemide, atenolol, mannitol, ranitidine, acyclovir, cimetidine, hydroclorothiazide	Rege et al., 2001
Ethanol	×			30	×		Hydrocortisone, propranolol, ketoprofen	Sugano et al., 2001
Ethanol		×		0.5	×		Quercetin, taxol, oestradiol 17-beta-D-glucuronide, propranolol	Walgren and Walle, 1999
Ethanol		×		5	×		Hydrochlorothiazide	Ginski et al., 2000
Ethanol		×		5	×		Dexamethasone	Yamashita et al., 2000
Ethanol		x^2		3	×		Mefenamic acid, danazol, phenytoin	Taub et al., 2002
Ethanol			×	1	×		Mannitol	Ungell, 2005
FaSSIF		×		100	×	×	Theophylline, phenylalanine, cyclosporin A, sodium fluorescein	Ingels et al., 2002
FaSSIF		×		100	×	×	20 compounds	Ingels, 2004
Gelucire 44/14		×		1	×		NCE Sch 56592, Sch X, Sch Y, PEG 4000, Mannitol, Cimetidine, Phenytoin, Diltiazem, Propranolol	Saha and Kou, 2000
Gelucire 44/14		×		1	×		UC-781	Deferme et al., 2002
HCO-40		×		10			Nadolol, propranolol, FITC-dextran, lucifer yellow	Takahashi et al., 2002
HPMC		×		0.01			Furosemide, atenolol, mannitol, ranitidine, acyclovir, cimetidine, hydroclorothiazide	Rege et al., 2001

(continued)

Excipients	Model			Concentration (Maximum tolerable, %)[1]		Purpose			Tested compounds	Reference
	Membrane	Cell	Tissue	Donor	Receptor	Adsorption	Solubility	Biorelevance		
Human plasma		×			100 (BL)	×		×	Quercetin, taxol, oestradiol 17-beta-D-glucuronide, propranolol	Walgren and Walle, 1999
Human plasma		×			100 (BL)	×		×	Furosemide, verapamil	Chung et al., 2001
Lactose		×		0.2					Furosemide, atenolol, mannitol, ranitidine, acyclovir, cimetidine, hydroclorothiazide	Rege et al., 2001
Methanol		×		1	1		×		SCH-A, SCH-B, SCH-E, mannitol, propranolol, progesterone	Krishna et al., 2001
Methyl cellulose		×		0.4					jNCE Sch 56592, Sch X, Sch Y, PEG 4000, Mannitol, Cimetidine, Phenytoin, Diltiazem, Propranolol	Saha and Kou, 2002
1-Methyl-2-pyrrolidone		×		2.5			×		Hydrochlorothiazide	Ginski et al., 2000
1-Methyl-2-pyrrolidone		×		0.5			×		Digoxin	Ingels, 2004
Myrj 52		×		200 μM			×		Epirubicin	Lo, 2003
PEG 300		×		20	20		×		Testosterone, Taxol, Doxorubicin	Huggeret al., 2002a
PEG 300		x[2]		20	20		×		Taxol	Hugger et al., 2002b

PEG 300	×	1	×	NCE Sch 56592, Sch X, Sch Y, PEG 4000, Mannitol, Cimetidine, Phenytoin, Diltiazem, Propranolol	Saha and Kou, 2003
PEG 400	×	30	×	Hydrocortisone, propranolol, ketoprofen	Sugano et al., 2001
PEG 400	×	1	×	Hydrochlorothiazide	Ginski et al., 2000
PEG 400	×	2	×	Dexamethasone	Yamashitaet al., 2000
PEG 400	×	1.5	×	Furosemide, atenolol, mannitol, ranitidine, acyclovir, cimetidine, hydroclorothiazide	Rege et al., 2001
PEG 400	×	1	×	Mannitol	Ungell, 2005
PEG 400	×	5	×	Nadolol, propranolol, FITC-dextran, lucifer yellow	Takahashi et al., 2002
PEG 600	×	10	×		Watanabe et al., 2000
Pluronic F68	×	1	×	NCE Sch 56592, Sch X, Sch Y, PEG 4000, Mannitol, Cimetidine, Phenytoin, Diltiazem, Propranolol	Saha and Kou, 2000
Pluronic F68	×	10	×	Nadolol, propranolol, FITC-dextran, lucifer yellow	Takahashi et al., 2002
Pluronic F68	×	<8	×	Cyclosporin A	Seelballucket al., 2003
Pluronic L81	×	<0.012	×	Cyclosporin A	Seelballuck et al., 2003

(continued)

Excipients	Model			Concentration (Maximum tolerable, %)[1]		Purpose			Tested compounds	Reference
	Membrane	Cell	Tissue	Donor	Receptor	Adsorption	Solubility	Biorelevance		
Pluronic P85		×		5			×		Fluorescein, doxorubicin, etoposide, taxol, valoproic acid, 3′-azido-3′-deoxythymidine (AZT), loperamide	Batrakovaet al., 1999
Pluronic P85		×		<0.06			×		Cyclosporin A	Seelballucket al., 2003
Polysorbate 20		×		0.1			×		Metformin	Dimitrijevic et al., 2000
Polysorbate 60		×		0.1			×		Metformin	Dimitrijevicet al., 2000
Polysorbate 85		×		0.1			×		Metformin	Dimitrijevic et al., 2000
Povidone		×		1			×		NCE Sch 56592, Sch X, Sch Y, PEG 4000, Mannitol, Cimetidine, Phenytoin, Diltiazem, Propranolol	Saha and Kou, 2000
Propylene carbonate		×		0.1			×		Hydrochlorothiazide	Ginski et al., 2000

Propylene glycol	×	1.5	×		Furosemide, atenolol, mannitol, ranitidine, acyclovir, cimetidine, hydroclorothiazide	Rege et al., 2001
Propylene glycol	×	20	×		Nadolol, propranolol, FITC-dextran, lucifer yellow	Takahashi et al., 2002
Propylene glycol	×	20	×		Nadolol, propranolol, FITC-dextran, lucifer yellow	Takahashi et al., 2002
Propylene glycol + PEG 400	×	1.5 + 2	×		Digoxin	Ingels, 2004
Propylene glycol/ Tween 80 3/2	×	1	×		NCE Sch 56592, Sch X, Sch Y, PEG 4000, Mannitol, Cimetidine, Phenytoin, Diltiazem, Propranolol	Saha and Kou, 2000
Sodium caprate	×	10 mM	×	×	Epirubicin	Lo and Huang, 2000
Sodium cholate	×	8 mM	×	×	Mannitol, PEG 4000, PEG 9000	Meaney and O'Driscoll, 2000
Sodium deoxycholate	×	1.2 mM	×	×	Epirubicin	Lo and Huang, 2000
Sodium lauryl sulphate	×	0.004	×		Furosemide, atenolol, mannitol, ranitidine, acyclovir, cimetidine, hydroclorothiazide	Rege et al., 2001
Sodium lauryl sulphate	×	0.003	×		Epirubicin	Lo, 2003
Sodium taurocholate	×	0.5	×	×	Mannitol, PEG 4000, PEG 9000	Meaney and O'Driscoll, 2000
Sodium taurocholate	x^2	0.25	×	×	Mefenamic acid, danazol, phenytoin	Taub et al., 2002

(continued)

Excipients	Model			Concentration (Maximum tolerable, %)[1]		Purpose			Tested compounds	Reference
	Membrane	Cell	Tissue	Donor	Receptor	Adsorption	Solubility	Biorelevance		
Solulan C24		×		0.005			×		Metformin	Dimitrijevicet al., 2000
Solulan 16		×		0.005			×		Metformin	Dimitrijevic et al., 2000
Solutol HS15		×		0.1			×		Digoxin	Ingels, 2004
Solutol HS15			×	0.3			×		Mannitol	Ungell, 2005
Tetraglycol		×		5			×		Hydrochlorothiazide	Ginski et al., 2000
Taurocholic acid		×		10 mM			×		Dexamethasone	Yamashita et al., 2000
Transcutol P		×		10			×		Nadolol, propranolol, FITC-dextran, lucifer yellow	Takahashi et al., 2002
Tween 20		×		200 µM			×		Epirubicin	Lo, 2003
Tween 80		×		0.5			×		NCE Sch 56592, Sch X, Sch Y, PEG 4000, Mannitol, Cimetidine, Phenytoin, Diltiazem, Propranolol	Saha and Kou, 2001
Tween 80		×		0.5			×		Taxol	Hugger et al., 2002b
Tween 80		x[2]		0.5			×		Taxol	Hugger et al., 2002b

Tween 80	×	5		×	Nadolol, propranolol, FITC-dextran, lucifer yellow	Takahashi et al., 2002
Tween 80	×	200 μM		×	Epirubicin	Lo, 2003
Tween 80	×	0.18		×	Furosemide, atenolol, mannitol, ranitidine, acyclovir, cimetidine, hydroclorothiazide	Rege et al., 2001
Vitamin E TPGS	×	0.05		×	Hydrochlorothiazide	Ginski et al., 2000
Vitamin E TPGS	×	1		×	UC-781	Deferme et al., 2002
Vitamin E TPGS	×		0.5		UC-781	Deferme et al., 2002
Vitamin E TPGS	×	0.2			Amprenavir	Yu et al., 1999

Table 3. Overview of the additives that have been proposed as solubility enhancers, to reduce adsorption and non-specific binding and/or to increase the biorelevance of *in vitro* absorption models. The table refers to the model used to evaluate the compatibility, the maximum tolerable concentration, the compartment in which the additive is included, the expected effect, the compounds tested in presence of the additive and the reference.

[1]as defined by paracellular marker compound, TEER measurements and/or enzymatical assays (MTT)

[2]MDCK cells

The concentrations mentioned should however be interpreted with care; the incubation time, pH, side of administration and the presence of other additives also have to be considered. The compatibility with the cell monolayer, and thus the acceptability of a certain additive, is usually based on an assessment of the transport of a hydrophilic marker (e.g. ^{14}C-mannitol or Lucifer yellow), TEER measurement and/or evaluation of the viability of the cells (e.g. MTT test) (Mosmann, 1983; Tada et al., 1986; Scudiero et al., 1988). More systematic studies of the impact of the different solubility enhancing additives on permeability are lacking and there is still a poor understanding of the possible impact of these additives on the different transport mechanisms of drugs. Various excipients have been demonstrated to interfere with active carrier mechanisms expressed in cell culture models. For instance, polyethylene glycol and vitamin E TPGS appeared to modulate (inhibition) the functionality of P-gp as observed by the impact of these excipients on the transport of taxol or doxorubicin (PEG 300) and rhodamine 123 and amprenavir (vitamin E TPGS) (Nerurkar et al., 1996; Dintaman and Silverman, 1999; Yu et al., 1999; Hugger et al., 2002b; Bogman et al., 2005; Brouwers et al., 2006). It is sometimes difficult to anticipate whether effects observed *in vitro* can be translated into an *in vivo* situation (Figure 3). Tween 80 and other surfactants have also been reported to have a modulatory effect on different efflux transporters, probably due to a disruption of the membrane structure (Oberle, 1995; Dimitrijevic et al., 2000; Rege et al., 2001). Inconsistency also exists with respect to the acceptability of bile salts as solubility enhancers. In a study by Meaney and O'Driscoll (2000), sodium taurocholate is shown to be compatible with the Caco-2 monolayer up to 10 mM, while in a study published by Udata and coworkers (2003), all of the tested bile salts (including sodium glycocholate, sodium taurocholate, sodium desoxycholate) had a negative effect on monolayer integrity (increase in the transport of mannitol), suggesting that these bile salts are toxic to the cell monolayer even at the lowest concentrations tested (10 mM, 10 mM and 5 mM, respectively).

Similar approaches to enhance solubility have been studied in the PAMPA system. For instance, Kansy et al. (2001) explored the use of glycocholic acid to solubilize compounds. As PAMPA is a completely artificial system, it is expected that, compared to cell-based models, higher concentrations of cosolvents can be used. Sugano and coworkers reported that DMSO, ethanol and PEG 400 could be used up to 30% without causing disruptions of the lipid layer (Sugano et al., 2001). An effect of these cosolvents on the physicochemical properties of the test compounds (e.g. impact on pKa) may however lead to an unpredictable effect on drug permeability (Sugano et al., 2001). Currently, DMSO is commonly used as a cosolvent in the PAMPA system at a concentration of 1 to 2%.

pH, Bile Salts and Proteins for Biorelevance

pH Adjustment

The importance of the pH to mimic the physiological conditions has been introduced in the section on "Physiological Relevance". It has been demonstrated that media buffered to a pH varying from 5 to 8 were compatible with the Caco-2 cell monolayer (Palm et al., 1999; Neuhoff et al., 2003, 2005). PAMPA models have even been shown to be compatible with a wider range of pH, i.e., from 4 to 10 (Bermejo et al., 2004).

Yamashita and coworkers studied the effect of medium pH on the transport rate of several passively and actively transported drugs in Caco-2 cells (including antipyrine, theophylline, hydrochlorothiazide, atenolol, terbutaline, nadolol, salicylic acid, furosemide and cephalexin). Transport was studied in the absence (apical and basolateral medium buffered at pH 7.4) or presence (apical medium buffered at pH 6.0, basolateral medium buffered at pH 7.4) of a pH-gradient. The observed differences in apparent permeability were attributed to both difference in drug partitioning and modification of carrier-mediated transport (Yamashita et al., 1997). A similar study was run in the PAMPA model (Sugano et al., 2001). The permeation of 30 model compounds, expected to be passively transcellularly transported, was assessed as a function of pH. For both models, it was concluded that a better prediction of the fraction absorbed in humans would be obtained under pH-gradient conditions. Although useful for standard screening tests, the value of this set-up can be questioned when performing mechanistic studies in the Caco-2 system. For instance, when performing bi-directional transport studies to explore the interaction with a carrier, the pH of the donor and receiver solvents should be similar in order to prevent bias created by the experimental conditions. Recently, the impact of a pH-gradient when performing bi-directional transport experiments of weak bases (Neuhoff et al., 2003) and weak acids (Neuhoff et al., 2005) has been discussed. The use of a system in which different pH values are maintained at the apical and basolateral side will lead to different concentrations of uncharged drug species, resulting in asymmetry in bi-directional transport and, for weak bases, a 'false' efflux component (Ungell et al., 2002; Neuhoff, 2005). This asymmetry in transport rate occurs independently of active transport mechanisms; it is therefore not advisable to use an efflux ratio obtained in a pH-gradient system as an indication of the involvement of an active transport system (Neuhoff et al., 2003; Volpe, 2004). Similarly, when transport of weak acids is studied, a pH gradient over the membrane will create a false asymmetry over the membrane which is not associated with interaction with transporters and can be interpreted as false uptake (Neuhoff et al., 2005). However, many acids are taken up actively via H^+-dependent systems (e.g. MCT carrier system) and, thus, when studying weak acids, two different pH systems have to be used: one with and another without a gradient in order to rule out an inappropriate interpretation (Neuhoff et al., 2005). A compromise pH of 7.0 at both sides of the monolayer (Yamashita et al., 1997) has been suggested when performing bi-directional mechanistic studies. However, when assessing pH-dependent carrier systems, the pH-gradient remains key for mechanistic studies [e.g. human peptide transporter hPEPT1, MCT, PAT1(imino) and thiamine transporter] (Steffansen et al. 2004). A possible pH-dependence should also be considered in studies of drug/drug-interactions involving P-glycoprotein or proton dependent uptake transporters (Ungell et al., 2002; Neuhoff et al.,

2003; Neuhoff et al., 2005). For these reasons, we suggest to work under pH-gradient conditions for the absorptive ranking of compounds. When performing more mechanistic polarity studies, we suggest to select the same pH on both sides during the first screening phase and elaborate on the differential pH conditions when the compounds are moving forward through the later development stages (lower throughput).

Bile Salts

Sodium taurocholate, sodium cholate, sodium taurodeoxycholate, sodium taurodihydrofusidate and other bile salts have been studied for their effect on the epithelial integrity of cell monolayers and on the transport of model compounds in Caco-2 cells (Anderberg et al., 1992; Lo and Huang, 2000). As described previously and summarized in Table 3, different bile salts were suggested as solubility enhancers (Lo and Huang, 2000, Meaney and O'Driscoll, 2000; Taub et al., 2001). More complex formulations containing bile salts as fasted state simulated intestinal fluid (FaSSIF) have also been evaluated in the Caco-2 model. It was demonstrated that this buffer was compatible with the Caco-2 monolayer for a period of at least two hours, without affecting the transport of theophyllin (passive diffusion) and phenylalanine (active absorptive transport). However, a concentration-dependent P-gp inhibitory effect of sodium taurocholate (present in FaSSIF) when assessing cyclosporin A transport was demonstrated (Ingels et al., 2002). Although the impact of using bile salts in Caco-2 experiments is not yet fully understood, we could expect that the use of bile salts included in the apical solvent could increase the solubilization of poorly water-soluble drugs in the Caco-2 cell culture model. In addition, it could also improve the physiological relevance of the model (Figure 3). Although we are convinced that the use of bile salts could become the preferred option in future especially for 'ranking' purposes, several aspects of this approach remain to be investigated, including the exact concentration and the type of bile salt(s) that should be used. In addition, the high cost of bile salts could also preclude the use of such buffers in high-throughput screening.

Proteins or Micellar Additives for the Creation of Sink Conditions

In order to preserve sink conditions in the static *in vitro* models, it has been proposed to include additives in the receiver compartment that are able to decrease the free drug concentration (Figure 5). Among all the proposed alternatives, the ideal option should be suitable for high-throughput screening and not increase the workload associated with the Caco-2 transport assay. The inclusion of serum albumin (used as such or as present in culture medium) in the receiver compartment has been demonstrated to modify the transport properties of drugs (Mathieu et al., 1999; Walgren and Walle, 1999; Aungst et al., 2000; Yamashita et al., 2000; Krishna et al., 2001; Deferme et al., 2002; Saha and Kou 2002; Demirbas and Stavchansky, 2003; Neuhoff, 2005). The presence of albumin in the basolateral compartment can promote drug partitioning from the cell monolayer into the basolateral compartment; in addition, it can also prevent

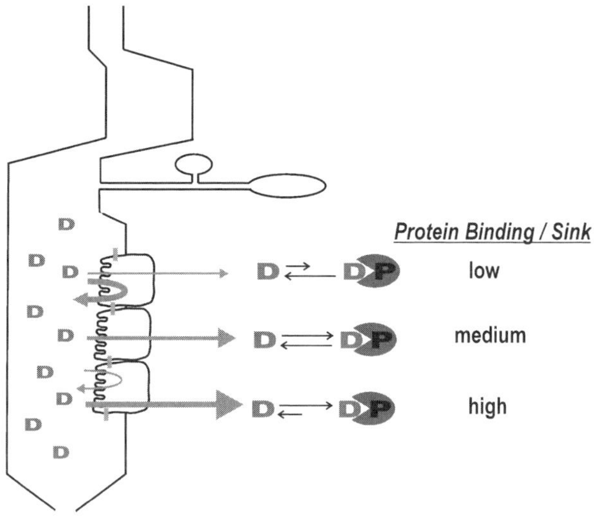

Figure 5. Impact of protein binding on drug transport.

the non-specific binding of the drug compounds to the plastic material. Inclusion of 4% bovine serum albumin (BSA) in the basolateral compartment has been proposed to be most relevant for *in vivo* conditions (Aungst et al., 2000, Saha and Kou, 2002; Demirbas and Stavchansky, 2003; Neuhoff, 2005). The inclusion of proteins in the receiver compartment can, however, have important consequences. It was for instance shown that the inclusion of BSA in the receiver compartment may affect the BCS permeability ranking of highly lipophilic new chemical entities (Saha and Kou, 2002). For protein-bound compounds, it was shown that failure to consider plasma binding could result in an overestimation of the basolateral to apical flux and so to a misleading net flux calculation. Under classic test conditions, furosemide and verapamil were shown to have a net apical secretion ratio of 4.2 and 1.3, respectively (Chung et al., 2001). In presence of human plasma in the basolateral compartment, the increase of the AP-to-BL and decrease of the BL-to-AP transport resulted in a reduction of the net apical secretion ratio, suggesting a much less significant efflux (Neuhoff, 2005). However, taking into account the low fraction of unbound drug in the BL compartment (e.g. low concentration of drug applied), the efficiency of the efflux transporter may increase, resulting in lower AP-to-BL and higher BL-to-AP transport (Neuhoff, 2005). The use of more physiologically relevant conditions is expected to be more in agreement with the *in vivo* absorption of many drugs and may partly explain why the efflux transporter in the intestine apparently has no or very little effect on limiting the *in vivo* oral absorption of many drugs (Chung and Chiou, 1999).

The studies mentioned so far all support the utilization of serum proteins at the basolateral side to increase the quality of the model. However, to our knowledge, no *in vitro/in vivo* correlation studies have been performed to assess the impact of the absence or presence of albumin. An important disadvantage of using protein containing media is the additional procedure needed for sample preparation, which is an important drawback for high-throughput screening assays. In addition, if transport studies are conducted in diffusion chambers, where stirring is provided by gas-bubbling, the surfactant property of BSA gives rise to an undesirable frothing and foaming effect (Saha and Kou, 2002). To overcome these issues, efforts have been made to find alternatives, so that sink conditions could be maintained without using materials which could interfere with the analytical method used. When assessing the transport of the antiviral compound UC-781 in Caco-2 cells, Deferme et al. demonstrated that vitamin E TPGS included in the basolateral compartment enhanced the transport rate of the compound (Deferme et al., 2002) while no additional sample preparation was needed. The use of analysis-friendly additives requires that their compatibility with the absorption model needs to be thoroughly studied. It is known that some of these additives can cause conformational changes in membrane bound proteins and that they may also alter membrane fluidity. Both effects may change drug permeation (Nerurkar et al., 1996; Rege et al., 2002). However, by only including these surfactants in the basolateral compartment, one might expect that the interaction with apically localized (efflux) carriers is limited (Deferme et al., 2002). A few studies indicate that non-ionic surfactants not only bind, adhere and incorporate into the lipid membrane (causing changes in membrane fluidity), but also cross the lipid bilayer into the cell interior and cause ATP depletion (Batrakova et al. 2001, Kabanov et al. 2003). This mechanism has been proposed to partly explain the P-gp inhibition by surfactants in chemotherapy (Kabanov et al., 2003). At present, no systematic studies comparing the different options and evaluating the impact of such additives on the *in vitro/in vivo* correlation, have been performed.

Also in the PAMPA model, alternatives to BSA have been explored to improve the biorelevance of the model. To overcome the adsorption and/or absence of sink conditions, different additives that do not require an additional step of sample preparation as compared with the addition of albumin, have been proposed. Recently, the use of Double-Sink PAMPA (DS-PAMPA) was proposed as biorelevant alternative to the classic PAMPA methodology (Avdeef, 2003). In DS-PAMPA, a non-specific binding agent (lipophilic sink) was included in the receiver compartment to create sink conditions.

Design of Permeability Screening Experiments

From the different examples reviewed, it is clear that when performing absorption studies, the experimental conditions will have a major impact on the generated data. The design of experimental conditions for PAMPA assays in early screening can probably be considered as straightforward. The biorelevant DS-PAMPA model may certainly be a good option for first screening experiments addressing passive transcellular absorption. The use of other physicochemical methods, such as calculation of lipophilicity or molecular descriptors or the use of chromatography systems, may be used as well for early ranking of passive membrane permeability, but their relevance in later stages of discovery is low. Co-solvents (at low concentrations) can be used to improve solubility of poorly soluble compounds. If cell-based systems are used for the evaluation of the absorption potential of drug compounds, it is necessary to carefully design the experiment based on the purpose of the assessment (absorption estimation, transport mechanism, BCS classification). A rational design of experiments will result in higher-quality data which will have more value for further decisions. Mechanistic approaches are in most cases not conceivable during early screening phases. Instead, a sequential approach is often used, involving the use of high or medium throughput, less resource demanding models for primary screens followed by low-throughput more predictive models for secondary screens and mechanistic studies (Figure 6). For the ranking of compounds during the early screening phase (based on their apical to basolateral transport), DMSO ($\leq$1%) can be used in standard screening procedures. This is currently the option in high-throughput transport screening of poorly water-soluble drugs. If compounds are not soluble in DMSO 1%, dimethylacetamide (DMA) (3%) or 1-methyl-2-pyrrolidone (2.5%) are good alternatives. In the "ranking mode", it is recommended to work under pH-gradient conditions to mimic physiological conditions. When performing more in-depth permeation studies, the experimental set-up will be based on the purpose of the study. In order to elucidate the mechanism of transport, a bi-directional study without a pH gradient may be considered. Additional assays evaluating the pH or concentration effect on the transport may be required. For the classification of drug compounds following the BCS guidelines (CDER, 2000), a thorough validation of the model is necessary. The United States Pharmacopeial Convention (USP) proposes the use of reference molecules to validate the system. An additional in-house evaluation of the model is needed to ensure the method suitability. Each laboratory should have a reference set of 10 to 20 different compounds (high and low permeability model drugs and model drugs for active transport carriers) and edit internal specifications and acceptance criteria for their model. If the physiological relevance of the media is taken into account, we suggest the use of biorelevant apical media [e.g. bile salt-containing solvent system buffered at pH 6.5 or FaSSIF] and analysis-friendly basolateral media [solvent buffered at pH 7.4 containing a sink condition creating additive (e.g. vitamin E TPGS or albumin)]. We have previously validated the use of FaSSIF as an apical solvent (Ingels et al., 2004). The practical applicability and biorelevance of such a set-up remains, however, somewhat unclear, mainly based on the poor understanding of potential side effects of bile salts on efflux carriers. The use of bile salts in the *in vitro* system could mask the effect of efflux carriers on total drug absorption; on the other hand, considering the presence of bile salts in intestinal fluids, the contribution of efflux carriers may be overestimated when using HBSS-like buffers without bile salts. The effect of vitamin E TPGS on apical efflux carriers should be minimal as this is only included in the basolateral compartment.

The inclusion of albumin in the receiver compartment is less advisable due to the additional sample preparation required; however, if a generic high

Development phase	PAMPA	Cell-based system		Tissue-based system
	Rapid Screening Mode			Rapid Screening Mode
Early Screening ("Ranking")	**Donor:** 2% DMSO*, pH 6.5 **Receiver:** Sink condition creating excipient**, pH 7.4	**Donor:** 1% DMSO*, pH 6.5 **Receiver:** 1% DMSO*, pH 7.4		**Donor:** 1% DMSO***, pH 7.4 **Receiver:** 1% DMSO***, pH 7.4
	In-depth Permeation Assessment		Transport Mechanism	In-depth Permeation Assessment
	pH-dependent permeation study	Biorelevant permeation study	Bidirectional study	All Studies
Extensive Permeability Assessment, Mechanistic Studies	**Donor:** 2% DMSO*, pH 4.0 - 10.0 **Receiver:** 2% DMSO*, pH 7.4	**Donor:** FaSSIF, pH 6.5 **Receiver:** Albumin 4 % or Vitamin E TPGS 0.5%	**Donor:** 1% DMSO*, pH 7.0 **Receiver:** 1% DMSO*, pH 7.0	**Donor:** 1% DMSO***, pH 7.4 **Receiver:** 1% DMSO***, pH 7.4 < 0,5% BSA can be used
			pH-dependent transport study	
			Donor: 1% DMSO*, pH 5.0 - 8.0 **Receiver:** 1% DMSO*, pH 7.4	

* DMSO 1 or 2 % can be replaced by dimethylacetamide 4-5 % or 1-methylpyrrolidone 2.5%

** One can consider to extend the use of e.g., vitamin E TPGS 0.5 % to the PAMPA technique. A ready-to-use solution for the receiver compartment is currently available (DS-PAMPA™, pIon Inc, MA, USA)

*** DMSO can be replaced by Solutol 0,3%

Figure 6. Guideline for assay design of membrane-, cell- and tissue-based systems, depending on the development phase and expected output.

throughput extraction procedure is available, the use of albumin remains a valuable option.

Conclusion

Membrane-, cell- and tissue-based systems have demonstrated their usefulness in the evaluation of drug permeation. Each method has its own characteristics and can be introduced at different stages of drug development for different applications. Artificial membrane-based systems have been promoted as a low-cost alternative to Caco-2 cells in early development for the evaluation of the passive component in drug permeation. The Caco-2 monolayer model is currently the standard in the pharmaceutical industry for high-throughput and/or multimechanistic assays. Tissue-based systems are more involved at later stages of development to evaluate site-specific or species-dependent absorption. From the examples presented in this study, it is apparent that the experimental conditions strongly affect the generated data. At later stages of development, for example for BCS classification, it is obvious that the choice of experimental conditions is essential. In this perspective, it is important that the conditions used for the official classification assay have been validated. The standardization of cell culturing conditions is currently well recognized (Ungell and Karlsson, 2003). This recognition of standardization to increase data quality should be extended towards the experimental conditions, to prevent that each compound would be analyzed and evaluated by its own specific experimental set-up (Ingels and Augustijns, 2003).

List of Abbreviations

ADMET Absorption, Distribution, Metabolism, Excretion, Toxicity
AP .... Apical
BAMPA .... Bio-mimetic Artificial Membrane Permeability Assay
BCS .... Biopharmaceutical Classification System
BL .... Basolateral
BSA .... Bovine Serum Albumin
CM .... Culture Medium
DMA .... Dimethylacetamide
DMEM .... Dulbecco's Modified Eagle Medium
DMSO .... Dimethylsulfoxide
FaSSIF .... Fasted State Simulated Intestinal Fluid
GI .... Gastrointestinal
GIT .... Gastrointestinal Tract
HBSS .... Hanks' Balanced Salt Solution
HDM .... Hexadecane Membrane
HEPES .... N-2-hydroxyethylpiperazine-N′-2-ethane sulphonic acid
IAM .... Immobilized Artificial Membrane
Isc .... Short–circuit current

KBR....................Krebs Bicarbonate Ringer's Solution
MCT....................Monocarboxylic acid transporter
MDCK....................Madin-Darby Canine Kidney
MES....................2-(N-morpholino) ethane sulphonic acid
MTT........3-(4,5-dimethylthiazol-2-yl)-2,5-diphenyl tetrazolium bromide
NMP....................N-methylpyrrolidone
PAMPA....................Parallel Artificial Membrane Permeability Assay
PAT1....................Proton-coupled amino acid transporter
PEG....................Polyethylene Glycol
PEPT1....................Oligopeptide Transporter
PD....................Potential difference
PG....................Propylene Glycol
PVP....................Polyvinyl Pyrrolidone
TEER....................Transepithelial Electrical Resistance
TM....................Transport Medium

References

Anderberg EK and Artursson P. Cell Cultures to Assess Drug ABsorption Enhancement. In: *Drug absorption enhancement. Concepts, possibilities, limitations and trends.* Ed: A. (Bert) G. de Boer. 1994. Harwood Academic publishers.

Anderberg EK, Nyström C, and Artursson P. Epithelial Transport of Drugs in Cell Culture. VII: Effects of Pharmaceutical Surfactant Excipients and Bile Acids on Transepithelial Permeability in Monolayers of Human Intestinal Epithelial (Caco-2) Cells. *J Pharm Sci* 1992; 81: 879–887.

Annaert P, Tukker JT, Van Gelder J, Naesens L, De Clercq E, Van den Mooter G, Kinget R, and Augustijns P. *In vitro, ex vivo* and *in situ* Intestinal Absorption Characteristics of the Antiviral Ester Prodrug Adefovir Dipivoxil. *J Pharm Sci* 2000; 89: 1054–1062.

Artursson P and Karlsson J. Correlation Between Oral Drug Absorption in Humans and Apparent Drug Permeability Coefficients in Human Intestinal Epithelial (Caco-2) Cells. *Biochem Biophys Res Commun* 1991; 175: 880–885.

Artursson P, Palm K, and Luthman K. Caco-2 Monolayers in Experimental and Theoretical Predictions of Drug Transport. *Adv Drug Del Rev* 2001; 46: 27–43.

Artursson P, Ungell AL and Löfroth JE. Selective Paracellular Permeability in Two Models of Intestinal Absorption: Cultured Monolayers of Human Intestinal Epithelial Cells and Rat Intestinal Segments. *Pharm Res* 1993; 10: 1123–1129.

Augustijns P, Bradshaw TP, Gan LSL, Henden RW, and Thakker DR. Evidence of a Polarized Efflux System in Caco-2 Cells Capable of Modulating Cyclosporin a Transport. *Biochem Biophys Res Commun* 1993; 197: 360–365.

Aungst BJ, Nguyen NH, Bulgarelli JP, and Oates-Lenz K. The Influence of Donor and Reservoir Additives on Caco-2 Permeability and Secretory Transport of HIV Protease Inhibitors and Other Lipophilic Compounds. *Pharm Res* 2000; 17: 1175–1180.

Avdeef A. Absorption and drug development. 2003-Wiley-Interscience NY, pp. 116–246.

Avdeef A, Strafford M, Block E, Balogh MP, Chambliss W, and Khan I. Drug Absorption in vitro Model: Filter-Immobilized Artificial Membranes 2. Studies of the Permeability Properties of Lactones in *Piper methysticum* Forst. *Eur J Pharm Sci* 2001; 14: 271–280.

Balimane PV and Chong S. Cell Culture-Based Models for Intestinal Permeability: A Critique. *Drug Discov Today* 2005; 10(5): 335–343.

Balimane PV, Chong S, and Morrison RA. Current Methodologies Used for Evaluation of Intestinal Permeability and Absorption. *J Pharmacol Toxicol Meth* 2000; 44: 301–312.

Balon K, Riebesehl BU, and Müller BW. Drug Liposome Partitioning as a Tool for the Prediction of Human Passive Intestinal Absorption. *Pharm Res* 1999; 16: 275–285.

Bangham AD. Liposomes: The Babraham Connection. *Chem Phys Lipids* 1993; 64: 275–285.

Bassolino D, Alper H, and Stouch TR. Drug-membrane Interactions Studied by Molecular Dynamics Simulation: Size Dependece of Diffusion. *Drug Des Discov* 1996; 13: 135–141.

Batrakova EV, Li S, Li Y, Alakhov V Yu, and Kabanov AV. Effect of Pluronic P85 on ATPase Activity of Drug Efflux Transporters. *Pharm Res* 2004; 21: 2226–2233.

Bermejo M, Avdeef A, Ruiz A, Nalda R, Ruell JA, Tsinman O, González I, Fernández C, Sánchez G, Garrigues TM, and Merino V. PAMPA – A Drug Absorption in vitro Model. 7. Comparing Rat in situ, Caco-2, and PAMPA Permeability of Fluoroquinolones. *Eur J Pharm Sci* 2004; 21: 429–441.

Bogman K, Zysset Y, Degen L, Hopfgartner G, Gutmann H, Alsenz J, and Drewe J. P-glycoprotein and Surfactants: Effects on Intestinal Talinolol Absorption. *Clin Pharmacol Ther* 2005; 77: 24–32.

Bohets H, Annaert P, Mannens G, van Beijsterveldt L, Anciaux K, Verboven P, Meuldermans W, and Lavrijsen K. Strategies for Absorption Screening in Drug Discovery and Development. *Cur Top Med Chem* 2001; 1: 367–383.

Brouwers J, Tack J, Lammert F, and Augustijns P. Intraluminal Drug and Formulation Behavior and Integration in in vitro Permeability Estimation: A Case Study with Amprenavir. *J Pharm Sci* 2006; 95: 372–383.

Center for drug evaluation and research (CDER), Food and drug administration (FDA). Waiver of *in vivo* bioavailibility and bioequivalence studies for immediate-release solid oral dosage forms based on a biopharmaceutics classification system. *Guidance for industry* 2000, pp. 1–13.

Chan OH, Schmid HL, Kuo B, Wright DS, Howson W, and Stewart BH. Absorption of Cam-2445, an NK1 Neurokinin Receptor Antagonist: *in vivo, in situ,* and *in vitro* evaluations. *J Pharm Sci* 1996; 85: 253–257.

Chung SM and Chiou WL. Potential Overestimation of the Role of Gastrointestinal Metabolism and p-glycoprotein Mediated Efflux in Contributing to Low Bioavailability of Drugs in Humans. *Pharm Sci Suppl* 1999; 1:presentation 4272.

Chung SM, Park EJ, Swanson SM, Wu TC, and Chiou WL. Profound Effect of Plasma Protein Binding on the Polarized Transport of Furosemide and Verapamil in the Caco-2 Model. *Pharm Res* 2001; 18: 544–547.

Conradi RA, Hilgers AR, Ho NFH, and Burton PS. The Influence of Peptide Structure on Transport Across Caco-2 cells. *Pharm Res* 1991; 8: 1453–1460.

Davies GR, Santa Ana CA, Morawski SG, and Fordtran JS. Permeability Characteristics of Human Jejunum, Proximal Colon and Distal Colon: Results of Potential Difference Measurements and Unidirectional Fluxes. *Gastroenterology* 1982; 83: 844–850.

Deferme S, Van Gelder J, Ingels F, Van den Mooter G, De Buck S, Balzarini J, Naesens L, De Clercq E, Kinget R, and Augustijns P. Intestinal Absorption Characteristics of the Low Solubility Thiocarboxanilide UC-781. *Int J Pharm* 2002; 234: 113–119.

Dimitrijevic D, Shaw AJ, and Florence AT. Effects of Some Non-Ionic Surfactants on Transepithelial Permeability in Caco-2 cells. *J Pharm Pharmacol* 2000; 52: 157–162.

Dintaman JM and Silverman JA. Inhibition of P-glycoprotein by D-α-tocopheryl Polyethylene Glycol 1000 Succinate (TPGS). *Pharm Res* 1999; 16: 1550–1556.

Egan WJ, Merz KM, and Baldwin JJ. Prediction of Drug Absorption Using Multivariate Statistics. *J Med Chem* 2000; 43: 3867–3877.

Fallingborg J, Christensen LA, Ingelman-Nielsen M, Jacobsen BA, Abildgaard K and Rasmussen HH. pH Profile and Regional Transit Times of the Normal Gut Measured by Radiotelemetry Device. *Aliment Pharmacol Ther* 1989; 3: 605–613.

Fix JA. Intestinal Rings and Isolated Intestinal Mucosal Cells.Chapter 4. In: Borchardt RT et al.(eds.), *Models for Assessing Drug Absorption and Metabolism.* Plenum Press, New York, 1996.

Friedrichsen GM, Chen W, Begtrup M, Lee C-P, Smith PL, and Borchardt RT. Synthesis of Analogs of L-valacyclovir and Determination of their Substrate

Activity for the Oligopeptide Transporter in Caco-2 Cells. *Eur J Pharm Sci* 2002; 16: 1–13.

Gan LSL and Thakker DR. Applications of the Caco-2 Model in the Design and Development of Orally Active Drugs: Elucidation of Biochemical and Physical Barriers Posed by the Intestinal Epithelium. *Adv Drug Del Rev* 1997; 23: 77–98.

Goodwin JT, Mao B, Vidmar TJ, Conradi RA, and Burton PS. Strategies Toward Predicting Peptide Cellular Permeability from Computed Molecular Descriptors. *J Peptide Res* 1999; 53: 355–369.

Grass GM and Sweetana SA. *In vitro* Measurement of Gastrointestinal Tissue Permeability Using a New Diffusion Cell. *Pharm Res* 1988; 5(6): 372–376.

Gray V and Dressman JB. Change of pH Requirements for Simulated Intestinal Fluid TS. *Pharmacopeial Forum* 1996; 22: 1943–1945.

Griffiths R, Lewis A, and Jeffrey P. Models of Drug Absorption *in situ* and in Conscious Animals. Chapter 5. In: Borchardt RT et al. (eds.), *Models for Assessing Drug Absorption and Metabolism.* Plenum Press, New York, 1996.

Grinius L, Stanton DT, Morris CM, Howard JM, and Curnow AW. Profiling of Drugs for Membrane Activity Using Liposomes as an in vitro Model System. *Drug Dev Ind Pharm* 2002; 28: 193–202.

Hanisch G, Utter L, Ungell A-L, and Lucas M. Surface pH Measurements During in vitro Transport Studies Across Intestinal Epithelia: Thickness of Acid Microclimate. AAPS annual meeting in San Diego, California, 6–10 Nov, 1994. *Pharm Res.*1994, vol 11:10 suppl.

Hidalgo IJ, Hillgren KM, Grass GM, and Borchardt RT. Characterization of the Unstirred Water Layer in Caco-2 cell Monolayers Using a Novel Diffusion Apparatus. *Pharm Res* 1991; 8: 222–227.

Hidalgo IJ and Li J. Carrier-mediated Transport and Efflux Mechanisms in Caco-2 Cells. *Adv Drug Del Rev* 1996; 22: 53–66.

Hidalgo IJ, Raub TJ, and Borchardt RT. Characterization of Human Colon Carcinoma Cell Line (Caco-2) as a Model System for Intestinal Epithelial Permeability. *Gastroenterology* 1989; 96: 736–749.

Hilgers AR, Conradi RA, and Burton PS. Caco-2 cell Monolayers as a Model for Drug Transport Across the Intestinal Mucosa. *Pharm Res* 1990; 7: 902–910.

Hugger ED, Audus KL, and Borchardt RT. Effects of Poly(ethylene glycol) on Efflux Transporter Activity in Caco-2 Cell Monolayers. *J Pharm Sci* 2002; 91: 1980–1990.

Ingels F, Deferme S, Destexhe E, Oth M, Van den Mooter G, and Augustijns P. Simulated Intestinal Fluid as Transport Medium in the Caco-2 Cell Culture Model. *Int J Pharm* 2002; 232: 183–192.

Ingels FM and Augustijns PF. Biological, Pharmaceutical, and Analytical Considerations with Respect to the Transport Media used in the Absorption Screening System, Caco-2. *J Pharm Sci* 2003; 92(8): 1545–1558.

Irvine JD, Takahashi L, Lockhart K, Cheong J, Tolan JW, Selick HE, and Grove JR. MDCK (Madin-Darby Canine Kidney) Cells: A Tool for Membrane Permeability Screening. *J Pharm Sci* 1999; 88(1): 28–33.

Jacobs RE and White SH. The Nature of Hydrophobic Binding of Small Peptides at the Bilayer Interface: Implications for the Insertion of Transbilayer Helices. *Biochemistry* 1989; 28: 3421–3437.

Kabanov AV, Batrakova EV, and Alakhov VYu. An Essential Relationship Between ATP Depletion and Chemosensitizing Activity of Pluronic® Block Copolymers. *J Control Release* 2003; 91: 75–83.

Kansy M, Avdeef A, and Fischer H. Advances in Screening for Membrane Permeability: High-Resolution PAMPA for Medicinal Chemists. *Drug Discov Today* 2004; 1: 349–355.

Kansy M, Senner F, and Gubernator K. Physico-chemical High Throughput Screening: Parallel Artificial Membrane Permeation Assay in the Description of Passive Absorption Process. *J Med Chem* 1998; 41: 1007–1010.

Kim DC, Burton PS, and Borchardt RT. A Correlation Between the Permeability Characteristics of a Series of Peptides Using an in vitro Cell Culture Model (Caco-2) and Those Using an in situ Perfused Rat Ileum Model of the Intestinal Mucosa. *Pharm Res* 1993; 10: 1710–1714.

Krishna G, Chen KJ, Lin CC, and Nomeir AA. Permeability of Lipophilic Compounds in Drug Discovery Using in-vitro Human Absorption Model, Caco-2. *Int J Pharm* 2001; 222: 77–89.

Lipinski CA, Lombardo F, Dominy BW, and Feeney PJ. Experimental and Computational Approaches to Estimate Solubility and Permeability in Drug Discovery and Development Settings. *Adv Drug Deliv Rev* 2001; 46: 3–26.

Lo Y-L and Huang J-F. Effects of Sodium Deoxycholate and Sodium Caprate on the Transport of Epirubicin in Human Intestinal Epithelial Caco-2 Cell Layers and Everted Gut Sacs of Rats. *Biochem Pharmacol* 2000; 59: 665–672.

Lucas M. Determination of Acid Surface pH in vivo in Rat Proximal Jejunum. *Gut* 1983; 24: 734–739.

Lundahl P and Beigi F. Immobilized Liposome Chromatography of Drugs for Model Analysis of Drug-Membrane Interactions. *Adv Drug Deliv Rev* 1997; 23: 221–227.

Makhey VD, Guo A, Norris DA, Hu P, Yan J, and Sinko PJ. Characterization of the Regional Intestinal Kinetics of Drug Efflux in Rat and Human Intestine and in Caco-2 Cells. *Pharm Res* 1998; 15: 1160–1166.

Mälkiä A, Murtomäki L, Urtti A, and Kontturi K. Drug Permeation in Biomembranes. *In vitro* and *in silico* Prediction and Influence of Physicochemical Properties. *Eur J Pharm Sci* 2004; 23: 13–47.

Marrink SJ and Berendsen HJC. Permeation Process of Small Molecules Across Lipid Membranes Studied by Molecular Dynamics Simulations. *J Phys Chem* 1994; 98: 4155–4168.

Martin YC. A practitioner's Perspective of the Role of Quantitative Structure-Activity Analysis in Medicinal Chemistry. *J Med Chem* 1981; 24: 229–237.

Mathieu F, Galmier M-J, Pognat J-F, Petit J, and Lartigue C. Transepithelial Transport of Bepiridil in the Human Intestinal Cell Line, Caco-2, Using Two Media, DMEMc and HBSS. *Int J Pharm* 1999; 181: 203–217.

Matsson P, Bergström CAS, Nagahara N, Tavelin S, Norinder U, and Artursson P. Exploring the Role of Different Drug Transport Routes in Permeability Screening. *J Med Chem* 2005; 48: 604–613.

Meaney C and O'Driscoll C. Mucus as a Barrier to the Permeability of Hydrophilic and Lipophilic Compounds in the Absence and Presence of Sodium Taurocholate Micellar Systems Using Cell Culture Models. *Eur J Pharm Sci* 1999; 8: 167–175.

Miret S, Abrahamse L, and de Groene EM. Comparison of in vitro Models for the Prediction of Compound Absorption Across the Human Intestinal Mucosa. *J Biomol Screen* 2004; 9: 598–606.

Mosmann T. Rapid Colorimetric Assay for Cellular Growth and Survival: Application to Proliferation and Cytotoxicity Assays. *J Immunol Met* 1983; 65: 55–63.

Narawane M, Podder SK, Bundgaard H, and Lee VHL. Segmental Differences in Drug Permeability, Esterase Activity and Ketone Reductase Activity in the Albino Rabbit Intestine. *J Drug Targeting* 1993; 1: 29–39.

Nerurkar MN, Burton PS, and Borchardt RT. The Use of Surfactants to Enhance the Permeability of Peptides Through Caco-2 Cells by Inhibition of an Apically Polarized Efflux System. *Pharm Res* 1996; 13: 528–534.

Neuhoff S. Refined *in vitro* Models for Prediction of Intestinal Drug Transport: Role of pH and Extracellular Additives in the Caco-2 Cell Model. Doctoral thesis, 2005.

http://publications.uu.se/abstract.xsql?dbid=5814 (accessed June 2005).

Neuhoff S, Ungell AL, Zamora I, and Artursson P. pH dependent Bidirectional Transport of Weakly Basic Drugs Across Caco-2 Cell Monolayers: Implications for Drug-Drug Interactions. *Pharm Res* 2003; 20: 1141–1148.

Neuhoff S, Ungell AL, Zamora I, and Artursson P. pH Dependent Passive and Active Transport of Acidic Drugs Across Caco-2 Cell Monolayers. *Eur J Pharm Sci* 2005; 25: 211–220.

Oberle RL, Moore TJ, and Krummel DAP. Evolution of Mucosal Damage of Surfactants in Rat Jejunum and Colon. *J Pharmacol and Toxicol Methods* 1995; 33: 75–81.

Ogihara T, Tamai I, and Tsuji A. Structural Characterization of Substrates for the Anion Exchange Transporter in Caco-2 Cells. *J Pharm Sci* 1999; 88: 1217–1221.

Örnskov E, Gottfries J, Erickson M, and Folestad S. Experimental Modelling of Drug Membrane Permeability by Capillary Electrophoresis Using Liposomes, Micelles and Microemulsions. *J Pharm Pharmacol* 2005; 57: 435–442.

Palm K, Luthman K, Ros J, Grasjo J, and Artursson P. Effect of Molecular Charge on Intestinal Epithelial Drug Transport: pH-dependent Transport of Cationic Drugs. *J Pharmacol and Exp Therap* 1999; 291: 435–443.

Palm K, Luthman K, Ungell AL, Strandlund G, Beigi F, Lundahl P, and Artursson P. Correlation of Drug Absorption with Molecular Surface Properties. *J Pharm Sci* 1996; 85: 32–39.

Palm K, Luthman K, Ungell AL, Strandlund G, Beigi F, Lundahl P, and Artursson P. Evaluation of Dynamic Polar Molecular Surface Area as Predictor of Drug Absorption: Comparison with Other Computational and Experimental Predictors. *J Med Chem* 1998; 41: 5382–5392.

Pidgeon C and Venkataram UV. Immobilized Artificial Membrane Chromatography: Supports Composed of Membrane Lipids. *Anal Chem* 1989; 176: 36–47.

Polentarutti B, Peterson A, Sjöberg Å, Anderberg EK, Utter L, and Ungell AL. Evaluation of Viability of Excised Rat Intestinal Segments in the Ussing Chamber: Investigation of Morphology, Electrical Parameters and Permeability Characteristics. *Pharm Res* 1999; 16: 446–454.

Putnam WS, Pan L, Tsutsui K, Takahashi L, and Benet LZ. Comparison of Bidirectional Cephalexin Transport Across MDCK and Caco-2 Cell Monolayers: Interactions with Peptide Transporters. *Pharm Res* 2002a; 19: 27–33.

Putnam WS, Ramanathan S, Pan L, Takahashi LH, and Benet LZ. Functional Characterization of Monocarboxylic Acid, Large Neutral Amino Acid, Bile Acid and Peptide Transporters, and P-glycoprotein in MDCK and Caco-2 Cells. *J Pharm Sci* 2002b; 91: 2622–2635.

Quadros E, Cassidy JP, and Leipold H. Buccal Tissues and Cell Culture. Chapter 7. In: Borchardt RT (ed.), *Models for Assessing Drug Absorption and Metabolism.* Plenum Press, New York, 1996.

Reardon PM. In Vitro Nasal Models. Chapter 16. In: Borchardt RT (ed.), *Models for Assessing Drug Absorption and Metabolism.* Plenum Press, New York, 1996.

Rege BD, Kao JPY, and Polli JE. Effects of Non-Ionic Surfactants on Membrane Transporters in Caco-2 Cell Monolayers. *Eur J Phar Sci* 2002; 16: 237–246.

Rege BD, Yu LX, Hussain AS, and Polli JE. Effect of Common Excipients on Caco-2 Transport of Low-permeability Drugs. *J Pharm Sci* 2001; 90: 1776–1786.

Rubas W, Jezyk N, and Grass GM. Comparison of the Permeability Characteristics of a Human Colonic Epithelial (Caco-2) Cell Line to Colon of Rabbit, Monkey and Dog Intestine and Human Drug Absorption. *Pharm Res* 1993; 10: 113–118.

Saha P and Kou JH. Effect of Bovine Serum Albumin on Drug Permeability Estimation Across Caco-2 Monolayers. *Eur J Pharm Biopharm* 2002; 54: 319–324.

Scudiero DA, Shoemaker RH, Paull KD, Monks A, Tierney S, Nofziger TH, Currens MJ, Seniff D, and Boyd MR. Evaluation of a Soluble Tetrazolium/formazan Assay for Cell Growth and Drug Sensitivity in Culture Using Human and Other Tumor Cell Lines. *Canc Res* 1988; 48: 4827–4833.

Sjöström M, Sjöberg Å, Utter L, Hyltander A, Karlsson J, Stockman E, and Ungell AL. Excised Human Intestinal Segments as a Mechanistic Tool for Verifying Transport Properties of Drug Candidates. Abstract AAPS 2000. *Pharm Sci* 2000; suppl.

Söderholm JD, Hedamn L, Artursson P, Franzen L, Larsson J, Pantzar N, Permert J, and Olaison G. Integrity and Metabolism of Human Ileal Mucosa in vitro in the Ussing Chamber. *Acta Physiol Scand* 1998; 162: 47–56.

Steffansen B, Nielsen CU, Brodin B, Eriksson AH, Andersen R, and Frokjaer S. Intestinal Solute Carriers: An Overview of Trends and Strategies for Improving Oral Drug Absorption. *Eur J Pharm Sci* 2004; 21: 3–16.

Stenberg P, Luthman K, and Artursson P. Prediction of Membrane Permeability to Peptides from Calculated Dynamic Molecular Surface Properties. *Pharm Res* 1999a; 16: 205–212.

Stenberg P, Luthman K, Ellens H, Lee CP, Smith PL, Lago A, Elliott JD, and Artursson P. Prediction of the Intestinal Absorption of Endothelin Receptor Antagonists Using Three Theoretical Methods of Increasing Complexity *Pharm Res* 1999b; 16: 1520–1526.

Stewart BH, Chan OH, Jezyk N, and Fleisher D. Discrimination Between Drug Candidates Using Models for Evaluation of Intestinal Absorption. *Adv Drug Del Rev* 1997; 23: 27–45.

Sugano K, Hamada H, Machida M, Ushio H, Saitoh K, and Terada K. Optimized Conditions of Bio-mimetic Artificial Membrane Permeation Assay. *Int J Pharm* 2001; 228: 181–188.

Sugano K, Nabuchi Y, Machida M, and Aso Y. Prediction of Human Intestinal Permeability Using Artificial Membrane Permeability. *Int J Pharm* 2003; 257: 245–251.

Sugano K, Takata N, Machida M, Saitoh K, and Terada K. Prediction of Passive Intestinal Absorption Using Bio-mimetic Artificial Membrane Permeation Assay and the Paracellular Pathway Model. *Int J Pharm* 2002; 241: 241–251.

Sugiyama T, Yamamoto A, Kawabe Y, Uchiyama T, Quan Y-S-, and Muranishi S. Effects of Various Absorption Enhancers on the Intestinal Absorption of Water Soluble Drugs by in vitro Ussing Chamber Method: Correlation with an in situ Absorption Experiment. *Biol Pharm Bull* 1997; 20: 812–814.

Tada H, Shiho O, Kuroshima KI, Koyama M, and Tsukamoto K. An Improved Colorimetric Assay for Interleukin 2. *J Immunol Met* 1986; 93: 157–165.

Taillardat-Bertschinger A, Carrupt PA, Barbato F, and Testa B. Immobilized Artificial Membrane HPLC in Drug Research. *J Med Chem* 2003; 46: 655–665.

Taylor DC, Pownall R, and Burke W. The Absorption Of Beta-adrenoceptor Antagonists in Rat in situ Small Intestine; The Effect of Lipophilicity. *J Pharm Pharmacol* 1985; 37: 280–283.

Tsuji A and Tamai I. Carrier-mediated Intestinal Transport of Drugs. *Pharm Res* 1996; 13: 963–977.

Udata C, Patel J, Pal D, Hejchman E, Cushman M, and Mitra AK. Enhanced Transport of a Novel Anti-HIV Agent –cosalane and its Congeners Across Human Intestinal Epithelial (Caco-2) Cell Monolayers. *Int J Pharm* 2003; 250: 157–168.

Ungell AL. *In vitro* Absorption Studies and Their Relevance to Absorption from the GI Tract. *Drug Develop and Indust Pharmacy* 1997; 23: 879–892.

Ungell AL. Transport Studies Using Intestinal Tissue Ex-vivo. Chapter 11. In: Lehr C-M (ed.), *Cell Culture Models of Biological Barriers; in vitro Test Systems for Drug Absorption and Delivery*. Taylor& Francis London, NY, 2002, pp. 164–188.

Ungell AL. Prediction of Human Drug Absorption Using in Silico and in vitro Techniques. Bullentin Technique Gattefosse No 98; 2005, pp. 19–30.

Ungell AL and Karlsson J. Cell Cultures in Drug Discovery: An Industrial Perspective. In: Van de Waterbeemd et al. (eds.),*Drug Bioavailability. Estimation of Solubility, Permeability, Absorption and Bioavailability*. P. Wiley WCH 2003, pp. 72–89.

Ungell AL, Nylander S, Bergstrand S, Sjöberg Å, and Lennernäs H. Membrane Transport of Drugs in Different Regions of the Intestinal Tract of the Rat. *J. Pharm Sci* 1997; 87: 360–366.

Ungell AL, Zamora I, Neuhoff S, and Artursson P. Real and False Efflux of Weak Basic Drugs Across Caco-2 Monolayers. AAPS Annual Meeting, Toronto, ON, Canada, 2002.

Ussing HH and Zerahn K. Active Transport of Sodium as the Source of Electric Current in the Short-circuited Isolated Frog Skin. *Acta Physiol Scand* 1951; 23: 110–127.

USP. USP Pharmacopeial Forum. *US Pharmacopeia* 1998; 24: 6015–6023.

Veber DF, Johnson SR, Cheng H-Y, Smith BR, Ward KW, and Kopple KD. Molecular Properties that Influence the oral Bioavailibility of Drug Candidates. *J Med Chem* 2002; 45: 2615–2623.

Volpe DA. Permeability Classification of Representative Fluoroquinolones by a Cell Culture Method. *AAPS Pharm Sci* 2004; 6,Article 13.

Walgren RA and Walle T. The Influence of Plasma Binding on Absorption/exsorption in the Caco-2 Model of Human Intestinal Absorption. *J Pharm Pharmacol* 1999; 51: 1037–1040.

Watanabe E, Sudo R, Takahashi M, and Hayashi M. Evaluation of Absorbability of Poorly Water-soluble Drugs: Validity of the Use of Additives. *Biol Pharm Bull* 2000; 23: 838–843.

Watanabe E, Takahashi M, and Hayashi M. A Possibility to Predict the Absorbability of Poorly Water-soluble Drugs in Humans Based on Rat Intestinal Permeability Assessed by an in vitro Chamber Method. *Eur J Pharm Biopharm* 2004; 58: 659–665.

Wiedmann TS and Kamel L. Examination of the Solubilization of Drugs by Bile Salt Micelles. *J Pharm Sci* 2002; 91: 1743–1764.

Wright LL and Painter GR. Role of Desolvatation Energy in the Nonfacilitated Membrane Permeability of Dideoxyribose Analogs of Thymidine. *Mol Pharmacol* 1992; 41: 957–962.

Yamashita S, Furubayashi T, Kataoka M, Sakane T, Sezaki H, and Tokuda H. Optimized Conditions for Prediction of Intestinal Drug Permeability Using Caco-2 Cells. *Eur J Pharm Sci* 2000; 10: 195–204.

Yamashita S, Tanaka Y, Endoh Y, Taki Y, Sakane T, Nadai T, and Sezaki H. Analysis of Drug Permeation Across Caco-2 Monolayer: Implication for Predicting *in vivo* Drug Absorption. *Pharm Res* 1997; 14: 486–491.

Yang CY, Cai SJ, Liu H, and Pidgeon C. Immobilized Artificial Membranes – Screens for Drug Membrane Interactions. *Adv Dr Del Rev* 1997; 23: 229–256.

Yee S. In Vitro Permeability Across Caco-2 cells (colonic) can Predict in vivo (small intestine) Absorption in Man – Fact or Myth. *Pharm Res* 1997; 14: 763–766.

Yu L, Bridgers A, Polli J, Vickers A, Long S, Roy A, Winnike R, and Coffin M. Vitamin E-TPGS Increases Absorption Flux of an HIV Protease Inhibitor by Enhancing its Solubility and Permeability. *Pharm Res* 1999; 16: 1813–1816.

Yu LX, Amidon GL, Polli JE, Zhao H, Mehta MU, Conner DP, Shah VP, Lesko LJ, Chen M-L, Lee VHL, and Hussain AS. Biopharmaceutics Classification System: The Scientific Basis for Biowaiver Extensions. *Pharm Res* 2002; 19: 921–925.

Zakelj S, Legen I, Veber M, and Kristl A. The Influence of Buffer Composition on Tissue Integrity During Permeability Experiments 'in vitro'. *Int J Pharm* 2004; 272: 173–180.

8

The Use of Solubilizing Excipients and Approaches to Generate Toxicology Vehicles for Contemporary Drug Pipelines

MARCUS E. BREWSTER

Chemical and Pharmaceutical Development, Johnson & Johnson, Pharmaceutical Research and Development, Beerse, BELGIUM

CLAIRE MACKIE AND MARC NOPPE

Research and Early Development, Johnson & Johnson, Pharmaceutical Research and Development, Beerse, BELGIUM

ANN LAMPO

Preclinical Development, Johnson & Johnson, Pharmaceutical Research and Development, Beerse, BELGIUM

THORSTEINN LOFTSSON

Faculty of Pharmacy, The University of Iceland, Reykjavik, ICELAND

Introduction

The purpose of this chapter is two-fold and includes approaches for identifying potentially problematic drug candidates with regard to formulation, in general, and for the preparation of toxicology vehicles, in particular. In addition, an attempt is made to provide insight as to what oral and parenteral excipients are appropriate for early human testing and, by extension, which of these materials can reasonably be used in GLP toxicology evaluation intended to support these Phase I human assessments. These considerations are becoming more visible in the drug development arena as evidenced by a number of recent symposia and congresses (Liu 2005; Van Gelder, 2006). The need for this type of guidance is rooted in the nature of contemporary drug pipelines and the fact that traditional decision trees for formulation are less and less relevant, prompting a search for new formulation adjuncts.

Within the pharmaceutical industry the nature of drug screening has evolved over the years such that high throughput screening techniques have become routine. These hit identification strategies put a type of evolutionary pressure on emerging drug candidates, which has led to a systematic increase in molecular weight, lipophilicity and a decrease in water solubility of lead compounds over time (Lipinski et al., 1997; Lipinski, 2001). This, in turn, has had a significant impact on what is required from formulators in that the number of formulation options and by extension, excipients, has to be increased to address the larger number of challenges being presented (Liu, 2000).

Biopharmaceutical Classification System (BCS) and the Developmental Classification System (DCS) and Their Implications

For a drug to be orally available, the compound must dissolve and be absorbed through the gut in such a way so as to generate adequate drug levels at the pharmacologically active site so that the desired action(s) is obtained in a reproducible manner (Yu, 1999). Retrospective studies in the late 1980's showed that >40% of drug failures in development could be traced to poor biopharmaceutical properties namely, poor dissolution or poor permeability (Prentis et al., 1988). This situation has been improved based on analysis completed in the 2000's, however poor solubility continues to impact the development of a large number of potential drug candidates (Kola and Landis, 2004). In recognition of the importance of these factors, the FDA and other world-wide boards of health have defined a Biopharmaceutical Classification System (BCS) in which drugs are divided into 4 types based on their solubility and permeability characteristics (Amidon et al., 1995; Dressman et al., 1998; Yu et al., 2002). Soluble, permeable drugs are termed Class I compounds with oral bioavailability only being limited by their ability to reach appropriate sites of absorption in the GI tract. Class II drugs are poorly soluble but permeable through the gut meaning that oral adsorption is limited by drug solubility and, as a consequence of the Noyes-Whitney equation, dissolution rate. Class III compounds are soluble but poorly permeable meaning that oral bioavailability is limited by the barrier properties of the GI tract. Finally, Class IV compounds are both insoluble and poorly permeable combining the limitations of both Class II and III materials. High throughput discovery methodologies are selecting, more and more, for difficult Class II compounds.

The BCS was initially developed to provide a scientific approach for the granting of biowaivers; that is, for allowing drug developers to replace certain bioequivalence clinical studies by accurate *in vitro* dissolution data (Yu et al., 2002; Kalantzi et al., 2005). The BCS was initially designed for immediate release products that are absorbed throughout the gastrointestinal tract. In addition to these applications, it has been reported that the scientific underpinnings of the BCS can be very helpful in suggesting the nature and amount of work that may be needed to generate appropriate formulations, be they oriented towards early animal work, toxicology or human clinical assessment (Dressman et al., 2001, Sun et al., 2004)). The application of the BCS approach to early drug candidate development is sometimes termed a developmental classification systems (DCS). These techniques have been shown to be very useful for identifying the rate-limiting step and predicting intestinal drug absorption based on primary biopharmaceutical properties such as solubility and permeability.

The BCS provides specific definitions for solubility and permeability such that for a soluble compound, the drug amount associated with the highest administered (oral) dose dissolves completely in 250 mL of water across a pH range of 1 to 7.5 (Amidon et al., 1995; Yu et al., 2002). A compound is considered permeable if it is >90% bioavailable or if >90% of an orally administered dose can be accounted for in the urine. Surrogates for these data have been suggested including the use of cell cultures (Caco-2), in situ intestinal perfusion and transport through tissue explants. An immediate release product is characterized as rapidly dissolving when not less than 85% of the labeled amount of drug substance is dissolved within 30 min (using a USP I apparatus at 100 rpm or a USP II apparatus at 50 rpm at pH 1, 4.5 and 6.8). These conservative definitions are useful in defining, from a regulatory standpoint, where dissolution data can be a surrogate for clinical pharmacokinetics, but are less helpful in pointing out where solubility and permeability may limit drug development. That is, there are many Class III and IV compounds with large markets and medical benefits (Cheng et al., 2004). A number of publications suggests the expansion of the number of classes in the BCS based on these concerns (Bergstrom et al., 2003).

The DCS is a more specific tool to help drug developers make risk decisions concerning potential pharmaceutical candidates (Figure 1) (Dressman et al., 2001; Sun et al., 2004). These factors are assessed across the development spectra

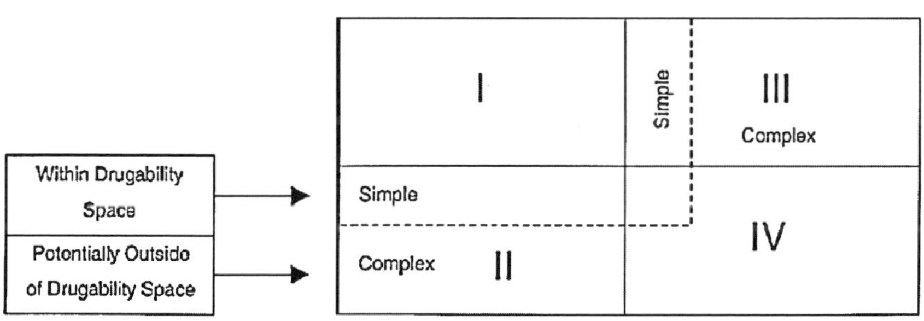

Figure 1. Biopharmaceutical Classification System and Development Classification System. Class I compounds are defined as soluble and permeable through the gastrointestinal tract, Class II as poorly soluble but permeable through the GI tract, Class III as soluble but poorly permeable and Class IV as both poorly soluble and permeable. The further classification of Class II and III (simple and complex) is intended to provide additional data on the develop-ability of the drug candidate.

from drugability in early Discovery to initial human clinical studies and, for our purposes, in the development of toxicology vehicles. The main driver for the application of the DCS is the nature of the information available, which in turn is related to the developmental phase and the amount of drug available for testing. As the nature of the data becomes more predictive, the DCS segments Class II and III into sub-classes such as simple (meaning that the limiting property can be addressed such that an acceptable fraction absorbed can be achieved) or complex (meaning that there is a significant risk that a significant fraction absorbed may not be achievable within usual resource and other constraints). For the purpose of this discussion, developmental phases in early drug evaluation are binned into library screening, hit-to-lead optimization, candidate selection/early animal pharmacokinetics and first-in-man pharmacokinetic assessment.

Library Screening/*In Silico* Evaluation

At very early stages, the amount of compound available is limited or nonexistent, meaning that screening falls in the purview of *in silico* prediction (Agoram et al., 2001; Zhao et al., 2001; Bergstrom et al., 2003; Johnson 2003; Dannenfelser et al., 2004; Keuntz et al., 2006; Johnson and Zheng, 2006; Jung et al., 2006). A number of commercially available and internally developed software packages have been developed to address predictions at this stage. The ADMET Predictor™ and GastroPlus™ systems provided by SimulationsPlus have been the topic of a number of publications and will be referred to in the current discussion (Figure 2) (Agoram et al., 2001; Dannenfelser et al., 2004; Keuntz et al., 2006). The software allows a number of predictions based solely on structural information although experimental data can also be added to improve the accuracy of the approach. At this stage, compounds or compound libraries can be screened producing solubility and permeability information as reported in a dimensionless Dose Number, Dissolution Number and Absorption Number. Based on appropriate threshold values (for example, for solubility 10 μg/mL and for predicted human effective jejunal permeability, 0.5×10^{-4} cm/sec), these parameters can suggest where absorption is limited by dose, dissolution rate and/or permeability. At this early stage, a preliminary DCS value can be assessed by comparing the Dose, Dissolution and Absorption numbers such that Class I compounds have acceptable Dose and Permeability Numbers, while Class II materials have low Dose and or low Dissolution Numbers but good Absorption Numbers. Class III compounds have good Dose/Dissolution Numbers but a low Absorption Numbers and Class IV candidates have both a low Absorption Number and poor Dose/Dissolution Numbers. While preliminary, these data can already give insight into many aspects of the formulation needs of the compound. These can be selectively assessed based on the functionality of the program to generate sensitivity analysis through the use of Spider plots (Kuentz et al., 2006). Thus, for a Class II candidate, a solubility target can be identified which, theoretically, would result in good absorption and this can in turn lead to a target solubility value which can then be addressed by appropriate decision trees.

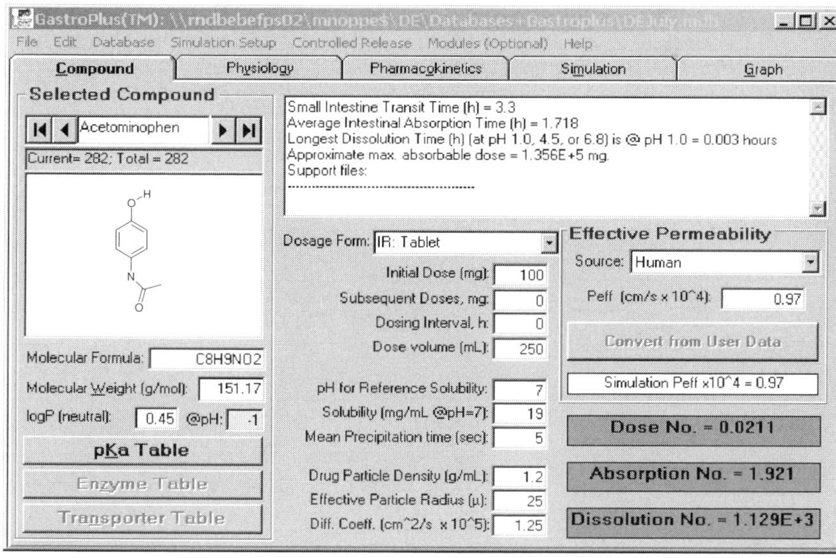

A

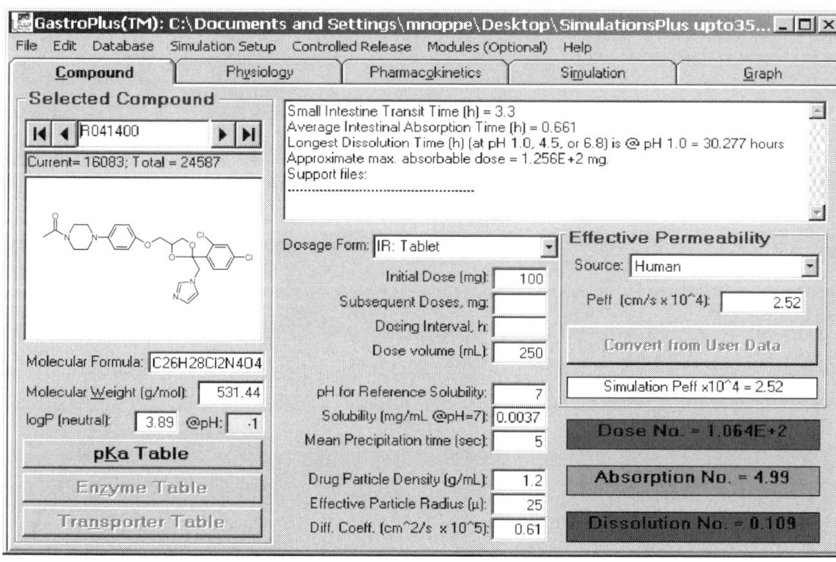

B

Figure 2. Estimation of DCS class based on the GastroPlus software of SimulationPlus. Based on the dimensionless Dose, Absorption and Dissolution Numbers, drugs can be defined as DCS Class I (A), II (B), III (C) and IV (D) based solely on structural information. The program gives predicted values for solubility, pKa and effective human intestinal permeability in order to suggest a fraction absorbed into the portal vein. (See color insert after Index.)

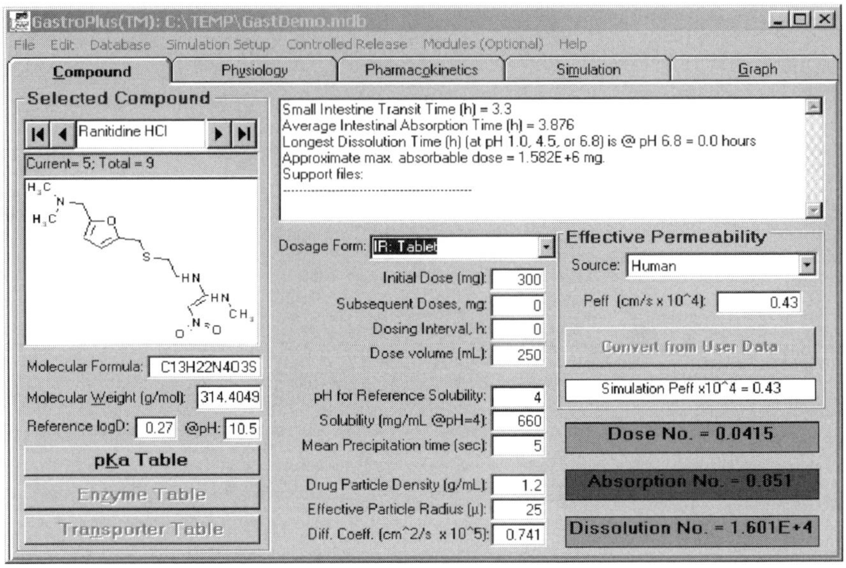

C

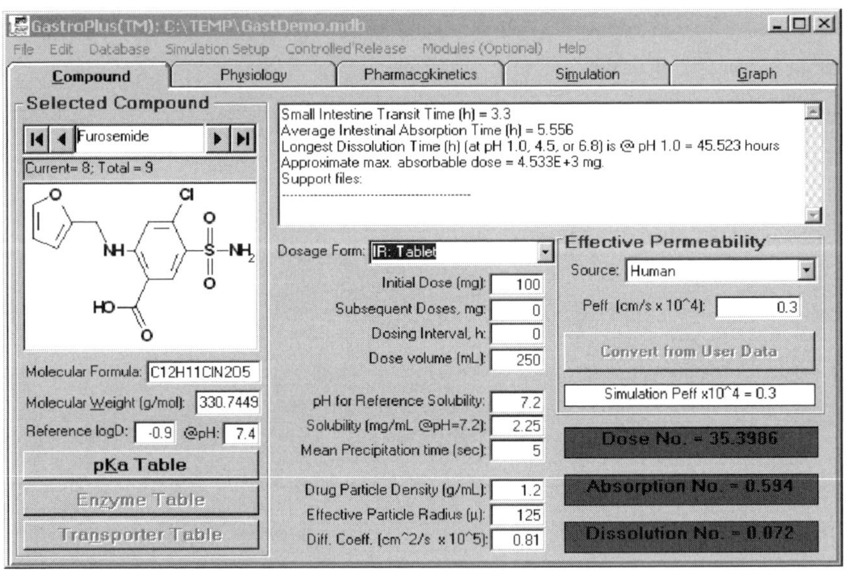

D

Figure 2. (*Continued*)

Itraconazole, a widely used broad-spectrum anti-fungal agent has been used as an example (De Beule and Van Gestel, 2001; Peeters et al., 2002). As indicated in Figure 3, the compound is predicted to have extremely low water solubility but useful intestinal permeability. The parameter sensitivity analysis indicated that fraction absorbed could be substantially increased if the apparent solubility of the drug could be increased to 100 μg/mL or greater. Such manipulation is possible through the use of hydrophilic cyclodextrins such as 2-hydroxypropyl-

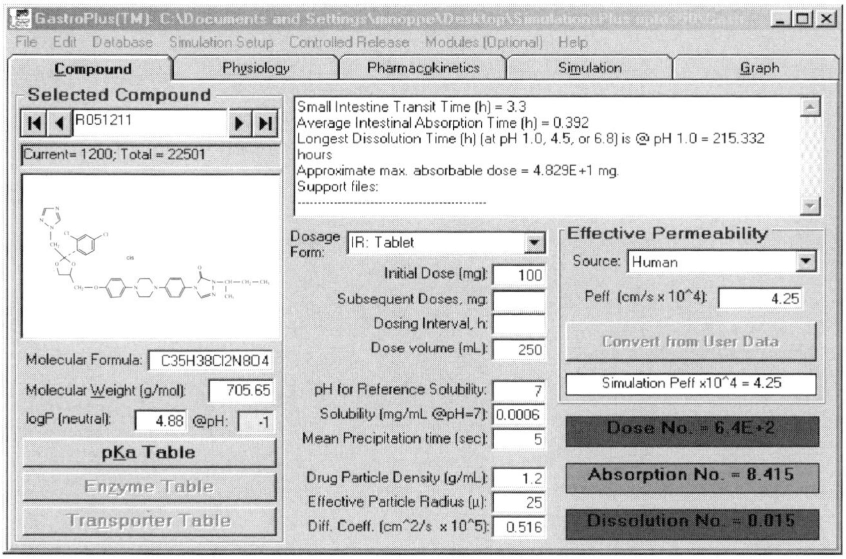

A

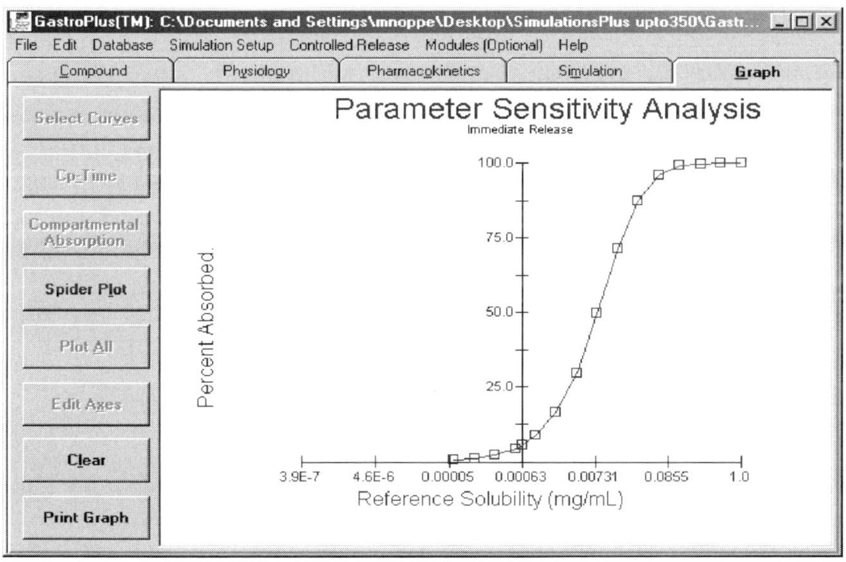

B

Figure 3. Predicted DCS classification of Itraconazole (A) as well as the effect of solubility on fraction absorbed (B). The Spider plot suggest good oral bioavailability at solubility values above 100 µg/mL. The solubility of itraconazole at neutral pH is estimated at 1 ng/mL. Itraconazole can be solubilized in 2-hydroxypropyl-β-cyclodextrin to levels in excess of 10 mg/mL which suggests Class I behavior (C). (See color insert after Index.)

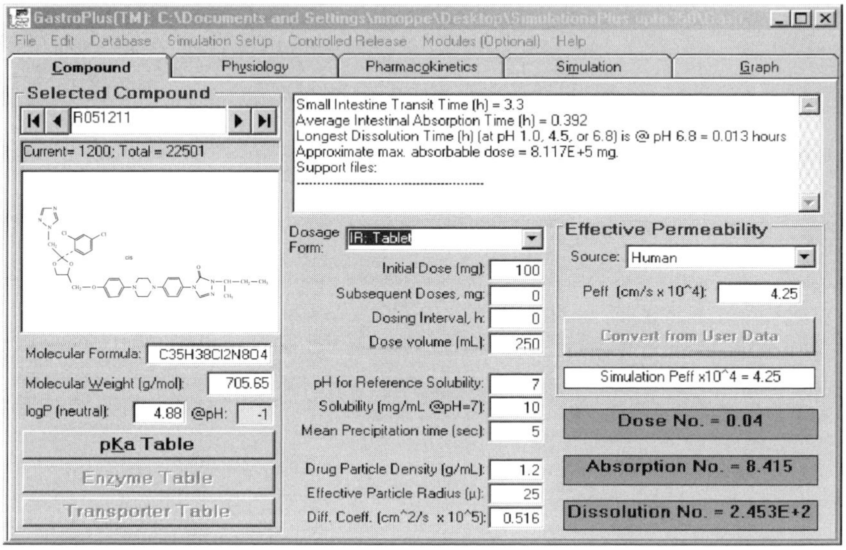

C

Figure 3. (*Continued*)

β-cyclodextrin (HPβCD) (Peeters et al., 2002; Davis and Brewster, 2004; Loftsson et al., 2004). Based on the ability to generate 10 mg/mL solutions of the drug in aqueous cyclodextrin solutions, GastroPlus predicts that this complex Class II compound can be formulated such that it behaves as a Class I material. This has been born out clinically (Koks et al., 2002). In addition, the effect of particle size and other variables on drug absorption can be modeled such that an *in silico* prediction of solubility-limited or dissolution rate limited pharmacokinetics can be suggested. At the initial phases of drug candidate screening, the ultimate human dose is not known, meaning the traditional Dose/Solubility ratio cannot be used. In the absence of an established dose range, a compound is assumed to be of average potency (about 1 mg/kg as suggested by Lipinski et al., 1997).

Lead Optimization/High Throughput Screening Analysis

If a candidate is selected for further assessment, it is produced in amounts sufficient for high throughput evaluation and other experimentation to screen for potency against a selected target. In addition, solubility and permeability can be assessed using the same DMSO-haystack solutions used to assess activity, based on kinetic solubility analysis and permeability through immobilized artificial membranes (Hidalgo, 2001; Zhu et al., 2002; Stoner et al., 2004; Dehring et al., 2004; Avdeef, 2005; Box et al., 2006). Kinetic solubility involves adding a small volume of a solution of the drug of interest in a water-miscible organic solvent (DMSO is generally used) to an aqueous buffer at a selected pH (Lipinski et al., 1997). The rate and extent (at a specified time) of precipitation of the compound gives insight into the critical supersaturation point of the compound in the defined system which can be compared in many instances to solubilities determined by traditional thermodynamic means (Dehring et al., 2004). The precipitation can be assessed by turbidity using either nephlemetric-based detectors or by modified flow cytometers. The approach is fast and has the advantage of discounting physical form which is often not optimized at early stages but which can dramatically affect thermodynamic solubilities. Typically, kinetic solubilities are binned into certain ranges as their exact value may not be particularly informative. To this end, scores of low, medium and high might be applied based on solubility ranges of:

Low Solubility (Score of 1) $<10\ \mu M$
Medium Solubility (Score of 2) $(10 < x < 50)\ \mu M$
High solubility (Score of 3) $>50\ \mu M$

Permeability can be assessed using parallel artificial membrane assays or PAMPA (Avdeef 2005). In the PAMPA assay, a sandwich is formed between two 96-well plates such that a donor and receptor cell is created, separated by a 125 μm filter coated with a dodecane solution of dioleoylphosphatidylcholine. Drug diffusion from the donor to receptor compartments is measured analytically over time using either a UV plate-reader or by LC/MS and effective permeability calculated. Data is then binned as high (a score of 3), medium (a score of 2) or low (a score of 1) based on the following criteria:

High Permeability (Score of 1) $> 1 \times 10^{-6}$ cm/sec
Medium Permeability (Score of 2) $(0.5 < x < 1) \times 10^{-6}$ cm/sec
Low Permeability (Score of 3) $< 0.5 \times 10^{-6}$ cm/sec

Another widely used permeability model in early drug development is the Caco-2 cell line (Yee, 1997; Camenisch et al., 1998; Ingels and Augustijns, 2003; Van Breemen et al., 2005). This cell culture, derived from a human colon carcinoma, assesses both diffusion and carrier-mediated transport of drugs. Comparisons of PAMPA and Caco-2 permeability models (Van Dijck et al., 2003; Kerns et al., 2004) suggest that the two provide similar information if simple diffusion is assessed; however, Caco-2 results are laboratory-, passage number- and culturing condition-dependent so that care should be taken in assessing permeability values obtained from different sources (Masungi et al., 2004). In our laboratory, characterization of permeability using Caco-2 suggests the following definitions:

High Permeability $> 10 \times 10^{-6}$ cm/sec
Medium Permeability $(1 < x < 10) \times 10^{-6}$ cm/sec
Low Permeability $< 1 \times 10^{-6}$ cm/sec

This suggests that DCS Class I compounds are those with high solubility and high permeability based on kinetic solubility and PAMPA or Caco-2 analyses. Class II compounds are those associated with high permeability and by medium or low solubility while Class III compounds are associated with high solubility and medium or low permeability. Class IV materials have both low solubility and low permeability (Figure 4).

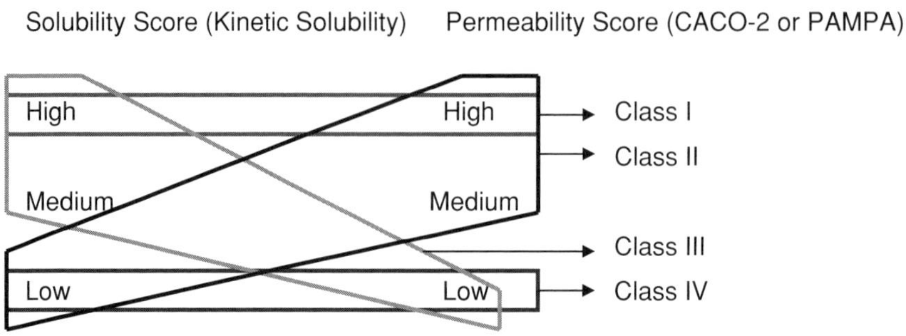

Figure 4. Estimation of DCS classification based on kinetic solubility and permeability (Caco-2 or PAMPA) information as binned by values cited in the text.

Compound Selection/Early Animal Pharmacokinetics

As a compound evolves, the formulateability of a drug candidate is derived from its behavior in *in vivo* systems. One of the most important aspects in this regard is the oral bioavailability of the material since this combines absorbability information from solubility and permeability together with other factors (Burton et al., 2002; Stoner et al., 2004; Parrott et al., 2005; Jolivette and Ward, 2005; Yang et al., 2005; Singh 2006). While a well-absorbed compound can have poor bioavailability, a poorly absorbed compound, by definition, will have a poor pharmacokinetic profile (Van de Waterbeemd and Gifford, 2003). A knowledge of the solubility and permeability of the drug candidate can suggest appropriate solubilization strategies resulting in adequate systems for proving principle and assessing toxicity.

As a compound moves further in development towards candidate selection, more drug substance is available, allowing better refinements to the provisional DCS score. At this stage, traditional thermodynamic solubilities are available using shake-flask and other techniques (Blasko et al., 2001; Fiese 2003; Balbach and Korn, 2004). Cut-off values are library- and company-specific, but relate to the likelihood that formulation problems will be encountered. Lipinski suggests that compounds with a solubility limit less than 50 μg/mL will cause formulation issues (Lipinski et al., 1997). Others have reported other limits. In the context of the current model, thermodynamic solubilities suggest the following classification (Figure 1):

Class I >1 mg/mL
Class II (simple) 10 μg/mL < x < 1 mg/mL
Class II (complex) < 10 μg/mL

The kinetics of absorption can be determined through a number of ways *in vivo* including deconvolution of blood levels obtained after oral dosing as well as in situ intestinal perfusion (Johnson and Swindell, 1996; Veng 1980; Veng 1980a). In the latter technique, the drug is formulated in a MES buffer at pH 6.5 and perfused through rat jejunum at a flow rate of 0.25 mL/min at drug concentrations of 20 to 100 μM. The amount lost during perfusion is assumed to be due to absorption from which a permeability estimate can be derived and, subsequently, an extrapolated maximum absorbable daily dose in man based on the following equation:

$$\text{MADD} = k_a \times S \times \text{SIWV} \times \text{SITT}$$

where MADD is the maximal absorbable daily dose, k_a is the rate of absorption based on a given perfusion rate, S is the solubility of the drug at pH 6.5, SIWV is the volume of water in the small intestine and SITT is the transit time through the small intestine (Johnson and Swindell, 1996; Curatolo 1998).

This system can give DCS classification based on the following values for permeability:

Class I $> 6 \times 10^{-5}$ cm/sec
Class II (simple) 3×10^{-5} cm/sec $< x < 6 \times 10^{-5}$ cm/sec
Class II (complex) $< 3 \times 10^{-5}$ cm/sec

A comparison of the DCS value estimated using *in silico*, HTS and rat in situ perfusion estimated for a given subset of clinical candidates is given in Table 1 based on internal historical data. As indicated, *in silico* data predicted DCS class correctly in 74% of cases and HTS data in 81% of cases using rat in situ perfusion as the reference data set. In many cases, errors were associated with the misclassification of DCS Class I or III with class II or IV (permeability errors). Misclassifications based on solubility errors were less common. Also some very poorly soluble Class II compounds can masquerade as Class IV candidates based on precipitation and adsorption to the filters and plastics associated with the permeability apparatus.

Human Pharmacokinetics

The ultimate judge of the BCS/DCS class is the behavior of the drug in man. Absorption rate is generally assessed by pharmacokinetic deconvolution of blood concentrations (Chan and Gibaldi, 1985; Mahmood 2004). As a first step and based on preformulation screening, i.v. dosing, drug-in-capsule or a simple suspension provides useful comparative data (Balbach and Korn, 2004). For a complex Class II compound, formulation modification is usually required and the comparison of the modified drug substance versus the unmodified material can provide useful data on what form the dosage design efforts should take.

Principles of Solubility, Dissolution Rate and Supersaturation

This evolving classification system is designed to give insight into the nature of the formulation challenge and, to some extent, what techniques will likely be useful in generating appropriate formulations including those needed for toxicology assessment. The limiting parameters for oral bioavailability for Class II

	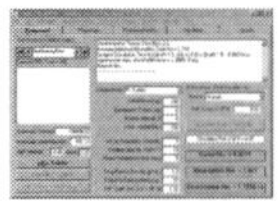		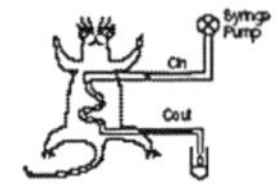
Compound	**GastroPlus DCS score**	**HTS DCS score**	**Rat in situ perfusion-based DCS score**
Compound 1	1	1	1
Compound 2	2	2	1
Compound 3	1	3	3
Compound 4	1	3	3
Compound 5	1	1	1
Compound 6	2	1	1
Compound 7	3	2	3
Compound 8	2	2	2
Compound 9	2	2	2
Compound 10	2	2	2
Compound 11	1	3	3
Compound 12	2	2	2
Compound 13	1	1	1
Compound 14	2	4	2
Compound 15	1	3	1
Compound 16	2	4	4
Compound 17	2	2	2
Compound 18	1	3	3
Compound 19	3	3	3
Compound 20	2	2	2
Compound 21	2	2	2
Compound 22	2	2	2
Compound 23	2	2	2
Compound 24	2	2	2
Compound 25	2	4	2
Compound 26	2	2	2
Compound 27	1	1	1
% Correct*	74%	81%	100%

Table 1. Sample DCS Associated with *in silico*, HTS and in situ rat intestinal perfusion.

compounds are solubility and dissolution rate. To that end, a short overview of these factors will be assessed before specific formulation suggestions are made.

Solubility

Solubility is a complex phenomenon related to the free energy of the solid being dissolved and the free energy of the formed solution (Liu, 2000, Fiese, 2003; Strickley, 2004). The free energy of the solid is an intrinsic property of the specific material assessed while the free energy of solution depends on the solvent and solute concentration. As long as the free energy of the solution is less than the free energy of the solid, the material will dissolve up to its saturation solubility (that is, the point at which the free energies of the solution and the solid are the same). Increasing the solubility of a solid can be completed by increasing the free energy of the solid (i.e., changing the salt or polymorph form or by reduction of particle size). A second approach is to decrease the chemical potential (μ) of the solute in solution by inclusion of cosolvents or solubilizing excipients (Strickley 2004). Cosolvents decrease the chemical potential of the bulk media meaning a higher solution concentration is needed to reach unity between solid and solution free energies. Complexing agents perform the same function but via specific interaction mechanisms. For ionizable compounds, pH adjustment can increase solubility since the ionized species is more water soluble than the corresponding unionized entity. In general, the aqueous solubility of a weak acid will increase 10-fold for every pH unit above its pKa and the solubility of a weak base will increase 10-fold for every pH unit below its pKa. pH adjustment is the easiest solubilization method explaining why this figures so highly in formulation decision trees for liquid systems.

One additional comment is that not all poorly soluble compounds (Class II and IV) have the same root causes for their limiting solubility. Solubility can be controlled by log P meaning that methods to improve wettability may be paramount. Alternatively, poor aqueous solubility may be associated with a high melting point meaning that crystal lattice forces are responsible. These two factors give rise to the terms "greaseballs" (compounds whose solubility is log P-limited) and "brick dust" (compounds whose solubility is melting point-limited). A semi-qualitative indication of whether a compound falls into the "greaseball" or "brick dust" category can be assessed using the empirical relationships of Yalkowsky such as:

$$\text{Log Sw} = 0.5 - 0.01(\text{Tm-25}) - \text{Log P}$$

where Log S_w is the log of the aqueous solubility, log P is the log of the octanol-water partition coefficient and Tm is the melting point for the drug (Yalkowski and Valvani, 1980; Valvani et al., 1981; Jain and Yalkowsky, 2001). These finding were subsequently interpreted in a theoretical context by Amidon and Williams (1982). These considerations suggest that different approaches may be useful for solubilizing compounds whose solubility is melting point-related (where increasing the chemical potential of the solid may be useful by changing the physical form of the drug (Gardner et al., 2004; Hilde and Morris, 2004)) versus those whose solubility is log P-related (where decreasing the chemical potential of the solution may be useful). To this end, application of aspects of both approaches may be advantageous.

Dissolution Rate

A significant aspect of pharmaceutical research aimed at improving the performance of poorly water-soluble drugs has been targeted to overcoming limitations imposed by the (modified) Noyes-Whitney equation which defines dissolution rate (dC/dt) as:

$$\frac{dC}{dt} = \frac{D \times A \times (C_s - C_t)}{h \times V}$$

where D is the diffusion coefficient, h, the diffusion layer thickness at the solid-liquid interface, A, the surface area of drug exposed to the dissolution media, V, the volume of the dissolution media, C_s, the saturation solubility of the drug and C_t, the drug concentration at time, t (Noyes and Whitney, 1897; Horter and Dressman, 2001; Dokoumetzidis et al., 2006). That is, dissolution rate can be increased by increasing the surface area of the drug (via micro- or nanosizing), by decreasing the diffusional layer thickness (through improving wettability by e.g. addition of surfactants or by hydrodynamic parameters) and by altering the saturation solubility of the drug (through formation of a supersaturated drug solution via solid dispersion, complexation approaches or by manipulation of the solid form to give more soluble salts, polymorphs or amorphous material).

Supersaturation

Supersaturated systems are metastable solutions containing the solute of interest at a concentration above its saturation solubility (Liu 2000; Kashchiev and Van Rosmalen, 2003). Depending on the physical stability and their tendency to precipitate, supersaturated systems are of significant interest in pharmaceutical dosage form design from both the topical and oral standpoints (Gao et al., 2004). While not directly applicable to oral dosing, lessons learned from the transdermal field can be useful in addressing oral and parenteral formulation. The optimal delivery of drugs through and to the skin is a complex balance of a number of formulation features such as the ability of the drug to penetrate the barrier layers of the skin including the stratum corneum (the pull properties) as well as the ability of the drug to leave the formulation (the push properties). The formulation push is improved not simply by increasing the concentration of drug in the formulation but by optimizing its thermodynamic activity (saturation state) in the vehicle (Pellet et al., 1994; Pellet et al., 1997; Pellet et al., 1997a). Supersaturated systems can play an important role in this context. For poorly water soluble drugs intended for oral use, the saturation solubility may be far too low to allow for efficient dissolution, absorption or exposure. One method that may be useful in improving the pharmaceutical performance of these drug candidates is to develop trial formulations or dosage forms that generate relatively stable supersaturated systems (Gao et al. 2003; Gao et al., 2004). A number of options are available for this task including the use of solid dispersions and solutions (Taylor and Zografi, 1997; Serajuddin 1999; Serajuddin 1999; Leuner and Dressman, 2000). Depending on their preparation, these formulations contain the drug substance in solution (dispersed at the molecular level) or dispersed at various aggregate levels in a carrier matrix. As the carrier matrix dissolves the drug is released at supersaturated concentrations. Generally, a nucleation inhibitor is needed to stabilize the formed supersaturation solution allowing for absorption and reasonable bioavailability (Krill et al., 2001; Gao et al., 2002; Goa et al., 2004; Crew et al., 2005; Appel et al., 2006). An example is itraconazole which was developed as a solid solution of the drug and hydroxypropylmethylcellulose coated on an inert sugar sphere (Sporanox® Oral Capsule) (De Conde et al., 1999; Peeters et al., 2002). In this dosage form, itraconazole is present in a molecularly dispersed solid solution which dissolves to give a supersaturated solution of the drug in the stomach with the co-dissolving HPMC acting to prevent precipitation. The supersaturated solution is sufficiently stable to allow for >85% fraction absorbed and 55% bioavailability (Brewster et al., 2004).

A number of approaches are available for screening for excipients which stabilize formed supersaturated systems. Peeters et al. (2003) suggested a method which involved solubilizing the drug of interest into a water-miscible organic solvent such as tetrahydrofuran (THF) or dimethylacetamide (DMA) to generate a concentrated solution (100 mg/mL). This solution was then added drop-wise into aqueous solution of 0.01 N HCl maintained at 37°C containing 250 mg of the excipient of interest until a precipitate formed. At various times after addition of the drug (5, 30, 60 and 120 min), a sample of the stirring suspension was withdrawn, filtered and analyzed. The extent of supersaturation and the physical stability of the solution was measured. Excipients that were found to be useful in stabilizing supersaturated solutions could be segregated into several classes:

- pharmaceutical polymers (which may act as nucleation inhibitors) including HMPC, HPC, PVP K30, PVP-VA 64, PEG 400
- complexing agents including HPβCD and SBEβCD
- surfactants, micelle-forming agents and S(M)EDDS-forming agents including cremophor RH40, sodium dodecyl sulfate, polysorbate 20 and 80, vitamin E TPGS

Excipients

In order to achieve appropriate levels of solubilization, dissolution rate and potentially supersaturation, excipients are required (Rowe et al., 2006; International Pharmaceutical Excipients Council 2006). Excipients are components of dosage forms which are included for a variety of reasons and are usually not directly associated with the pharmacological activity of the selected drug substance. Historically, excipients have comprised ingredients e.g., to mask or improve taste, to optimize binding properties of a granulate or to color the final formulation. The reasons for choosing excipients have changed over the years based on the fact that the drug substance needs more and more help to dissolve, be absorbed and reach its active site. Thus, excipients such as preservatives, solubilizers and disintegration agents (i.e., functional excipients) are now commonly used. The International Pharmaceutical Excipient Council (2006) defines an excipient as "any substance other than the active drug or prodrug which has been appropriately evaluated for safety and is included in a drug delivery system to either:

- aid processing of the system during manufacture
- protect, support or enhance stability, bioavailability or patient acceptability
- aid in product identification
- provide any other attribute of the overall safety and effectiveness"

Excipients, unlike the active pharmaceutical ingredient (API), have no regulatory status in and of themselves. In the US, their use is generally supported by widespread experience with these materials based on:

- Being recognized as "generally regarded as safe (GRAS)" by the FDA and other agencies (USA 21CFR182, 184 and 186). Note that use of GRAS listed materials and food additives are specifically related to oral products.
- Being approved as a food additive by the FDA (21CFR171) or other agencies
- Being present in an approved drug product. In this instance, the use of the excipient, without additional data, is limited to circumstances where the same material is used in the same administration route, at the same level of exposure, for the same duration and in the same patient population. The FDA maintains a list of "Inactive Pharmaceutical Ingredients" which contains the product in which they are formulated, their route and the amount in the dosage form (FDA, 2006).

The "Guidance for Industry—Drug Product, Chemistry, Manufacturing and Controls Information" draft issued in January of 2003 (FDA 2003) bin excipients into several types:

- Compendial, Non-Novel Excipients
- Noncompendial, Non-Novel Excipients
- Novel Excipients
- Excipients of human or animal origin
- Excipients that can impart their own pharmacology activity

Compendial, non-novel excipients are listed in national formularies such as the USP/NF, the European Pharmacopeia or the Japanese Pharmacopeia which legally dictates specifications and analytical testing. Their use is usually of low risk but is dependent on dose, route, duration and patient population. Noncompendial, non-novel excipients may require additional testing before use but that again depends on dose, route, duration and patient population. Sources for these excipients may include the USA FDA GRAS list and equivalent indices in Europe (the "E" list) and Japan. Novel excipients are those that either have not been used before or have not been used by the suggested route of administration. Novel excipients require additional testing for NDA approval at a level similar to that of the drug product. The nature and extent of toxicology needed to use an excipient is included in "Guidance for Industry—Nonclinical Studies for the Safety Evaluation of Pharmaceutical Excipients" (FDA 2005). In addition, the guidance does segregate the duration of excipient use into three categories:

- short term (<14 days)
- intermediate term (2 weeks to three months)
- long term (>three months)

The information presented therefore provides a kind of flow chart to assess the risk of using a particular excipient. In increasing order, these would seem to be:

- Use of a compendial excipient in a manner similar to that in a marketed product
- Use of a noncompendial, non-novel excipient in a manner similar to that in a marketed product
- Use of a compendial excipient by the same route but at different doses
- Use of a noncompendial excipient by the same route but at different doses
- Use of a compendial or noncompendial excipient by another administration route
- Use of a novel excipient

Assessing specifically which excipient can be used will then rely on the appropriate database. For the purposes of this overview the following information sources were used:

- The US Pharmacopoeia (USP26/NF21) (USP 2006)
- The European Pharmacopoeia (5th Edition) (EP 2006)
- FDA's Inactive Ingredient Database (FDA 2006)
- FDA GRAS list (21CRF184) (FDA GRAS 2006)
- Monographs and Evaluation of the Joint WHO/FAO Expert Committee on Food Additives (JECFA 2006)
- Guidance for Industry—Q3C Impurities: Residual Solvents (FDA 1997) and Tables and Lists (2003) (FDA 2003a)
- Handbook of Pharmaceutical Additives (Ash and Ash 2005)
- Handbook on Injectable Drugs (Tissel 2001)
- Handbook of Pharmaceutical Excipients (Rowe et al., 2006)
- Adverse Reactions to Drug Formulation Agents, A Handbook of Excipients (Weiner and Bernstein 1989)
- Excipient Toxicity and Safety (Weiner and Kotkoskie 2000)

Toxicological data given for each source can be reported in different formats and using different terminology. For example, International Program on Chemical Safety (IPCS) used the notation of Tolerable Daily Intake (TDI), the World Heath Organization gives an Acceptable Daily Intake (ADI) and the FDA uses the term Permitted Daily Exposure (PDE). A summary of excipients and their potential acceptability for toxicology and first-in-human use is given in Table 2.

Excipient	USP or EP monograph	GRAS	Residual solvent (PDE mg/d)	FDA inactive ingredient list or marketed product
Water Soluble Excipients				
Dimethylacetamide	NO	NO	Class 2 (10.9)	Teniposide (iv)
Dimethyl sulfoxide	YES	NO	Class 3 (50)	iv, sc, topical
Ethanol	YES	YES	Class 3 (50)	Many
Glycerin	YES	YES	NO	Many
N-Methylpyrrolidone	NO	YES	Class 2 (5.3)	Leuprolide (sc)
PEG 300	YES	YES	NO	Many
PEG 400	YES	YES	NO	Many
Poloxamer 407	YES	NO	NO	po, topical
Propylene glycol	YES	YES	NO	Many
HPβCD	EP	NO	NO	Sporanox (po, iv)
SBEβCD	NO	NO	NO	Voraconazole (iv)
α–Cyclodextrin	EP	YES	NO	Alprostadil (ic)
-Phospholipids-				
Phosphatidylcholine	YES*	YES	NO	iv—Doxil, Ambisome
Distearoylphosphatidylglycerol	YES		NO	iv—8.4%
Dimyristoylphosphatidylcholine			NO	Visudyne
Dimyristoylphosphatidylglycerol	YES		NO	0.15%
Water Insoluble Excipients				
Beeswax	YES	YES	NO	Tigan Suppo
Oleic Acid	YES	YES	NO	Ketoprofen Topical
Soy fatty acids			NO	Rapamune
Vitamin E	YES	YES	NO	Amprenavir
Corn oil glycerides	YES		NO	Gynecure Vaginal
-Medium chain mono/di/tri gly.-				

Long chain triglycerides	YES	YES	NO	Propofol
Castor Oil	YES	YES	NO	Tetrazepam
Corn Oil	YES	YES	NO	Sandimmune
Peanut Oil	YES	YES	NO	Prometrium
Peppermint Oil	YES	YES	NO	Nifedipine
Safflower Oil			NO	Propofol
Hydrogenated soybean oil	YES	YES	NO	Vesondid
Hydrogenated vegetable oil	YES		NO	Rondec
Surfactants				
Cremophor EL	YES		NO	Miconazole
Cremophor RH 40	YES		NO	Tegratol
Cremophor RH 60	YES		NO	
Tween 20	YES	YES	NO	Ketoprofen
Tween 80	YES	YES	NO	Docetaxel
Vit E TPGS	YES		NO	Halofantrine
Solutol HS-15	YES		NO	Vitamin A
Span 20	YES		NO	Pevaryl
Softigen 767	YES		NO	Penetran lotion
Labrasol	YES		NO	Prograf
Labrafil M1944CS	YES		NO	
Labrafil M-2125CS	YES		NO	
Polyoxyl 8 stearare	YES		NO	Riopan
Polyoxyl 40 stearate	YES	YES	NO	Solian

Table 2. Potentailly useful excipient for toxicology and first-in-human formulations.

* As Lecithin

Potential Formulations for Toxicology and First-in-Human Testing

The BCS approach relates directly to solid oral dosage forms but the scientific basis of this system can also be applied to liquid formulations intended for parenteral or oral use. For liquids, the two main drivers for appropriate dosage form design is sufficient drug solubility and stability (Strickley, 2004). For parenteral systems, the solubility of drug candidates of interest is often insufficient in simple aqueous medium (Sweetana and Akers, 1996; Strickley, 1999; Akers, 2002). In addition to their commercial utility, liquid formulations are beneficial in the prosecution of early toxicology and clinical trials based on the ability to meter the desired dose as well as the ease of administration. A number of decision trees for the selection of appropriate liquid formulations have been published in the literature focussing on various technologies and excipients (Lee et al., 2003; Strickley 2004). Such approaches are important to minimize the risk of formulation effects which may mask the pharmacological properties of a drug candidate by altering its bioavailability or pharmacokinetics (Bittner and Mountfield, 2002; Bittner and Mountfield, 2002a; Bittner and Mountfield, 2002b). One approach for assessing liquid-based formulations is given in Figure 5. This approach assesses solubility and dosage form complexity but is arguably somewhat pipeline specific. In the first branch of the decision tree, oral liquids can be solutions, where a solubilizing agent may be needed, or a suspension wherein the drug is milled and dosed in an appropriate suspending agent such as hydroxpropylmethylcellulose (HPMC or Methocel). The use of suspension or solution for toxicology evaluation or as clinical trial materials can be driven by a variety of factors as described below:

Solutions

- Gives best opportunity for good oral bioavailability
- Only option (with a few exceptions, i.e., nanosuspensions) for intravenous dosing (Akkar and Muller, 2003)
- Physical form and state of the drug is unimportant meaning that physical form optimization is not essential at an early stage
- Content uniformity is not an issue
- Convenient control of dosing through dilution and easy metering

On the other hand

- The number of poorly soluble drugs is increasing
- There is no universal approach, meaning that each formulation needs to be tailored to the drug. This is inherently time-consuming and labor-intensive
- Depending on their type, solubilizers can exert their pharmacological own actions which may be toxic
- Co-solvents are limited by their mechanism of solubilization (Rao and Stella, 2003)

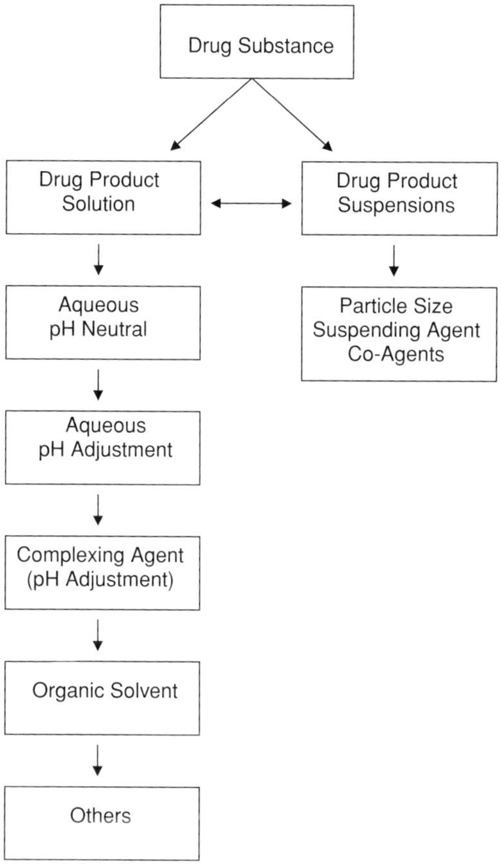

Figure 5. Decision tree of selection of aqueous formulation for toxicology and early clinical use.

- Precipitation from vehicles containing organic modifiers is often problematic
- Chemical stability can be limiting

Suspensions

- They are generally easy to preparc and characterize
- They can be highly standardized (a useful attribute in both GLP and GMP environments)
- The likelihood that consistency between toxicological and clinical formulation can be conserved is reasonably high
- Development time is limited and dosage form configuration is generally straightforward
- Stability data can be collected in an early phase (at the time of preparation of the toxicology supplies) meaning that they are not limiting with regard to clinical dosing

- Dissolution rate is such that these formulations are useful for most drug substances except those associated with very poor water solubility and wettability
- These formulations are, in principle, better mimics of eventual solid dosage forms
- Chemical stability (to light, temperature, etc.) is usually good

On the other hand

- A significant number of oral bioavailability failures are observed with suspensions
- Physical form concerns are important meaning that these features need to be optimized early in the development process
- Particle size can be important especially for very poorly water-soluble drugs. In some cases nanodispersions are required which have different requirements in terms of processing and stabilization
- Suspensions can be supersaturated sometimes giving artifactually higher exposure
- Exact dose is sometimes difficult to determine (soluble fractions versus suspended drug versus the amount of drug that will eventually dissolve)
- Resuspension can be difficult and can affect content uniformity
- Physical form and particle size growth (Oswald Ripening) may not be stable as a function of time

For solution-based systems, a simple system based on dissolving the drug candidate in water or a pH 7.4 aqueous buffer should be attempted. Failing in this, pH adjustment is usually the easiest and safest manner to improve the solubility of an acidic or basic drug in aqueous media. For parenteral products, the range of pH values found in marketed products is reasonable wide (2-12) (Sweetana and Akers 1996; Strickley 2004). For biocompatibility reasons, it is recommended that the pH of the vehicle be maintained between 4 and 8, however values outside of this range may be considered. Similar limits for oral products should be considered although wider latitude may be justified.

The second branch of decision tree involves the use of cyclodextrin complexation (Loftsson and Brewster, 1996; Loftsson and Brewster, 1997; Davis and Brewster, 2004; Loftsson et al., 2004). Cyclodextrins are cyclic sugar oligomers derived from starch containing various numbers of α-1,4-lined glucose residues [(α = hexomer (α-CD), β = heptomer (β-CD), γ = octomer (γ-CD), δ = nonamer (δ-CD))] (Brewster et al., 1989). These oligomaltose systems take the form of a truncated cone wherein the primary hydroxyl functions of the sugar groups are oriented to the narrower end of the torus while the secondary groups are oriented to the wider face. This architecture generates a hydrophilic external surface which provides for aqueous solubility and a lipophilic cavity into which appropriately sized molecules or sub-structures can include and form water-soluble complexes. No covalent bonds are formed or broken during guest/host (i.e., drug/cyclodextrin) complex formation and in solution the guest and host molecules are in a dynamic equilibrium with the complex (Loftsson and Brewster, 1996). In addition to increasing the water solubility of the guest substrates, this equilibrium can increase their chemical stability (based on insulating a sensitive portion of the chemical guest from the environment), reduce unaesthetic smells and tastes, convert oils and liquids to powders and reduce various formulation-based incompatibilities (Brewster et al., 1989; Challa et al., 2005).

Unfortunately, the cyclodextrin which is of the best size to form complexes with the majority of drugs, i.e., β-cyclodextrin (β-CD), is itself only poorly water-soluble. This results from intramolecular hydrogen bonding and associated high crystal lattice energy. To overcome this limitation, β-CD has been chemically altered to generate derivatives which are more soluble but which retain the beneficial complexation characteristics of the parent compound (Uekama et al., 1998; Uekama 1999). Of the chemical derivatives of β-CD, the most successful to date are 2-hydroxypropyl-β-cyclodextrin (HPβCD) and sulfobutylether-β-cyclodextrin (SBEβCD) (Thompson 1997). HPβCD is prepared by treating base-solubilized β-CD with propylene oxide followed by purification. As with other chemically modified pharmaceutical starches, HPβCD is a statistical mixture of various hydroxypropylated isomers with a defined substitution average and substitution distribution. These materials can be made reproducibly and in large quantities. Importantly, HPβCD and SBEβCD are not toxic even at high doses and interact with lipophiles in a quantitatively similar manner as does β-CD (Rajewski et al., 1995; Gould and Scott, 2005).

The use of cyclodextrins is often preferred to organic solvents. This is related to two features of solubilization with these functional excipients, (1) the toxicology profile of HPβCD and SBEβCD is well understood and is superior to many organic solvents (Gould and Scott, 2005) and (2) the nature of solubilization. Cyclodextrins solubilize lipophiles by forming dynamic inclusion complexes in solution (Loftsson and Brewster, 1996). Based on this mechanism, cyclodextrins often solubilize compounds as a linear function of their concentration. This means that as a solution is administered, both the drug and cyclodextrin concentration are reduced in a linear manner suggesting that precipitation is, theoretically, not likely after either oral or i.v. dosing. Organic solvents, on the other hand, solubilize solutes as a log function of their concentration as described by the extended Hildebrand equation (Pitha and Teruhiko 1992; Pitha et al., 1992). This log-linear relationship means that as an organic solvent is introduced into an aqueous environment, the solubilizing power of the formulation is rapidly lost and precipitation can occur. Depending on the nature of this precipitation, the drug can be effectively removed from the equilibria required for absorption from the GI tract. For parenteral liquids, precipitation can occur at the site of injection or at other loci in the body. Once included in the cyclodextrin cavity, the guest molecule may be released through complex dilution, by replacement of the included guest by some other suitably sized molecule such as dietary lipids or, if the complex is located in close approximation to a lipophilic biological membrane such as the oral mucosa, the guest may be transferred to the matrix for which it has the highest affinity.

The regulatory status of the cyclodextrins continues to evolve. β-Cyclodextrin is used in a number of formulations in numerous countries throughout the world. The excipient is compendial in the US (USP24/NF19), Europe (Ph. Eur. 3rd Ed.) and Japan (Japanese Pharmaceutical Codex) and is generally recognized as safe by the FDA (FDA GRAS 2006). HPβCD is available in approved products in the US, Europe and the rest of the world including Sporanox® (itraconazole) oral and i.v. solution, a hydrocortisone-based mouth wash (Iceland) and an indomethacin-based eye drop (France) (Davis and Brewster 2004; Loftsson et al., 2004). A monograph is available in the European Pharmacopoeia (Ph. Eur. 4.6) and HPβCD is cited in the FDA's list of Inactive Pharmaceutical Ingredients (FDA 2006). Recently, an i.m. formulation of Zeldox® (ziprasidone) and an i.v. formulation for voriconazole (Vfend®) which contains SBEβCD has been approved (SBEβCD is not yet compendial but is cited in the FDA's List of Inactive Pharmaceutical Ingredients) while an HPγCD eye drop solution for Voltaren (diclofenac) is available in France. These properties of cyclodextrins, including their proven safety in man, make them useful excipients at all phases of the development process as well as in marketed products.

Extensive toxicology testing has been completed for HPβCD as well as for SBEβCD (Rajewski et al., 1995; Gould and Scott, 2005). With regard to dosing, Sporanox oral and i.v. products contain 10 mg of itraconazole solubilized in 400 mg HPβCD/mL of product (PRD, 2003; Davis and Brewster, 2004). Oral dosing (200 mg) of the active means an excipient dose of 8g HPβCD. Intravenous dosing involves a 400 mg loading dose (days 1 and 2) and a 200 mg dose daily thereafter. This means that 16g of HPβCD are administered for the first two days, followed by 8 g daily dose. The 40% HPβCD solution is diluted two-fold with saline prior to administration. Parenteral (iv) dosing of voriconizole is made possible through the use of SBEβCD. Voriconazole (10 mg/mL) is solubilized in a solution containing 16% w/v of the cyclodextrin. At a drug dosage of 3-6 mg drug/Kg BW, a SBEβCD dose of between 3 and 7 grams are expected (PDR 2003; Davis and Brewster, 2004). As with Sporanox, the i.v. solution is diluted two-fold with saline prior to drug dosing.

The third branch of the decision tree suggests the use of organic solvents and co-solvents. As defined by Strickley (2004), organic cosolvents are generally divided into water-soluble and water-insoluble varieties. Water-soluble vehicles include polyethylene glycol (especially PEG 400), ethanol, propylene glycol and glycerin and can be used as such, in combination with water or in more complex ternary mixtures. Cosolvents can be the best solution in formulating poorly water-soluble drug substances since solubility in these systems can often be orders of magnitude higher than that in water. Co-solvents suffer from two main drawbacks including local and or systemic toxicity and the tendency for the drug to precipitate on dilution. Of the most common co-solvents, PEG 400 is widely used based on its high degree of water miscibility related to extensive hydrogen-bond formation and dipole-dipole interactions. The amount of PEG 400 in oral products varies from a small percentage in multi-component mixtures to high doses as a single excipient. Toxicity data on PEG is available from various sources.

The JECFA (2006) suggest a limit of 10 mg/Kg or about 500-1000 mg in man. These data are extrapolated from animal toxicology and the limit is clearly conservative based on what is in marketed products. The FDA inactive ingredient list suggests formulations containing 18% PEG can be injected and oral solution or suspension containing upwards of 60% PEG are commercially available. The Agenerase® Oral Solution of amprenavir delivers 15.6 g of PEG 400 for the requisite drug dose (Strickley 2004). PEG 400 can also alter gastrointestinal drug transit time in some species which may affect pharmaceutical performance (Schulze et al., 2003). Ethanol is another common solubilizer and is used in oral, topical and parenteral products. Its safe use is suggested in a number of databases including the ICH guidelines on residual solvents (FDA 1997, FDA 2003a). In this registry, ethanol is considered a class 3 material with a generic PDE of 50 mg/day. This suggest a limit of 5000 ppm in the drug product. However, significantly higher amounts are incorporated in marketed drug products and over-the-counter preparations. The FDA inactive ingredient (FDA 2006) list includes parenteral formulations containing almost 50% ethanol and oral solutions or elixirs with greater than 90% ethanol in some cases at significant volumes. Glycerin is also widely available in both parenteral and oral products with percentages of greater than 20% in products designed for i.v. or i.m. dosing and greater than 90% for oral administration. Finally, for propylene glycol the FDA inactive ingredient list suggests formulations containing 50% PG can be injected and oral solution or suspensions containing upwards of 90% PG are commercially available. The JECFA (2006) suggests a limit of 25 mg/Kg or about 2 g in man and suggest that levels of 2500 mg/kg in rat and dog are without toxicological effects. Again, the Agenerase® Oral Solution of amprenavir delivers 50.6 g of propylene glycol for the requisite drug dose. The disconnect which sometimes exist between published safety limits and levels that are already being dosed in approved products complicates a straightforward interpretation of excipient toxicity data. This can result from many reasons including using chronic toxicity limits applied to acute or subacute dosing, using animal data wherein the species tested is more sensitive to the excipient than man or comparing different administration routes. For example, N-methylpyrollidone (NMP, Pharmasolv®) is cited in the residual solvent guidance as having a PDE of 5.3 mg (FDA 2003a). The Q3C ICH figure for NMP is based on inhalation toxicity in the rat where some toxicity was seen in off-springs. NMP is present in marketed formulations far in excess of the 5.3 mg PDE including for the delivery of an in-situ forming gel of leuprolide acetate. This s.c. formulation delivers 160 mg of NMP. An NMP-based formulation for subgingival application of tetracycline-containing gels is also approved.

After cyclodextrins and co-solvents, other approaches can be applied including the use of surfactants and micelle forming agents. Surfactants can be classified as amphoteric (lecithin), non-ionic (Tween 80 or Cremophor EL) or ionic (sodium lauryl sulfate or sodium palmitate). Cremophor is a polyoxyethylenated castor oil derivative which is a common solubilizing excipient in a number of formulations including those for paclitaxel, propofol, teniposide and clanfenur

(Gelderblom et al., 2001). The compound is a useful solubilizer but is also associated with allergic responses in sensitive species and individuals. The material cannot be used in dogs due to anaphylacotoid reactions thought to be associated with histamine release from mast cells. In man, reactions ranging from rash to shock have been observed and appear to be individual and drug product dependent. Even though patients receiving cremophor-containing medication may be pretreated with steroids as well as H1/H2 blockers, some reaction has been noted in ~40% of patients and significant reactions occur in 1.5–3% of treated individuals in the case of paclitaxel (Gelderblom et al., 2001). In addition, cremophor-related dosing is thought to be associated with lipoprotein alterations, with neurotoxicity and reversal of P-glycoprotein activity. The WHO guidelines suggest an ADI of up to 7.5 mg/kg bw (JECFA 2006). Vitamin E-TPGS is gaining widespread use as a safe solubilizer (Wu and Hopkins, 1999). It is used orally in marketed products at doses in excess of 10 g/d.

Conclusions: How Will Excipients Be Evaluated for Potential Use in GLP, GMP and GCP Evaluation

The collected data suggest that historical decision trees for the use of excipient in toxicology investigation are rapidly becoming obsolete. This has prompted a search for additional approaches and solubilizing formulation adjuncts. A possible way forward in trying to assess what excipients are useful may include: (1) Determine the type of excipient based on novelty, compendial status and expected duration of dosing. (2) Assess whether the excipient is GRAS listed, is present in a marketed product or has been evaluated by the JECFA or other agencies. Assess dose, duration of administration, route and specific patient population to see if this is consistent with its intended use and if this is different from the suggested Phase I investigation. (3) Assess whether the excipient is considered a residual solvent and note the class and if appropriate the PDE. Critically assess the nature of the data provided and compare this with the use of these solvent in marketed formulations with regard to dose, duration, route and patient population. (4) Design the GLP toxicology studies in a balanced way using data generated to support human clinical testing. While risk increases with the novelty of the excipient, it should not be a matter of policy to automatically exclude an excipient. (5) If risk-benefit analysis deems it necessary, a novel excipient may be characterized in keeping with published guidelines and monographs (Steinberg et al., 1996).

Acknowledgements

The authors are grateful to Prof. Jennifer Dressman for her kind review of the chapter as well as scientific suggestions.

List of Abbreviations

αCD ... α-Cyclodextrin
ADI ... Acceptable Daily Intake
API ... Active Pharmaceutical Ingredient
βCD ... β-Cyclodextrin
BCS ... Biopharmaceutical Classification System
CFR ... United States Code of Federal Regulations
CMC ... Chemistry, Manufacture and Control
DCS ... Developmental Classification System
DMA ... Dimethylacetamide
DMSO ... Dimethyl sulfoxide
EP ... European Pharmacopoeia
FDA ... United States Food and Drug Administration
γCD ... γ-Cyclodextrin
GCP ... Good Clinical Practice
GI ... Gastrointestinal
GLP ... Good Laboratory Practice
GMP ... Good Manufacturing Practice
GRAS ... Generally Regarded as Safe
HPβCD ... 2-Hydroxypropyl-β-cyclodextrin
HPγCD ... 2-Hydroxypropyl-γ-cyclodextrin
HPC ... Hydroxypropylcelluose
HPMC ... Hydroxypropylmethylcellulose
IPCS ... International Program on Chemical Safety
IPEC ... International Pharmaceutical Excipient Council
JECFA ... Joint WHO/FAO Expert Committee on Food Additives
JP ... Japanese Pharmacopoeia
MADD ... Maximal Absorbable Daily Dose
NF ... United States National Formulary
NMP ... N-Methylpyrollidone
PAMPA ... Parallel Artificial Membrane Permeability Assay
PDE ... Permitted Daily Exposure
PEG ... Polyethylene glycol
PVP ... Polyvinylpyrollidone
PVP-VA ... Polyvinylpyrollidone-co-vinyl acetate
SBEβCD ... Sulfobutylether-β-cyclodextrin
TDI ... Total Daily Intake
THF ... Tetrahydrofuran
USP ... United States Pharmacopeia

References

Agoram B, Woltosz W, and Bolger M. Predicting the Impact of Physiological and Biochemical Processes on Oral Drug Absorption. *Adv Drug Deliv Rev* 2001; 50: S41–S67.

Akers MJ. Excipient-Drug Interactions in Parenteral Formulations. *J Pharm Sci* 2002; 91: 2283–2300.

Akkar A and Müller RH. Intravenous Itraconazole Emulsions Produced by SolEmuls Technology. *Eur J Pharm Biopharm* 2003; 56: 29–36.

Amidon GL, Lennernås H, Shah VP, and Crison JR. Theoretical Basis for a Biopharmaceutical Drug Classification: The Correlation of *in vitro* Drug Product Dissolution and *in vivo* Bioavailability.*Pharm Res* 1995; 12: 413–420.

Amidon G and Williams NA. A Solubility Equation for Non-Electrolytes in Water. *Int J Pharm* 1982; 11: 249–256.

Apple LE, Babcock WC, Friesen DT, Ray RJ, Shamblin RM, and Smithey DT. Pharmaceutical Dosage Forms Comprising a Low Solubility Drug and Polymer. International Patent Appl 2006: WO2006024944 A2.

Ash M and Ash I. *Handbook of Pharmaceutical Additives*. Gower Publishing Ltd, Hampshire, UK, 1997.

Avdeef A. The Rise of PAMPA. *Expert Opin Drug Metab Toxicol* 2005; 1: 325–342.

Balbach S and Korn C. Pharmaceutical Evaluation of Early Developmental Candidates "The 100 mg Approach." *Int J Pharm* 2004; 275: 1–12.

Bergstrom CAS, Strafford M, Lazorova L, Avdeef A, Luthman K, and Artursson P. Absorption Classification of Oral Drugs Based on Molecular Surface Properties. *J Med Chem* 2003; 46: 558–570.

Bittner B and Mountfield RJ. Formulation and Related Activities for the Oral Administration of Poorly Water-Soluble Compounds in Early Discovery Animal Studies: An Overview of Frequently Applied Approaches Part 1. *Pharmazeut Indust* 2002; 64: 800–807.

Bittner B and Mountfield RJ. Formulation and Related Activities for the Oral Administration of Poorly Water-Soluble Compounds in Early Discovery Animal Studies: An Overview of Frequently Applied Approaches Part 2. *Pharmazeut Indust* 2002a; 64: 985–991.

Bittner B and Mountfield RJ. Intravenous Administration of Poorly Soluble New Drug Entities in Early Drug Discovery: The Potential Impact of Formulation on Pharmacokinetic Parameters. *Curr Opin Drug Disc Dev* 2002b; 5: 59–71.

Blasko A, Leahy-Dios A, Nelson W, Austin S, Killion R, Visor G, and Massey I. Revisiting the Solubility Concept of Pharmaceutical Compounds. *Monat Chem* 2001; 132: 789–798.

Box KJ, Volgyi G, Baka E, Stuart M, Takacs-Novak K, and Comer J. Equilibration Versus Kinetic Measurements of Aqueous Solubility and the Ability of

Compounds to Supersaturate in Solution—A Validation Study. *J Pharm Sci* 2006; 95: 1298–1307.

Brewster ME, Simpkins JW, Hora MS, Stern WC, and Bodor N. Review: Potential Use of Cyclodextrins in Parenteral Formulations. *J Parent Sci Technol* 1989; 43: 231–240.

Brewster ME, Verreck G, Chun I, Rosenblatt J, Mensch J, Van Dijck A, Noppe M, Arien T, Bruining M, and Peeters J. The Use of Polymer-Based Electrospun Nanofibers Containing Amorphous Drug Dispersions in the Delivery of Poorly Water-Soluble Pharmaceuticals. *Die Pharmazie* 2004; 59: 387–391.

Burton P, Goodwin JT, Vidmar TJ, and Amore BM. Predicting Drug Absorption: How Nature Made it a Difficult Problem. *J Pharm Exp Therap* 2002; 303: 889–895.

Camenisch G, Alsenz J, Van de Waterbeemd H, and Folkers G. Estimation of Permeability by Passive Diffusion through Caco-2 Cell Monolayers Using the Drug's Lipophilicity and Molecular Weight. *Eur J Pharm Sci* 1998; 6: 313–319.

Challa R, Ahuja A, Ali J, and Khar RK. Cyclodextrin in Drug Delivery: An Updated Review. *AAPS PharmSciTech* 2005; 6: E329–E357.

Chan K and Gibaldi M. Assessment of Drug Absorption after Oral Administration. *J Pharm Sci* 1985; 74: 388–393.

Cheng C, Yu L, Lee H, Yang C, Lue C, and Chou C. Biowaiver Extension Potential to BCS Class III High Solubility-Low Permeability Drugs: Bridging Evidence for Metformin Immediate-release Tablet. *Eur J Pharm Sci* 2004; 22: 297–304.

Crew MD, Shanker RM, Smithey DT, Miller WK, and Friesen DT. Stabilized Pharmaceutical Solid Compositions of Low Solubility Drugs, Poloxamers and Stabilizing Polymers. *International Patent Appl* 2005: WO 2005065656 A2.

Curatolo W, Physical Chemical Properties of Oral Drug Candidates in the Discovery and Exploratory Development Setting. 1998; PSTT 1: 387–393.

Dannenfelser R, He H, Joshi Y, Bateman S, and Serajuddin ATM. Development of Clinical Dosage Forms for a Poorly Water Soluble Drug I: Application of Polyethylene Glycol-Polysorbate 80 Solid Dispersion Carrier System. *J Pharm Sci* 2004; 93: 1165–1175.

Davis ME and Brewster ME. Cyclodextrin-Based Pharmaceutics: Past, Present, Future. *Nature Rev Drug Discov* 2004; 3: 1023–1035.

De Beule K and Van Gestel J. Pharmacology of Itraconazole. *Drugs* 2001; 61(Suppl. 1): 27–33.

De Conde V, Gilis P, and Putteman P. Pellets having a Core Coated with a Lipid Lowering Agent and a Polymer. International Patent 1999: WO 9955313.

Dehring KA, Workman H, Miller K, Mandagere A, and Poole S. Automated Robotic Liquid Handling/Laser-based System for High Throughput Measurement of Kinetic Aqueous Solubility. *J Pharm Biomed Anal* 2004; 35: 447–456.

Dokoumetzidis A, Papadopoulou V, and Macheras P. Analysis of Dissolution Rate Using Modified Versions of Noyes-Whitney Equation and the Weibull Function. *Pharm Res* 2006; 23: 256–261.

Dressman JB, Amidon G, Reppas C, and Shah VP. Dissolution Testing as a Prognostic Tool for Oral Drug Absorption: Immediate Release Dosage Forms. *Pharm Res* 1998; 15: 11–22.

Dressman J, Butler J, Hempenstall J, and Reppas C. The BCS: Where do We go from Here? *Pharm Tech* 2001; 25: 68–76.

FDA, Guidance for Industry. Q3C Impurities: Residual Solvents. http://www.fda.gov/cder/ guidance/Q3Cfinal.htm (Dated 1997).

FDA, Guidance for Industry. Drug product—Chemistry, Manufacturing and Control Information. http://www.fda.gov/cder/guidance/1215dft.pdf (Dated 2003).

FDA, Guidance for Industry. Q3C—Tables and List. http://www.fda.gov/cder/guidance/Q3CT&Lrev1.pdf (Dated 2003a).

FDA, Guidance for Industry. Nonclinical Studies for the Safety Evaluation of Pharmaceutical Excipients. http://www.fda.gov/cder/guidance/5544fnl.htm (Dated 2005).

FDA, Center for Drug Evaluation and Research, Inactive Ingredient Database. http://www.accessdata.fda.gov/scripts/cder/iig/index.cfm (2006).

FDA GRAS, Summary of All GRAS Notices. http://www.cfsan.fda.gov/~rdb/opa-gras. html (May 2006).

Fiese EF. General Pharmaceutics—The New Physical Pharmacy. *J Pharm Sci* 2003; 92: 1331–1342.

Gao P, Guyton ME, Huang T, Bauer J, Stefanski K, and Lu L. Enhanced Oral Bioavailability of a Poorly Water Soluble Drug PNU-91325 by Supersaturable Formulations. *Drug Dev Ind Pharm* 2004; 30: 221–229.

Gao P, Hageman MJ, Morozowich W, Dalga RJ, Stefanski KJ, Huang T, Karim A, Hassan F, and Forbes, JC. Pharmaceutical Compositions Having Reduced Tendency for Drug Crystallization. *International Patent Appl* 2002: WO 02056878 A2.

Gao P, Rush BD, Pfund WP, Huang T, Bauer JM, Morozowich W, KUO M, and Hageman MJ. Development of a Supersaturable SEDDS (S-SEDDS) Formulation of Paclitaxel with Improved Oral Bioavailability. *J Pharm Sci* 2003; 92: 2386–2398.

Gardner CR, Walsh CT, and Almarsson O. Drugs as Materials: Valuing Physical Form in Drug Discovery. *Nature Rev Drug Discov* 2004; 3: 926–934.

Gelderblom H, Verweij J, Nooter K, and Sparreboom A. Chrmophor EL: The Drawbacks and Advantages of Vehicle Selection for Drug Formulation. *Eur J Cancer* 2001; 37: 1591–1598.

Gould S and Scott RC. 2-Hydroxypropyl-β-cyclodextrin (HP-β-CD): A Toxicological Review. *Food Chem Toxicol* 2005; 43: 1451–1459.

Hidalgo I. Assessing the Absorption of New Pharmaceuticals. *Curr Top Med Chem* 2001; 1: 385–401.

Hilde LR and Morris KR, Physics of Amorphous Solids. *J Pharm Sci* 2004; 93: 3–12.

Horter D and Dressman JB. Influence of Physicochemical Properties on Dissolution of Drugs in the Gastrointestinal Tract. *Adv Drug Deliv Rev* 2001; 46: 75–87.

Ingels F and Augustijns P. Biological, Pharmaceutical and Analytical Considerations with Respect to the Transport Media Used in the Absorption Screening System, Caco-2. *J Pharm Sci* 2003; 92: 1545–1558.

International Pharmaceutical Excipients Council. IPEC, the International Pharmaceutical Excipients Council is the World-wide Trade Association of Manufacturers and Users of Pharmaceutical Excipients. http://www.ipec.org (2006).

Jain N and Yalkowsky SH. Estimation of Aqueous Solubility I: Application to Organic Non-electrolytes. *J Pharm Sci* 2001; 9: 234–252.

JECFA. Monographs, and Evaluation of the Joint WHO/FAO Expert Committee on Food Additives. http://www.inchem.org/pages/jecfa.html (2006).

Johnson KC. Dissolution and Absorption Modeling: Model Expansion to Simulate the Effects of Precipitation, Water Absorption, Longitudinally Changing Intestinal Permeability, and Controlled Release on Drug Absorption. *Drug Devlop Ind Pharm* 2003; 29: 833–842.

Johnson KC and Swindell AC. Guidance in the Setting of Drug Particle Size Specifications to Minimize Variability in Absorption. *Pharm Res* 1996; 1795–1798.

Johnson S and Zheng W. Recent Progress in the Computational Prediction of Aqueous Solubility and Absorption. *AAPS J* 2006; 8: E27–E40.

Jolivette L and Ward K. Extrapolation of Human Pharmacokinetic Parameters from Rat, Dog and Monkey Data: Molecular Properties Associated with Extrapolative Success or Failure. *J Pharm Sci* 2005; 94: 1467–1483.

Jung S, Choi S, Um S, Kim J, Choo H, Choi S, and Chung S. Prediction of the Permeability of Drug through Study on Quantitative-Permeability Relationship. *J Pharm Biomed Anal* 2006; 41: 469–475.

Kalantzi L, Reppas C, Dressman JB, Amidon G, Junginger HE, Midha KK, Shah VP, Stavchansky SA, and Barends DM, Biowaiver Monograph for Immediate Release Solid Dosage Forms: Acetominophen (Paracetamol). *J Pharm Sci* 2005; 95: 4–14.

Kashchiev D and Van Rosmalen GM. Review: Nucleation in Solutions Revisited. *Cryst Res Technol* 2003; 38: 555–574.

Kerns EH, Li C, Petusky S, Farris M, Ley R, and Jupp P. Combined Application of Parallel Artificial Membrane Permeability Assay and Caco-2 Permeability Assays in Drug Discovery. *J Pharm Sci* 2004; 93: 1440–1453.

Koks C, Meenhorst P, Bult A, and Beijnen J. Itraconazole Solution: Summary of Pharmacokinetic Features and Review of Activity in the Treatment of Fluconazole-resistant Oral Candidosis in HIV-infected Persons. *Pharmacol Res* 2002; 46: 195–201.

Kola I and Landis J. Can the Pharmaceutical Industry Reduce Attrition Rates? *Nature Rev Drug Discov* 2004; 2: 711–715.

Krill SK, Fort JJ, Law D, Schmitt EA, and Qiu Y. Inhibitors of Crystallization in a Solid Dispersion. *International Patent Appl* 2001: WO 0134119 A2.

Kuentz M, Nick S, Parrott N, and Roethlisberger D. A Strategy for Preclinical Formulation Development Using GastroPlus as Pharmacokinetic Simulation Tool and a Statistical Screening Design Applied to a Dog Study.*Eur J Pharm Sci* 2006; 27: 91–99.

Lee Y, Zocharski PD, and Samas B. An Intravenous Formulation Decision Tree for Discovery Compound Formulation Development. *Int J Pharm* 2003; 253: 111–119.

Leuner C and Dressman J. Improving Drug Solubility for Oral Delivery using Solid Dispersion. *Eur J Pharm Biopharm* 2000; 50: 47–60.

Lipinski CA. Avoiding Investment in Doomed Drugs. *Curr Drug Discov* 2001; 1: 17–19.

Lipinski CA, Lombardo F, Dominy BW, and Feeney PJ. Experimental and Computational Approaches to Estimate Solubility and Permeability in Drug Discovery and Development Settings. *Adv Drug Deliv Rev* 1997; 23: 3–25.

Liu R. *Water-insoluble Drug Formulation*. Interpharm Press, Englewood, CO, 2000.

Liu R. Meeting of the American Association of Pharmaceutical Scientists (AAPS)—Roundtable: Challenges and New Approaches to the Formulation of Poorly Water-Soluble Compounds for the Support of Toxicology Studies, Meeting of the American Association of Pharmaceutical Scientists (AAPS), Nashville, Tennessee, November 6–10, 2005.

Loftsson T and Brewster M. Pharmaceutical Applications of Cyclodextrins. I. Drug Solubilization and Stabilization. *J Pharm Sci* 1996; 85: 1017–1025.

Loftsson T and Brewster M. Cyclodextrins as Pharmaceutical Excipients. *Pharmaceut Technol Europe* 1997; 9: 26–34.

Loftsson T, Brewster ME, and Masson M. Role of Cyclodextrins in Improving Oral Drug Delivery. *Am J Drug Deliv* 2004; 2: 261–275.

Mahmood I. Estimation of Absorption Rate Constant (Ka) following Oral Administration by Wagner-Nelson, Loo-Riegelman, and Statistical Moments in the Presence of a Second Peak. *Drug Metab Drug Interact* 2004; 20: 85–100.

Masungi C, Borremans C, Willems B, Mensch J, Van Dijck A, Augustijns P, Brewster M, and Noppe M. Usefulness of a Novel CACO-2 Cell Perfusion System I. In Vitro Prediction of the Absorption Potential of Passively Diffused Compounds. *J Pharm Sci* 2004; 93: 2507–2521.

Noyes AA and Whitney WR. The Rate of Solution of Solid Substances in their Own Solutions. *J Am Chem Soc* 1897; 19: 930–934.

Parrott N, Paquereau N, Coassolo P, and Lave T. An Evaluation of the Utility of Physiologically Based Models of Pharmacokinetics in Early Discovery. *J Pharm Sci* 2005; 94: 2327–2343.

PDR. *Physician's Desk Reference, Fifty-seventh Edition.* Medical Economics Company, Montvale, NJ, 2003.

Peeters J, Neeskens P, Tollenaere J, and Van Remoortere P. Characterization of the Interaction of 2-Hydroxypropyl-β-cyclodextrin with Itraconazole at pH 2, 4 and 7. *J Pharm Sci* 2002; 91: 1414–1422.

Peeters J, Vandecruys R, and Brewster M. The Use of Supersaturation Studies in Early Pharmaceutical Development. American Association of Pharmaceutical Scientists Annual Meeting and Exposition, Salt Lake City, Utah, USA, October 26–30, 2003.

Pellet MA, Castellano S, Hadgraft J, and Davis AF. The Penetration of Supersaturated Solutions of Piroxicam across Silicon Membranes and Human Skin In Vitro. *J Control Rel* 1997; 46: 205–214.

Pellet MA, Davis AF, and Hadgraft J. Effect of Supersaturation on Membrane Transport: 2 Piroxicam. *Int J Pharm* 1994; 111: 1–6.

Pellet MA, Roberts MS, and Hadgraft J. Supersaturated Solutions Evaluated with an In Vitro Stratum Corneum Tape Stripping Technique. 1997; 151: 91–98.

Pitha J and Teruhiko H. Effect of Ethanol on Formation of Inclusion Complexes of Hydroxypropyl Cyclodextrins with Testosterone or with Methyl Orange.*Int J Pharm* 1992; 243–251.

Pitha J, Teruhiko H, Torres-Labandeira J, and Irie T. Preparation of Drug-Hydroxypropyl Cyclodextrin Complexes by a Method Using Ethanol or Aqueous Ammonium Hydroxide as Cosolubilizers. *Int J Pharm* 1992; 80: 253–258.

Prentis, R A, Lis, Y, and Walker, SR. Pharmaceutical Innovation by Seven UK-owned Pharmaceutical Companies.*Br J Clin Pharmacol* 1988; 25: 387–396.

Rajewski RA, Traiger G, Bresnahan J, Jaberaboansari P, Stella VJ, and Thompson DO. Preliminary Safety Evaluation of Parenterally Administered Sulfoalkyl Ether β-Cyclodextrin Derivatives.*J Pharm Sci* 1995; 84: 927–932.

Rao VM and Stella VJ. When Can Cyclodextrins be Considered for Solubilization Purposes? *J Pharm Sci* 2003; 92: 927–932.

Rowe RC, Sheskey PJ, and Owen S (Eds). *Handbook of Pharmaceutical Excipients, Fifth Edition.* Pharmaceutical Press and the American Pharmsists Association, London, 2006.

Schulze JDR, Waddington WA, Ell, PJ, Parsons GE, Coffin MD, and Basit AW. Concentration-Dependent Effects of Polyethylene Glycol 400 on Gastrointestinal Transit and Drug Absorption.*Pharm Res* 2003; 20: 1984–1888.

Serajuddin AT. Solid Dispersion of Poorly Water-soluble drugs. Early Promises, Subsequent Problems and Recent Breakthroughs. *J Pharm Sci* 1999; 88: 1058–1066.

Singh S. Preclinical Pharmacokinetics: An Approach Towards Safer and Efficacious Drugs.*Curr Drug Metab* 2006; 7: 165–182.

Steinberg M, Borzelleca JF, Enters EK, Kinoshita FK, Loper A, Mitchell DB, Tamulinas CB, and Weiner ML. A New Approach to the Safety Assessment of Pharmaceutical Excipients.*Reg Toxicol Pharmacol* 1996; 24: 149–154.

Stoner C, Cleton A, Johnson K, Oh D, Hallak H, Brodfuehrer J, Surendran N, and Han H. Integrated Oral Bioavailability Projection Using In Vitro Screening Data as a Selection Tool. *Int J Pharm* 2004; 269: 241–249.

Stoner C, Gifford E, Stankoic C, Lepsy C, Brodfuehrer J, Prasad J, and Surendran N. Implementation of an ADME Enabling Selection and Visualization Tool for Drug Discovery.*J Pharm Sci* 2004; 93: 1131–1141.

Strickley RG. Parenteral Formulations of Small Molecule Therapeutics Marketed in the United States (1999)—Part I. *PDA J Pharm Sci Technol* 1999; 53: 324–349.

Strickley RG. Solubilizing Excipients in Oral and Liquid Formulations. *Pharm Res* 2004; 21: 201–230.

Sun D, Yu L, Hussain M, Wall D, Smith R, and Amidon G. In Vitro Testing of Drug Absorption for Drug 'Developability' Assessment: Forming an Interface between In Vitro Preclinical Data and Clinical Outcome. *Curr Opin Drug Discov Develop* 2004; 7: 75–85.

Sweetana S and Akers MJ. Solubility Principles and Practices for Parenteral Drug Dosage Form Development. *PDA J Pharm Sci Technol* 1996; 50: 330–342.

Taylor LS and Zografi G. Spectroscopic Characterization of Interactions between PVP and Indomethacin in Amorphous Molecular Dispersions. *Pharm Res* 1997; 14: 1691–1698.

Thompson DO. Cyclodextrins—Enabling Excipients: Their Present and Future Use in Pharmaceuticals. *Crit Rev Therap Drug Carrier Syst* 1997; 14: 1–104.

Tissel AL. *Handbook on Injectable Drugs, Eleventh Edition.* American Society of Heath System Pharmacists, Bethesda, MD, 2001.

Uekama K. Cyclodextrins in Drug Delivery System.*Adv Drug Del Rev* 1999; 36: 1–2.

Uekama K, Hirayama F, and Irie T. Cyclodextrin Drug Carrier Systems. *Chem Rev* 1998; 98: 2045–2076.

Valvani SC, Yalkowsky SH, and Roseman TJ. Solubility and Partitioning: IV Aqueous Solubility and Octanol-Water Partition Coefficients of Liquid Non-Electrolytes.*J Pharm Sci* 1981; 70: 502–507.

Van Breemen R and Li Y. Caco-2 Cell Permeability Assays to Measure Drug Absorption. *Expert Opin Drug Metab Toxicol* 2005; 1: 175–185.

Van de Waterbeemd H, and Gifford E. ADMET In Silico Modelling: Towards Prediction Paradise? *Nature Rev Drug Discov* 2003; 192–204.

Van Dijck A, Masungi C, Mensch J, Borremans C, Willems B, Mackie C, Brewster M, and Noppe, M. Parallel Artificial Membrane Permeability Assay (PAMPA) Combined with a 10-Day Multiscreen CACO-2 Cell Culture for Prediction of Passive and Active Absorption of Drugs. 2003 American Association of Pharmaceutical Scientists Annual Meeting and Exposition, Salt Lake City, Utah, USA, October 26–30, 2003.

Van Gelder J. EUFEPS Conference: When Poor Water Solubility Becomes an Issue: From Early Stage to Proof of Principle, Verona, Italy, April 26–27, 2006.

Veng PP. Novel Deconvolution Method for Linear Pharmacokinetic Systems with Polyexponential Impulse Response.*J Pharm Sci* 1980; 69: 312–318.

Veng PP. Novel Approach to Bioavailability Testing: Statistical Method for Comparing Drug Input Calculated by a Least-square Deconvolution Technique. *J Pharm Sci* 1980a; 69: 318–324.

Weiner M and Bernstein IL. *Adverse Reactions to Drug Formulation Agents: A Handbook of Excipients.* Marcel Dekker, New York, NY, 1989.

Weiner ML and Kotkoskie LA. *Excipient Toxicity and Safety.* Marcel Dekker, New York, NY, 2000.

Wu S and Hopkins W. Characteristics of D-α-Tocopheryl PEG 100 Succinate for Applications as an Absorption Enhancer in Drug Delivery Systems. *Pharm Tech* 1999; 23: 52–68.

Yalkowsky SH and Valvani S. Solubility and Partitioning I: Solubility of Non-Electrolytes in Water. *J Pharm Sci* 1980; 69: 912–922.

Yang Z, Zadjura L, D'Arienzo C, Marino A, Santone K, Klunk L, Greene D, Lin P, Colonno R, Wang T, Meanwell N, and Hansel S. Preclinical Pharmacokinetics of a Novel HIV-1 Attachment Inhibitor BMS-378806 and Prediction of its Human Pharmacokinetics. *Biopharm Drug Disp* 2005; 26: 387–402.

Yee S. In Vitro Permeability Across Caco-2 Cells (Colonic) Can Predict In Vivo (Small Intestinal) Absorption in Man—Fact or Myth. *Pharm Res* 1997; 14: 763–766.

Yu L. An Integrated Model for Determining Causes for Poor Oral Drug Absorption. *Pharm Res* 1999; 16: 1883–1887.

Yu L, Amidon G, Polli J, Zhao H, Mehta M, Conner D, Shah VP, Lesko L, Chen M, Lee VL, and Hussain AS. Biopharmaceutics Classification System: The Scientific Basis for Biowaiver Extensions. *Pharm Res* 2002; 19: 921–925.

Zhao YH, Le J, Abraham MH, Hersey A, Eddershaw PJ, Luscombe CN, Boutina D, Beck G, Sherborne B, Cooper J, and Platts JA. Evaluation of Human Intestinal Data and Subsequent Derivation of a Quantitative Structure-Activity Relationship (QSAR) with the Abraham Descriptors. *J Pharm Sci* 2001; 90: 749–784.

Zhu C, Jiang L, Chen T, and Hwang K. A Comparative Study of Artificial Membrane Permeability Assay for High Throughput Profiling of Drug Absorption Potential. *Eur J Med Chem* 2002; 37: 399–407.

9

Solubilizing Vehicles for Oral Formulation Development

ROBERT G. STRICKLEY AND REZA OLIYAI

Formulation & Process Development, Gilead Sciences, Foster City, CA

This chapter focuses on solubilizing excipients in oral formulations with selected examples of commercially available over-the-counter and prescription human pharmaceutical products. The intent is to educate the reader on the need for and the chemical contents of solubilized oral formulations, and to act as a practical guide in assisting the development scientist in choosing a solubilizing vehicle for oral administration. The formulation philosophy is to minimize the excipients and the suggested development approach is from "simple to complex" (i.e., keep the formulation as simple as possible). This chapter is organized in four sections. The first section is an introduction with basic concepts. The second section addresses solubilizing excipients and mixtures of excipients. The third section deals with liquid-filled capsules. The fourth section focuses on oral solutions. The last three sections are further divided into water-soluble vehicles and lipid-containing vehicles.

Introduction

Drug molecules that are poorly water-soluble can be difficult to effectively administer *in vivo* due to solubility limitations. Fortunately, there is a wide selection of solubilizing excipients that are generally regarded as safe that can be judicially used to safely and effectively administer drugs with a wide variety of physiochemical properties and chemical structures. Information on solubilizing vehicles in pharmaceutical formulations can be obtained in the official prescribing information for a specific commercial product (Physician's Desk Reference, 2006), useful internet websites for products in the United States of America (Internet websites 1–3), United Kingdom (Internet website 4), Japan (Internet website 5), and other countries (Internet websites 6–9), published review articles (Wang and Kowal, 1980; Sweetana and Akers, 1996; Powell et al., 1998; Strickley, 1999;

Oral solution	Liquid-filled capsule
Pediatric	Increase oral bioavailability
Adult dose modification	Low dosage form
Ease of swallowing	Faster onset of action
Oral topical solution	Increase therapeutic effect
Early clinical	Decrease food effect on oral bioavailability
	Drug molecule is an oil

Table 1. Reasons for developing an oral solution or a liquid-filled capsule.

Stuchlik and Zak, 2001; Strickley, 2004), and various company brochures (Internet websites 10–21). For most oral products, only the qualitative listing of excipients is provided and not the exact amount(s), but the United States Food and Drug Administration inactive ingredient database (Internet website 22) does provide the maximum amounts of excipients used in commercial products by the various routes of administration.

Solubilized oral formulations include oral solutions and solutions filled into soft or hard capsules composed of gelatin or hydroxypropylmethylcellulose. The reasons to develop a solubilized formulation in a capsule for oral administration include increasing the oral bioavailability of a poorly water-soluble molecule, a low dosage strength (e.g., 2.5 μg Hectorol® soft gelatin capsules), a drug that occurs as an oil, a rapid onset of action by decreasing the time for maximum plasma concentration, an increased therapeutic effect by increasing the maximum plasma concentration, and to decrease the effect of food on oral bioavailability (Table 1). Oral solutions are easily metered dosage forms which are intended for pediatric administration, adult dose modification, patients who cannot swallow tablets/capsules, and oral topical solutions (Table 1). Oral solutions are also used in early clinical trials as a means to more rapidly evaluate a new chemical entity for Phase I safety and/or pharmacokinetics, since a simple oral solution can be developed and manufactured more rapidly than a solid oral dosage form.

Solubility of an orally administered drug should be viewed as relative to its dose and absorption. In the case where the maximum absorbable dose of the drug is less than its dose, then the limitation is either solubility or permeability of the drug. Permeability is difficult to alter without affecting other physiological processes, but temporary inhibition of intestinal efflux can increase the oral bioavailability of drugs that are substrates for the P-glycoprotein transporter when co-administered with a P-glycoprotein inhibitor such as ritonavir. However, if solubility is the limiting factor then formulation design can often significantly improve *in vivo* performance by using solubilizing excipients that: a) create thermodynamically stable solutions thereby eliminating the solid disintegration and dissolution processes, b) create supersaturated solutions upon dilution, or c) create solubilizing micelles during *in vivo* digestion.

The required solubility of an orally administered drug is usually much higher in a capsule than in a bulk oral solution, since a capsule is limited to approximately one milliliter (mL) in fill volume, whereas an oral solution can be administered in a dose volume of 10's of milliliters. Thus, one of the main challenges in developing a capsule formulation is to achieve the required solubility. The concentration of a drug in an oral solution is a critical variable, not only due to dose relative to solubility, but also taste since most drugs have poor organoleptic properties: the higher the drug concentration the poorer the taste. Many drugs have pH-dependent solubility and the ideal situation in developing an aqueous-based oral solution is when the pH of maximum stability is also the pH of maximum (or acceptable) solubility for a given drug. The balance between drug concentration and dose volume is such that: 1) In an early clinical trial the dose is dissolved in a volume as low as five milliliters to as high as 10's of milliliters; and 2) In an intended commercial oral solution the full adult dose should be contained within a reasonable upper volume, such as thirty milliliters, but also such that the lowest pediatric dose is contained within a measurable lower volume, such a 0.25–1.0 mL. For example, Sustiva® Oral Solution contains 30 mg/mL of efavirenz in which the adults dose is rather high at 600 mg once-a-day, thus the full adult dose is twenty milliliters of the oral solution. The lowest volume that can be administered accurately orally is ~0.05 mL (50 μL). For example, Rocaltrol® oral solution contains calcitriol at 1 μg/mL and the pediatric dose in patients over three years of age is 0.25–0.5 μg, which is 0.25–0.5 mL daily, but for children less than three years of age the dose is 10–15 ng/kg, which is only 0.01–0.015 mL/kg daily.

Bulk solution formulations require solubilizing excipients when the dose of active ingredient is large relative to its aqueous solubility and thus allowing for a convenient dose volume. For water-insoluble drugs, the solubilizing vehicles used in simple oral solutions include aqueous-based cosolvents at pH 2–10, or organic-containing solutions using polyethylene glycol (PEG) 400, propylene glycol (PG), medium-chain-triglyceride, or a mixture such as PEG 400/PG, PEG 400/d-α-tocopherol polyethylene glycol 1000 succinate (TPGS), or PEG 400/PG/ethanol. Many commercial over-the-counter oral solution formulations contain polyethylene glycol, propylene glycol, glycerin, polysorbate 20, or poloxamer 407. The upper %'s of solvent used in oral solutions is up to 100% medium-chain triglyceride, 100% PEG 400, 55% propylene glycol (the higher %'s are contraindicated in children younger than four years of age), 42% ethanol, 40% hydroxypropyl-β-cyclodextrin (HP-ß-CD), and 12% TPGS.

The largest daily amount of an excipient administered from a commercial product is often rather difficult to estimate since both the dose and actual % of excipient are required, but the actual % is usually not reported. However from the available information some estimates can be attempted and Table 2 is a listing of the estimated maximum amount of certain excipients administered orally. It is estimated that the maximum daily amount administered by the

Excipient	Estimated maximum amount administered orally	Product, Drug
Cremophor EL	620 mg, b.i.d.	Norvir® capsule (Ritonavir)
Cremophor RH 40	≥400 mg	(Internet website 22)
Ethanol	3.1 mL, b.i.d. 4.2 mL q.d.	Norvir® Oral Solution (Ritonavir) Kaletra® Oral Solution (Lopinavir/Ritonavir)
Gelucire® 44/14	1.5 grams/day	Solufen-200 mg capsule (Ibuprofen)
Hydroxypropyl-ß-cyclodextrin	8.0 grams, q.d.	Sporanax® Oral Solution (Itraconazole)
Labrasol	1.8 grams/day	Cyclosporin A capsule (generic)
Medium-chain triglyceride	20 mL, q.d.	Sustiva® 30 mg/mL Oral Solution (Efavirenz)
Oleic acid	3.5 grams, b.i.d.	Norvir® capsule (Ritonavir)
PEG 400	16 grams, b.i.d.	Agenerase® Oral Solution (Amprenavir)
Propylene glycol	51 grams, b.i.d.	Agenerase® Oral Solution (Amprenavir)
TPGS	11 grams, b.i.d.	Agenerase® Oral Solution (Amprenavir)

Table 2. Estimated maximum amount of selected excipients administered from oral formulations.

oral route is twenty milliliters of medium-chain triglyceride (Efavirenz, Sustiva® 30 mg/mL Oral Solution), 102 grams of propylene glycol, thirty-two grams of PEG 400, twenty-two grams of TPGS (Amprenavir, Agenerase® Oral Solution), eight grams of HP-ß-CD (Itraconazole, Sporanox® Oral Solution), ~seven grams of oleic acid (Norvir® capsules), ~six mL of ethanol (Ritonavir, Norvir® Oral Solution), ~1.8 grams of Labrasol (Cyclosporin A, generic capsule), ~1.5 grams of Gelucire® 44/14 (Ibuprofen, Solufen-200 mg capsules), and ~1.2 grams of cremophor EL (Ritonavir, Norvir® capsules).

Chemical stability is critical to establishing a shelf-life which can be defined as 5–10% degradation of the active ingredient, and the usual goal for a commercial product is to have a shelf-life of two years at 25°C/60% R.H., and six months at 40°C/75% R.H. The chemical stability requirements for an oral solution intended only for early clinical trials is less stringent than a commercial oral solution, and a shelf-life of six months is normally sufficient. The range in pH for oral solutions is 2–10, and to maximize chemical stability of an aqueous-based oral solution, the ideal solution pH is the pH of maximum stability of the active ingredient. The mechanism of chemical degradation can also affect shelf-life. Chemical stability of a drug in solution will not be concentration dependent if the degradation reaction is first-order with respect to the drug as in a cyclization reaction, or in a pseudo-first-order reaction such as hydrolysis. However, chemical stability of a drug in solution will be concentration dependent if the degradation reaction is second-order such as dimerization or another bimolecular reaction.

When using low or high pH solutions, the choice of buffer capacity is based on a balance between *in vivo* performance (oral bioavailability and local irritation to the mouth and esophagus) and *in vitro* physical/chemical stability. The buffer capacity is minimized to a concentration(s) needed to maintain solution pH during extended shelf storage, but at the same time high enough to keep the drug in solution once the oral solution comes in contact with the stomach. Co-administration of a high pH oral solution may affect the oral bioavailability of a weak basic drug since an elevated stomach pH could potentially reduce the *in vivo* solubility of a weak base.

The topic of solid-state solubilization techniques is beyond the scope of this chapter, but some solubilizing excipients are used in solid-state. Solubilizing techniques in solid-state include established commercially successful approaches using micronized particles (Kim and Park, 2004), and/or surfactants such as sodium lauryl sulfate, to approaches that are promising such as solid solutions, solid dispersions, solid lipid nanoparticles (Bummer, 2004) that have received much research and development attention since the 1960's, and currently are in many clinical trials but there are still few commercial successes (Serajuddin, 1999; Dannenfelser et al., 2004). Manufacturing of solid dispersions using supercritical fluid technology has received considerable attention (York, 2004; Won et al., 2005). SkyePharma PLC in collaboration with First Horizon Pharmaceuticals in 2005 received approval to market a solid-state formulation of Fenofibrate (Triglide™) that can be taken with or without food. Triglide™ is the first approved product that utilizes SkyePharma's proprietary IDD®-P solubilization technology (Internet website 23) and contains micronized particles, egg lecithin, and sodium lauryl sulfate. Entocort® EC capsules were approved in 2005 and contain 3 mg of micronized budesonide, a water-insoluble corticosteroid, and also contain Tween 80 in enteric coated sugar spheres that dissolve at pH > 5.5, from which the oral bioavailability is 9–21% due to 80–90% first-pass metabolism. In 2005 Kaletra® (lopinavir/ritonavir) became available as a solid tablet to replace the soft gelatin capsule, and the solid tablet is made by a melt extrusion process that contains a fixed dose combination of 200 mg of lopinavir and 50 mg of ritonavir and includes the excipients copovidone, sorbitan monolaurate, silicon dioxide, and sodium stearyl fumarate.

Solubilizing Excipients and Mixtures

There is a wide selection of solubilizing excipients that can be generally categorized into various types including water-soluble organic solvent/excipients, water-insoluble organic solvents/excipients, triglycerides, semi-solids, surfactants, phospholipids, and cyclodextrins (Table 3). The question as to which solubilizing vehicle to choose for a particular drug molecule depends on many factors, but in general the formulation philosophy and approach is from simple to complex, meaning to minimize the formulation components. The reasons to minimize the excipients include overall simplicity, cost of goods, ease of procurement, ease of manufacture, and very importantly to minimize toxicity. Table 4 is a flow chart of a suggested order of solubilization approaches for oral formulations arranged in a "simple to more complex" manner. Tables 5 and 6 are listings of solubilizing vehicles used in oral formulation filled into capsules and in oral solutions, respectively. The salt form of the drug can have an impact on the solubility but is not included in this discussion.

Water-Soluble Organic Solvents

The water-soluble organic solvents in commercially available solubilized oral formulations are polyethylene glycol 400 (PEG 400), ethanol, propylene glycol, glycerin, and Transcutol® HP. The solvent Transcutol® is purified diethylene glycol monoethyl ether and is a powerful solubilizer that is used in various European products.

PEG 400 is commonly used in prescription and over-the-counter liquid-filled capsules. Mixtures containing PEG 400 are also commonly used such as PEG 400 and propylene glycol, PEG 400 and medium-chain triglycerides, PEG 400 and peppermint oil, as well as the ternary mixture of PEG 400, propylene glycol, and 8% ethanol. The amount of PEG 400 can affect oral bioavailability, and in humans administration of 150 milliliters of a 1 mg/mL aqueous solution of ranitidine along with one gram of PEG 400 resulted in approximately 48% absorption of ranitidine, while the same aqueous solution but with 2.5 grams or 5 grams of PEG 400 resulted in approximately 21% absorption of ranitidine. The increase in the amount of administered PEG 400 was also associated with a decrease in small intestinal transit time of 270 minutes to 230–240 minutes (Schulze, 2003).

Surfactants

Water-miscible surfactant molecules contain both a hydrophobic and hydrophilic portion, and can solubilize many poorly water-soluble drugs. Surfactants can also self-assemble to form micelles once the surfactant monomer concentration reaches the critical micelle concentration. Thus surfactants can solubilize drug molecules by either a direct cosolvent effect or by uptake into micelles. The non-ionic surfactants in commercially available solubilized oral formulations include polyoxyl 35 castor oil (cremophor EL), polyoxyl 40 hydrogenated

Water-soluble	Water-insoluble excipients	Triglycerides	Surfactants
Solvents	Beeswax	*Long-chain triglycerides*	Polyoxyl 35 castor oil (Cremophor EL)
Ethanol	Corn oil mono-di-triglycerides	Castor oil	Polyoxyl 40 hydrogenated castor oil (Cremophor RH40)
Glycerin	Glyceryl monolinoleate (Maisine™ 35-1)	Corn oil	Polyoxyl 60 hydrogenated castor oil (Cremophor RH60)
PEG 300	Glyceryl monooleate (Peceol™)	Cottonseed oil	Polysorbate 20 (Tween 20)
PEG 400	Glycerol esters of fatty acids (Gelucire® 39/01)	Olive oil	Polysorbate 80 (Tween 80)
Poloxamer 407	Medium chain monoglycerides (Capmul MCM)	Peanut oil	d-α-tocopheryl polyethylene glycol 1000 succinate (TPGS)
Propylene glycol	Medium chain diglycerides	Peppermint oil	Sorbitan monooleate (Span 20)
Transcutol® HP	Oleic acid	Safflower oil	PEG 300 caprylic/capric glycerides (Softigen® 767)
Cyclodextrin	PEG 300 caprylic/capric propylene glycol diesters (Captex 200)	Rapseed oil	PEG 300 linoleic glycerides (Labrafil® M-2125CS)
Hydroxypropyl-β-cyclodextrin	Polyglyceryl oleate (Plurol® Oleique)	Sesame oil	PEG 300 oleic glycerides (Labrafil® M-1944CS)
Phospholipids	Propylene glycol monolaurate (Lauroglycol™ FCC)	Soybean oil	PEG 400 caprylic/capric glycerides (Labrasol®)
Phosphatidylcholine (lecithin)	Propylene glycol dicaprylocaprate (Labrafac™ PG)	Hydrogenated soybean oil	PEG 1500 lauric glycerides (Gelucire® 44/14)
	Soy fatty acids	Hydrogenated vegetable oils	PEG 1500 stearic glycerides (Gelucire® 50/13)
	d-α-tocopherol (Vitamin E)	*Medium-chain triglycerides*	Polyoxyl 8 stearate (PEG 400 monostearate)
		Caprylic/capric triglycerides derived from coconut oil or palm seed oil	Polyoxyl 40 stearate (PEG 1750 monostearate)

Table 3. Solubilizing excipients used in commercially available solubilized oral formulations.

Capsule	Oral solution
Water-soluble organic solvent ↓ Long-chain triglyceride ↓ Medium-chain triglyceride ↓ Water-insoluble organic solvent ↓ Organic solvents and surfactant ↓ Triglyceride and surfactant ↓ Microemulsion ↓ SEDDS (self-emulsifying drug delivery system)	Aqueous, pH 2–10 ↓ Cosolvent (Aqueous/organic solvent), pH 2–10 ↓ Organic solvent(s) (100%) ↓ Aqueous with complexation, pH 2–10 ↓ Oil-in-water emulsion ↓ Microemulsion ↓ SEDDS (self-emulsifying drug delivery system)

Table 4. Flow chart of suggested order of solubilization approaches for oral liquid formulations: "simple to complex".

castor oil (cremophor RH 40), polyoxyl 60 hydrogenated castor oil (cremophor RH 60), polysorbate 20 (Tween 20), polysorbate 80 (Tween 80), d-α-tocopherol polyethylene glycol 1000 succinate (TPGS), sorbitan monooleate (Span 80), polyoxyl 40 stearate, and various polyglycolyzed glycerides including Labrafil® M-1944CS, Labrafil® M-2125CS, Labrasol®, Gelucire® 44/14, Gelucire® 50/13 and Softigen® 767. The ionic phospholipid surfactant egg phosphatidylcholine (lecithin) is present in various commercial and clinical products.

Cremophors are pegylated castor oil or hydrogenated castor oil, and are complex mixtures of relatively hydrophobic and hydrophilic molecules, and are synthesized by reacting either castor oil or hydrogenated castor oil with varying amounts of ethylene oxide. Cremophors are quite effective at solubilizing very hydrophobic drugs and are normally not used alone. Polysorbates are pegylated derivatives of sorbitan monoesters (i.e., monolaurate, monopalmitate, monostearate, monooleate, or monoisostearate) or sorbitan triesters (i.e., tristearate or trioleate) and are available in many grades. Polysorbate 80, also known as Tween 80, is a mixture of polyoxyethylene 20 sorbitan monooleates and is a yellow viscous liquid at room temperature. Sorbitan monooleate, also known as Span 80, is a yellow viscous liquid at room temperature and is the ester of oleic acid and the cyclic hexitol of sorbitol.

There are many different polyglycolyzed glycerides and they are generally used to formulate water-insoluble drugs in lipid based formulations such as self-emulsifying drug delivery systems (SEDDS) in order to improve oral

Number of solubilizing excipients	Solvent mixture	Product, drug
1	Corn Oil	Valproic acid/Depakene®
	Gelucire® 44/14	Ibuprofen/Solufen 200 mg
	Medium chain mono-diglycerides	Dutasteride/Avodart™
	Medium-chain triglycerides	Calcitrol/Rocaltrol®
	Oleic acid	Testosterone Undecanoate/ Restandol 40 mg
	Peanut oil	Progesterone/Prometrium®
	PEG 400	Ibuprofen/Advil® Migrane
	Peppermint oil	Simethicone (over-the-counter)
	Sesame oil	Dronabinol/Marinol®
	Soybean oil	Vitamin A (over-the-counter)
	$\alpha-$tocopherol	Ethyl Icosapentate/Epadel® Capsules 300
2-Binary	PEG 400/Glycerin	Etoposide (Generic)
	PEG 400/Tween 20	Bexarotene/Targretin®
	PEG 400/Peppermint oil	Nimodipine/Nimotop®
	PEG 400/PG	Sudafed® Cold & Cough
	Medium-chain triglyceride/ Ethanol	Paricalcitrol/Zemplarl®
	Medium-chain triglyceride/ Glycol esters of fatty acids	Tocopherol nicotinate/ Juvela® N Soft capsules
	Medium chain mono-diglycerides/α-tocopherol	Saquinavir/Fortovase®
	Glyceryl monooleate/Propylene glycol esters of fatty acid	Menatetrenone/Glakay® Capsules 15 mg
	Sesame oil/α-tocopherol	Alfacalcidol/One-Alpha Capsules
	Gelucire® 44/14 and PEG 20,000	Fenofibrate/Lipirex
3-Ternary	PEG400/Ethanol 8%/PG	Digoxin/Lanoxin®
	Medium-chain triglyceride/ PEG/PG	Robitussin® Cough, Cold & Flu
	Oleic acid/Cremophor EL/ Ethanol	Ritonavir/Norvir®
	Oleic acid/Cremophor EL/PG	Lopinavir and Ritonavir/ Kaletra®
	TPGS/PEG 400/PG	Amprenavir/Agenerase®

(*continued*)

Number of solubilizing excipients	Solvent mixture	Product, drug
	Cremophor RH 60/ Hydrogenated oil/Glyceryl monooleate	Indomethacin farnesil/ Infree® Capsules
	Soybean oil/Hydrogenated vegetable oils/ Hydrogenated soybean oil	Isotretinoin/Accutane®
4-Quatenary	Ethanol/PG/Cremophor EL/ Medium chain mono-diglycerides	Tipranavir/Aptivus®
	Rapseed oil/Hydrogenated soybean oil/Partially hydrogenated plant oils/PG	Clofazimine/Lamprene Capsules 100 mg
5	Ethanol 11.9%/PG/Cremophor RH 40/Corn oil mono-di-triglycerides/α-tocopherol	Cyclosporin A/Neoral®
6	Ethanol/PG/PEG/Cremophor EL/Sorbitan monooleate/ Tween 80	Cyclosporin A/Gengraf®

Table 5. Solubilizing vehicles in oral formulations—capsules.

bioavailability. Polyglycolyzed glycerides are synthesized by the partial pegylation of vegetable oils, and consist of mono-, di-, and triglycerides and mono- and di-fatty acid esters of polyethylene glycol.

The micelle-forming molecule d-α-tocopherol polyethylene glycol 1000 succinate (TPGS) is an effective vehicle for lipid-based drug delivery, and is also a water-soluble source of the water-insoluble oil Vitamin E. TPGS was first synthesized in 1950 by the Eastman Chemical Company as a water-dispersible form of Vitamin E with improved gastrointestinal absorption (Cawley and Stern, 1954). TPGS is self-affirmed as generally recognized as safe, is manufactured to meet the specifications in the United States National Formulary, and is recognized by formulators as 'an effective oral absorption enhancer for improving the bioavailability of poorly absorbed drugs and as a vehicle for lipid-based drug delivery' (Wu and Hopkins, 1999). TPGS forms micelles at concentrations greater than or equal to 0.2 mg/mL in water and improves the aqueous solubility of some drugs (Yu et al., 1993; Constantinides et al., 2006). TPGS is a potent inhibitor of active intestinal efflux even at concentrations ten-fold below the critical micelle concentration, suggesting that monomeric TPGS is capable of inhibiting the efflux mechanism. Therefore, TPGS not only improves *in vivo* performance by solubility-enhancing micelle formation, but also by increasing the overall intestinal permeability via inhibiting an efflux mechanism.

Number of solubilizing excipients	Solvent mixture	Product, drug(s)
1	100% Medium chain mono-diglycerides Water with 40% hydroxypropyl-β-cyclodextrin Water with 23% ethanol	Sustiva® 30 mg/mL Oral Solution (Efavirenz) Sporanox® Oral Solution (Itraconazole) Donnatal® Elixir (Phenobarbital)
2-Binary	40% Propylene glycol and 10% Ethanol PEG/propylene glycol	Lanoxin® Elixir Pediatric (Digoxin) Children's Tylenol® Cold (Acetaminophen, chlorpheniramine maleate, and dextromethorphan)
3-ternary	~12% TPGS, ~17% PEG 400 and ~55% propylene glycol 12.5% Ethanol, olive oil, and Labrafil M-1944CS.	Agenerase® oral solution (Amprenavir) Sandimmune® oral solution (Cyclosporin A)
4-Quatenary	Cremophor EL, propylene glycol, ethanol, peppermint oil, water	Norvir® oral solution (Ritonavir)
5	Propylene glycol, 42% ethanol, glycerin, Cremophor RH 40, peppermint oil, water. 11.9% Ethanol, dl-α-tocopherol, corn oil-mono-di-triglycerides, Cremophor RH 40, propylene glycol.	Kaletra® oral solution (Lopinavir and ritonavir) Neoral® oral solution (Cyclosporin A)
7	Polysorbate 80, phosphatidylcholine, propylene glycol, mono- and diglycerides, 1.5–2.5% ethanol, soy fatty acids, ascorbyl palmitate.	Rapamune® Oral Solution (Sirolimus)

Table 6. Solubilizing vehicles in oral formulations—oral solutions.

Cyclodextrins

Cyclodextrins are cyclic (α-1,4)-linked oligosaccharides of α-D-glucopyranose containing a relatively hydrophobic central cavity and a hydrophilic outer surface. Cyclodextrins are designated α-, β-, or γ-corresponding to six, seven or eight glucopyranose units, with cavity diameters of 4.7–5.3, 6.0–6.5 and 7.5–8.3 Å, respectively (Irie and Uekama, 1997). The central cavity is lined with methylene groups (-CH_2-) and ethereal oxygens of the glucopyranose units, and is estimated to have a polarity similar to that of aqueous ethanolic solutions (Loftsson and Brewster, 1996). Cyclodextrins can increase the equilibrium solubility of some hydrophobic molecules by forming a non-covalent inclusion complex if the molecule or a portion of the molecule (i.e., non-polar side-chain, or an aromatic or heterocyclic ring) is of the appropriate size to fit inside the central cavity. The two most common and preferred water-soluble β-cyclodextrin derivatives are hydroxypropyl-β-cyclodextrin and sulfobutylether-β-cyclodextrin (Captisol®) which have a degree of substitution between four and eight surface modifications per β-cyclodextrin molecule.

Water-Insoluble Organic Solvents and Solids

A lipid-based oral formulation is used for water-insoluble drugs in cases where typical formulation approaches (i.e., solid wet granulation, solid dry granulation, water-soluble liquid in a capsule) do not provide the required bioavailability, or when the drug itself is an oily substance (e.g., dronabinol, ethyl icosapentate, indomethacin farnesil, teprenone, and tocopherol nicotinate). The water-insoluble solvents used in commercially available solubilized oral formulations include oleic acid, dl-α-tocopherol (Vitamin E), medium-chain mono- and diglycerides, long-chain triglycerides (peanut oil, corn oil, soybean oil, sesame oil, olive oil, peppermint oil, and castor oil), medium-chain triglycerides derived from coconut oil and palm seed oil, mono- and diesters of propylene glycol, or monoesters of glycerol.

Oleic acid is the common name for (Z)-9-octadecenoic acid and is a nearly colorless liquid at room temperature. Vitamin E is the common name for d-α-tocopherol, is an oily liquid at room temperature and is also an antioxidant (Constantinides et al., 2004; Constantinides et al., 2006). Propylene glycol esters of fatty acids such as propylene glycol laurate and propylene glycol monolaurate (Lauroglycol™FCC) can be used alone or as a co-surfactant. The medium-chain diester of propylene glycol (Labrafac™PG) is a clear liquid for soft gelatin capsule formulations. The glycerol monoesters of oleic acid (Peceol™) and linoleic acid (Maisne™35-1) are liquid solvents that can be the oily phase of SEDDS formulations.

Medium-chain triglycerides are commonly used to solubilize water-insoluble drugs. The mixture of ethanol and fractionated medium chain triglyceride of coconut oil is sometimes used as a solubilizing vehicle. A mixture of medium-chain triglycerides, polyethyleneglycol, and propylene glycol is also used as a solubilizing vehicle. Long-chain triglycerides are also commonly used to solubilize lipid soluble drugs in over-the-counter soft gelatin capsules such as soybean oil solubilizes vitamin A, and corn oil solubilizes 6 mg lutein, which is a natural carotenoid dietary supplement. Peppermint oil is used for both solubility and also to impart a favorable taste, such as in over-the-counter soft gelatin capsules containing 125 mg of simethicone for gas relief.

The water-insoluble solids include hydrogenated soybean oil, hydrogenated vegetable oil, beeswax, and soy fatty acids and are usually used in a mixture of solvents to solubilize lipophilic drugs.

Microemulsion Oral Formulations

Microemulsions are thermodynamically stable isotropically clear dispersions composed of a polar solvent, oil, and a surfactant(s). Labrafil® and Gelucire® 44/14 are all-in-one self-emulsifying surfactants which are in many oral products throughout the world. Microemulsions have much potential for drug-delivery since very hydrophobic molecules can be solubilized and formulated for oral administration (Tenjarla, 1999). All of the commercial products are actually nonaqueous microemulsions, also known as microemulsion preconcentrates or self-emulsifying drug delivery systems (SEDDS), since the polar solvent is not water. Upon contact with aqueous media, such as gastrointestinal fluids, a SEDDS formulation spontaneously forms a fine dispersion or aqueous microemulsion.

Capsules: Soft and Hard

Table 7 is a listing of selected commercially available solubilized oral formulations in the United States of America arranged alphabetically by drug name and also showing the drug's chemical structure, the dose, the marketed formulation, the list of excipients, and the estimated maximum amount of each excipient administered if possible to determine. Tables 8 and 9 are lists of selected commercially available lipid-based oral formulations in the United Kingdom and Japan, respectively, arranged alphabetically by drug name and also showing the drug's chemical structure, indication, dose, marketed formulation, list of excipients, storage conditions, and the year of initial marketing. The limited geographical focus is due to the availability of formulation information of commercial pharmaceutical products in these countries.

Capsules typically vary in volume from 0.5–1.5 mL, but the ideal volume filled into a capsule is approximately 0.7 mL to allow for an easily swallowed capsule. Thus ideally the entire dose of active ingredient should be solubilized in a total volume of less than or equal to 0.7 mL to allow for a dose regimen of 1 capsule per dose (Liu, 2000), otherwise a larger capsule is required, or a dose regimen of multiple capsules per dose must be considered. The dose range of active ingredients in capsule formulations is from 0.25 μg to 500 mg, and the upper mass of an entire liquid-filled capsule is about 1650 mg. Both soft and hard gelatin capsules are used for solubilized oral formulations. Liquid formulations can be filled into either soft or hard capsules, whereas semisolids or solid formulations are filled into hard capsules.

The range of complexity of solubilized oral formulations filled into a capsule varies from a simple one-excipient formulation such as PEG 400 or a long-chain triglyceride, to complex microemulsion preconcentrates which contain oil, cosolvent, and surfactant excipients. The preferred water-soluble organic solvents

Molecule(s)/ trade name/ company/ indication	Chemical structure	Dose	Commercial oral formulation	Excipients	Estimated maximum amount of excipient administered[1]	Storage
Acetaminophen, Dextromethorphan HBr, Guaifenesin, Pseudoephedrine HCl/ 1) Robitussin® Cough, Cold & Flu and 2) Sudafed® Cold & Cough/over-the-counter	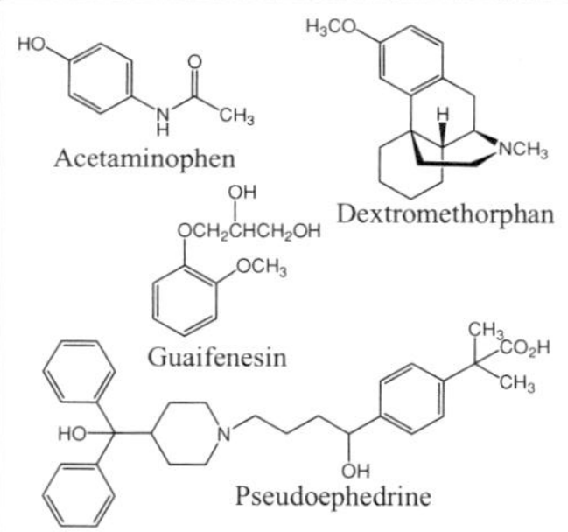	2 capsules every 4 hours, up to 8 per day	Soft gelatin capsule 250, 10, 100, 30 mg	1) Medium-chain triglyceride (fractionated coconut oil) PEG PG Total mass is 1410 mg 2) PEG 400 PG Total mass is 1450 mg		RT
Acetaminophen, Chlorpheniramine maleate, Dextromethorphan HBr, Pseudoephedrine HCl/Children's Tylenol® Cold plus Cough/over-the-counter	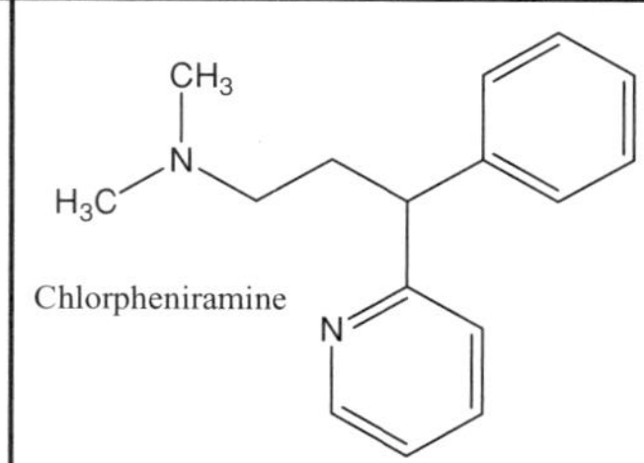	Up to 10 mL q.i.d.	Oral Solution Each mL has 32, 0.2, 1, and 3 mg	Citric acid Colors Corn syrup Flavors PEG PG Sodium benzoate Sodium carboxy-methylcellulose Sorbitol Water pH 3.5		RT

Alfacalcidol/ One-Alpha Capsules/ Leo Laboratories/ Calcium Regulator	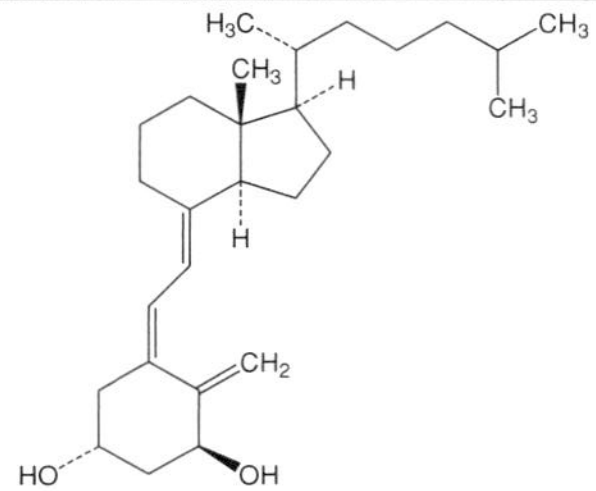	0.5–1 μg (1 capsule) q.d.	Soft gelatin capsule 0.25, 0.5, 1.0 μg	Sesame oil dl-α-tocopherol		
Amprenavir/ Agenerase®/Glaxo SmithKline/HIV	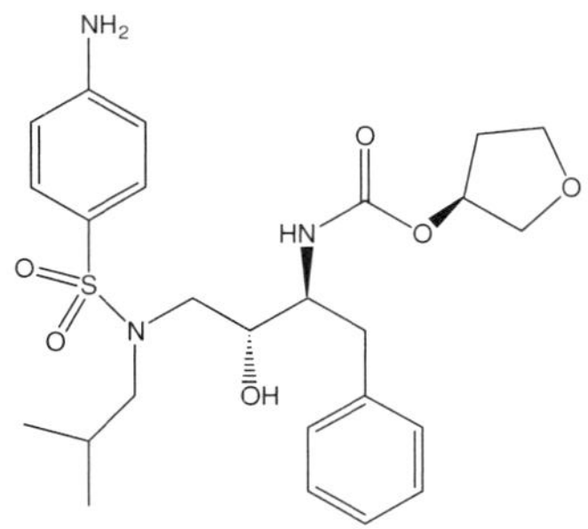	1) 1200 mg (8 capsules) b.i.d.	1) Soft gelatin capsule 50, 150 mg	1) TPGS (280 mg in the 150 mg capsule) PEG 400 (247, 740 mg) Propylene glycol (19, 57 mg)	1) TPGS 2240 mg b.i.d 4480 daily; PEG 400 5920 mg b.i.d 11,840 mg daily Propylene glycol 456 mg b.i.d 912 mg daily	RT

(*continued*)

Molecule(s)/ trade name/ company/ indication	Chemical structure	Dose	Commercial oral formulation	Excipients	Estimated maximum amount of excipient administered[1]	Storage
		2) Pediatric: 17 mg/kg (1.1 mL) t.i.d Adults: 1400 mg (92 mL) b.i.d	2) Oral Solution 15 mg/mL	2) TPGS (~12%) PEG 400 (~17%) Propylene glycol (~55%) Sodium chloride Sodium citrate Citric acid Flavors/ sweeteners	2) Adult TPGS 11,000 mg b.i.d 22,000 daily PEG 400 15,640 mg b.i.d 31,280 mg daily Propylene glycol 50,600 mg b.i.d 101,000 mg daily	RT
Bexarotene/ Targretin®/ Ligand/ Antineoplastic	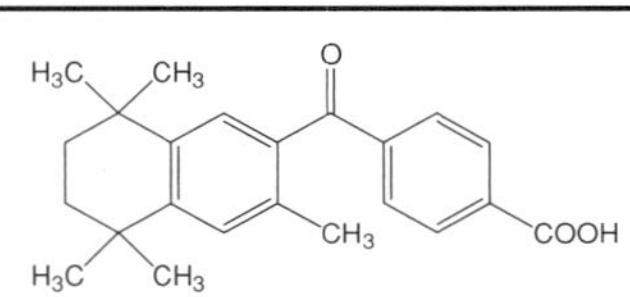	300 mg/m^2/ day ~500 mg (6 capsules)	Soft gelatin capsule 75 mg	PEG 400 Polysorbate 20 Povidone BHA		RT Avoid high temperature, humidity and light

Calcitrol/Rocaltrol®/ Roche/Calcium Regulator	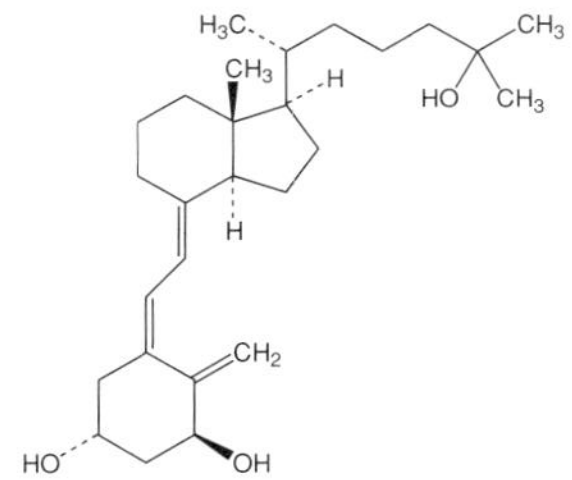	0.25–1 μg/day (1 capsule)	1) Soft gelatin capsule 0.25, 0.5 μg 2) Oral Solution 1 μg/mL	1) Fractionated triglyceride of coconut oil (medium chain triglyceride) 2) Fractionated triglyceride of palm seed oil (medium chain triglyceride)	Medium chain triglyceride 1 mL	15°–30°C. Protect from light
Clomethiazole edisilate/ Heminevrin Capsules/ AstraZeneca/ Sedative	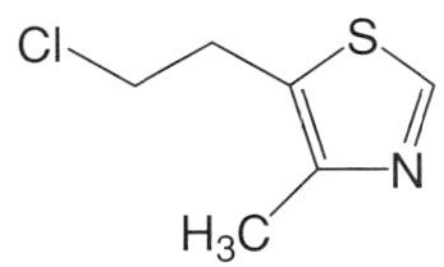	1–4 capsules as needed	Soft gelatin capsule 192 mg	Medium-chain triglycerides from fractionated coconut oil		
Cyclosporin A/ I. Neoral®/ Novartis/ Immunosuppressant Prophylaxis for organ transplant rejection	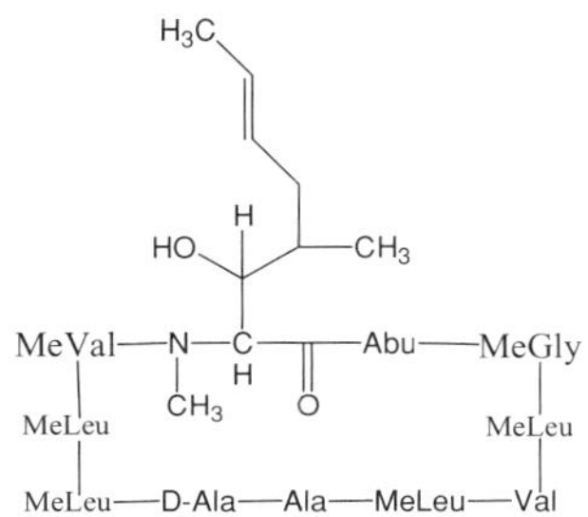 $C_{62}H_{117}N_{11}O_{12}$, MW = 1202.64	1–9 mg/kg/ day 70– 700 mg (1–7 capsules, up to 7 mL)	1) Soft gelatin capsule 25, 100 mg 2) Oral Solution 100 mg/mL	1) Ethanol 11.9% Corn oil-mono-di-triglycerides Cremophor RH 40 Propylene glycol dl-α-tocopherol	Ethanol 0.83 mL	RT, do not store Oral Solution in the refrigerator.

(*continued*)

Molecule(s)/ trade name/ company/ indication	Chemical structure	Dose	Commercial oral formulation	Excipients	Estimated maximum amount of excipient administered[1]	Storage
			Oral bioavailability is 20–50%	2) Ethanol 11.9% Corn oil-mono-di-triglycerides Cremophor RH 40 dl-α-tocopherol Propylene glycol		
Cyclosporin A/ II. Sandimmune®/ Novartis		2–14 mg/kg/ day 150–1000 mg (2–10 capsules, up to 10 mL)	1) Soft gelatin capsule 25, 50, 100 mg 2) Oral Solution 100 mg/mL Oral bioavailability is <10–89%	1) Ethanol 12.7% Corn oil Glycerol Polyoxyethylated linoleic glycerides (Labrafil M-2125CS) 2) Ethanol 12.5% Olive oil Polyoxyethylated oleic glycerides (Labrafil M-1944CS)	Ethanol 1.2 mL	

Cyclosporin A/ III. Gengraf®/ Abbott		1–9 mg/kg/day 70–700 mg (1–7 capsules)	Hard gelatin capsule 25, 100 mg Orally bioequivalent to Neoral®	Ethanol 12.8% Polyethylene glycol Cremophor EL Polysorbate 80 Propylene glycol Sorbitan monooleate		RT
Cyclosporin A/ IV. Cyclosporin Capsules/Sidmak		1–9 mg/kg/day 70–700 mg (1–7 capsules)	Soft gelatin capsule 100 mg	Caprylic/capric triglycerides (Labrafac) dl-a-tocopherol Glyceryl caprylate PEG-8 caprylic/ capric glycerides (Labrasol) PEG-35 castor oil		
Digoxin/Lanoxin®/ Glaxo SmithKline/ Treatment of mild to moderate heart failure	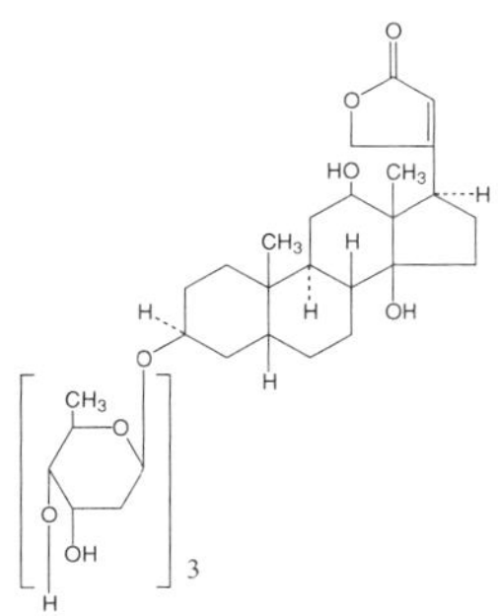	1) 50–350 μg (1–3 capsules) 2) 10–35 μg/kg or <250 ug (< 5 mL)	1) Soft gelatin capsule 50, 100, 200 μg 2) Elixir Pediatric 50 μg/mL	1) PEG 400 Ethanol 8% Propylene glycol Water 2) Ethanol 10% Methyl paraben 0.1% Citric acid Flavor Propylene glycol 40% Sodium phosphate Sucrose	Propylene glycol <2.0 mL; Ethanol < 0.5 mL	

(*continued*)

Molecule(s)/ trade name/ company/ indication	Chemical structure	Dose	Commercial oral formulation	Excipients	Estimated maximum amount of excipient administered[1]	Storage
Doxercalciferol/ Hectorol®/Bone Care/Management of secondary hyperpara-thyroidism associated with chronic renal dialysis.	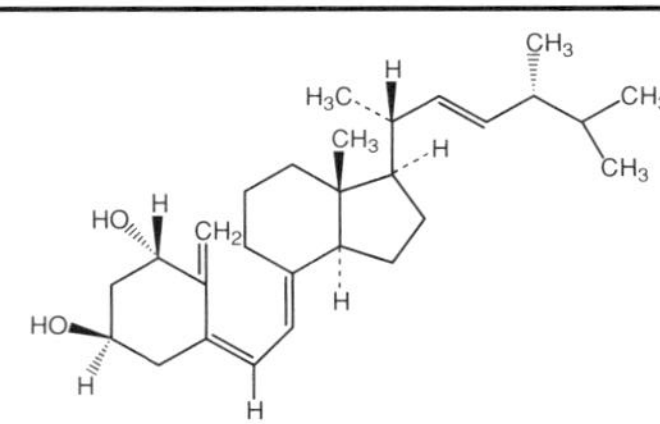	<200 μg (up to 8 capsules) t.i.d.	Soft gelatin capsule 2.5 μg	BHA Ethanol Fractionated triglyceride of coconut oil (medium chain triglyceride)		RT
Dronabinol/ Marinol®/Roxane and Unimed/ Anorexia or nausea	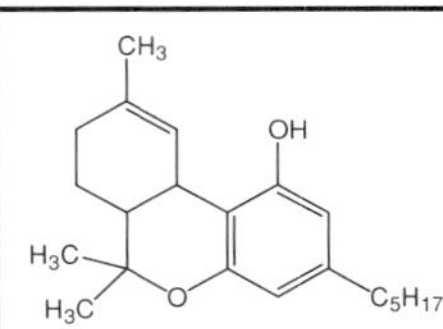 pKa = 10.6	2.5–10 mg 1 capsule	Soft gelatin capsule 2.5, 5, 10 mg	Sesame oil		8°–15°C. Protect from freezing
Dutasteride/ Avodart™/ GlaxoSmithKline/ Treatment of benign prostrate hyperplasia	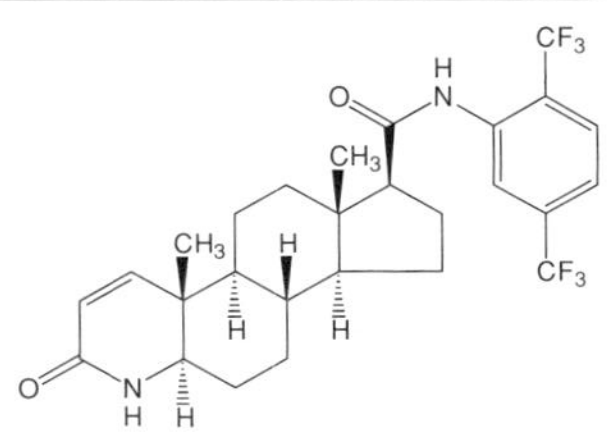	0.5 mg 1 capsule	Soft gelatin capsule 0.5 mg Oral bioavailability is 40–94%	Mixture of mono- and diglycerides of caprylic/ capric acid BHT		RT

Etoposide/(Generic)/ Mylan/ Antineoplastic	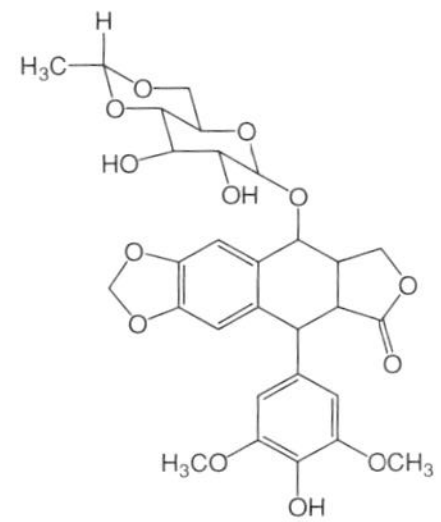		Soft gelatin capsule 50 mg Oral bioavailability is 25–75%	Citric acid Glycerin PEG 400 Water		
Ethyl Icosapentate/ Epadel® Capsules 300/Mochida Pharmaceuticals/ Hyperlipidemia	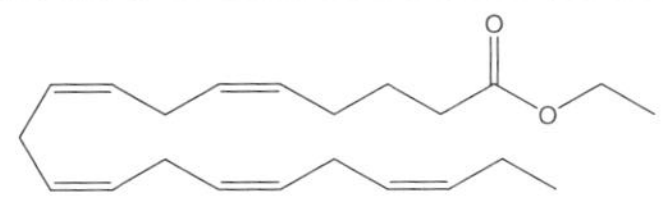	600 mg (2 capsules) t.i.d.	Soft gelatin capsule 300 mg	α-tocopherol		
Ibuprofen/ 1) Advil® Migrane Liquidgels/ (Generic)/and 2) Solufen 200 mg/ Irexsynthelabo/ Analgesic	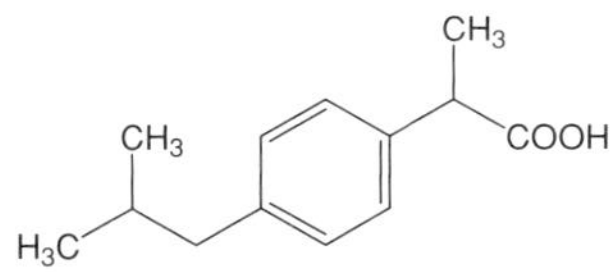	200 mg up to 6 capsules per day	Soft gelatin capsule 200 mg	1) PEG 400 and 2) Gelucire® 44/14 (244 mg)	Gelucire® 44/14 1950 mg per day	RT
Indomethacin farnesil/Infree® Capsules/Eisai Co./ Anti-inflammatory and analgesic	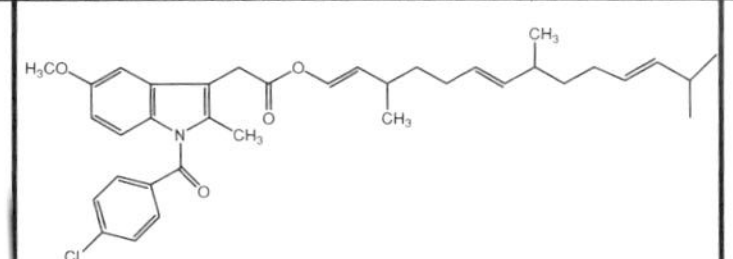	200 mg (1–2 capsules) b.i.d.	Soft gelatin capsule 200 mg	Cremophor RH 60 Hydrogenated oil Glyceryl monooleate		

(continued)

Molecule(s)/ trade name/ company/ indication	Chemical structure	Dose	Commercial oral formulation	Excipients	Estimated maximum amount of excipient administered[1]	Storage
Isotretinoin/ Accutane®/Roche/ Antiachne	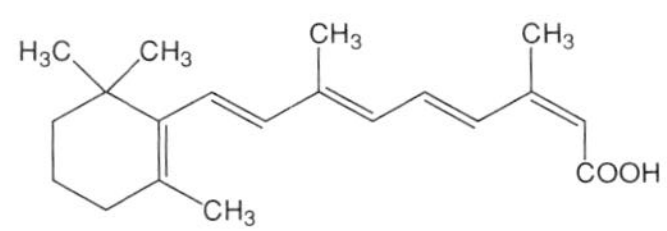	0.5–2 mg/ kg/day (1–4 capsules)	Soft gelatin capsule 10, 20, 40 mg	Beeswax BHA EDTA Hydrogenated soybean oil flakes Hydrogenated vegetable oils Soybean oil		RT, Protect from light
Itraconazole/ Sporanox®/ Ortho Biotech and Janssen/ Antifungal	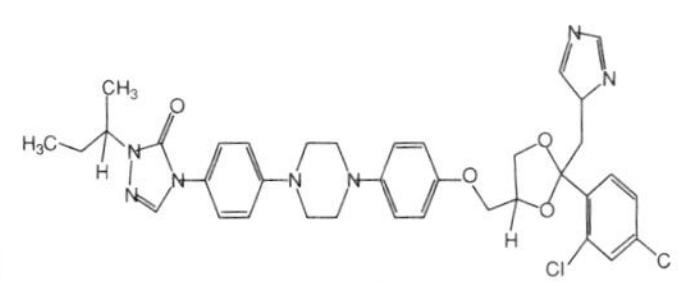 pKa =3.7	200 mg (20 mL) q.d.	Oral Solution 10 mg/mL pH 2	Water HP-β-CD 40% Propylene glycol 2.5% Sodium saccharin Sorbitol Flavors	HP-β-CD 8.0 grams PG 0.5 mL	

Lopinavir and Ritonavir/ Kaletra®/Abbott/ HIV	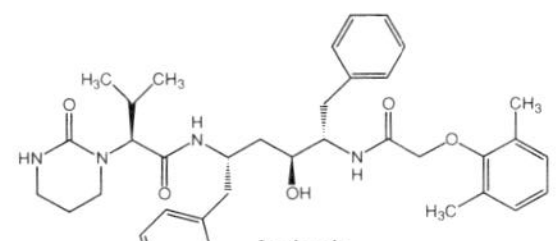 Lopinavir 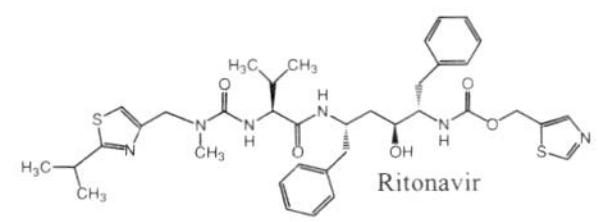 Ritonavir	1) 400 mg/ 100 mg (2 tablets) b.i.d, 800 mg/ 200 mg (4 tablets) q.d. 2) 400 mg/ 100 mg (3 capsules) b.i.d 3) Up to 5 mL	1) Tablet 200 mg lopinavir and 50 mg ritonavir 2) Soft gelatin capsule 133.3 mg lopinavir and 33.3 mg ritonavir 3) Oral solution 80 mg/mL lopinavir and 20 mg/mL ritonavir	1) Sorbitan monolaurate Sodium stearyl fumarate Copovidone Silicon dioxide 2) Oleic acid Cremophor EL Propylene glycol 3) Ethanol (42.2% v/v) Glycerin Cremophor RH 40 Propylene glycol Sodium chloride Sodium citrate/ Citric acid Flavors/sweeteners Water	4.2 mL q.d. 2.1 mL b.i.d. Ethanol	1) Room temperature 2,3) 2–8°C, or at room temperature for <2 months
Lubiprostone/ Amitiza™/ Sucampo/ Treatment of chronic idiopathic constipation in adults	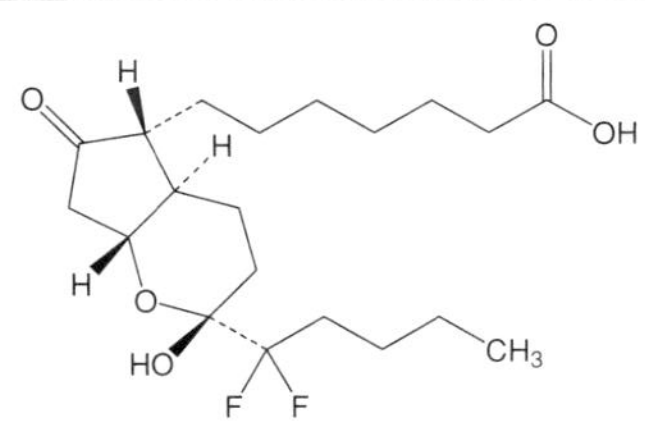	24 μg b.i.d.	Soft gelatin capsule, 24 μg	Medium chain triglyceride		RT

(*continued*)

Molecule(s)/ trade name/ company/ indication	Chemical structure	Dose	Commercial oral formulation	Excipients	Estimated maximum amount of excipient administered[1]	Storage
Menatetrenone/ Glakay® Capsules 15 mg/Eisai Co./ Osteoporosis	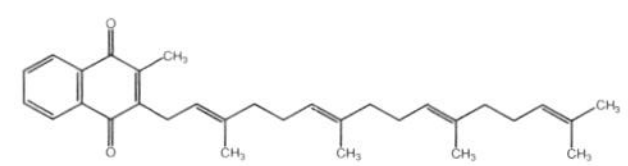	45 mg (3 capsules) t.i.d.	Soft gelatin capsule 15 mg	Propylene glycol esters of fatty acid, Glyceryl monooleate		
Nimodipine/ Nimotop®/Bayer/ Vasodilator	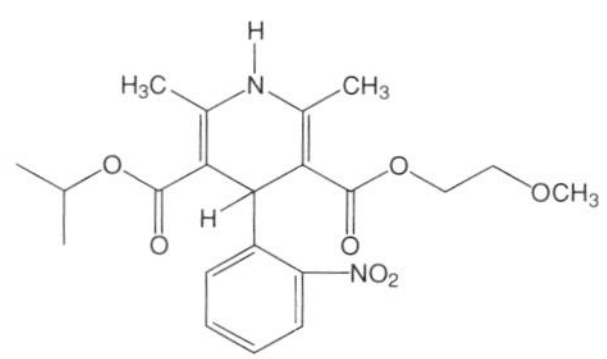	60 mg (2 capsules) every 4 hours	Soft gelatin capsule 30 mg	Peppermint oil PEG 400 Water		
Paricalcitrol/ Zemplarl®/Abbott/ Calcium Regulator	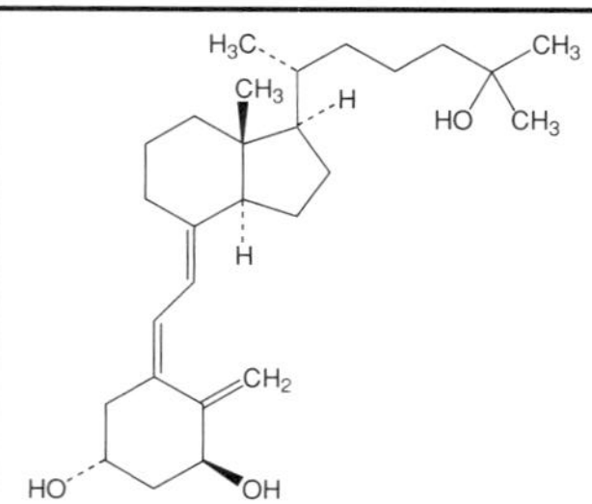	1–4 μg/day or 1–4 ug/ t.i.w. (1 capsule)	Soft gelatin capsule 1,2,4 μg	Medium chain triglyceride (coconut and palm kernel oil) Ethanol BHT		RT
Phenobarbital Elixir/ (Generic)/ Pharmaceutical Associates/ Anticonvulsant and sedative	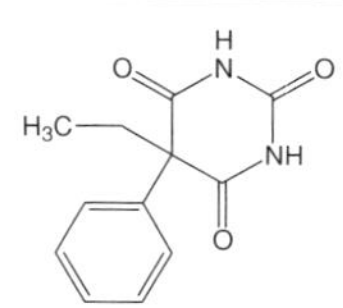		Elixir 4 mg/mL	Ethanol 23% Glucose Sodium saccharin Water		

Progesterone/ Prometrium®/ Solvay/Hormone	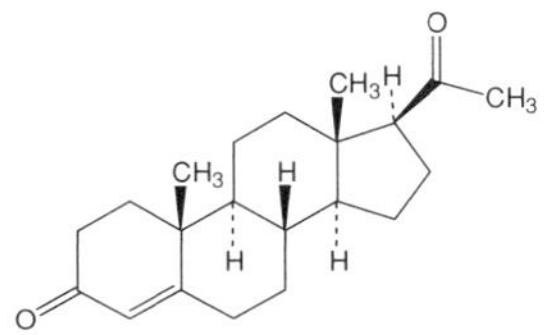	200 mg (1 capsule) q.d.	Soft gelatin capsule 100 mg micronized	Peanut oil		RT Protect from light and excessive moisture
Ritonavir/Norvir®/ Abbott/HIV	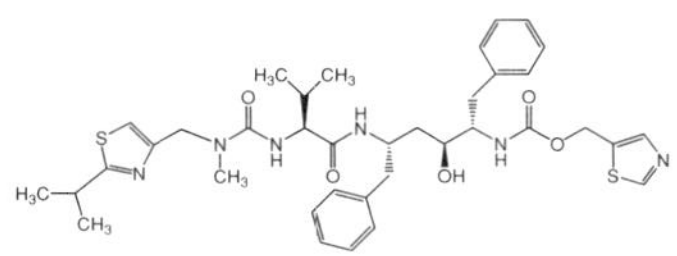	1) 600 mg (6 capsules) b.i.d 2) up to 600 mg (7.5 mL)	1) Soft gelatin capsule 100 mg 2) Oral Solution 80 mg/mL	1) BHT Ethanol Oleic acid ~75% Cremophor EL ~13% 2) Ethanol (42%) Water Cremophor EL Propylene glycol Citric acid Flavors/ sweetener/dye	Ethanol 3.1 mL b.i.d.	Capsule 2–8°C, or at room temperature for <1 month Oral Solution RT
Saquinavir/ Fortovase®/Roche/ HIV Protease inhibitor. Discontinued in 2006	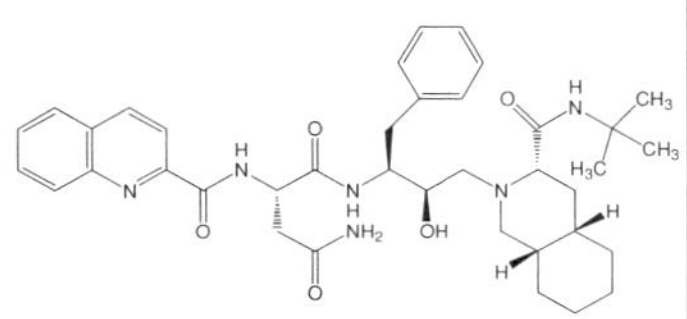	1200 mg (6 capsules) t.i.d	Soft gelatin capsule 200 mg	Medium chain mono- and diglycerides Povidone DL-α-tocopherol		2–8°C, or at room temperature for <3 months

(*continued*)

Molecule(s)/ trade name/ company/ indication	Chemical structure	Dose	Commercial oral formulation	Excipients	Estimated maximum amount of excipient administered[1]	Storage
Sirolimus/ Rapamune®/ Wyeth-Ayerst/2001 in the UK/ Immunosuppressant	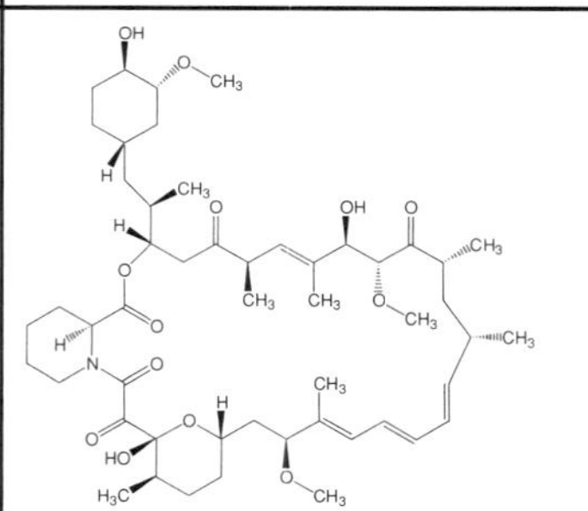	6 mg (6 mL) loading dose followed by 2 mg (2 mL) q.d.	Oral solution 1 mg/mL	Phosal 50 PG® (Ascorbyl palmitate Ethanol 1.5–2.5% Mono- and diglycerides Polysorbate 80 Phosphatidyl-choline Propylene glycol Soy fatty acids)		2–8°C, or at room tempera-ture for <15 days
Tipranavir/Aptivus®/ Boehringer Ingelheim/ HIV Protease inhibitor	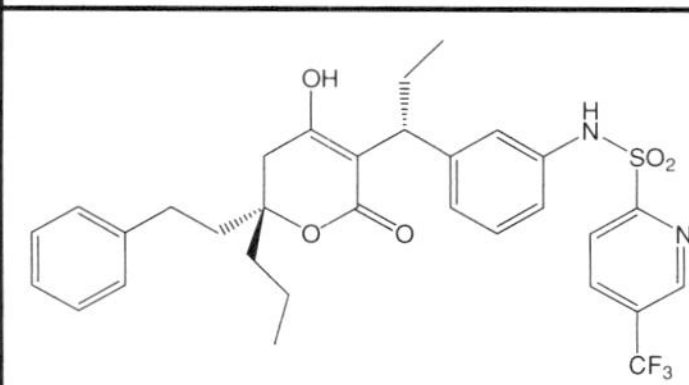	500 mg b.i.d.	Soft gelatin capsule 250 mg	Ethanol (7% w/w) Cremophor EL Propylene glycol Medium chain mono- and diglycerides		2–8°C, or at room tempera-ture for <2 months

Tocopherol nicotinate/ Juvela® N Soft capsules/ Eisai Co./ Hypertension, hyper-lipidemia	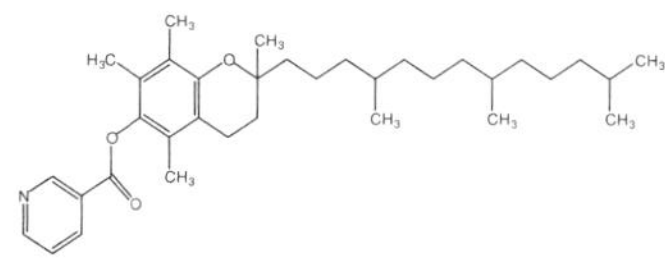	200 mg (1 capsule) t.i.d.	Soft gelatin capsule 200 mg	Medium-chain triglycerides Glycol esters of fatty acid Aspartic acid		
Tretinoin/ Vesanoid®/Roche/ Antineoplastic	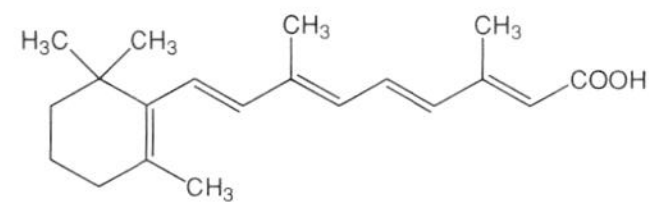	45 mg/m^2/ day (5–10 capsules)	Soft gelatin capsule 10 mg	Beeswax BHA EDTA Hydrogenated soybean oil flakes Hydrogenated vegetable oils Soybean oil		RT
Valproic acid/ Depakene®/ Abbott/Antiepileptic	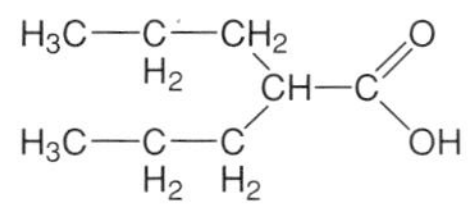	250–1000 mg (1–5 capsules)	Capsule 250 mg	Corn oil		RT

Table 7. List of selected commercially available solubilized formulations for oral administration in the United States of America.

[1]The estimated maximum amount of each excipient is calculated by knowing the exact amount or % of the excipient and the dose.

Molecule/ trade name/ company/ date of initial marketing	Chemical structure	Indication	Dose	Commercial oral formulation	Excipients	Storage
Alfacalcidol/ One-Alpha Capsules/ Leo Laboratories/ 2000	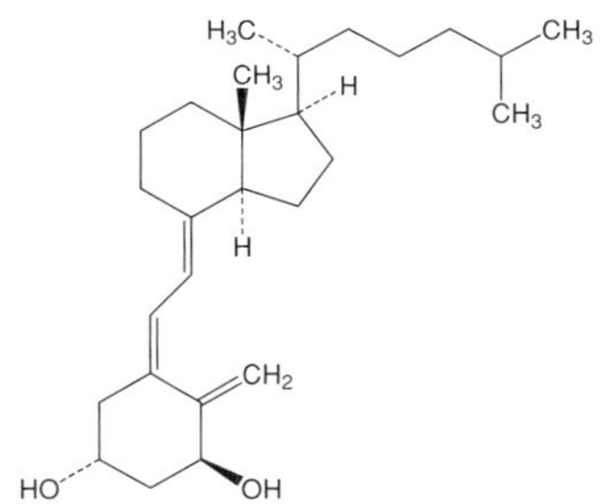	Calcium Regulator	0.5–1 μg q.d (1 capsule)	Soft gelatin capsule 0.25, 0.5, 1.0 μg	Sesame oil dl-α-tocopherol	RT
Clofazimine/ Lamprene Capsules 100 mg/Alliance Pharmaceuticals/1998	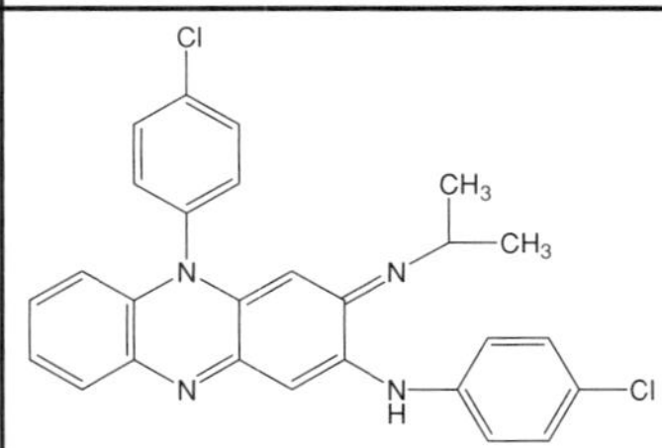	Treatment of leprosy in combination with dapsone and rifampicin	Maximum 300 mg q.d. for up to 3 months (3 capsules)	Soft gelatin capsule 100 mg (micronized suspension in an oil-wax base)	Rapseed oil, wax blend (Beeswax, Hydrogenated soybean oil, Partially hydrogenated plant oils) BHT Citric acid PG	Below 25°C

Clomethiazole edisilate/Heminevrin Capsules/ AstraZeneca/2002	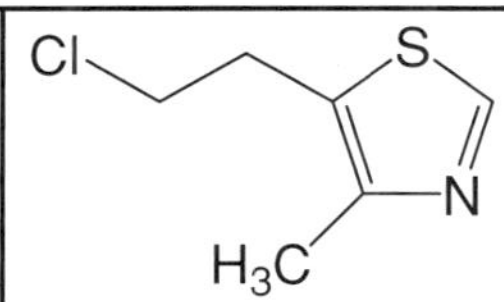	Sedative	1–4 capsules as needed	Soft gelatin capsule 192 mg	Medium-chain triglycerides from fractionated coconut oil	RT
Efavirenz/Sustiva® 30 mg/mL Oral Solution/ Bristol-Meyers Squibb/1999	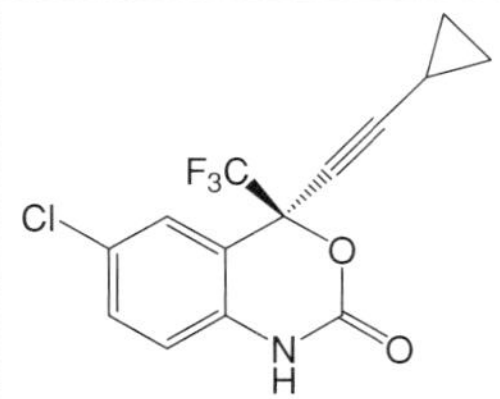	HIV antiviral	Adults: 600 mg q.d. (up to 20 mL); Pediatrics: 270–600 mg (9–20 mL)	Oral solution 30 mg/mL	Medium-chain triglycerides Benzoic acid Strawberry and Mint flavor	RT
Fenofibrate/Fenogal/ Genus/2002	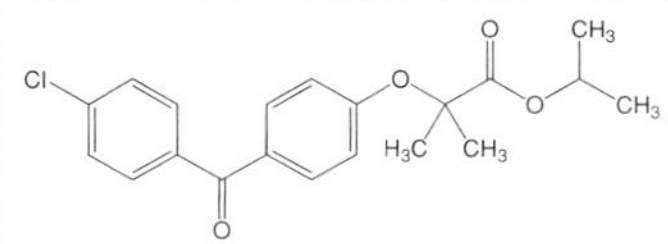	Antihyper- lipoproteine mic	200 mg (1 capsule) q.d.	Hard gelatin capsule 200 mg	Lauryl macro- golglycerides (Gelucire® 44/14) Macrogel 20,000 (PEG 20,000), HPC	

(*continued*)

Molecule/ trade name/ company/ date of initial marketing	Chemical structure	Indication	Dose	Commercial oral formulation	Excipients	Storage
Testosterone Undecanoate/ Restandol 40 mg/ Organon Laboratories/1981	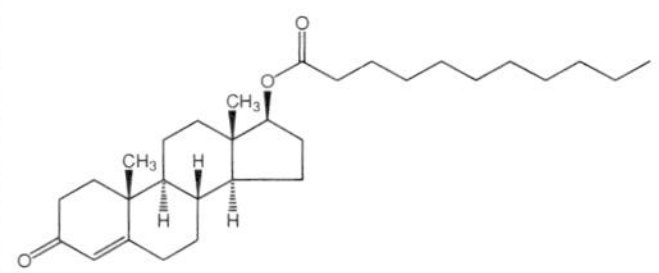	Hormone replacement therapy	40–160 mg q.d. (1–4 capsules)	Soft gelatin capsule 40 mg HSE (equivalents of testosterone, 61 mg of testosterone undecanoate	Oleic acid	Wholesaler in the refrigerator (2–8°C), and the patient at RT. Protect from light and heat.
Valproic acid/ Convulex 100 mg, 200 mg, 500 mg/ Pharmacia/1991	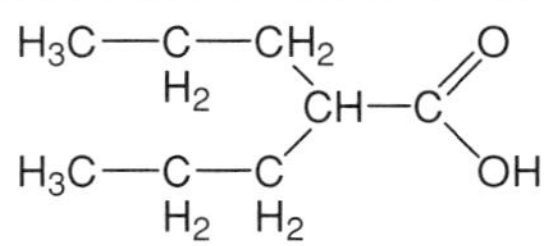	Antiepileptic	10–60 mg/ kg/day up to 2500 mg per day (1–5 capsules)	Soft gelatin capsule 100, 200 and 500 mg	Medium-chain triglycerides, Coating is hydrozypropy-methylcellulose phthalate, and dibutyl-phthalate	RT

Table 8. List of selected commercially available solubilized lipid-based formulations for oral administration in the United Kingdom.

for soft gelatin capsules are PEG 400 and propylene glycol. Ethanol can also be used in capsules, but the amount is limited to less than 15% since ethanol can diffuse out of a capsule due to its volatility, and can potentially dissolve the capsule shell. If solubility and/or oral bioavailability is still not sufficient, the next level of complexity is to add a surfactant such as TPGS, cremophor EL, cremophor RH 40, Tween 80, Span 80, Softigen® 767, Labrasol®, Labrafil® M-1944CS, or Labrafil® M-2125CS. If solubility and/or oral bioavailability are still not sufficient, the next level of complexity is a non-aqueous self-emulsifying drug delivery system (SEDDS) which contains an oil, a surfactant and a co-surfactant.

Physical stability and capsule shell integrity are critical as water and solvents permeate into and from the capsule shell and the enclosed solution formulation. Capsule shells will dissolve or swell when exposed to excess water, thus solutions filled into capsules are composed of a high percentage of an organic excipient(s), with minimum amount of water to maintain capsule integrity. Soft gelatin capsule shells contain up to 50% water and immediately after manufacturing over a period of days, water can migrate from the capsule shell into the liquid and also the reverse until an equilibrium is reached. Water migration is critical to physical stability not only because of capsule shell integrity, but also water-insoluble drugs can precipitate with increasing water content. Soft gelatin capsules contain a plasticizer and are thus more amenable to the more hygroscopic solvents than are hard gelatin capsules, which do not contain any plasticizer. Hard gelatin capsule shells become brittle as water is leached into the liquid, thus hygroscopic solvents that extract water from the capsule shell such as PEG 400, propylene glycol, and medium chain monoglycerides are typically used in mixtures or with a small amount of water.

As a guide to solubilized oral formulation drug development, the following examples illustrate the formulation philosophy of "simple to complex". Starting with one-excipient formulations to complex microemulsion preconcentrates.

Water-Soluble Solubilizing Vehicles in Capsules—One Solubilizing Excipient

PEG 400 is used to solubilize many drugs in highly concentrated solutions that are filled into capsules.

1) **PEG 400** solubilizes 200 mg of ibuprofen in over-the-counter liquid gel-caps Advil® Migrane and Liqui-Gels for a fast onset of action. PEG 400 has the highest solubilization power for ibuprofen and was selectively higher for the active S-isomer compared to the R-isomer (Nerurkar et al., 2005). PEG 400 solubilizes 50 mg of the water-insoluble etoposide in VePesid® (etoposide, 50 mg), which is now a generic product.

Molecule/ trade name/ company/ date of initial marketing	Chemical structure	Indication	Dose	Commercial oral formulation	Excipients	Storage
Ethyl Icosapentate/ Epadel® Capsules 300/ Mochida Pharmaceuticals/ 1990	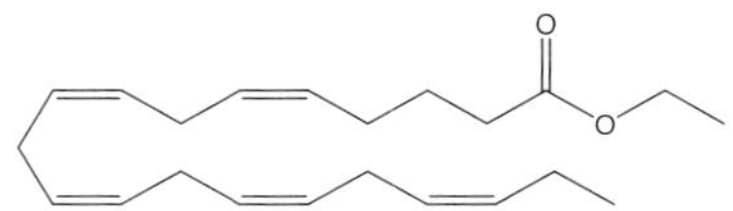	Hyper-lipidemia	600 mg (2 capsules) t.i.d.	Soft gelatin capsules 300 mg	α-tocopherol	RT
Indomethacin farnesil/Infree® Capsules/ Eisai Co./1991	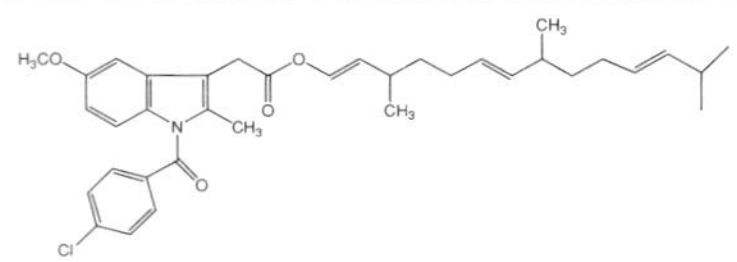	Anti-inflammatory and analgesic	200 mg (1–2 capsules) b.i.d.	Hard gelatin capsule 100 mg (solid)	α-tocopherol, Hydrated silicon dioxide, MCC, Tartaric acid, PEG 6000	RT
				Soft gelatin capsules 200 mg (liquid)	Aspartic acid, Cremophor RH 60, Hydrogenated oil, Glyceryl monooleate	RT
Menatetrenone/ Glakay® Capsules 15 mg/Eisai Co./1995	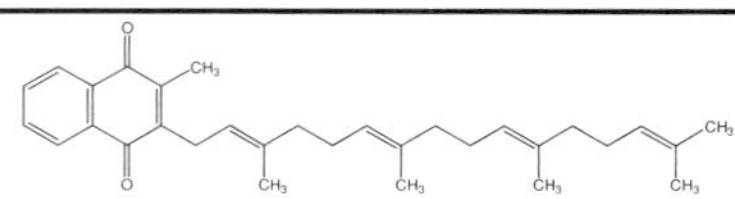	Osteoporosis	45 mg (3 capsules) t.i.d.	Soft gelatin capsule 15 mg (contents of capsule is a viscous liquid or semisolid)	Propylene glycol esters of fatty acid, Glyceryl monooleate	RT, protect from light

Teprenone/ Selbex® Capsules 50 mg And Fine Granules 10%/ Eisai Co./1984	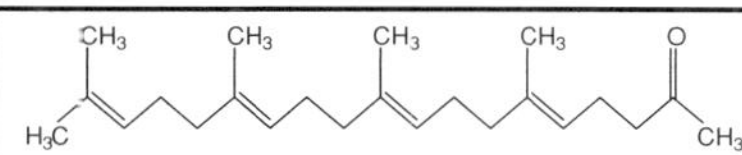	Acute gastritis	150 mg (3 capsules) t.i.d	Hard capsule 50 mg (contents of capsule are granules or powder)	α-tocopherol, Hydrated silicon dioxide, Talc, Mannitol, PEG 6000 Lactose	RT
			150 mg (1.5 grams of granules) t.i.d.	Fine granules, 10% w/w	α-tocopherol, Hydrated silicon dioxide, Talc, HPC, Mannitol, Lactose	RT
Tocopherol nicotinate/ Juvela® N Soft capsules/Eisai Co./1984	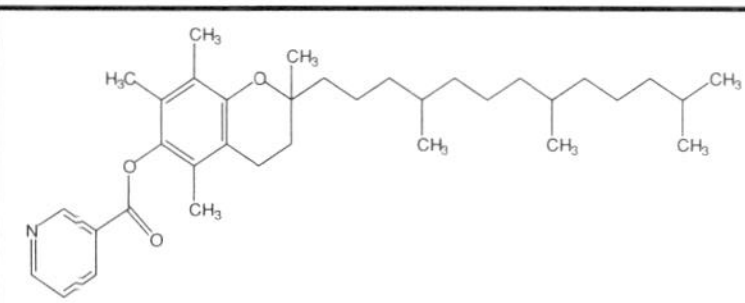	Hypertension, hyper-lipidemia	200 mg (1 capsule) t.i.d.	Soft gelatin capsule, 200 mg (contents of capsule are viscous suspension or semisolid)	Aspartic acid, Medium-chain triglycerides, Glycol esters of fatty acid	RT

Table 9. List of selected commercially available solubilized lipid-based formulations for oral administration in Japan.

Water-Soluble Solubilizing Vehicles in Capsules—Two Solubilizing Excipients

A mixture of PEG 400 and propylene glycol is commonly used in over-the-counter soft gelatin capsules, as is a mixture of PEG 400 and medium-chain triglycerides. Some prescription products contain a mixture of peppermint oil and PEG 400.

1) A mixture of **PEG 400 and propylene glycol** in over-the-counter products solubilizes: 1) 100 mg of the laxative docusate sodium in soft gelatin capsules; 2) 200 mg of guaifenesin, 20 mg of pseudoephedrine HCl, and 10 mg of dextromethorphan HBr in Robitussin® Cold & Cough capsules; 3) 325 mg of acetaminophen, 15 mg of dextromethorphan, and 6.25 mg of doxylamine succinate in Vicks® Nyquil® LiquiCaps; and 4) 250 mg of acetaminophen, 15 mg of dextromethorphan HBr, 100 mg of guaifenesin, and 30 mg of pseudoephedrine HCl in Sudafed® Cold & Cough capsules.
2) A mixture of **PEG 400 and medium-chain triglycerides** solubilizes 200 mg of ibuprofen and 30 mg of pseudoephedrine in over-the-counter liquid Advil® Cold & Sinus liquid gel-caps.
3) A mixture of **peppermint oil and PEG 400** solubilizes 30 mg of nimodipine in Nimotop® soft gelatin capsules.
4) A mixture of **polysorbate 20 and PEG 400** in combination with povidone and butylated hydroxy anisole (BHA) solubilizes 75 mg bexarotene, a benzoic acid derivative that is a selective activator of the retinoid X receptor indicated for the treatment of T-cell lymphoma, in Targretin® soft gelatin capsules. The dose of bexarotene is 300–750 mg (four to ten capsules) once daily and Targretin® soft gelatin capsules should be stored at controlled room temperature. Targretin® has been available in the United Kingdom (UK) since 2001, and is also available in the United States of America (USA).

Water-Soluble Solubilizing Vehicles in Capsules—Three Solubilizing Excipients

A mixture of PEG 400, propylene glycol, and 8% ethanol can be used in soft gelatin capsules.

1) A mixture of **PEG 400, propylene glycol, and 8% ethanol** solubilizes 50, 100 or 200 μg of digoxin in Lanoxicap® soft gelatin capsules. The absolute oral bioavailability of digoxin from the Lanoxicap® soft gelatin capsules is 90–100%, compared to 60–80% for the tablets. The dose of digoxin is 50–350 μg per day, which is one to three capsules.

Lipid-Containing Solubilizing Vehicles in Capsules—One Solubilizing Excipient

The simplest lipid-based formulations contain only one excipient such as Gelucire® 44/14, oleic acid, α-tocopherol, corn oil, peanut oil, sesame oil, medium-chain triglyceride, or medium-chain mono- and diglycerides. There are at least 10 commercially available one-lipid excipient oral formulations, and they all happen to be in soft capsules.

1) **Gelucire®44/14** solubilizes 200 mg of ibuprofen in Solufen®-Gé capsules and is marketed in Europe. Each Solufen®-Gé 200 mg capsule contains 244 mg of Gelucire® 44/14 and up to six capsules per day are administered, which corresponds to a daily intake of 1464 mg of Gelucire® 44/14 (Table 2).
2) **Medium-chain triglycerides** solubilize 192 mg of clomethizole edisilate in Heminevrin Capsules® soft gelatin capsules (60 mg equivalent of clomethiazole). Clomethiazole is an oily viscous liquid as the free base but the edisilate (ethanedisulfonate) salt is a solid. Clomethiazole is a short-acting hypnotic and sedative with anticonvulsant effects, and Heminevrin is indicated for treatment of restlessness and insomnia that accompany alcohol withdrawal. Clomethiazole has a short half-life and low bioavailability due to extensive hepatic first-pass metabolism. The dose of Heminevrin is one to four capsules as needed, and the capsules are stored at room temperature. Heminevrin Capsules® have been available in the United Kingdom since 1998.

   Calcitriol is a non-ionizable and water-insoluble calcium-regulator intended for the treatment of hypocalcemia. Calcitriol is formulated as a solution in a fractionated medium-chain triglyceride of coconut oil, along with the antioxidants BHA and butylated hydroxy toluene (BHT), as 0.25 μg and 0.5 μg Rocaltrol® soft gelatin capsules, which have been available in the UK since 1996, and are also available in the USA. Rocaltrol® is also available (USA only) as a 1 μg/mL oral solution formulated in a fractionated triglyceride of palm seed oil. Calcitriol is not available in Japan, but a similar molecule, falecalcitriol, is available and marketed as a conventional solid tablet formulation. The dose of calcitriol is 0.25–0.5 μg (one capsule) and the product should be stored at controlled room temperature.

   Lubiprostone is a water-insoluble heptanoic acid derivative used in the treatment of chronic idiopathic constipation in adults, and is solubilized in medium-chain triglyceride in 24 μg Amitiza™ soft gelatin capsules. The dose of lubiprostone is 24 μg (one capsule) twice daily with food. Amitiza™ should be stored at controlled room temperature and was approved in the USA in 2006.

3) **Sesame oil** solubilizes 2.5 mg, 5 mg or 10 mg dronabinol, also known as Δ-9-tetrahydrocannabinol, in Marinol® soft gelatin capsules. Dronabinol is the principal psychoactive component of cannabis and finds therapeutic application as an anti-emetic for treatment of the nausea and vomiting associated with cancer chemotherapy or as an appetite stimulant to treat AIDS wasting syndrome. Dronabinol exists as a hydrophobic oil at ambient room temperature, and after oral administration is almost completely absorbed (90–95%). However, the bioavailability of dronabinol is only about 10–20% due to extensive hepatic first-pass metabolism. The dose of dronabinol is 2.5–10 mg (one capsule) twice daily. Marinol® soft gelatin capsules should be stored at 8–15°C and must be protected from freezing. Marinol® is available in the USA.
4) **Corn oil** solubilizes 250 mg of valproic acid, an anticonvulsant with an intrinsic water solubility of 1.3 mg/mL, in Depakene® soft gelatin capsules available in the USA. In the UK, valproic acid is formulated as a solution in **medium-chain triglycerides** as 100-mg, 200-mg and 500-mg Convulex enteric coated soft gelatin capsules, where it has been available since 1991. The dose of valproic acid is 600 mg to 2500 mg (two to ten capsules) daily, and both products should be stored at controlled room temperature.
5) **Peanut oil** partially solubilizes 100 mg of progesterone, a water-insoluble steroid that is sparingly soluble in vegetable oils, in Prometrium® soft gelatin capsules which are available in the USA. The absolute oral bioavailability of progesterone has not been determined, but the systemic exposure is increased following micronization. The dose of progesterone is 200–400 mg (two to four capsules) once daily and the product should be stored at controlled room temperature in moisture-proof containers protected from light. Progesterone, formulated in vegetable fat, is available in the UK as Cyclogest® 200 mg and 400 mg pessaries, where it is indicated for the treatment of premenstrual syndrome.
6) **Oleic acid** solubilizes 61 mg of testosterone undecanoate in Restandol 40 mg soft gelatin capsules. Testosterone undecanoate is an ester prodrug of testosterone intended for oral administration in hormone replacement therapy. Free testosterone is inactive following oral administration due to virtually complete hepatic first-pass extraction. However, the undecanoate ester prodrug is transported via the intestinal lymphatic system, thereby circumventing the hepatic portal circulation and the associated presystemic first-pass metabolism. The oral dose of testosterone undecanoate is 40–160 mg equivalents of testosterone (one to four capsules) once daily. Restandol 40 mg soft

gelatin capsules must be stored refrigerated prior to dispensing, after which it may be stored at controlled room temperature. Restandol has been available in the UK since 1981.

7) **Capric and caprylic mono- and diglycerides** solubilizes 0.5 mg of dutasteride, a water-insoluble steroid derivative used in the treatment of benign prostatic hyperplasia, in Avodart® soft gelatin capsules. The average bioavailability of dutasteride from Avodart® is 60%, with a range of 40–94%. The dose of dutasteride is 0.5 mg (one capsule) once daily and the product should be stored at controlled room temperature. Avodart® has been available in the UK since 2003, and is also available in the USA.
8) **α-tocopherol** solubilizes 300 mg of ethyl icosapentate, which exists as a liquid at ambient room temperature, in Epadel® soft gelatin capsules. Ethyl icosapentate is used in the treatment of hyperlipidemia and arteriosclerosis obliterans. The dose of ethyl icosapentate is 600 mg (two capsules) three times daily and the product should be stored at controlled room temperature in moisture-proof containers protected from light. Epadel® has been available in Japan since 1990.

Lipid-Containing Solubilizing Vehicles in Capsules—Two Solubilizing Excipients

The next level of complexity in lipid-based formulations is those that contain two excipients. Some typical combinations are sesame oil with α-tocopherol, medium-chain triglyceride with ethanol, and propylene glycol esters of fatty acids with glyceryl monooleate. There are at least 6 commercially available oral formulations with two lipid excipients.

1) The mixture of **lauryl macrogolglycerides (Gelucire® 44/14) and PEG 20,000** solubilizes 200 mg of fenofibrate, which is a water-insoluble prodrug of fenofibric acid that is a lipid regulating agent, and filled into hard gelatin capsules in various European products: Fenogal, Lipirex 200 mg, CiL® 200 mg Kapselin, and Fenofibrat AZU® 200 mg Kapselin.
2) The mixture of **sesame oil and α-tocopherol** solubilizes 0.25 μg, 0.5 μg, and 1.0 μg of alfacalcidol, a de-hydroxy derivative of calcitriol, in One-Alpha® soft gelatin capsules which have been available in the UK since 2000. The dose of alfacalcidol is 0.25–1.0 μg (one capsule) daily and the product should be stored at controlled room temperature. In Japan, alfacalcidol is available in capsule, solution and powder formulations (Alfarol®) that employ potassium sorbate as the major excipient.
3) The mixture of **ethanol** and a fractionated **medium-chain triglyceride** of coconut oil solubilizes 2.5 μg of doxercalciferol in Hectorol® soft gelatin capsules. The dose of doxercalciferol

is 10–20 μg (four to eight capsules) three times weekly and the product should be stored at controlled room temperature. Hectorol® is available in the USA.

4) The mixture of **propylene glycol esters of fatty acids and glyceryl monooleate** solubilizes 15 mg of menatetrenone (also known as vitamin K_2), which is a water-insoluble prenylated napththoquinone used in the treatment of osteoporosis, in Glakay® soft gelatin capsules. The bioavailability of menatetrenone is highly dependent on the dietary status of the patient at the time of dosing, with an approximate sevenfold increase in the systemic exposure (as determined from the plasma concentration area-under-the-curve) being observed following post-prandial dosing as compared to the fasted state . The role of dietary fat on the bioavailability of menatetrenone was influenced by the quantity of fat consumed at the time of dosing, with a threefold increase in exposure occurring when the amount of fat was increased from nine grams to 35–54 grams. The dose of menatetrenone is 45 mg (three capsules) three times daily after meals and the product should be stored at ambient room temperature protected from light. Glakay® has been available in Japan since 1995.

5) A mixture of **α-tocopherol and medium-chain mono- and diglycerides** solubilizes 200 mg of saquinavir in Fortovase® soft gelatin capsules. Saquinavir is a water-insoluble human immunodeficiency virus (HIV) protease inhibitor that was first introduced in 1996 as a conventional oral solid dosage form (Invirase®). Compared to the original Invirase® formulation, the mean bioavailability of saquinavir is increased by approximately 3.3-fold (range: 2.1–5.3-fold) by the Fortovase® formulation. Although the absolute bioavailability of saquinavir from the Fortovase® formulation has not been reported, comparison of the systemic exposure to that following a similar dose of Invirase®, from which the mean bioavailability has been reported to be 4% (range: 1–9%), suggests that the bioavailability of saquinavir from the Fortovase® formulation is approximately 15%. The bioavailability of saquinavir from the Invirase® formulation was found to be subject to considerable food effect, increasing approximately sevenfold when administered following a heavy breakfast as compared to the fasted state. Thus, it is recommended that saquinavir be administered within two hours of a meal. The dose of Fortovase® is 1200 mg (six capsules) three times daily or, if taken in combination with ritonavir, which is known to increase the bioavailability of saquinavir, 1000 mg (five capsules) twice daily. Fortovase® capsules must be stored refrigerated at 2–8°C or at room temperature for no more than

three months. However, Fortovase® was discontinued in February 2006 (Internet website 24).

6) A mixture of **medium-chain triglycerides and glycol esters of fatty acids** solubilizes 200 mg of tocopherol nicotinate, which exists as an orange-yellow liquid or solid that is practically insoluble in water, in Juvela® N soft gelatin capsules. Tocopherol nicotinate is the nicotinic acid ester of α-tocopherol and is indicated in the treatment of hypertension, hyperlipidemia and peripheral circulatory disturbances. The bioavailability of tocopherol nicotinate is subject to a considerable food effect, increasing some 30-fold following post-prandial administration as compared to the fasted state. The dose of Juvela® N is 200 mg (one capsule) three times daily and the product may be stored at controlled room temperature. Juvela® N has been available in Japan since 1984.

Lipid-Containing Solubilizing Vehicles in Capsules—Three Solubilizing Excipients

Ascending the level of complexity of lipid-based formulations are those that contain mixtures of three excipients. Typical examples of such combinations include: 1) medium-chain triglycerides, PEG, and PG; 2) TPGS, PEG 400, and PG; 3) oleic acid, cremophor EL, and ethanol or PG; 4) Polysorbate 20, PEG 400, and povidone; 5) medium-chain mono and diglycerides, α-tocopherol, and povidone; 6) medium-chain triglycerides, glycol esters of fatty acids, and aspartic acid. There are at least seven commercially available, three-excipient lipid-based oral formulations, and all are delivered in soft gelatin capsules.

1) A mixture of **medium-chain triglycerides, polyethyleneglycol, and propylene glycol** solubilizes the combination of 250 mg of acetaminophen, 10 mg of dextromethorphan HBr, 100 mg of guaifenesin, and 30 mg of pseudoephedrine HCl in over-the-counter Robitussin® Cold, Cough & Flu liquid-filled capsules.
2) A mixture of (approximate proportions) 23% **d-α-tocopherol polyethylene glycol 1000 succinate (TPGS)**, 60% **PEG 400** and 5% **propylene glycol** solubilizes 50 mg and 150 mg of **Amprenavir,** a poorly water-soluble HIV protease inhibitor, in Agenerase® soft gelatin capsules. TPGS forms micelles and improves the aqueous solubility of amprenavir from 36 μg/mL to 720 μg/mL (Yu et al., 1999). The bioavailability of amprenavir in conventional capsule or tablet formulations is near zero, but has been shown to increase to 69 ± 8% following oral administration of 25 mg/kg to beagle dogs in formulations containing 20% TPGS, with a further increase in bioavailability to 80 ± 16% occurring when the content of TPGS was increased

to 50%. (Wu and Hopkins, 1999). The dose of amprenavir is 1200 mg (eight capsules) twice daily and the product should be stored at controlled room temperature. Agenerase® soft gelatin capsules and oral solution have been available in the UK since 2000 and are also available in the USA.

3) A mixture of **oleic acid, cremophor EL and propylene glycol** solubilizes 133.3 mg of lopinavir and 33.3 mg of ritonavir, which are water-insoluble HIV protease inhibitors used in the treatment of HIV infection, in the fixed-dose combination Kaletra® soft gelatin capsules. The dose of Kaletra® is three capsules twice daily and the product must be stored refrigerated at 2–8°C or at controlled room temperature for no more than two months. Kaletra® has been available in the UK since 2001, and is also available in the USA.
4) A mixture of **cremophor RH 60, hydrogenated oil, and glyceryl monooleate** solubilizes 200 mg of indomethacin farnesil, which occurs as an oily liquid that is practically insoluble in water, in Infree® soft gelatin capsules. Indomethacin farnesil is an ester prodrug of the anti-inflammatory and analgesic indomethacin. The oral bioavailability of indomethacin farnesil is decreased when administered in the fasted state, but absorption is improved following an ordinary meal containing ten grams of fat. The dose of indomethacin farnesil is 200 mg (1 capsule) twice daily and the product may be stored at controlled room temperature. Indomethacin farnesil has been available in Japan since 1991.
5) A combination of beeswax, **soybean oil, hydrogenated vegetable oils and hydrogenated soybean oil** solubilizes 10 mg, 20 mg and 40 mg isotretinoin, an isomer of tretinoin that is an anti-comedogenic indicated in the treatment of severe cystic acne, in Accutane® soft gelatin capsules. Isotretinoin demonstrates rapid, but erratic absorption that is approximately doubled following post-prandial administration, as compared to the fasted state. The absolute bioavailability of isotretinoin has not been reported. The dose of isotretinoin is 0.5–1.0 mg/kg/day (one to two capsules) and the product should be stored at controlled room temperature protected from light. Isotretinoin is available in the USA, and has been available in the UK since 1983. There are currently two generic versions of this product available, both having formulations nearly identical to that of Acutane®.
6) A mixture of **oleic acid, cremophor EL, and ethanol** solubilizes 100 mg of ritonavir in Norvir® soft gelatin capsules. Ritonavir is an HIV protease inhibitor with a peptide-like structure that has an intrinsic water solubility of 1.0 μg/mL and two weakly basic thiazole groups with pKa's of 1.8 and 2.6, which preclude

the possibility of solubilization through manipulation of the formulation pH. The initial hard gelatin capsule formulation of ritonavir (Norvir®) was marketed as an amorphous, semi-solid dispersion containing 100 mg of ritonavir solubilized in a mixture of caprylic/capric medium-chain triglycerides, polyoxyl 35 castor oil (cremophor EL), citric acid, ethanol, polyglycolyzed glycerides, polysorbate 80 and propylene glycol (Strickley, 2004). However, unexpected precipitation of ritonavir as a less soluble crystalline form in the excipient matrix negatively impacted both the drug dissolution rate and bioavailability, leading to a temporary withdrawal of the product from the market in 1998. Norvir® was reintroduced in 1999 after re-formulation as a thermodynamically stable solution in soft gelatin capsules. The dose of ritonavir is 600 mg (six capsules) twice daily and the product must be stored refrigerated at 2–8°C or at room temperature for no more than one month. The amount of oleic acid in each Norvir® capsule is approximately 590 mg, thus up to 3.5 grams of oleic acid are consumed twice-a-day (Table 2). The amount of cremophor EL in each Norvir® capsule is approximately 105 mg, thus up to 0.62 grams of cremophor EL are consumed twice-a-day (Table 2). Norvir® is available in the UK and the USA, but is being replaced by the combination product, Kaletra®.

7) A combination of beeswax, **soybean oil, and hydrogenated vegetable and soybean oil** solubilizes 10 mg of tretinoin, a water-insoluble antineoplastic agent, in Vesanoid® soft gelatin capsules. The oral bioavailability of tretinoin has been reported to be >60% based on recovery of approximately 2/3 of an administered dose in the urine. The dose of tretinoin is 45 mg/m^2/day (eight capsules) twice daily and the product should be stored at room temperature protected from light. Tretinoin is available in the USA, and has been available in the UK since 2001.

Lipid-Containing Solubilizing Vehicles in Capsules—Four Solubilizing Excipients

The next level of complexity of lipid-based formulations currently marketed contain mixtures of four excipients, which can be SEDDS formulations or mixtures such as beeswax, soybean oil, hydrogenated vegetable oils and hydrogenated soybean oil. There are at least two commercially available four-excipient lipid-based oral formulations and both are delivered in soft gelatin capsules. Aptivus® soft gelatin capsule (tipranavir) is a SEDDS formulation with ethanol and PG as the polar solvents, cremophor EL as the surfactant, and medium-chain mono-and-diglycerides as the oily phase.

1) An oil-wax matrix composed of **rapseed oil, beeswax, hydrogenated soybean oil and partially hydrogenated plant oils**

partially solubilizes 100 mg of clofazimine to a suspension in Lamprene® soft gelatin capsules. Clofazimine is practically insoluble in water and used in the treatment of leprosy. The oral bioavailability of clofazimine from Lamprene® is 70%, increasing further when administered with food. The dose of clofazimine is up to 300 mg (three capsules) daily and the product must be stored below 25°C and protected from moisture. Clofazimine has been available in the UK since 1998.

2) A mixture of **medium-chain mono-and-diglycerides, cremophor EL, ethanol (7% w/w), and propylene glycol** solubilizes 250 mg of tipranavir, a water-insoluble non-peptide protease inhibitor of HIV in Aptivus® soft gelatin capsules (Physician's Desk Reference, 2006). Tipranavir is a substrate for both the intestinal efflux P-glycoprotein and also the hepatic metabolic enzyme CYP3A4, and thus must be co-administered with 200 mg of ritonavir in order to achieve effective plasma concentrations. The absolute oral bioavailability has not been reported, but co-administration of ritonavir increases the overnight trough plasma concentrations of tipranavir by 29-fold. The dose of tipranavir is 500 mg (two capsules) twice daily. Aptivus® should be stored at 2–8°C prior to opening the bottle and at room temperature for less than two months after opening the bottle. Aptivus® was approved by the United States Food and Drug Administration in June 2005.

Lipid-Containing Solubilizing Vehicles in Capsules—Five and Six Solubilizing Excipients

The highest level of complexity of lipid-based formulations in currently marketed products can be illustrated by three cyclosporin A products, which contain mixtures of five or six excipients and are microemulsion preconcentrates. Gengraf®, Neoral®, and Sandimmune® soft gelatin capsules are all stored at room temperature.

1) A mixture of **ethanol (11.9%), PG, corn oil mono-di-triglycerides, cremophor RH 40, and α-tocopherol** solubilizes 25 or 100 mg of cyclosporin A in Neoral® soft gelatin capsules. Cyclosporin A is a sparingly water-soluble lipophilic cyclic peptide with a molecular weight of 1201 Da used in preventing rejection of transplanted kidneys, livers and hearts, and is commercially available in multiple solubilized oral formulations. Cyclosporin A was originally formulated as Sandimmune® in 25 mg, 50 mg and 100 mg soft gelatin capsules as well as a 100 mg/mL oral solution. The Sandimmune® soft gelatin capsules contain cyclosporin A dissolved in 12.7% ethanol, corn oil, glycerol, and Labrafil M-2125CS, while the oral solution contains 12.5% ethanol, olive oil, and Labrafil M-1944CS. The absolute

oral bioavailability of cyclosporin A as Sandimmune® is erratic and is <10% in liver transplant patients, and as high as 89% in some renal transplant patients. To improve the oral bioavailability, cyclosporin A was reformulated as a microemulsion preconcentrate and solubilized in 11.9% ethanol, propylene glycol, corn oil-mono-di-triglycerides, polyoxyl 40 hydrogenated castor oil (cremophor RH 40), and DL-α-tocopherol in 25 mg or 100 mg Neoral® soft gelatin capsules as well as a 100 mg/mL oral solution. Neoral® forms a nanoemulsion once dispersed in aqueous media (Bekerman et al., 2004), and after oral administration of Neoral® the cyclosporin A area-under-the-curve of the plasma concentration versus time profile is 20% to 50% higher than with Sandimmune®. Also, the peak plasma concentrations of cyclosporin A are 40% to 106% higher after oral administration of Neoral® compared to Sandimmune®.

2) A mixture of **ethanol (12.8%), PEG, cremophor EL, Tween 80, propylene glycol, and sorbitan monooleate** solubilizes 25 or 100 mg of Cyclosporin A in Gengraf® hard gelatin capsules. Gengraf® and Neoral® are bioequivalent with virtually identical pharmacokinetics.
3) A mixture of **medium-chain triglycerides (Labrafac), dl-α-tocopherol, Labrasol, and cremophor EL solubilizes** 100 mg of Cyclosporin A in soft gelatin capsules in a generic by Sidmak.
4) A clinical formulation of cyclosporin that was shown to be bioequivalent to Neoral® used a liquid-filled capsule formulation that spontaneously forms a nanosuspension when mixed in an aqueous media (Bekerman et al., 2004). The liquid vehicle is composed of 200 mg of cyclosporin dissolved in a mixture of 800 μL of organic solvent [**NMP** (N-methylpyrrolidone), PG and/or PEG], 140 mg of **egg phosphatidylcholine**, 540 μL of **Tween 80**, 540 μL of **Span 80**, 400 μL of **cremophor RH 40**, and 260 μL of **medium-chain triglyceride** and filled into soft gelatin capsules. The formulation composition was carefully selected and was chosen since *in vitro* it resulted in the smallest particle size of the nanosuspension (25 nm), and in the *in vivo* clinical trial the selected formulation resulted in the highest plasma maximum concentration and area-under-the-curve.

Oral Solutions and Elixirs

Oral solutions can be developed for either early Phase I clinical trials of a new chemical entity (then further develop a solid oral dosage form), or for intended commercial manufacture. Oral solutions for early clinical trials are developed to rapidly introduce a new chemical entity into human trials, because an oral solution can be the simplest formulation to develop and manufacture within a short period of time. For a water-insoluble new chemical entity the solubilizing vehicles used in simple oral solutions for early clinical trials include PEG 400, PG or a mixture of PEG 400/PG; PEG 400/TPGS; PEG 400/ethanol; or PEG 400/PG/ethanol. If the new chemical entity has poor taste at the concentrations used, then a small amount of peppermint oil can be added. The volume used in an early clinical trial is typically 10's of milliliters, thus the concentrations used depend on the dose such that a high dose drug of 500 mg can be administered using fifty milliliters of a 10 mg/mL oral solution, or five milliliters of a 100 mg/mL oral solution. A commonly used Phase I oral solution formulation is 20/80 ethanol/PEG 400 administered up to ten milliliters twice-a-day (Lamarre, 2003).

Commercial oral solutions are inherently more complex than capsule-filled formulations, and the challenges in developing an oral solution include solubility, chemical stability, physical stability, preservation, and taste-masking. Oral solutions range in complexity from formulations such as one solvent with a preservative to complex multi-excipient solutions containing flavors, sweeteners, colors, buffers, salts, preservatives, surfactants, and organic solvents. Elixirs are sweetened hydroalcoholic oral solutions that are specially formulated for oral use in infants and children. A mixture of polyethylene glycol and propylene glycol, or propylene glycol and ethanol are commonly used in over-the-counter and prescription oral solutions for pediatric cold, cough and/or flu remedies

Many oral solutions are intended for pediatric administration, of which oral solution formulations are a subset of a larger choice of formulation type such as suspension, syrup, powder or microcapsules for constitution to a suspension, powder for reconstitution to a solution or suspension, solid particles (powder, coated particles, extended release, enteric-coated granules, beads) in packets or capsules to be sprinkled on food, oral powders, and chewable tablets. The broader topic of pediatric formulation development is beyond the scope of this chapter, but this chapter will cover selected oral solutions for pediatric administration.

Aqueous-Based Oral Solution

The water-soluble solubilizing excipients in oral solutions include ethanol, propylene glycol, polyethylene glycol, glycerin, and hydroxypropyl-β-cyclodextrin.

1) Digoxin, a non-ionizable cardiotonic glycoside, is practically insoluble in water and is solubilized in **40% propylene glycol, 10% ethanol**, flavor, sweetener, preservative and buffers to 50 μg/mL in Lanoxin® Elixir Pediatric. The package configuration is a 60-mL bottle with a 1-mL calibrated dropper marked at 0.1 mL divisions. The absolute oral bioavailability of digoxin from the elixir is 70–85%. The pediatric dose is 10–35 μg/kg with a maximum dose of 250 μg using the elixir, which is bioequivalent to the upper adult dose of 200 μg using the capsule formulation.

Thus the maximum volume of the elixir is five milliliters which corresponds to two milliliters of PG, and 0.5 mL of ethanol per day.

2) Phenobarbital, an anticonvulsant and sedative with an intrinsic water solubility of 1 mg/mL, is solubilized in water, **23% ethanol**, glucose, sodium saccharin and flavors to 3.5 mg/mL in Donnatal® Elixir.

3) A mixture of **polyethylene glycol, propylene glycol**, citric acid, and water at pH 3.5 dissolves 32 mg/mL of acetaminophen, 0.2 mg/mL of chlorpheniramine maleate, 1 mg/mL of dextromethorphan HBr, and 3 mg/mL of pseudoephedrine HCl in over-the-counter Children's Tylenol® Cold plus Cough oral solution.

4) Itraconazole, a weakly basic ($pK_a \sim 3.7$) water-insoluble antifungal drug, is solubilized to 10 mg/mL using a combination of non-covalent complexation with **40% hydroxypropyl-β-cyclodextrin** (i.e., ~400 mg/mL) in water and pH adjustment to ~2 in Sporanox® oral solution (Peeters, et al., 2002). The relative oral bioavailability of itraconazole from the oral solution is 149% ± 68% compared to capsules from which the oral bioavailability is 55%. Therefore the oral bioavailability of itraconazole from the oral solution can be estimated to be 45–82%. The dose of Sporanox® oral solution is up to twenty milliliters once-a-day, which is eight grams of hydroxypropyl-β-cyclodextrin per dose representing the estimated maximum amount administered orally per dose (Table 2).

Lipid-Containing Oral Solutions

While the majority of oral solutions are aqueous based, at least seven are lipid-based or contain a significant amount of a lipid as a critical formulation component.

1) Calcitriol, in addition to being available as Rocaltrol® soft gelatin capsules, is available in the USA as a 1 μg/mL oral solution in a **fractionated triglyceride of palm seed oil** along with the antioxidants BHA and BHT. The pediatric dose in patients over three years of age is 0.25–0.5 μg (0.25–0.5 mL) daily. For children less than three years of age, the dose is 10–15 ng/kg (0.01–0.15 mL/kg) daily. Rocaltrol® oral solution is dispensed in a 15-mL multi-dose container and is supplied with twenty single-use graduated oral dispensers. Rocaltrol® oral solution should be stored at controlled room temperature protected from light.

2) Efavirenz is a water-insoluble non-nucleoside reverse transcriptase inhibitor widely used in the treatment of HIV infection. Efavirenz is available as Sustiva® 600 mg tablets in both the UK

and the USA, but has also been available in the UK since 1999 as a 30 mg/mL oral solution in **medium-chain triglycerides** in combination with benzoic acid and strawberry/mint flavor. The daily dose of efavirenz is 600 mg (twenty milliliters) for adults and 270–600 mg for pediatrics, which is nine to twenty milliliters and this upper dosing regimen delivers the estimated maximum amount of medium chain triglycerides per unit dose of any currently marketed oral lipid-based formulation (Table 2). The colorless Sustiva® 30 mg/mL Oral Solution is packaged in a 180-mL multi-dose container that should be stored at controlled room temperature.

3) Amprenavir is available in both Agenerase® soft gelatin capsule and Agenerase® oral solution formulations. As an oral solution, amprenavir is solubilized to the extent of 15 mg/mL in a combination of (approximate percentages) **12% TPGS, ~17% PEG 400 and ~55% propylene glycol** and flavored with grape, bubblegum and peppermint. In children over four years of age, the dose of amprenavir is 17 mg/kg (1.1 mL/kg) three times daily which delivers a total daily amount of eight grams of TPGS, thirty-six grams of propylene glycol, and eleven grams of PEG 400 assuming a total patient body weight of twenty kilograms. Due to the potential toxicity of the large dose of co-administered propylene glycol (~1650 mg/kg per day), Agenerase® oral solution is contraindicated in infants and children below the age of four years. The oral bioavailability of amprenavir from the oral solution is ~14% less than that from the capsule formulation, thus requiring the maximum adult dose of the oral solution to be adjusted to 1400 mg, which is approximately ninety-two milliliters, twice a day. The total excipient amounts co-administered in conjunction with a twice daily dose of ninety-two milliliters of Agenerase® oral solution is twenty-two grams of TPGS, 102 grams of propylene glycol, and thirty-two grams of PEG 400, representing the estimated highest amounts of PEG 400, TPGS and propylene glycol given orally (Table 2). Agenerase® oral solution is packaged in a 240-mL multi-dose container and should be stored at room temperature. Agenerase® oral solution has been available in the UK since 2000, and is also available in the USA.

4) Ritonavir is formulated as both Norvir® soft gelatin capsules and Norvir® oral solution, which contains 80 mg/mL of ritonavir solubilized in a mixture of **cremophor EL, propylene glycol, 42% ethanol, water, and peppermint oil**. The pediatric dose of ritonavir is 250–450 mg/m^2 or up to a maximum of 600 mg (7.5 mL) twice daily. The total amount of ethanol administered in conjunction with a daily adult dose of ritonavir as the oral solution

is 3.2 mL twice daily, representing the highest estimated amount of ethanol given orally (Table 2). The Norvir® oral solution is packaged in a 240-mL multi-dose container and should be stored at room temperature. Norvir® oral solution has been available in the UK since 1996 and is also available in the USA.

5) Lopinavir and ritonavir are co-formulated as both Kaletra® soft gelatin capsules and Kaletra® oral solution. Kaletra® fixed-dose combination oral solution contains 80 mg/mL of lopinavir and 20 mg/mL of ritonavir solubilized in **propylene glycol, 42% ethanol, water, glycerin, cremophor RH 40 and peppermint oil**. Kaletra® oral solution is packaged in a 160-mL multi-dose container and should be stored refrigerated at 2–8°C or at room temperature for no more than two months. Kaletra® oral solution has been available in the UK since 2001 and is also available in the USA.

6) Cyclosporin A is available as Neoral® and Sandimmune® oral solutions, as well as Neoral®, Sandimmune® and Gengraf® soft gelatin capsules. Sandimmune® oral solution contains 100 mg/mL of cyclosporin A dissolved in 12.5% **ethanol, olive oil, and Labrafil M-1944CS**. The dose of Sandimmune® oral solution ranges from one to seven milliliters daily and is to be administered after dilution with milk or orange juice. Grapefruit and grapefruit juice inhibit the metabolism of cyclosporin and should be avoided. Neoral® oral solution is a microemulsion preconcentrate containing 100 mg/mL of cyclosporin A dissolved in 11.9% **ethanol, dl-α-tocopherol, corn oil-mono-di-triglycerides, cremophor RH 40, and propylene glycol**. The dose of Neoral® oral solution ranges from one to seven milliliters daily and should be further diluted with orange or apple juice at room temperature and administered immediately. Both Neoral® and Sandimmune® are packaged in 50-mL multi-dose bottles and once opened, the contents should be consumed within two months. Both Neoral® and Sandimmune® oral solution should be stored at room temperature, and should not be stored refrigerated since at less than 20°C the solution may gel or form sediment, which does not affect product performance or dosing. Neoral® has been available in the UK since 1995, and both Neoral® and Sandimmune® oral solutions are available in the USA.

7) Sirolimus is a non-ionizable and water-insoluble immunosuppressant that is formulated in Rapamune® as a 1 mg/mL solution in **polysorbate 80** and the proprietary excipient combination, Phosal 50 PG®, which is composed of **phosphatidylcholine, propylene glycol, mono- and diglycerides, 1.5–2.5% ethanol, soy fatty acids and ascorbyl palmitate**. The oral bioavailability of

sirolimus from Rapamune® oral solution is approximately 14% when dosed in the fasted state, increasing to approximately 20% when given with a high-fat meal. Rapamune® is also available in a nanoparticulate tablet formulation from which a bioavailability of 27% is achieved. The loading dose of sirolimus is 6 mg (six milliliters) followed by a maintenance dose of 2 mg (two milliliters) daily. Rapamune® is supplied in a 60-mL multi-dose container and should be stored refrigerated at 2–8°C and used within thirty days of opening. Alternatively, Rapamune® may be stored at controlled room temperature for up to fifteen days. Rapamune® is available in the USA and has been available in the UK since 2001.

Conclusions

There is a wide selection of solubilizing excipients that can be used in oral formulations either by themselves or in combination with other excipients. The solubilizing excipients are generally categorized into various types including water-soluble organic solvent/excipients, water-insoluble organic solvents/excipients, triglycerides, surfactants, phospholipids, and cyclodextrins. The water-soluble organic solvents in commercially available solubilized oral formulations are ethanol, glycerin, polyethylene glycol 300 (PEG 300), PEG 400, propylene glycol, and Transcutol® HP. The non-ionic surfactants in commercially available solubilized oral formulations include polyoxyl 35 castor oil (cremophor EL), polyoxyl 40 hydrogenated castor oil (cremophor RH 40), polysorbate 20, polysorbate 80, d-α-tocopherol polyethylene glycol 1000 succinate (TPGS), sorbitan monooleate, polyoxyl 40 stearate, and various polyglycolyzed glycerides including Labrafil® M-1944CS, Labrafil® M-2125CS, Labrasol®, Gelucire® 44/14, Gelucire® 50/13 and Softigen® 767. The water-insoluble solvents/excipients used in commercially available solubilized oral formulations include oleic acid, dl-α-tocopherol (Vitamin E), long-chain triglycerides, medium-chain triglycerides, mono and diglycerides, propylene glycol esters, and hydrogenated oils. The only cyclodextrin in oral pharmaceutical formulation is hydroxypropyl-β-cyclodextrin. The phospholipid in oral formulations is currently limited to phosphatidylcholine, but other phospholipids are sure to be included in future oral formulations such as solid dispersions and solid lipid nanoparticles.

Common solubilizing vehicles in oral formulation are mixtures of solubilizing excipients including binary mixtures of: PEG 400 and propylene glycol; PEG 400 and medium-chain triglycerides; PEG 400 and peppermint oil; PEG 400 and Tween 20; and medium-chain triglycerides and ethanol. Ternary mixtures include PEG 400, propylene glycol, and ethanol, medium-chain triglycerides, PEG, and propylene glycol; oleic acid, cremophor and ethanol; oleic acid, cremophor, and propylene glycol; and TPGS, PEG 400, and propylene glycol. More complex mixtures include the microemulsion preconcentrates being composed of: medium chain mono-diglycerides, cremophor, ethanol, and propylene glycol; ethanol, propylene glycol, corn oil mono-di-triglycerides, cremophor, and α-tocopherol; and ethanol, propylene glycol, PEG, cremophor, sorbitan monooleate, and Tween 80.

The amount of active drug substance contained in solubilized oral formulations ranges from 0.25 μg to 500 mg in liquid-filled capsules, and from 1 μg/mL to 100 mg/mL for oral solutions. The total amount of lipid excipient administered in a single dose of a capsule formulation ranges from 0.5–5 grams, but for oral solution products can range from as low as 0.1 milliliters to as high as twenty milliliters. Lipid-based formulations range in complexity from simple, one-excipient solutions (e.g., sesame or corn oil) to multi-excipient, self-emulsifying drug delivery systems (SEDDS). Some solubilized oral formulations and oral lipid-based products require long-term storage at 2–8°C, and can tolerate room temperature storage for only brief periods due to chemical and/or physical stability issues. While in some cases just the opposite is recommended in that some oral solutions are stored at room temperature and not refrigerated due to possible gelling or precipitation upon cooling.

Future prospects will include advances in scientific understanding of the interactions between drugs and excipients with endogenous *in vivo* fluids and physiological processes, technological innovations in new manufacturing techniques such as nanotechnology, and improvements in existing methods such as solid dispersions, as well as new excipients.

List of Abbreviations

BHA .... butylated hydroxy anisole
BHT .... butylated hydroxytoluene
b.i.d .... twice-a-day
EDTA .... ethylenediaminetetraacetic acid
HIV .... human immunodeficiency virus
HP-β-CD .... hydroxypropyl-β-cyclodextrin
HEC .... hydroxyethylcellulose
HPC .... hydroxypropylcellulose
HSPC .... hydrogenated soy phosphatidylcholine
MCC .... microcrystalline cellulose
PEG .... polyethylene glycol
PG .... propylene glycol
q.d. .... once-a-day
q.i.d .... four times-a-day
t.i.d .... three times-a-day
t.i.w .... three times-a-week
RT .... room temperature
SEDDS .... self-emulsifying drug delivery system
TPGS .... d-α tocopheryl polyethylene glycol 1000 succinate

References

Bekerman T, Golenser J, and Domb A. Cyclosporin Nanoparticulate Lipospheres for Oral Administration. *J Pharm Sci* 2004; 93: 1264–1270.

Bummer PM. Physical Chemical Considerations of Lipid-Based Oral Drug Delivery-Solid Lipid Nanoparticles. *Crit Reviews in Ther Drug Carrier Sys* 2004; 21: 1–19.

Cawley D and Stern MH. Water-Soluble Tocopherol Derivatives. US Patent 1954: 2,680,749.

Constantinides PP, Han J, and Davis SS. Advances in the Use of Tocols as Drug Delivery Vehicles. *Pharm Res* 2006; 23: 243–255.

Constantinides PP, Tustian A, and Kessler DR. Toco Emulsions for Drug Solubilization and Parenteral Delivery. *Adv Drug Del Rev* 2004; 56: 1243–1255.

Dannenfelser R-M, He H, Joshi Y, Bateman S, and Serajuddin ATM. Development of Clinical Dosage Forms for a Poorly Water-Soluble Drug I: Applications of Polyethylene Glycol-Polysorbate 80 Solid Dispersion Carrier System. *J Pharm Sci* 2004; 93: 1165–1175.

Internet websites accessed February 2007:

1) http://www.rxlist.com
2) http://www.fda.gov/cder
3) http://www.accessdata.fda.gov/scripts/cder/iig/index.cfm
4) http://emc.medicines.org.uk
5) http://www.e-search.ne.jp/~jpr/jpr_db/eindex.html
6) for the European Union: http://www.emea.eu.int/index/indexh1.htm
7) for Sweden: http://www.mpa.se/eng/spc/index.shtml
8) for New Zealand: http://www.medsafe.govt.nz/DatasheetPage.htm
9) for South Africa: http://home.intekom.com/pharm/
10) http://www.gattefosse.com
11) http://www.croda.com
12) http://www.sasol.com
13) http://www.corporate.basf.com
14) http://www.cydexinc.com
15) http://www.gaylordchemical.com
16) http://www.ispcorp.com
17) http://www.lipoid.com
18) http://www.spectrumchemical.com
19) http://www.cognis.com
20) http://www.pharmaline.com
21) http://www.abitec.com
22) http://www.accessdata.fda.gov/scripts/cder/iig/index.cfm
23) http://www.skyepharma.com
24) http://www.rocheusa.com/products/FTVDearDoctorFINAL.pdf

Irie T and Uekama K. Pharmaceutical Applications of Cyclodextrins. III. Toxicology Issues and Safety Evaluation. *J Pharm Sci* 1997; 86: 147–162.

Lamarre D. An NS3 Protease Inhibitor with Antiviral Effects in Humans Infected with Hepatitis C Virus. *Nature* 2003; 426: 186–189.

Liu R., *Water-Insoluble Drug Formulation*, Interpharm Press, Denver, Colorado, 2000.

Loftsson TL and Brewster ME. Pharmaceutical Applications of Cyclodextrins. I. Drug Solubilization and Stabilization. *J Pharm Sci* 1996; 85: 1017–1024.

Nerurkar J, Beach JW, Park MO, and Jun HW. Solubility of (±)-Ibuprofen and the S (+)-Ibuprofen in the Presence of Cosolvents and Cyclodextrins. *Pharm Dev and Tech* 2005; 10:413–421.

Peeters J, Neeskens P, Tollenaere JP, Remoortere PV, and Brewster ME. Characterization of the Interaction of 2-Hydroxypropyl-ß-Cyclodextrin with Itraconazole at pH 2, 4, and 7. *J Pharm Sci* 2002; 91: 1414–1422.

Physician's Desk Reference, 60th ed., Thomson Healthcare, Inc. , Montvale, NJ, 2006.

Powell MF, Nguyen T, and Baloian L. Compendium of Excipients for Parenteral Formulations. *J Parent Sci Technol* 1998; 52: 238–411.

Schulze JDR, Waddington WA, Ell PJE, Parsons GE, Coffin MD, and Basit AW. Concentration-Dependent Effects of Polyethylene Glycol 400 on Gastrointestinal Transit and Drug Absorption. *Pharm Res* 2003; 20: 1984–1988.

Serajuddin ATM. Solid Dispersion of Poorly Water-Soluble Drugs: Early Promises, Subsequent Problems, and Recent Breakthroughs. *J Pharm Sci* 1999; 88: 1058–1066.

Strickley RG. Parenteral Formulations of Small Molecule Therapeutics Marketed in the United States (1999)-Part I. *J Parent Sci Technol* 1999; 53: 324–349.

Strickley RG. Solubilizing Excipients in Oral and Injectable Formulations. *Pharm Res* 2004; 21: 201–230.

Stuchlik M and Zak, S. Lipid-Based Vehicle for Oral Delivery. *Biomed Papers* 2001; 145: 17–26.

Sweetana S and Akers MJ. Solubility Principles and Practices for Parenteral Dosage Form Development. *J Parent Sci Technol* 1996; 50: 330–342.

Tenjarla S. Microemulsions: An Overview and Pharmaceutical Applications. *Crit Reviews in Ther Drug Carrier Sys* 1999; 16: 461–521.

Wang Y-C J and Kowal RR. Review of Excipients and pH's for Parenteral Products Used in the United States. *J Parent Sci Technol* 1980; 34: 452–462.

Won D-H, Kim M-S, Lee S, Park J-S, and Hwang S-J. Improved Physicochemical Characteristics of Felodipine Solid Dispersion Particles by Supercritical Anti-Solvent Precipitation Process. *Intl J Pharm* 2005; 301: 199–208.

Wu SH-W and Hopkins WK. Characteristics of D-α-Tocopheryl PEG 1000 Succinate for Applications as an Absorption Enhancer in Drug Delivery Systems*Pharm Tech* 1999; 23: 52–68.

York P, Kompella UB, and Shekunov BY. *Supercritical Fluid Technology for Drug Product Development.* Marcel Dekker, Inc., New York, NY, 2004.

Yu L, Bridgers A, Polli J, Vickers A, Long S, Roy A, Winnike R, and Coffin M. Vitamin-E-TPGS Increases Absorption Flux of an HIV Protease Inhibitor by Enhancing its Solubility and Permeability. *Pharm Res* 1999; 16: 1812–1817.

10

Solubilizing Systems for Parenteral Formulation Development—Small Molecules

JAMES E. KIPP

BioPharma Solutions, Baxter Healthcare Corporation, Round Lake, IL, USA

Introduction

"Parenteral" is defined[1] as "situated or occurring outside the intestine, and especially introduced otherwise than by way of the intestines"—pertaining to essentially any administration route other than enteral. This field is obviously too broad for an adequate focus in one book, let alone one chapter. Many have nonetheless used the term synonymously with injectable drug delivery. We restrict ourselves to this latter usage. This would thus include intravenous, intramuscular, subcutaneous, intrathecal, and subdural injection. In this chapter we discuss the theoretical and practical aspects of solubilizing small molecules for injectable formulation development and will examine the role of surfactants and other excipients in more recent parenteral delivery systems such as liposomes, solid-drug nanoparticles and particulate carriers.

Today, pharmaceutical additives have adopted a more prominent and specialized role in parenteral drug delivery. The need to formulate and deliver water-insoluble compounds is clearly recognized. In a competitive marketplace, there is an obvious need for rapid drug discovery and formulation development. Lipinski (2000) has suggested that increased reliance on high-throughput screening has led to the investigation of drug candidates with less than optimal physicochemical properties, such as aqueous solubility, and hence to more difficulty in formulation and clinical development. As a result of high-throughput screening using universal solvents such as dimethyl sulfoxide, promising candidates often fail bioavailability requirements, in large part due to solubility issues (Lipinski et al., 1997). The deliverable quantity of a drug that can be dissolved may be limited by its toxicity at the higher concentration and that of the solubilizing agents themselves (e.g., Cremophor EL, polysorbate 80). Second, targeted drug delivery is becoming important for selective activity, high therapeutic efficacy and low toxicity. The need to provide sustained delivery has led to the use of more elaborate injectable platforms, for example liposomal and nanoparticulate drug delivery, and this has fostered the use of new polymeric surfactants and carriers.

The use of solubilizing excipients in the delivery of injectable drugs carries a greater number of restrictions than in oral applications. Unquestionably, injectable dosage forms must be held to the strictest standards of purity, sterility, and product consistency. Risks associated with injection usually necessitate the intermediacy of a healthcare professional, although cost containment in the healthcare industry has contributed to an increasing number of injectable medications that are administered at home. This evolution is certain to continue, making the design of safe, easy-to-use parenteral drugs of paramount importance. It follows that the choice of excipients must assist in optimizing this safety.

In intravenous drug delivery, constituents of the formulation interact immediately with tissues near the injection site, with blood and vasculature of the circulatory system, the heart, and with highly vascularized organs—the liver, spleen, lungs and kidneys. A large variety of immediate, acute complications may arise. Many drugs and formulation excipients may elicit an immune or pseudoallergic response, as in, for example, antibody-mediated (Type II) or antibody-complex mediated (Type III) hypersensitivity reactions (Coico et al., 2003). In some of these reactions, the drug or excipient acts as a hapten and may combine with circulating blood cells to induce antibody formation, some of which involves the complement cascade. Blood cell lysis and thrombosis may also occur, particularly near the site of injection. Phlebitis or embolism may subsequently occur elsewhere in the body, particularly in the lungs (pulmonary embolism). Simple chemical properties of the drug or excipients may also cause adverse reactions. Solutions at high pH or high osmolality, for example, may be too irritating. Poorly water-soluble drugs may precipitate upon injection. Cytotoxic agents may extravasate and cause necrosis of the surrounding tissues. In the case of intramuscular delivery, care must be exercised that a vein or artery is not entered if the desire is to limit distribution of a toxic drug. On the other hand, muscle damage may occur if the toxicity threshold is exceeded. Subcutaneous delivery, in which the formulation is injected into the loose connective tissue underneath the dermis, shares with other injection routes some of the associated risks such as immunologic response. However, subcutaneous delivery may carry a higher infection and inflammation risk because the drug formulation is absorbed more slowly from the injection site than one that is intravenously delivered.

Theoretical Aspects of Solubilizer Choice in Parenteral Formulation Design

Most of the currently marketed injectable drugs rely on complete dissolution in the formulation medium. In choosing a solubilizing vehicle the maxim "like dissolves like" is often quoted but inadequately explained. Often this is taken to mean that two substances with "similar intermolecular forces" will mix more readily than those with dissimilar interaction forces. These forces consist of those between the same compounds (cohesive: solute-solute and solvent-solvent) and between different compounds (adhesive: solute-solvent). Specifically, what are these forces and which are important?

Molecular shape and electronic charge distribution (including that induced by interaction with surrounding molecules) affect intermolecular attraction and thus solubility. This charge polarization may be permanent, or induced by interactions with neighboring molecular dipoles (that are permanent or induced). The induced dipole-induced dipole component that is instantaneous as compared to molecular motion, is termed the dispersion force, and is fairly constant per unit surface area for all molecules (Adamson, 1973). The dispersion force contributes to all intermolecular interactions, even between polar molecules. The vector sum of all induced and permanent dipolar interactions (excluding hydrogen bonding) determines the so-called van der Waals force. These dipolar forces, permanent and induced, decrease with the inverse of the intermolecular separation taken to the sixth power.

Hydrogen bonding is a rapidly reversible, highly directional, dipole-dipole interaction. The energy of an individual hydrogen bond usually ranges from 1 kcal/mol to 5 kcal/mol. A relationship between intermolecular separation and hydrogen bond strength is difficult to derive because of its directional nature. Furthermore, hydrogen bonds may not form between two molecules because of internal (intramolecular) hydrogen bonding. An important effect on stabilization is the number of hydrogen bonds that can form between molecules. The number of molecules coordinated by hydrogen bonding to surrounding molecules (solute and solvent) has both enthalpic and entropic contributions (Eisenberg and Kauzmann, 1969; Wu et al., 1982; Gaballa and Neilson, 1983; Gorbaty and Demianets, 1985; Sciortino et al., 1992). It is the high strength of the O-H hydrogen bond combined with high coordination number that results in the large cohesive energy of water, as reflected in low vapor pressure and high boiling point. Given this cohesiveness, the ability of the solute to disrupt hydrogen bonding between water molecules by competition for coordination sites is a very important factor in dissolution. The contribution of hydrogen bonding interactions (water-water, water-solute, and solute-solute) depends on the nature of the donor and acceptor (e.g., electronegativity of the atoms involved), the environment (e.g., dielectric strength of the medium), and the average coordination number.

Because dispersion forces are relatively weak as compared to hydrogen bonding, a solute with a large surface area and few available sites for hydrogen bonding will require more chemical work to dissolve in water than molecules that can form multiple hydrogen bonds relative to the total molecular surface area. Water molecules that cannot interact with the solute surface by hydrogen bonding must orient themselves so that such bonding interactions face away from the interface and toward neighboring water molecules. This restriction in water mobility results in a significant, unfavorable negative entropy change. Molecules with apolar surfaces thus tend to aggregate in water so that this hydrophobic surface area is reduced—the so-called "hydrophobic effect". The tendency of non-polar surfaces to avoid contact with water is often ascribed to strong self-association of water at this interface ("iceberg effect"); simulations have failed however to prove this (Lynden-Bell et al., 2001). The hydrophobic effect is a major determinant of the solubility of many organic compounds in water, and excipients that enhance this effect lower the solubility of organic compounds in water whereas agents that have been termed "antihydrophobic" may increase their solubility. The latter include amphiphiles that possess both polar and nonpolar regions, and thus can lower surface tension at a hydrophobe-water interface. Antihydrophobic ions, which are often used as protein denaturants, function by bridging between nonpolar surfaces and water (Tanford, 1980; Breslow, 1980; Breslow, 1984; Breslow, 1991; Breslow, 1994). Prohydrophobic additives are generally simple salts with cations of small ionic radii (e.g., lithium or sodium); antihydrophobic additives include salts of large ("soft") ions such as guanidinium perchlorate.

For solute and solvent molecules that are similar in shape, size and polarity, the solubility of liquids is essentially driven by entropy of mixing. Liquids that are freely soluble in water have two important properties—miscibility, resulting in a large positive entropy change, and sufficiently low cohesive energy as reflected by the vapor pressure, boiling point, or melting point of the material. To achieve dissolution of a solid, lattice forces must be overcome, and the melting point cannot be prohibitively high if dissolution is to occur. The propensity for the melt to mix with water, driven by enthalpy and entropy, is expressed through the solute activity coefficient. Methods have been published for estimation of the activity coefficient (Ornektekin et al., 1996; Gmehling et al., 1998; Lohmann et al., 2001; Hwang et al., 2001). One of the more practical is the partitioning model of Yalkowsky, in which the activity coefficient may be approximated by the octanol-water partition coefficient. Log P can be easily estimated by group contribution methods (Leo et al., 1971; Hansch and Leo, 1979; Leo, 1993; Wang et al., 1997). The following semi-empirical equation, which expresses the logarithm of molar solubility at temperature T as a function of melting point (T_m) and *Log P*, has been derived by Yalkowsky and co-workers, and assumes a fixed entropy of fusion (Jain and Yalkowsky, 2001), that of a semi-rigid, small organic molecule:

$$Log\,S = -0.01(T_m - T) - Log\,P + 0.5 \tag{1}$$

Polarity may be qualitatively defined as the ability of a solute to dissolve in a polar solvent, which results from interaction with surrounding molecules by dipolar, non-dispersive forces. By this definition, hydrocarbons are nonpolar because they possesses no permanent dipole moments, and the entire molecular surface must solely interact with its environment via dispersion forces. Thus methanol is more polar than octanol because the surface area of methanol that interacts only via dispersion forces (hydrophobic surface area) is much less than that of octanol. For liquids, increasing solute polarity generally causes an increase in water solubility. This is not necessarily true for solids because polarity

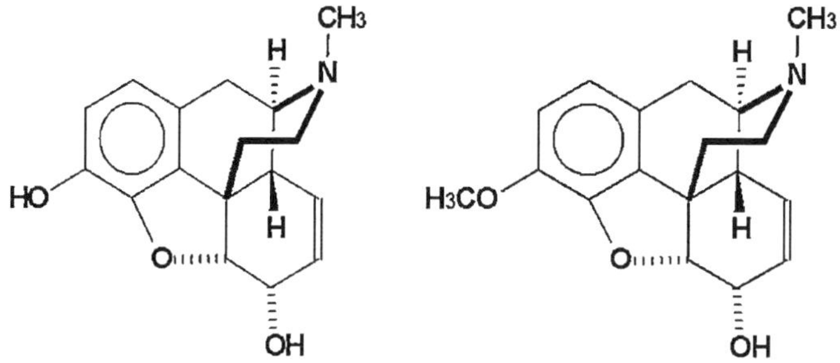

morphine

Melting point >254°C (dec)
Solubility = 1 g in 5000 mL
# H-bond donors = 2
# H-bond acceptors = 4
Log P = 0.89

codeine

Melting point = 154-156°C
Solubility = 1 g in 120 mL
# H-bond donors = 1
# H-bond acceptors = 4
Log P = 1.19

Figure 1. Comparison of morphine and codeine (melting points and solubilities from endnote 13, octanol-water partition coefficients from Avdeef (2003)).

enhancement may also increase lattice energy. Figure 1 compares the structures of morphine and codeine. Morphine has an extra hydroxy group that is available for hydrogen bonding, and accordingly has a higher melting point than codeine (Figure 1). Substitution of the OH group with OCH_3 increases hydrophobicity as expected, yet despite a significantly higher Log P (1.2 versus 0.72), codeine is more soluble because of reduced lattice energy.

Strategies for the Formulation of Parenteral Drugs

Adjustment of pH and Salt Formation

Figure 2 is a flowchart that highlights various strategic paths in parenteral formulation development as a function of basic physical properties (melting point, log P, and solute polarity). The vertical bar on the right-hand side associates a spectrum of solute polarity, from highest at the top and lowest at the bottom, with the most fruitful formulation strategy that will generally apply. Compounds at the top right-hand column are those that are the most polar. Compounds of this type will possess hydrogen bond donors and/or acceptors with highly electronegative heteroatoms (oxygen or nitrogen) that form strong hydrogen bonds. As a consequence, the molecule can lose or gain protons to form solvated ions (e.g., carboxylate, or ammonium ion). Adjustment of pH to form ionic compounds is the easiest and by far the most prevalent approach to solubilization. This strategy is limited however, when the solubility of the uncharged drug molecule (S_0) is very low, or if the solubility product (K_{SP}) of the salt generated by addition of acid or base is very low. As shown in equation 2, an increase of one pH unit above the pKa of a monoprotic acid theoretically results in a tenfold increase in solubility.

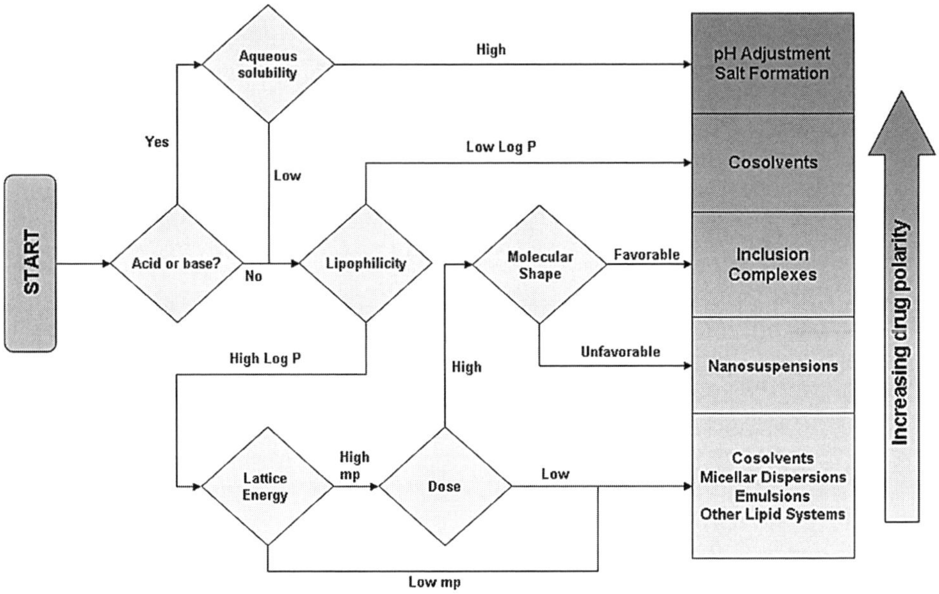

Figure 2. Flowchart for solubilization of parenteral drugs. (See color insert after Index.)

Likewise, a decrease of one pH unit below the pKa of a monobasic drug causes a ten-fold solubility increase. S_0 is the solubility of the uncharged species.

$$S = S_0(1 + 10^{pH-pKa}),\ \textit{for organic acids} \qquad S = S_0(1 + 10^{pKa-pH}),\ \textit{for bases} \qquad (2)$$

In regions where the ion product for a 1:1 salt (M^+X^-) exceeds K_{sp}, and solubility is limited by that of the salt, solubility is approximately equal to $\sqrt{K_{SP}}$. The effects of the degree of acid or base ionization and K_{SP} on the solubility of a very poorly soluble (3×10^{-4} mM) organic acid (HA) and organic base (B) are illustrated in Figure 3. It is observed that a significant solubility enhancement is not realized until the adjusted pH is at least 4 units removed from the pKa of either compound.

Examples abound of drug formulations that need to be adjusted to extreme pH in order to dissolve the drug. Thiopental has a pKa of 7.6 at 20°C (McLeish, 1992) and the solubility of the neutral species is approximately 80 μg/mL (Yalkowsky and He, 2003). Because of this low solubility, the formulation pH is almost 4 log units above the pKa, approximately 10 to 11. Another example is phenytoin, with a pKa of 8.3 (Larsen, 1989), and an intrinsic solubility of 26 μg/mL (Yalkowsky and He, 2003). The commercial preparation is adjusted to pH 12.[2] At physiologic pH (7.4), the solubility of these compounds is quite low, and injection of these high pH formulations, when diluted in blood, may cause precipitation near the injection site, especially upon rapid injection. Soft tissue irritation and inflammation has occurred at the site of injection with and without extravasation of intravenous phenytoin. Reactions may vary from slight tenderness to necrosis (Wheless, 1998).

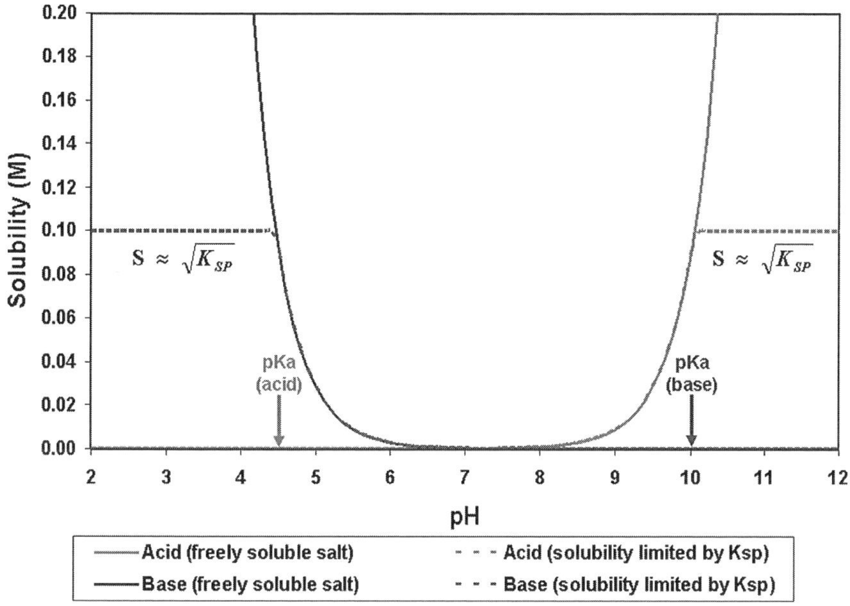

Figure 3. Solubility of organic acid (pKa = 4.5) and base (pKa = 10), illustrating the effect of solubility product, $[M^+][X^-]$, where M^+ and X^- are the counterions of the acid and base respectively.

Table 1 presents examples of marketed drugs that are poorly soluble in their uncharged state, yet which have been successfully formulated by pH adjustment.

Use of Organic Cosolvents

Injectable formulations are severely restricted as compared to orals with respect to the types and concentrations of cosolvents or surfactants permissible in the vehicle. The cytosol of tissue is largely water, and thus high osmotic pressure associated with injection of cosolvent mixtures is a major formulation impediment that may be ameliorated in oral formulations by the mucosal barrier of the enteral tract.

Progressing down the far right column of Figure 2, one finds compounds of lower polarity that either cannot exhibit acid or base behavior over a reasonable pH range (e.g., 3 to 12), or which have intrinsic solubilities that are too low to achieve the desired solubility enhancement by pH adjustment alone. These compounds usually contain polar groups (e.g., alcohols, or amines), capable of hydrogen bonding, and thus these compounds have a low octanol-water partition constant (e.g., less than 2). Water miscibility of the cosolvent requires some degree of hydrogen bonding between the cosolvent and water, and yet the solute must also dissolve in the solvent mixture. This ambivalent interaction between cosolvent and solute and cosolvent with water generally requires that the solute have some intermediate degree of polarity. Drugs that are too hydrophobic (high Log P) may therefore not be amenable to formulation using simple cosolvency.

Drug (generic name)	Brand name	Form	pH of solution (reconstituted, if powder)	Solution components after reconstitution (mg/mL)	Total volume/ vial (mL)	Administration routes
doxycycline hyclate (hydrochloride hemiethanolate hemihydrate)	Vibramycin IV (Pfizer)	sterile powder	1.8–3.3	drug (10); ascorbic acid (48)	10	IV infusion after further dilution
dolasetron mesylate monohydrate	Anzemet (Aventis)	solution	3.2 to 3.8	drug (20); mannitol (38.2); acetic acid for pH adjustment	0.625	IV
doxorubicin hydrochloride	Adriamycin PFS (Pfizer)	lyophilized powder	3.0	drug (10); sodium chloride; pH adjusted with HCl	5	IV infusion, piggyback

acylovir sodium	Zovirax (GlaxoSmith-Kline)	lyophilized powder	11	drug (50)	10	IV infusion after further dilution
allopurinol sodium	Aloprim (Nabi)	lyophilized powder	11.1 to 11.8		30	IV infusion after further dilution
phenytoin sodium	Dilantin (Pfizer, Parke-Davis)	solution	12	50 mg/mL	2	IM (prophylactic dosage, neurosurgery)
warfarin sodium	Coumadin (Bristol-Myers Squibb)	lyophilized powder	8.1 to 8.3	drug (2); sodium phosphate dibasic heptahydrate (4.98); sodium phosphate monobasic monohydrate (0.194); sodium chloride (0.1); mannitol (38); sodium hydroxide for pH adjustment	2	IV (slow peripheral bolus)

Table 1. Poorly soluble drugs that have been solubilized by pH adjustment.[12]

An enhancement in solubility of several orders of magnitude may be obtained by using a pure organic solvent substantially less polar than water, but which is nonetheless water-miscible (e.g., propylene glycol). On the other hand, because of the nonlinear relationship between solubility and cosolvent mole fraction, attaining the desired solubility enhancement may require a high concentration of cosolvent. This would result in solutions with very high osmolality, thereby limiting injection volume or applicability of cosolvents. Only a small fraction (roughly 10% in 1996) of marketed injectables use cosolvents to solubilize drug (Sweetana, 1996). Following the pathway from "start" in Figure 2 to the first branching point, the decision "Acid or base?" must first be addressed. If the answer is negative, then a logical next decision is whether drug hydrophobicity (Log P) is too high for the use of cosolvents. Dissolution in water-miscible cosolvents, such as ethanol and propylene glycol, requires some solute hydrophilicity; and drugs with high Log P may thus require either dispersed systems such as emulsions or suspensions, formulation by inclusion complexation (e.g., cyclodextrins), micellization by surfactants, or a combination of cosolvents and surfactants (as shown at the bottom of the right hand column). Table 2 clearly illustrates that drugs that rely only upon the use of protic cosolvents (e.g., alcohols) for solubilization tend to have low to moderate log P values (less than 3). Common solvents include ethanol, propylene glycol, polyethylene glycol, tetrahydrofurfuryl alcohol, polyethyleneglycol ether, and glycerin. Amsacrine hydrochloride (log P = 1.12), is solubilized in the dipolar, aprotic solvent N,N-dimethylacetamide. These solvents, characterized by the presence of hydrogen-bonding acceptors (nitrogen or oxygen atoms) but no donor groups, tend to dissolve a wide array of insoluble materials, and include N-methyl-2-pyrrolidone (NMP), dimethylsulfoxide (DMSO), and N,N-dimethylformamide (DMF). Solubilization by these cosolvents may be primarily due to their ability to disrupt intermolecular hydrogen bonding between water molecules.

Melphalan Hydrochloride for Injection (Alkeran®) exemplifies the application of cosolvency. The drug is supplied as a lyophilized powder with a sterile diluent for reconstitution that is comprised of sodium citrate (0.2 g), propylene glycol (6.0 mL), ethanol (96% grade, 0.52 mL), and Water for Injection, diluted to a total volume of 10 mL.[3] ALKERAN for Injection is administered intravenously.

Another example in clinical development was formulation of the antitumor agent, VNP40101M (1,2-bis(methylsulfonyl)-1-(2-chloroethyl)-2-[(2-methylamino)carbonyl] hydrazine) (Krishna, 2001). VNP40101M is an N-sulfonylhydrazine with low aqueous solubility. It possesses no acidic or basic functional groups, and thus pH adjustment was not an option. However, the drug has a moderately low Log P (0.75), which should immediately suggest to the formulator that dissolution using a water-miscible cosolvent might be possible. This is also attractive from the standpoint of stability, as the drug is easily hydrolyzed in an aqueous medium, particularly under alkaline conditions. Accordingly, 10 mg/mL of drug concentrate could be formulated using a solvent system consisting of 30% ethyl alcohol and 70% polyethylene glycol-300

Generic name (Originator)	Brand name	Indications	Solubilizing components (v/v)	Melting point of neutral form of active (°C)[13]	Log P (Octanol-water partition coefficient) of active
Carmustine	BiCNU (Bristol-Myers Squibb)	antineoplastic	100% ethanol	30–32	1.53[14]
Chlordiazepoxide HCl	Librium (Roche)	sedative	20% propylene glycol; 1.5% benzyl alcohol	236–236.5	1.0 (Rytting et al., 2005)
Diazepam	Valium (Roche)	sedative	40% propylene glycol; 10% ethanol	125–126	2.82[14]
Digoxin	Lanoxin (GlaxoSmith-Kline)	cardiovascular agent	40% propylene glycol; 10% alcohol	230–265	1.26[14]
Lorazepam	Ativan (Wyeth)	hypnotic, anticonvulsant	18% PEG 400 in propylene glycol, with 2% benzyl alcohol as preservative	166–168	2.48 (Avdeef, 2003)
Nitroglycerin	Tridil (Faulding, US) NitroBid IV (Aventis)	cardiovascular	30% propylene glycol, 30% ethanol 45% propylene glycol	13.5	1.62[14]
Amsacrine	Amsidine (Pfizer)	antineoplastic	100% dimethyl-acetamide, formulated as HCl salt	HCl salt: 231–232 (drug pKa = 7.43)[16]	1.12 (Cornford et al., 1992)
Pentobarbital sodium	Nembutal	hypnotic	drug (50 mg/mL), 40% propylene glycol, 10% alcohol, 10% pH 9.5 (approx.)	122–123	2.08 (Avdeef, 2003)
Phenobarbital sodium	Luminal	anticonvulsant	68% propylene glycol, 10% ethanol	174–178	1.53 (Avdeef, 2003)
Phenytoin sodium	Dilantin (Pfizer)	anticonvulsant	40% propylene glycol, 10% ethanol (pH adjusted to 12)	295–298	2.24 (Avdeef, 2003)

Table 2. List of cosolvent systems employed in parenteral product development.

(PEG-300). Citric acid (0.6%) was added to lower the pH and minimize base-catalyzed hydrolysis. This solution could be diluted to 1 mg/mL for infusion using 5% dextrose (D5W) or normal saline (NS) and stored for 24 hours without incurring visible precipitation or degradation.

Micellar Solubilization Using Surfactants

For compounds with large hydrophobic groups, and thus high Log P (e.g., >3), the use of cosolvents in combination with surfactants (e.g., polysorbate 80, Cremophor EL) may generally be required. Surfactants are amphiphiles that are characterized by the presence of polar regions and sizeable non-polar domains. These molecules are entropically forced to accumulate at the boundaries of phases having different polarity. The same thermodynamic factors cause micellization of the surfactant to occur above its critical micelle concentration (CMC). Hydrophobic molecules can escape interaction with water (hydrophobic effect) by residing within the hydrophobic interior of the micelle. When a water-miscible cosolvent and a surfactant are combined, many effects on solubilization may compete with each other and the outcome may be difficult to predict. In general, the cosolvent increases the CMC of the surfactant and increases solubility of the drug in the continuous phase. Molecules of the organic cosolvent may also be contained within the micelle. This may reduce the hydrophobicity of the micellar interior. Surfactants may, however, be combined with non-aqueous solvents to prepare concentrated solutions, which when reconstituted with water form aqueous micellar dispersions. Formulation of paclitaxel (Taxol®) is one example. Paclitaxel has a large hydrophobic molecular surface, and high log P (3.96).[4] Accordingly, the drug cannot be formulated in an aqueous medium using cosolvents alone. The current formulation is a clear nonaqueous solution that is intended for dilution with a suitable parenteral fluid prior to intravenous infusion. Each mL contains 6 mg paclitaxel, 527 mg of purified Cremophor® EL (polyoxyethylated castor oil) and 49.7% (v/v) dehydrated alcohol, USP.[5]

The use of surfactants in parenteral products is limited by their potent side-effects. All surfactants, especially those that are ionic may cause hemolysis. Anaphylactoid reactions have been noted, particularly in association with the destruction of lymphocytes, resulting in histamine release. Phospholipids are generally well-tolerated, and are frequently used in the preparation of emulsions. Nonionic surfactants such as Cremophor EL and polysorbate 80 (Tween 80) are known to cause hypersensitivity reactions (Eschalier et al., 1988; Mounier et al., 1995; Theis et al., 1995; Munoz et al., 1998; Volcheck and Van Dellen, 1998), including hypotension, largely via histamine release mechanisms. Because of potential side-effects, it is now traditional to pretreat the patient with an antihistamine and possibly a corticosteroid (e.g., prednisone) before administration of formulations that contain Cremophor EL.

Table 3 is a list of three representative injectable formulations that contain surfactants or surfactants in non-aqueous media for aqueous dilution.

Generic name (originator)	Brand name	Form	Formulation components (mg/mL)	Melting point of active[13]	Log K(octanol-water) of active
amphotericin B	Fungizone	lyophilized powder	amphotericin B (2); sodium deoxycholate (4.1); sodium phosphates (2.02)—further diluted for infusion	170 °C (decomposes)	3.04[4]
cyclosporin A	Sandimmune (Novartis)	solution	drug (50); Cremophor EL (650); alcohol (32.9%, v/v)—further diluted before infusion	148–151 °C	3.49[4]
paclitaxel	Taxol (Bristol Myers Squibb)	solution	paclitaxel (6); Cremophor® EL (527); dehydrated alcohol, USP (49.7%, v/v)	213 °C (dec)	3.31[14]

Table 3. List of parenterals that are formulated as colloidal dispersions by use of surfactants.

Inclusion Complexes

Application of cosolvents and surfactants may often not achieve the required solubility enhancement, or ratio between the desired dose and the intrinsic solubility of the pharmaceutical agent. If the desired solubility enhancement is too high, then dispersed systems (e.g., suspensions, emulsions) may be the only option. Cyclodextrins may offer an alternative route. The only cyclodextrins that are generally recognized for use in injectable products are 2-hydropropyl-beta-cyclodextrin (HP-BCD) and sulfobutylether beta-cyclodextrin (SBE-BCD). HP-BCD is found in itraconazole for injection (Sporanox® IV). Each mL of Sporanox IV contains 10 mg (1%, w/v) of itraconazole, solubilized by hydroxypropyl-beta-cyclodextrin (400 mg, 40% w/v), 2.5% (v/v) propylene glycol, and pH adjusted to 4.5, in water for injection. The product is packaged in 25-mL colorless glass ampoules, each containing 250 mg of itraconazole, the contents of which are diluted in 50 mL 0.9% Sodium Chloride Injection, USP (normal saline) prior to infusion. After final dilution, the infusion contains approximately 13% HP-BCD (w/v). As shown by this example however, the amount of cyclodextrin required on a per weight basis may be quite large. A molar excess of cyclodextrin relative to drug is required for complete complexation, and the high molecular weight of the cyclodextrin derivative (MW, HP-BCD = 1440 g/mol) results in a large mass. Such large quantities may limit the route of administration. For example, the use of the cyclodextrin in an oral capsule or tablet might be impractical because of excipient volume.

Sulfobutyl ether beta-cyclodextrin (SBE-BCD) is another cyclodextrin derivative that is being used in commercial parenteral products. This compound is made available by Cydex[6], and is currently used in at least two marketed products: ziprasidone mesylate (Geodon®) and voraconazole (Vfend®), both by Pfizer[7]. Geodon for Injection is administered intramuscularly upon reconstitution and contains a lyophilized form of ziprasidone mesylate trihydrate. After reconstitution, each mL of ziprasidone mesylate for injection contains 20 mg of ziprasidone, 4.7 mg of methanesulfonic acid, and 294 mg (29.4%, w/v) of sulfobutylether beta-cyclodextrin sodium (SBE-BCD). Vfend I.V. is intended for reconstitution and administration by intravenous infusion, and is supplied as a sterile lyophilized powder in a 30-mL single-use vial containing 200 mg of active drug and 3200 mg of sulfobutyl ether beta-cyclodextrin sodium. Lyophilized drug is reconstituted with Water for Injection to produce a solution containing 10 mg/mL Vfend and 160 mg/mL (16%, w/v) of sulfobutyl ether beta-cyclodextrin sodium. This solution is diluted to 5 mg/mL or less, and therefore a maximal cyclodextrin concentration of 8%, w/v, is reached. Based on average degree of substitution, SBE-CD is supplied by Cydex as the heptasodium salt. For some parenteral products, this may pose issues with hyperosmolality because one equivalent of SBE-CD dissociates into eight osmolar equivalents in solution.

Cyclodextrin complexation also depends on a suitable molecular topology that allows the guest molecule to fit within the hydrophobic host cavity. Cyclodextrins form inclusion compounds with hydrophobic guest molecules in aqueous solutions largely because of the hydrophobic effect. Complex formation reduces interaction of water molecules with the hydrophobic interior surface of the cyclodextrin and the hydrophobic surface of the guest molecule. Water molecules that are restricted in their mobility within the cyclodextrin cavity are liberated upon complexation. The fact that polar as well as nonpolar compounds can effectively form complexes shows that complexation entails both enthalpic and entropic contributions. Thus cyclodextrins are known to form complexes with acids, amides, and small ions (Cramer and Henglein, 1956; Wojcik and Rohrbach, 1975).

Table 4 lists representative cyclodextrin formulations that are currently available for parenteral use.

Dispersed Systems

Emulsions

Dispersed systems, such as emulsions, have also been used to achieve high drug loading for parenteral administration. Emulsions generally consist of a vegetable oil (e.g., soybean), a phospholipid surfactant (e.g., lecithin), and glycerol added for isotonicity. The surfactant (emulsifier) is necessary to provide an energy barrier to agglomeration of the emulsion droplets. Unlike, micellar solutions that are thermodynamically stable, emulsions and suspensions are thermodynamically unstable. Stabilization thus hinges upon the ability to kinetically impede coalescence of droplets. The interaction energy for a system of like-charged colloidal particles comprises an attractive van der Waals interaction and a repulsive electrostatic double layer interaction (Derjaguin and Landau, 1941; Verwey and Overbeek, 1948; Crocker and Grier, 1996).

The anesthetic, propofol, has been formulated as an emulsion, Diprivan®. Propofol is a liquid at room temperature, has a high solubility in vegetable oils, and therefore a large octanol-water partition coefficient (Log P = 3.83, pH 6 to 8.5[8]). Because drug concentration in the continuous aqueous phase affects pain on injection, further benefit is realized by keeping the drug in an oil phase. The pKa of propofol is 11, and therefore not amenable to salt formation. In addition to the active component, the formulation also contains soybean oil (10%, w/v), glycerol (2.25%, w/v), egg lecithin (1.2%, w/v), and disodium edetate (0.005%, w/v). Solution pH is 7–8.5. Disodium edetate acts as an antimicrobial agent (Bobey et al., 2001).

The number of solid drugs with moderate to high melting point (approximately 200°C and above) that can be formulated as emulsions may be limited because of water and oil insolubility due to high lattice energy. Drugs that are soluble in oil principally interact via hydrophobic forces (dispersion forces), which are weaker than dipole-dipole interactions, including hydrogen bonding. Lattice energy, however, is largely determined by polarity. Thus a solute with a high melting point must possess some degree of polarity (i.e., presence of permanent dipoles and ability to hydrogen bond), and these strong intermolecular

Drug (generic name)	Brand name	Form	Administration routes	Market
α-cyclodextrin				
alprostadil (PGE_1)	Prostandin, Prostavasin Caverject	Solution Powder with custom diluent	intravenous intracavernosal injection	Europe, Japan Europe, US
2-hydroxypropyl-β-cyclodextrin				
itraconazole	Sporanox IV (Janssen)	Solution	oral, intravenous	Europe, US
mitomycin	Mitozytrex, Mitoextra	Powder for reconstitution as solution	intravenous	US (SuperGen) Europe (Novartis)
sulfobutylether-β-cyclodextrin (Captisol(R))				
voriconazole	Vfend (Pfizer)	Powder for reconstitution as solution	intravenous	Europe, US
ziprasidone mesylate	Geodon (Pfizer) Zeldox (Pfizer)	Powder for reconstitution as solution	intramuscular intramuscular	US Europe

Table 4. List of parenterals that are formulated by use of cyclodextrins.

forces cannot be as readily overcome by the weaker dispersion forces between solute and oil. Thus most drugs that can be formulated in emulsions are generally liquids or low melting solids that have high octanol-water partition coefficients (high log P).

Table 5 lists marketed parenterals that are formulated as emulsions and the corresponding melting points and Log P values of the active principal component.

Liposomes

Only a few liposomal formulations have gained acceptance, possibly because of the high cost of large-scale manufacturing. Liposomes have been reported in the research literature for decades but have only gained recent acceptance for lipophilic antifungals such as amphotericin B and a handful of antineoplastic agents such as doxorubicin. Liposomes are built from a multimolecular assemblage of phospholipids arranged in bilayered structures, or lamellae, with one or several concentric lamellae surrounding an aqueous core. Water-soluble drugs (e.g., doxorubicin) may be carried within the aqueous core of the liposome, whereas hydrophobic drugs such as amphotericin B may interact with the lipid membrane. By compartmental loading of the drug, the amount of nonsequestered drug is reduced and potential toxicity reduction may be realized. As compared with emulsions, the ability to load lipophilic drugs into liposomes is significantly more limited because of the relatively low volume of the hydrophobic compartment as compared with the aqueous core. Another disadvantage is that liposomes are generally less physically stable than emulsions, and must often be lyophilized.

Table 6 presents currently marketed liposomal formulations for parenteral administration.

Suspensions

Compared with solutions, suspensions afford superior loading. As with emulsions, suspended particles must be kinetically stabilized with surfactants to prevent aggregation. The Ostwald-Freundlich equation,

$$\ln \frac{S}{S_0} = \frac{2v\gamma}{rRT} = \frac{2M\gamma}{\rho rRT}, \tag{3}$$

which pertains to spherical particles, defines the effects of particle radius (r), molar volume (v), density (ρ), and interfacial tension (γ) on solubility, S, at temperature T. S_0 is the solubility of a flat, solid sheet ($r \rightarrow \infty$), M is the molecular weight of the solid, and R is the ideal gas constant. Reducing the particle size increases apparent, or local drug solubility, all other factors being constant. As with other dispersions such as emulsions, stabilization depends on the ability to kinetically impede particle coalescence. Instability can also result from a shift in size distribution to larger particles (Ostwald ripening). This is a consequence of equation 3, in which smaller particles must have higher local solubility than large particles. This concentration gradient causes growth of large particles at

Generic name (originator)	Brand name	Indications	Formulation components (mg/mL)	Melting point of active[13]	Log K(octanol-water) of active[16]
Diazpam (Baxter)	Dizac(R)	Anxiety, epilepsy (I.V. administration only)	drug (2), fractionated soybean oil (150), diacetylated monoglycerides (50), egg phospholipids (12), glycerin (22), sodium hydroxide, pH = 8	125–126 °C	2.82 (experimental)
Felbinac ethyl ester (Baxter)	Daitac	Anti-inflammatory		liquid near RT[15]	4.34 (calculated)
Flurbiprofen axetil (Kaken)	Ropion Lipfen	Anti-inflammatory, post-operative pain.		110–111 °C	4.96 (experimental)[18]
Alprostadil, Prostaglandin E_1 (Taisho)	Palux, Liple, Englandin	1) Venous insufficiency 2) Diabetic skin ulcers 3) Peripheral vasodilator		115–116 °C	3.20 (experimental)[17]
Dexamethasone palmitate	Limethason, Limetazon	Chronic articular rheumatism		60–65 °C[b]	9.00 (calculated)[18]
Menaquinone-4, Vitamin K_2 (Eisai)	Glakay	Vitamin K deficiency		oil at room temperature	16.04 (calculated)
Perflenapent (Sonus)	EchoGen	1) Cardiology contrast agent. 2) Radiology contrast agent.		liquid at room temperature	4.40 (experimental)

Perflubron-based emulsion (Alliance)	Imagent BP	Detection of brain tumor with CT.		liquid at room temperature	8.35 (calculated)
Perfluorodecalin and perfluo-rotripropylamine mixture (Yoshitomi)	Fluosol	Oxygen-carrying fluid to reduce myocardial ischemia		liquid at room temperature (mp = −10 °C, mixture cis- and trans-isomers)[19]	7.8 (calculated)
Propofol (AstraZeneca)	Diprivan	Anesthetic	soybean oil (100), glycerol (22.5), egg lecithin (12), and disodium edetate (0.005%); with sodium hydroxide to adjust pH to 7–8.5.	oil at RT (mp = 18°C)	3.79 (experimental)
Vitamin A (retinol) (Pfizer, Kabi)	Vitalipid	Parenteral nutrition		62–64°C	5.68 (experimental)
Vitamin D_2	Vitalipid	Parenteral nutrition		115–118°C	10.44 (calculated)
Vitamin E	Vitalipid	Parenteral nutrition		oil at RT (mp = 2.5–3.5°C)	12.18 (calculated)
Vitamin K_1	Vitalipid	Parenteral nutrition		oil at RT	11.71 (calculated)

Table 5. Parenteral drugs marketed as emulsions.

Drug (generic name)	Brand name	Form	pH of solution (reconstituted, if powder)	Solution components after reconstitution (mg/mL)	Total volume/vial (mL)	Administration routes
daunorubicin citrate	DaunoXome (Gilead)	unilamellar liposomes, core-loaded	4.9–6.0	daunorubicin citrate (2, base equivalent), encapsulated in liposomes consisting of distearoylphosphatidylcholine (28.2), cholesterol (6.72). and dispersed in an aqueous medium containing sucrose (85), 94 glycine (3.76), calcium chloride dihydrate (0.28)	25 (diluted before infusion)	infusion
doxorubicin HCl	Doxil (ALZA, Ortho-Biotech)	PEG calix-unilamellar liposomes, core-loaded (STEALTH® liposomes)	6.5	doxorubicin HCl (2), liposome carriers are composed of N-(carbonyl-methoxypolyethylene glycol 2000)-1,2-distearoyl-*sn*-glycero-3-phosphoethanolamine sodium salt (MPEG-DSPE), (3.19); fully hydrogenated soy phosphatidylcholine (HSPC) (9.58); cholesterol, (3.19), dispersed in a solution containing ammonium sulfate (2); histidine, hydrochloric acid and/or sodium hydroxide for pH control; and sucrose.	10, 30 (diluted before infusion)	iv infusion

amphotericin B	Ambisome (Gilead Sciences, marketed by Astellas Pharma US)	unilamellar liposomes, drug residing in bilayer membrane	5 to 6	Assumes final dilution to 2 mg/mL drug: amphotericin B (2); hydrogenated soy phosphatidylcholine (8.5); cholesterol (2.1); distearoylphosphatidylglycerol (3.4); alpha-tocopherol (0.03); sucrose (36); disodium succinate hexahydrate (1.1)	12 (after first dilution with water for injection)	iv infusion
cytarabine	DepoCyt (Enzon/ Skye-Pharma)	multivesicular lipid vesicles	5.5–8.5	cytarabine (10); cholesterol (4.1); triolein (1.2) dioleoylphosphatidylcholine (DOPC) (5.7); dipalmitoylphosphatidylglycerol (DPPG) (1.0)	5 (direct injection of vial)	intrathecal

Table 6. Representative marketed liposomal formulations.[12]

the expense of dissolving smaller ones. In order to obtain a stable suspension, the considerable potential energy created by the large interface between the solid and the surrounding medium must be reduced by adding surface-active agents. Surface stabilization may be achieved by using amphiphiles that migrate to the solid-liquid interface and provide a barrier to particle agglomeration. Non-ionic polymeric surfactants such as poloxamer 188, a triblock co-polymer of ethylene glycol and propylene glycol, are very effective stabilizers because of multiple attachment of hydrophobic domains at the particle surface. Entropically, the probability of detachment of all of these hydrophobic moieties is very low at room temperature, thus providing a strong surface affinity (Alexandridis and Hatton, 1995). Non-ionic surfactants may also create a hydration zone, a layer of tightly bound water molecules around each particle. When two particles meet, work is required to dislodge this water layer because of osmotic forces. Other entropic factors are also involved. The hydrophobic domains of the surfactant associate with the particle surface, with pendant hydrophilic domains extending into the aqueous medium. Attraction between particles necessitates the intertwining of these pendant chains leading to a restriction in chain mobility, and hence an unfavorable lowering of entropy (Lee et al., 2000). The combination of the entropic and enthalpic factors comprises so-called "steric" stabilization, and may provide an effective barrier to aggregation. A combination of steric and electrostatic stabilization is often required to achieve desired shelf life. Glycol copolymers such as poloxamers or polyethylene glycols, suffer however from reduced solubility in water at high temperatures, which may lead to particle aggregation. This results from thermally induced cleavage of hydrogen bonds between the hydrated polymer and water, leading to formation of visible polymer aggregates ("cloud point"). The ability to autoclave such formulations is limited if the cloud point lies below the sterilization temperature (121°C). Addition of cloud-point modifiers, usually anionic surfactants such as sodium dodecyl sulfate, may raise the cloud point and enhance stability at high temperature (Lee et al., 2000). Polysorbates (Tweens), poloxamines and poloxamers have been used as non-ionic surfactants. Bile salts (e.g., sodium cholate) and alkyl sulfonates (sodium dodecyl sulfate, sodium dioctylsulfosuccinate, and sodium lauryl sulfate, for example) have been effectively used as anionic surfactants. A number of steroids have been available for years as course suspensions for intramuscular and intra-articular delivery. Examples include Depo-Medrol® Sterile Aqueous Suspension[9]. The marketed product contains methylprednisolone acetate, a white crystalline powder that melts at about 208°C and is practically insoluble in water. Because of its high melting point and moderate Log P (2.56)[14], it cannot be formulated in oil as an emulsion. Each mL contains active drug (40 or 80 mg/mL), polyethylene glycol 3350 (29 or 28 mg/mL), myristyl gamma-picolinium chloride (0.195 mg or 0.189 mg/mL), and sodium chloride added to adjust tonicity. Solution pH is within 3.5 to 7.0.[10] Myristyl gamma-picolinium chloride is a cationic surfactant, added in small quantities as a preservative.

Table 7 is a current listing of parenteral suspension products.

Drug (generic name)	Brand name	Formulation pH	Solution components after reconstitution (mg/mL)[12]	Melting point (°C)/ log P(neutral)	Administration routes/dosage
Aurothioglucose	Solganal (Schering AG)	Not applicable; drug decomposes in water	drug: 50 sesame oil: 5%; aluminum monostearate: 2%; propylparaben	Not available (drug is unstable in aqueous media)	intramuscular
Azacitidine	Vidaza (Pharmion)	6.0 to 7.5	drug: 25; mannitol: 25	235–237 (hydrolyzes in water)[20]/2.17[14]	subcutaneous
Betamethazone acetate + betamethasone sodium phosphate	Celestone Soluspan (Schering-Plough)	6.8 to 7.2	drug: 3 (acetate), 3 (sodium phosphate); dibasic sodium phosphate: 7.1; monobasic sodium phosphate: 3.4; edetate disodium: 0.1; benzalkonium chloride (preservative): 0.2[21]	betamethasone acetate: 200–220 (decomposes)[21]/ 2.46[21]	intramuscular
Colfosceril palmitate (dipalmitoyl-phosphatidylcholine (DPPC))	Exosurf Neonatal (Glaxo Wellcome)	5 to 7	drug: 13.5; cetyl alcohol: 1.5; tyloxapol: 1; sodium chloride (0.1N)	235–236 (R-form)[15]/ 9.81[14]	intratracheal

(continued)

Drug (generic name)	Brand name	Formulation pH	Solution components after reconstitution (mg/mL)[12]	Melting point (°C)/ log P(neutral)	Administration routes/dosage
Cortisone acetate (cortisone-21-acetate)	Cortone acetate (Merck)	5.0 to 7.0	drug: 50; sodium chloride: 9; polysorbate 80: 4; sodium carboxymethylcellulose: 5; benzyl alcohol: 9	235–238[22]/2.1[14]	intramuscular
Triamcinolone diacetate	Aristocort Parenteral (Fujisawa)	6	drug: 25; polysorbate 80 (0.2%) polyethylene glycol 4000 (3%); sodium chloride: 0.85%; benzyl alcohol: 0.9%	235[22] / 1.92[14]	intralesional
	Aristocort Forte (Fujisawa)	6	drug: 40; polysorbate 80 (0.2%); polyethylene glycol 4000 (3%); sodium chloride: 0.85%; benzyl alcohol: 0.9%		intramuscular, intrasynovial or intra-articular
Triamcinolone hexacetonide	Aristospan (Fujisawa)		drug: 20; polysorbate 80 (0.4%); benzyl alcohol (0.9%); sorbitol	295–296[22] / 4.76[14]	intralesional, intraarticular

Dexamethasone acetate (dexamethasone -21-acetate monohydrate)	Decadron-LA (Merck)	Not found	drug: 8 (equivalent to dexamethasone); sodium chloride: 6.67; creatinine: 5; disodium edetate: 0.5; sodium carboxymethylcellulose: 5; polysorbate 80: 0.75; benzyl alcohol (preservative): 9; sodium bisulfite: 1	229–231[22]/2.46[14]	intramuscular, intra-articular, intrale-sional, soft-tissue
Methylprednisolone acetate	Depo-Medrol (Pfizer)	3.5 to 7.0	drug: 40; polyethylene glycol 3350: 29; sodium chloride: 8.7; Myristyl-gamma-picolinium chloride (preservative): 0.195	205–208[22] / 2.56[14]	intramuscular, intrasyn-ovial, intralesional
Medroxyprogesterone acetate (medroxy-progesterone-17-acetate)	Depo-Provera (Pfizer)		drug: 150; propylene glycol 3350: 28.9; polysorbate 80: 2.41; sodium chloride: 8.68; methylparaben: 1.37; propylparaben: 0.15	207–209[22]/ 4.09[14]	intramuscular
Hydrocortisone acetate (hydrocortisone-21-acetate)	Hydrocortone Acetate (Merck)	5.0 to 7.0	drug: 50; sodium chloride: 9; polysorbate 80: 4; sodium carboxymethylcellulose: 5; benzyl alcohol (preservative): 9	223(dec)[22]/2.19[14]	Intraarticular, intrale-sional, 25 and 50 mg/mL

Table 7. Suspensions marketed for parenteral use (excluding nasal, otic, ophthalmic or pulmonary).

Particle Carrier Systems

Next generation drugs may require a higher therapeutic index than conventional methodologies can afford. As a result, new excipients for extended release such as polymeric carriers, which as of 20 years ago only appeared in the research literature, are now reaching the market. The advent of Abraxane®[11], a paclitaxel formulation in which the drug is formulated as core-loaded albumin microparticulate, promises improved efficacy and safety (McNeil, 2003). Development of highly potent macromolecular drugs such as growth factors and hypothalamic releasing hormones that may have irreversible hormonal effects, has led to newer techniques to deliver such potent drugs more effectively over a prolonged period. Leuprolide acetate, for example, is a synthetic nonapeptide analog of naturally occurring gonadotropin-releasing hormone (GnRH or LH-RH). The acetate analog, which possesses greater potency than the natural hormone, was first developed in 1986 as a solution for the palliative treatment of advanced prostate cancer (DiMasi and Paquette, 2004). The solution, Lupron® Injection, is still being marketed (TAP Pharmaceuticals, Inc.) for daily administration by the subcutaneous route. Leuprolide acetate has also been developed in poly(lactide-co-glycolide) (PLGA) microspheres for subcutaneous or intramuscular delivery. Lupron® Depot (TAP Pharmaceuticals, Inc.) is available in a prefilled dual-chamber syringe containing sterile lyophilized PLGA microspheres. When the microspheres are mixed with diluent, a suspension is formed. The suspension may be administered monthly by intramuscular injection. Epic Therapeutics, a subsidiary of Baxter Healthcare Corporation, has been developing a leuprolide depot formulation (LeuProMaxx®) that is in clinical Phase II. Alza (Johnson & Johnson) was developing 3- and 6-month sustained release formulations (preclinical) that employ Alza's ALZAmer® depot platform. Nektar Therapeutics (previously Inhale Therapeutic Systems) has been developing a dry powder inhaled formulation of leuprolide. West Pharmaceutical Services was developing an intranasal delivery system (ChiSys®) based on the use of chitosan as a mucoadhesive to prolong contact with the nasal mucosa, and has sold this technology to Archimedes Pharmaceuticals. Newer formulations for 3-month sustained release are now on the market (e.g., Lupron Depot®-3 Month, by TAP Pharmaceuticals). Clearly, the pharmaceutical market is more receptive to excipients that facilitate sustained delivery via parenteral routes, enabling less frequent dosing of highly potent drugs, and encouraging greater patient compliance.

Conclusion

We have reviewed major solubilization strategies for the formulation of parenteral drug products, and have presented a theoretical framework for the qualitative assessment of these strategies, based on solute hydrophobicity versus polarity (as gauged by log P) and lattice energy of the solid, as reflected by melting point. In general, as solute polarity decreases and solid lattice energy increases, reduced solubility limits the number of formulation options and dispersed systems (liquid-liquid, solid-liquid) should seriously be considered.

List of Abbreviations

USP........United States Pharmacopeia
w/v........Weight-to-volume ratio
dec........Decomposes
RT........Room temperature (25 °C)

Endnotes

1. Webster's New Collegiate Dictionary, G. & C. Merriam Co., Springfield, Massachusetts, U.S.A., 1973.
2. Prescribing Information for Dilantin® (Phenytoin Sodium Injection, U.S.P.), Warner-Lambert, 1994.
3. Prescribing Information for Alkeran® (Melphalan HCl for Injection), GlaxoSmithKline.
4. Calculated using HyperChem version 7.5 (Hypercube, Inc., Gainesville, FL), Ghose-Crippen method.
5. Prescribing Information, TAXOL® (paclitaxel) INJECTION, Bristol-Myers Squibb Company, Princeton, NJ, 08543, March 2003.
6. Cydex, Inc., Lenexa, KS 66214 USA.
7. Geodon and Vfend are both registered trademarks of Pfizer Inc.
8. Diprivan® (propofol for injection) package insert, AstraZeneca 2001, 2004.
9. Depo-Medrol is a registered trademark of Pfizer Inc.
10. Package insert, Depo-Medrol®, methylprednisolone acetate injectable suspension, USP, Pharmacia & Upjohn Company, a subsidiary of Pharmacia Corporation, Kalamazoo, Michigan 49001, USA. Revised March 2003.
11. Abraxane is a registered trademark of American Pharmaceutical Partners, Inc.
12. Unless otherwise indicated, all information is from Physician's Desk Reference, 58th Ed., Medical Economics Co., NJ, 2004.
13. Unless otherwise indicated, all melting points cited from The Merck Index, 10th Edition. Rahway, NJ: Merck & Co.; 1983.
14. http://www.syrres.com/esc/est_kowdemo.htm, and references therein.
15. Chapman and Hall Chemical Database Online (File 303). DIALOG, Thompson Co; 1997.
16. Unless otherwise indicated, all Log P values obtained from: http://www.syrres.com/esc/est_kowdemo.htm, and references therein.
17. http://www.ps.toyaku.ac.jp/dobashi/database/structure/f_group/flurbiprofen_axetil.html.

18. http://www.ps.toyaku.ac.jp/dobashi/database/structure/d_group/dexamethasone_palmitate.html.
19. Web site (ChemExper): http://www.chemexper.com/chemicals/supplier/cas/306-94-5.html.
20. IARC. *Pharmaceutical Drugs. IARC Monographs on the Evaluation of Carcinogenic Risk of Chemicals to Humans, vol. 50.* Lyon, France: International Agency for Research on Cancer; 1990.
21. Package insert (Schering Corporation Kenilworth, NJ 07033 USA).
22. Merck Index, 11th Edition. Rahway, NJ: Merck & Co; 1989.

References

Adamson AW. *A Textbook of Physical Chemistry*. Academic Press, New York, NY, 1973, pp. 323–328.

Alexandridis P and Hatton TA. Poly(ethylene oxide)-poly(propylene oxide)-poly(ethylene oxide) Block Copolymer Surfactants in Aqueous Solutions and at Interfaces: Thermodynamics, Structure, Dynamics, and Modeling (Review). *Colloids and Surfaces A: Physicochem Eng Aspects* 1995; 96: 1–46.

Avdeef A. *Absorption and Drug Development: Solubility, Permeability, and Charge State.* Wiley, Hoboken, NJ, 2003, pp. 59–66.

Bobey DG, Russell AD, and Jones CB. Selection of Disodium Edetate as the Optimal Antimicrobial Additive for Use in Propofol Emulsion. *Pharm Ther* 2001; 11: 335–341.

Breslow R. Hydrophobic Effects on Simple Organic Reactions in Water. *Acc Chem Res* 1991; 24: 159–164.

Breslow R. Hydrophobic and Antihydrophobic Effects on Organic Reactions in Aqueous Solution. In: Cramer CJ, Truhlar DG (eds.), *Structure and Reactivity in Aqueous Solution: Characterization of Chemical and Biological Systems (ACS Symposium Series).* American Chemical Society, Washington DC, 1994, pp. 291–302.

Breslow R and Maitra U. On the Origin of Product Selectivity in Aqueous Diels-Alder Reactions. *Tetrahedron Lett* 1984; 25: 1239–1240.

Breslow R and Rideout DC. Hydrophobic Acceleration of Diels-Alder Reactions. *J Am Chem Soc* 1980; 102: 7817.

Coico R, Sunshine G, and Benjamini E. Hypersensitivity Reactions: Antibody-mediated (Type II) Cytotoxic Reactions and Immune Complex (Type III) Reactions. In: *Immunology, A Short Course (5th Edition).* Wiley, Hoboken, NJ, 2003, pp. 215–224.

Cornford EM, Young D, and Paxton JW. Comparison of the Blood-brain Barrier and Liver Penetration of Acridine Antitumor Drugs. *Cancer Chemother Pharmacol* 1992; 29(6): 439–444.

Cramer F and Henglein FM. Einschlussverbindugen der Cyclodextrin mit Gasen. *Angew Chem* 1956; 68: 649.

Crocker JC and Grier DG. When like charges attract: The Effects of Geometrical Confinement on Long-range Colloidal Interactions. *Phys Rev Letters* 1996; 77(9): 1897–1900.

Derjaguin BV and Landau L. *Acta Physicochimica (USSR)* 1941; 14: 633.

DiMasi JA and Paquette C. The Economics of Follow-on Drug Research and Development Trends in Entry Rates and the Timing of Development. *Pharmacoeconomics* 2004; 22(Supplement):2: 1–14.

Eisenberg D and Kauzmann W. *The Structure and Properties of Water*, Oxford University Press, Oxford, UK, 1969.

Eschalier A, Lavarenne J, Burtin C, Renoux M, Chapuy E, and Rodriguez M. Study of Histamine Release Induced by Acute Administration of Antitumor Agents in Dogs. *Cancer Chemother Pharmacol* 1988; 21: 246–250.

Gaballa GA and Neilson GW. The Effect of Pressure on the Structure of Light and Heavy Water. *Mol Phys* 1983; 50: 97–111.

Gmehling J, Lohmann J, Jakob A, Li J, and Joh RA. Modified UNIFAC (Dortmund) Model. 3. Revision and Extension. *Ind Eng Chem Res* 1998; 37: 4876–4882.

Gorbaty YE and Demianets YN. An X-Ray Study on the Effect of Pressure on the Structure of Liquid Water. *Mol Phys* 1985; 55: 571–588.

Hansch C and Leo A. *Substituent Constants for Correlation Analysis in Chemistry and Biology*. Wiley, New York, NY, 1979.

Hwang SM, Lee JM, and Lin H. New Group-interaction Parameters of the UNIFAC Model: Aromatic Methoxyl Binaries. *Ind Eng Chem Res* 2001; 40: 1740–1747.

Jain N and Yalkowsky SH. Estimation of the Aqueous Solubility I: Application to Organic Nonelectrolytes. *J Pharm Sci* 2001; 90(2): 234–252.

Krishna G, Hodnick WF, Lang W, Lin X, Karra S, Mao J, and Almassian B. Pharmaceutical Development and Manufacturing of a Parenteral Formulation of a Novel Antitumor Agent, VNP40101M. *AAPS PharmsciTech* 2001; 2 (3), article 14 (accessed June 2006).

Larsen JR and Larsen LS. Clinical Features and Management of Poisoning Due to Phenytoin. *Med Toxicol Adverse Drug Exp* 1989; 4(1): 229–245.

Lee RW, Shaw JM, McShane J, and Wood RW. Particle Size Reduction. In: Liu R (ed.), *Water-insoluble drug formulation.* Interpharm, Denver, CO, 2000, p. 478.

Lee RW, Shaw JM, McShane J, and Wood RW. Particle Size Reduction. In: Liu R (ed.), *Water-insoluble drug formulation.* Interpharm, Denver, CO, 2000, pp. 487–488.

Leo A. Calculating Log Poct From Structures. *Chem Rev* 1993; 93: 1281–1306.

Leo A, Hansch C, and Elkins D. Partition Coefficients and Their Uses. *Chem Rev* 1971; 71: 525–616.

Lipinski CA. Drug-like Properties and the Causes of Poor Solubility and Poor Permeability. *J Pharmacol Toxicol Methods* 2000; 44: 235–249.

Lipinski CA, Lombardo F, Dominy BW, and Feeney PJ. Experimental and Computational Approaches to Estimate Solubility and Permeability in Drug Discovery and Development Settings. *Adv Drug Del Rev* 1997; 23: 3–25.

Lohmann J, Joh R, and Gmehling J. From UNIFAC to Modified UNIFAC (Dortmund). *Ind Eng Chem Res* 2001; 40: 957–964.

Lynden-Bell RM, Rasaiah JC, and Noworyta JP. Using Simulation to Study Solvation in Water. *Pure Appl Chem* 2001; 73(11): 1721–1731.

MacNeil JS. Abraxane Comes Through with Proof of Superiority, Scientific Endorsements. *NanoBioTech News* 2003; 1(19): 1–5.

McLeish MJ. Thiopental Sodium. In: Brittain HG (ed), *Analytical Profiles of Drug Substances and Excipients (Vol. 21).* Academic Press, New York, NY, 1992, p. 540.

Mounier P, Laroche D, Divanon F, Mosquet B, Vergnaud MC, Esse-Comlan A, Piquet MA, and Bricard H. Anaphylactoid Reactions to an Injectable Solution of a Cremophor-containing Solution of Multivitamins. *Therapie* 1995; 50: 571–573.

Munoz A, Karila P, Gallay P, Zettelmeier F, Messner P, Mery M, and Grolleau R. A Randomized Hemodynamic Comparison of Intravenous Amiodarone with and without Tween-80. *Eur Heart J* 1998; 9: 142–148.

Ornektekin S, Paksoy H, and Demirel Y. The Performance of UNIFAC and Related Group Contribution Models Part II. Prediction of Henry's Law Constant. *Thermochimica Acta* 1996; 287: 251–259.

Rytting E, Lentz KA, Chen X-Q, Qian F, and Venkatesh S. Aqueous and Cosolvent Solubility Data for Drug-like Compounds. AAPS Journal 2005; 7(1) Article 10.

Sciortino F, Geiger A, and Stanley HE. Network Defects and Molecular Mobility in Liquid Water. *J Chem Phys* 1992; 96: 3857–3865.

Sweetana S and Akers MJ. Solubility Principles and Practices for Parenteral Drug Dosage Form Development. *PDA J Pharm Sci Tech* 1996; 50(2): 330–342.

Tanford C. The Hydrophobic Effect: Formation of Micelles and Biological Membranes, 2nd ed. Wiley, New York, NY, 1980.

Theis JG, Liau-Chu M, Chan HS, Doyle J, Greenberg ML, and Koren G. Anaphylactoid Reactions in Children Receiving High-dose Intravenous Cyclosporine for Reversal of Tumor Resistance: The Causative Role of Improper Dissolution of Cremophor EL. *J Clin Oncol* 1995; 13: 2508–2516.

Verwey EJ and Overbeek JTG. *Theory of the Stability of Lyophobic Colloids.* Elsevier, Amsterdam, 1948.

Volcheck GW and Van Dellen RG. Anaphylaxis to Intravenous Cyclosporine and Tolerance to Oral Cyclosporine: Case Report and Review. *Ann Allergy Asthma Immunol* 1998; 80: 159–163.

Wang R, Fu Y, and Lai L. A New Atom-additive Method for Calculating Partition Coefficients. *J Chem Inf Comput Sci* 1997; 37: 615–621.

Wheless JW. Pediatric Use of Intravenous and Intramuscular Phenytoin: Lessons Learned. *J Child Neurol* 1998; 13: S11–S14, discussion S30–S32.

Wojcik JF and Rohrbach RP. Small Anion Binding to Cycloamylose. Equilibrium Constants. *J Phys Chem* 1975; 79: 2251.

Wu AY, Whalley E, and Dolling G. Neutron Diffraction of Heavy Water to 15.6 Kbar. *Mol Phys* 1982; 47: 603–628.

Yalkowsky SW and He Y. *Handbook of Aqueous Solubility.* CRC Press, Boca Raton, FL, 2003, p. 764.

Yalkowsky SW and He Y. *Handbook of Aqueous Solubility.* CRC Press, Boca Raton, FL, 2003, p. 998.

11

Aqueous Solubilizing Systems for Parenteral Formulation Development—Proteins

ZAHRA SHAHROKH

Shire Human Genetic Therapies, Cambridge, Massachusetts

Introduction

Protein solubility ranges from low micrograms per ml to several hundreds of milligrams per ml, and is very compound-specific. Most antibodies are known to reach solubilities of hundreds of mg per ml whereas beta amyloid protein has very low solubility. Small structural changes could lead to significant changes in solubility, for example, cryo-immunoglobulins may be almost insoluble. The dose and route of administration may demand a higher concentration than possible in simple formulations, posing a challenge to the development of a clinically or commercially viable product. One important challenge is that proteins are typically administered via injections due to poor bioavailability by other delivery modes (See review articles in book edited by Audus and Raub, 1993; Pharmaceutical Business Review website, 2005), which restricts the types and levels of excipients (FDA website; Powell et al., 1998; Strickley, 1999, 2000). Further constraints are imparted by the small volume of administration appropriate for subcutaneous and intramuscular delivery routes which need to be consistent with patient compliance and ease of delivery. This can be very different from the volume/concentration constraints of intravenous administration. For therapeutic doses in the mg/kg levels, the less than approximately 1.2 mL acceptable volume for subcutaneous delivery may necessitate formulations containing hundreds of mg/mL protein. Moreover, toxicological studies may assess approximately 10-fold higher doses than those planned for clinical studies in order to establish a safety window. This necessitates even higher concentrations for non-clinical formulations than for clinical formulations.

This chapter reviews the approaches for addressing the solubility challenges of protein formulations in aqueous solutions using conditions that are suitable for parenteral administration and practices that are acceptable for biopharmaceuticals.

Definition of Protein Solubility

Protein solubility is a thermodynamic characteristic of the protein/solvent system defined as the concentration of soluble protein in equilibrium with the solid phase at a given pH, temperature, and solvent composition (Flynn, 1984; Arakawa and Timasheff, 1985; Middaugh and Volkin, 1992). For practical purposes, solubility of proteins can be defined as the maximum amount of protein that remains in a visibly clear solution (i.e. does not show protein precipitates, crystals, gels, or hazy soluble aggregates), or does not sediment at 30,000 g centrifugation for 30 min (Schein, 1990; Ducruix and Reis-Kautt, 1990).

Methods for Predicting Protein Solubility

While software programs exist to examine the solubility of small molecules based on their structure (see ACD and Osiris websites), current approaches are not used to assess the solubility of the protein, and the parameters described in the section above make the prediction of protein solubility based on structure difficult. The complexity arises from the changes in protein conformation that could occur by changes in experimental conditions, as well as the possibility of supersaturation or phase changes, not easily predictable from the primary sequence or crystal structure of the protein. An approach for predicting protein solubility based on hydrophobic interaction chromatography has been suggested (Gagnon et al., 1997). The theoretical basis for this observation has been described (Melander and Horvath, 1997) and is related to the correlation between the hydrophobic surface area of proteins that governs its interaction with chromatographic resins, and the salting out order, which correlates with solubility.

Methods for Measuring Protein Solubility

While the solubility of most small molecules can be determined by dissolving the solid bulk in volatile solvents, drying, and dissolving in various solutions until no further dissolution is observed, it is risky to expose protein solids to volatile solvents without protein denaturation or to generate pure protein solids in their native conformation that lack other excipients. Thus, protein solubility is generally determined by concentrating the protein solution until a phase change (precipitates, flocculants, gels, or other solid forms) is observed. The soluble phase is filtered or centrifuged to remove any precipitates, and the amount of protein in the soluble phase is determined.

Methods for concentrating proteins include membrane filtration, osmotic removal of solvent, and solvent evaporation (reviewed by Shire et al., 2004). Early in preformulation studies when material is limited, small-scale methods are needed to achieve high protein concentrations. Osmotic pressure dependent microdialysis (Saul and Don, 1984; Ceschini et al., 1996; Zhang and Hjerten, 1997), dialysis against a hygroscopic material such as Sephadex or polyethylene glycols (Middaugh and Volkin, 1992), solvent evaporation by vacuum or nitrogen flow (Rothstein, 1994), and precipitation by salts or solvents are among the procedures used. Selective protein precipitation with an inert agent such as polyethylene glycol 8000 provides a measure of the thermodynamic activity of the protein in equilibrium with PEG precipitates, and extrapolation of the logarithmic plot of the amount of soluble protein versus PEG concentration to zero PEG yields an estimate of the aqueous solubility of proteins in a quick and simple way (Middaugh et al., 1997). Though these micromethods have the advantage of requiring small amounts of protein, they may not be scalable or may lead to protein destabilization or phase changes that could give misleading results. These approaches are generally useful for rank ordering of solubility in various formulations during preformulation studies when materials are limited, but the results should be verified by representative manufacturing processes, which primarily involve filtration (van Reis and Zydney, 2001; also see Shire et al., 2004 for process examples). The pros and cons of the various methods for concentrating proteins, their limitations, and their application for determining protein solubility are listed in **Table 1**. Detailed discussions of the manufacturing considerations for high concentration proteins are provided by Shire et al. (2004).

An exception to the approach for determining protein solubility, that is based on concentrating the protein, occurs when the protein can be crystallized, in which case, the solubility of the protein is established by the protein concentration in the solution phase that is in equilibrium with the crystals. Though the crystals could also be dried and dissolved in other solvent systems to determine solubility in those systems, protein crystals carry salt and other ions, and may render different solubilities depending on the method of preparation.

For determination of protein solubility, after the solution phase is separated from the solid phase, the amount of protein in the solution phase is usually measured by spectrophotometry using an absorbance at 280 nm. After subtracting the blank signal from that of the protein, protein concentration is calculated using a predetermined extinction coefficient as follows:

$$\text{Protein conc. (mg/ml)} = \frac{\text{Measured Absorbance at 280 nm} \times \text{Sample Dilution}}{\text{Extinction coefficient (mL/mg.cm) at 280 nm} \times \text{Path length (cm)}}$$

Another approach to measuring protein concentration in the solution phase is the use of chromatographic methods, such as size exclusion or reverse phase HPLC. The peak area of a given injected volume of a protein solution is converted to protein amount using a standard curve that is generated by plotting peak area versus amount of injected standard. This approach is necessary in the rare event that a protein has few aromatic residues that are required for the 280 nm absorbance signal. Protein concentration in solution could also be determined

Concentrating method	Scale	Benefits	Limitations	Used for solubility determination
Osmotic driven (dialysis) methods: Against solutions	Few μLs to few mLs	Small-scale; Mechanically gentle; Rapid Simultaneous buffer exchange	Concentrations achieved may not translate to manufacturing scale	Yes
Against water absorbing materials			Have to buffer exchange later	
Solvent Evaporation: Vacuum, N_2 flow	Few μLs to few mLs	Small-scale	Slow (may lead to degradation); concentrates excipients	Yes, only if solution composition reached at end of process
		Pharmaceutically relevant and scalable		
Lyophilization	mLs		Must protect against drying induced degradation; concentrates excipients	
Precipitation (salt or solvent; supercritical Fluid)	Few μLs to Liters	Quick and scalable	Solution composition not pharmaceutically relevant	No[2]
			May lead to irreversible degradation	

Freezing	Few mLs to L	Ten fold concentration is achieved quickly	Also concentrates excipients	No
Chromatographic bind and elute	Few μLs to many mL	Wide range of scale	Solution may damage protein Solution composition not pharmaceutically relevant	No
Filtration Systems: Centrifugal (Centricon) Pressure (Amicon)	Hundreds of μLs to tens of mLs	Moderate scale; May include buffer exchange	Adsorption losses; Shear induced degradation Findings may not translate to manufacturing scale Large scale impractical for early development phases	Yes, if not material limited
Tangential Flow Filtration	Hundreds of mLs to Liters	Manufacturing scale; May include buffer exchange	High viscosity or pore clogging may limit concentrations reached	

Table 1. Commonly used methods for concentrating proteins[1].

[1] From Shire et al., 2004 with copyright permission.

[2] PEG precipitation has been used for solubility determination.

by Keldjal nitrogen measurements or amino acid analysis. These approaches are more time consuming and require more sophisticated analytical capabilities than the spectrophotometric method.

Solubility Principles for Proteins and Practical Parameters Influencing Protein Solubility

The solubility of a protein is a complex parameter since it is dependent not only on the amino acid composition and surface exposure of charged versus hydrophobic amino acids, it is also influenced by the formulation composition, the method and history of manufacture, and subsequent handling and storage conditions.

Effect of pH

pH is a key parameter that influences protein solubility (Rupley, 1968). Increasing ionization of the charged amino acids in a protein theoretically yields a more soluble protein due to increased intermolecular repulsion and increased solvation by water. At very low ionic strengths, where the salt does not mask electrostatic protein interactions, proteins often have their lowest solubility near their isoelectric point (pI) where the net charge is zero. Solubility then increases at pH values below or above the pI due to the increased charge density following ionization of amino and carboxylic groups, respectively. When proteins bind salts, the lowest solubility is reached when the net charge on the protein is the same as the number of bound ions (Retailleau et al., 1997), but the general trend with pH remains similar.

Effect of Temperature

The temperature effect on protein solubility is protein-dependent. Protein solubility generally increases with increasing temperature until protein unfolding occurs. However, there are examples of complex thermal behaviors of solubility (see examples in Bull, 1971; Middaugh and Volkin, 1992), such that the heat of solution changes sign as a function of temperature. An example is hemoglobin S which shows an initial decrease in solubility with increasing temperature, then increases at temperatures higher than 37°C (Middaugh et al., 1997). Other exceptions are hydrophobic compounds that show a decrease in solubility with increasing temperatures (Ismailos et al., 1991).

Effect of Ion Strength and Osmolytes

Protein solubility has been mechanistically explained by changes in bulk water surface tension and binding to water or ions versus protein-protein (Frommenhagen, 1965; Melander and Horvath, 1977; Bull and Breeze, 1980; Arakawa and

Timasheff, 1985; Schein, 1990; Leberman, 1991). Binding of proteins to specific excipients or ions influences solubility through changes in protein conformation or masking certain amino acids involved in self-association. The effect of salts on protein solubility has been studied for over a century and is the basis for protein crystallization. The effect of salts falls in two categories: at very low ionic strengths, solubility usually increases with increasing salt concentration because of shielding of electrostatic repulsions (salting-in); at higher ionic strengths, solubility decreases with increasing salt concentration because both the protein and the salt compete for hydration, and protein self-association is increased via hydrophobic effects (salting-out) (Melander and Horvath, 1977; Leberman, 1991). The salting-out and salting-in effect of the lyotropic (Hofmeister) salt series has been described (Bull and Breese, 1980; Leberman, 1991). Certain nonionic osmolytes (e.g. sugars) are also preferentially excluded from the protein surface, a phenomenon that leads to preferential hydration and stabilization of the protein into more compact conformations (Arakawa et al., 1991; Shimizu and Smith, 2004), but the impact on protein solubility depends on the protein and the level of the osmolyte (Middaugh, 1992, Jenkins, 1998).

Solid State Considerations

Since solubility is the concentration of protein in solution at equilibrium with the solid phase, the state of protein in the solid phase affects the solubility in the solution phase. Theoretical treatment of the protein solubility problem has often ignored solid phase interactions of the protein due to its complexity. A crystalline solid phase is expected to render a lower solubility than the amorphous solid phase. However the complexity and the heterogeneity of the protein in the solid state (e.g. amorphous, gel, or crystalline, or precipitates of native or denatured forms) makes it difficult to directly assess solid state effects.

Effect of Process and Handling

Proteins often bind ions, lipids, or other components to which they are exposed during their manufacture, which might consequently affect their solubility. Moreover, due to their conformational lability, proteins might partially denature or precipitate or form higher oligomers as a result of manufacturing manipulations, which might result in variable solubility levels of different preparations. The lability of protein conformation and the possibility of interacting with self, with surfaces, and with specific solutes lead to a protein solubility that depends on the history of sample treatment. One such manipulation is the method of concentrating. Concentrating a protein solution at small scale by centrifugal filtration might give a solubility that does not translate at an industrial scale due to shear-induced destabilization by stirring or by the pump systems used at larger scale, or due to instability from larger surface area interactions. Thus, studies done at small scale need to be verified at the manufacturing scale. Other interactions, for example with plastic surfaces (e.g. polystyrene) may lead to surface adsorption or denaturation (see for example, Tzannis et al., 1996; Mollman et al., 2006) and subsequently altered apparent solubility. During filtration used for aseptic processing, care should be taken to avoid excessive foaming which creates surfaces at which protein denaturation and apparent loss of solubility can occur. For a more detailed discussion of manufacturing considerations for generating high protein concentrations, see Shire et al. (2004).

Impact of Heterogeneity

Protein solubility can also depend on micro-heterogeneity and the form purified, for example the level of glycosylation might impact solubility (see for example Schmoekle et al., 2004). It is conceivable that depending on the consistency of the preparation with respect to glycosylation, different solubilities might be observed for preparations made near their solubility limit. Though less likely to occur for a pharmaceutical protein prepared under current industry standards, protein solubility might also vary depending on the level of purity of the preparation. For example, fibrinogen solubility was found to be linearly dependent on the initial amount of protein. This was attributed to either impurities or protein heterogeneity in the preparation (Leavis and Rothstein, 1974). Thus, it is evident that the solubility of a protein is a complex phenomenon and is affected by process parameters which could generate various native or non-native forms of the protein.

The following section describes approaches successfully used for formulating proteins with enhanced solubility.

Strategies for Enhancing Protein Solubility Useful in Pharmaceutically Acceptable Formulations

For a pharmaceutically viable product, approaches to increase protein solubility should render a stable product with excipients suitable for parenteral administration. Identification of a formulation composition optimal for protein solubility and also for protein stability requires an iterative process of evaluating the impact of various parameters on solubility and on stability, independently, and then jointly.

pH, Temperature, Ionic Strength

Enhancing the solubility of a protein therapeutic should begin by assessing the pH dependence of solubility, at various temperatures. A consideration for parenteral protein formulations is that pH is generally within a limited range of 4 to 8, below which injection site reactions are more prevalent (Napaporn et al., 2000) and above which protein degradation is common (Shahrokh, 1997; Wang, 1999). pH-solubility studies should thus be combined with an assessment of pH-dependence of protein structure and bioactivity, since an increased solubility might be due to formation of a structurally altered protein. An example is the TNF family of proteins that have higher solubility in the acidic pH range, but these species do not have native conformation and are inactive (Shahrokh, unpublished results). Moreover, since solubility near the pI is usually at a minimum, formulation pH within 0.5 unit of the pI of the protein is often avoided. Since solubility is dependent on the ionic strength of the formulation, pH-solubility studies are followed by evaluating the effect of ionic strength. Commonly used salts to adjust ionic strength are sodium chloride and sodium sulfate; other divalent cations such as Zn, Ca, and Mg are used when specifically required for protein structure or function. The limitations on tonicity of formulations for parenteral administration might limit the concentration of salt required to increase the solubility of the protein. Isotonic solutions are physiologically most suitable for parenteral administration, but up to 2-fold hypertonic formulations have been used (Strickley, 2000). This should be considered the upper limit of product tonicity following dilutions preparatory to administration.

Once a range of suitable pH and ionic strength are selected, the effect on protein stability is evaluated for the final selection of an optimal formulation pH. Only when the optimal pH and addition of common salts do not render the desired solubility are other additives considered. This adds to the complexity of the formulation and the challenge of maintaining stability. Recently, an empirical approach to determine protein phase diagrams using various biophysical techniques has been used to facilitate identification of optimal formulation conditions (Fan et al., 1995).

Amino Acids and Polyelectrolytes

One of the most common approaches to increase protein solubility is the addition of amino acids and their derivatives (see examples in Middaugh and Volkin, 1992). Specifically, charged amino acids such as arginine, lysine, aspartic and glutamic acid have been used to increase protein solubility in a concentration-dependent manner (Isaacs and Patel, 1990; Isaacs et al., 1990; Flores et al., 2001). In a study of several proteins ranging from 10–70 kDa and pI of 5.3 to 10.0, a 1:1 molar mixture of the charged amino acids arginine and glutamic acid at 50 mM each, increased the solubility by 2–9 fold (Golovanov et al., 2004). In this system, the individual amino acids did not impart such solubilization and stabilization. The broader utility of mixtures of amino acids to increase protein solubility could potentially reflect the greater surface interaction which minimizes the self-association by binding the differently charged amino acids to the protein. Interestingly, a simultaneous increase in the stability of these proteins against proteolytic degradation was also noted.

In a study of a cytokine formulation, sulfates and a range of arginine salts were found to be effective in generating more than a 10-fold increase in solubility (Flores et al., 2001). Unexpectedly, solubility increased with increasing sulfate concentrations, and the protein crystallized in sulfate formulations at low temperatures. These studies also indicated that counter ions of arginine with two or three charges (e.g. malate or phosphate) form complexes with arginine that do not dissociate in solution and result in an osmolality that is the same as the concentration of arginine itself, rather than the sum of the two ions. In contrast, the mono-charged salt forms of arginine (e.g. chloride or glutamate) dissociate in solution and give an unacceptably high osmolality for parenteral administration.

Arginine has also been used to increase the yield of soluble proteins in cell culture expression systems by facilitating protein refolding. In the case of lysozyme, increased efficiency of formation of soluble refolded protein by arginine has been attributed to the stabilization of the partially folded intermediate, rather than stabilization of the native form (Reddy et al., 2005). Stabilization of the partially folded intermediate of lysozyme by arginine prevented its aggregation and made it available to form the native protein. In another example, arginine, other amino acids, citrate, and polyelectrolytes such as polyphosphate and sulfated polysaccharides have all been shown to enhance refolding and solubility of human tissue factor pathway inhibitor (TFPI), a hydrophobic protein with very limited aqueous solubility (Dorin et al., 1996). Arginine increased TFPI solubility by 100-fold. Moreover, polyphosphate chain length influenced the yield of soluble refolded protein, with $n = 25$ repeating units being optimal, $n = 5$ being ineffective, and $n = 75$ increasing aggregation. The mechanism was speculated to be binding of the polyelectrolyte to the protein, thereby increasing the effective charge density and solvation. This is consistent with the observation that for the same ionic strength, the compounds tested differed in their solubility enhancement characteristics. These studies provide suggestions for compounds that might enhance protein solubility, even when refolding is not involved.

Polyanions have yielded different effects on the solubility of proteins with high pI. Polyanions such as sulfated polysaccharides have limited effects on the solubility of basic fibroblast growth factor, a protein with a high pI, since specific interactions yield reversible precipitates containing bioactive and native protein (Sommer and Rifkin, 1989; Shahrokh et al., 1994). By contrast, sulfated polysaccharides were most effective in enhancing the solubility of bone morphogenic protein, a very low solubility, high pI protein (Zhu, 1999).

Complexing Agents

Cyclodextrins have been used for enhancing the solubility of poorly water-soluble small molecules (Shimpi et al., 2005). The mechanism of solubilization is formation of reversible complexes between the hydrophobic moiety of the drug molecule and the hydrophilic cyclodextrin that is highly water soluble. A similar observation has been made with some proteins and peptides (Brewster et al., 1991; Johnson et al., 1994; Flores et al., 2001). Sulfated cyclodextrins and 2-hydroxypropyl-beta-cyclodextrins are approved excipients for parenteral administration, however, use of these compounds should take into consideration the cost involved.

Other Approaches

Sugars and polyols at low concentrations have been shown to increase the solubility of certain proteins, but at high concentrations decrease solubility due to preferential exclusion from the protein surface and competition with water (Arakawa and Timasheff, 1985; Middaugh and Volkin, 1992; Jenkins, 1998).

Though not of practical use for the formulation scientist, protein engineering could also be used for increasing the solubility of some proteins (Rollema et al., 1995; Murby et al., 1995; Avramopoulou et al., 2004; De Marco et al., 2004; Nallamsetty and Waugh, 2006). This strategy might be considered early in the development program once it is understood that achieving the desired dose necessitates redesign of the protein candidate for further development.

Organic solvents that are generally used for enhancing solubility of small molecules can denature protein structure and result in further aggregation or inactivation of the protein (Middaugh and Volkin, 1992). Chapter XII in this book addresses the use of organic solvents to generate soluble proteins for parenteral administration (C. Stevenson). Glycerol and/or nonionic surfactants have been claimed to increase the solubility of β-interferon (Shaked et al., 2001) and kahalalide polypeptide (Juijen et al., 2001).

Conclusions

Formulating protein therapeutics may be challenged by their limited solubility which complicates attaining the desired concentration for toxicological or clinical dose ranging studies, specifically for subcutaneous and intramuscular delivery routes where delivery volume is limited. Achieving high protein concentrations can be influenced by formulation parameters such as pH and ionic strength, and may also require addition of solubilizing agents or control of temperature. Pharmaceutically acceptable solubilizing agents for proteins are currently limited to certain classes of compounds, which include amino acids and their derivatives, polyionic salts and polyelectrolytes, cyclodextrins, and to a lesser extent, polar cosolvents, sugars, polyols, and surfactants. These compounds enhance protein solubility in aqueous solutions through different mechanisms such as the masking of intermolecular electrostatic interactions by increasing the surface charge of the protein, masking hydrophobic moieties through formation of inclusion complexes, and stabilization of the native folded form. In addition to formulation composition and temperature, parameters such as the process for concentrating the protein, the manufacturing history, and the conditions for handling and storage of the protein that may impact protein conformation and self-association are sometimes critical for achieving and maintaining high concentrations.

Optimization of protein solubility is done empirically and alongside protein stability and structure-activity determinations. Excipients suitable for enhancing protein solubility and parenteral administration are limited. Table 2 lists those

Parameter	Impact	Mechanism(s)	Limitations
Temperature	Usually solubility increases with increasing temperature until denaturation occurs (there are exceptions)	Dependent on heat of solution	Destabilization/precipitation at high temperatures; cold denaturation for some proteins
pH	Solubility is minimum at pI (at low ionic strength), and increases above and below pI	Increased charge density or conformational change	Destabilization at pH extremes
Ionic Strength	Usually "salting-in" at low ionic strengths and "salting-out" at high ionic strengths (there are exceptions)	Charge shielding at low ionic strength and excluded volume effect at high ionic strength	Limited range of ionic strength suitable for parenteral administration
Amino Acids	Charged amino acids may increase solubility	Increased charge density by interaction with the protein	Usually high concentration excipient is needed, which affects tonicity
Polyelectrolytes	Binding to the protein may increase solubility; high molecular weight protein complexes may also form which could decrease solubility	Increased charge density or conformational change	Potential pharmacological effect or toxicity of the excipients
Polyols	At low levels may increase solubility; at high levels may precipitate proteins	Increased hydrophilicity through interaction with the protein or inducing conformational change	Impact on tonicity; *in vivo* tolerability
Surfactants	At low levels may increase solubility; at high levels may denature the protein	Masking hydrophobic residues; inducing conformational change	Protein denaturation; *in vivo* tolerability at high concentrations
Cyclodextrins	May increase solubility	Masking of hydrophobic residues by forming inclusion complexes	High concentration excipient needed; affects tonicity; is costly
Organic Solvents	May increase solubility at low levels	Solubilize hydrophobic residues	Protein denaturation

Table 2. Parameters influencing protein solubility.

parameters that impact protein solubility together with their mechanisms and limitations.

Acknowledgements

The author is grateful for critical review of the manuscript by Drs. Marcio Voloch, Gaozhong Zhu, and Kathy Taylor at Shire HGT, Inc.

References

ACD website: http://www.acdlabs.com/products/phys_chem_lab/aqsol/new.html

Arakawa T and Timasheff SN. Theory of Protein Solubility. *Methods Enzymol.* 1985; 114: 49-77.

Arakawa T, Kita Y, and Carpenter JF. Protein-solvent Interactions in Pharmaceutical Formulations. *Pharm Res* 1991; 8(3): 285–291.

Audus K and Raub T (ed). Biological Barriers to Protein Delivery. *Pharm. Biotech.* Vol. 4. (KL Audus, TJ Raub, Eds), 1993.

Avramopoulou V, Mamalaki A, Tzartos, and Socrates J. Soluble, Oligomeric, and Ligand-binding Extracellular Domain of the Human alpha 7 Acetylcholine Receptor Expressed in Yeast : Replacement of the Hydrophobic Cysteine Loop by the Hydrophilic Loop of the ACh-binding Protein Enhances Protein Solubility. *J Biol Chem* 2004; 279(37): 38287–38293.

Brewster ME, Hora MS, Simpkins JW, and Bodor N. Use of 2-Hydroxypropyl-Cyclodextrin as a Solubilizing and Stabilizing Agent for Protein Drugs. *Pharm Res* 1991; 8: 792–795.

Bull HB. *An Introduction to Physical Biochemistry.* FA Davis Company, Philadelphia, Pa, 1971, pp. 150–152.

Bull HB and Breese K. Protein Solubility and the Lyotropic Series of Ions. *Arch Biochem Biophys* 1980; 202(1): 116–120.

Ceschini S, Pietroni P, and Angeletti M. A Multi-Sample Microdialysis Apparatus for Proteins and Nucleic Acids. *Prep Biochem Biotech* 1996; 26(3&4): 189 199.

De Marco V, Stier G, Blandin S, and de Marco A. The Solubility and Stability of Recombinant Proteins are Increased by Their Fusion to NusA. *Biochem Biophys Res Commun* 2004; 322(3): 766–771.

Dorin G, Arve BH, Pattison GI, Halenbeck RF, Johnson K, Chen B-L, Rana RK, Hoba MS, Madani H, Tsang M, Gustafson ME, Bild GS, and Johnson GV. Method of Solubilizing, Purifying, and Refolding Proteins. 1996. Patent No. PCT/US96/09980.

Ducruix AF and Reis-Kautt MM. Solubility Diagram Analysis and the Relative Effectiveness of Different Ions on Protein Crystal Growth. *Methods: A Companion to Methods in Enzymology.* 1990; 1(1): 25–30.

Fan H, Ralston J, Dibiase M, Faulkner E, and Middaugh R. Solution Behavior of IFN-β-1a: An Empirical Phase Diagram Based Approach. *J Pharm Sci* 2005; 94(9): 1893–1911.

FDA website, http://www.cfsan.fda.gov/~dms/eafus.html

Flores H, Lin T, Pai R, Matthews T, and Shahrokh Z. Apo2 ligand/TRAIL Formulations. US and International Patent 2001;PCT/US02/36251, WO 03042344.

Flynn GL. Solubility Concepts and Their Applications to the Formulation of Pharmaceutical Systems. *J Parent Sci Technol* 1984; 38(5): 202–209.

Frommhagen LH. The Solubility and Other Physicochemical Properties of human γ-Globulin Labeled with Fluorescein Isothiocyanate. *J Immunol* 1965; 95(3): 442–445.

Gagnon P, Mayes T, and Danielsson A. An Adaptation of Hydrophobic Interaction Chromatography for Estimation of Protein Solubility Optima. *J Pharm Biomed Anal* 1997; 16(4): 587–92.

Golovanov AP, Haulbergue GM, Wilson SA, and Lian L-Y. A Simple Method for Improving Protein Solubility and Long Term Stability. *J Am Chem Soc* 2004; 126: 8933–8939.

Isaacs BS and Patel H. Pharmaceutical Formulations of Plasminogen Activator Proteins. US Patent 1990;04980165.

Ismailos G, Reppas C, Dressman JB, and Macheras P. Unusual Solubility Behavior of Cyclosporine A in Aqueous Media. *J Pharm Pharmacol* 1991; 43: 287–289.

Jenkins WT. Three Solutions of the Protein Solubility Problem. *Prot Sci* 1998; 7(2): 376–382.

Johnson MD, Hoesterey BL, and Anderson BD. Solubilization of a Tripeptide HIV Protease Inhibitor Uing a Combination of Ionization and Complexation with Chemically Modified Cyclodextrins. *J Pharm Sci* 1994; 83(8): 1142–1146.

Leavis PC, and Rothstein F. The Solubility of Fibrinogen in Dilute Salt Solutions. *Arch Biochem Biophys* 1974; 161(2): 671–682.

Leberman R. The Hofmeister Series and Ionic Strength. *FEBS Letters* 1991; 284(2): 293–294

Melander W and Horvath C. Salt Effect on Hydrophobic Interactions in Pprecipitation and Chromatography of Proteins: An Interpretation of the Lyotropic Series. *Arch Biochem Biophys* 1977; 183(1): 200–215.

Middaugh CR, Tisel WA, Haire RN, and Rosenberg A. Determination of the Apparent Thermodynamic Activities of Saturated Protein Solutions. *J Biol Chem* 1997; 254: 367–370.

Middaugh CR and Volkin DB. Protein Solubility. In: Ahern TJ, and Manning MC (eds.), *Stability of Protein Pharmaceuticals*. Plenum Press, New York, 1992, pp. 109–134.

Mollmann SH, Jorgensen L, Bukrinsky JT, Elofsson U, Norde W, and Frokjaer S. Interfacial Adsorption of Insulin Conformational Changes and Reversibility of Adsorption. *Eur J Pharm Sci* 2006; 27(2-3): 194–204.

Murby M, Samuelsson E, Nguyen TN, Mignard L, Power U, Binz H, UhleUn M, and Stehl S. Hydrophobicity Engineering to Increase Solubility and Stability of a Recombinant Protein from Respiratory Syncytial Virus. *Eur J Biochem* 1995; 230(1): 38–44.

Nallamsetty S and Waugh DS. Solubility-Enhancing Proteins MBP and NusA Play a Passive Role in the Folding of Their Fusion Partners. *Protein Expr Purif* 2006; 45(1): 175–182.

Napaporn J, Thomas M, Brazeau G, and Shahrokh Z. Assessment of the Myotoxicity of Pharmaceutical Buffers Using an *in vivo* Muscle Model: Effect of pH, Buffer Capacity, Tonicity, and Buffer Type. *Pharm Develop Technol* 2000; 5(1): 123–130.

Nuijen B, Bouma M, Talsma H, Manada C, Jimeno JM, Lopez-Lazaro L, Bult A, and Beijnen JH. Development of a Lyophilized Parenteral Pharmaceutical Formulation of the Investigational Polypeptide Marine Anticancer Agent Kahalalide F. *Drug Dev Ind Pharm* 2001; 27(8): 767–780.

Osiris website: http://www.organic-chemistry.org/prog/peo/logS.html

Powell MF, Nguyen T, and Baloian L. Compendial of Excipients for Parenteral Formulations. *PDA J Pharm Sci & Technol* 1998; 52(5): 238–311.

Protein Drug Delivery: Penetrating a Growth Market. Datamonitor publisher. 2005. http://www.pharmaceutical-business-review.com

Reddy KRC, Rudolph LH, and Lange C. L-Arginine Increases the Solubility of Unfolded Species of Hen Egg White Lysozyme. *Prot Sci* 2005; 14(4): 929–935.

Retailleau P, Reis-Kautt M, and Ducruix A. No Salting-in of Lysozyme Chloride Observed at Low Ionic Strength Over a Large Rrange of pH. *Biophys J* 1997; 73: 2156–2163.

Rollema HS, Kulpers OP, Both P, de Vos WM, and Siezen RJ. Improvement of Solubility and Stability of the Antimicrobial Peptide Nisin by Pprotein Engineering. *Appl Environ Microbiol* 1995; 61(8): 2873–2878.

Rothstein F. Differential Precipiation of Proteins. Science and technology. *Bioprocess Technol* 1994; 18: 115–208.

Rupley JA, Comparison of Protein Structure in the Crystal and in Solution IV. Protein Solubility. *J Mol Biol* 1968; 35: 455–476.

Saul A and Don M. A Rapid Method of Concentrating Proteins in Small Volumes With High Recovery Using Sephadex G-25. *Analyt Biochem* 1984; 138(2): 451–453.

Schein CH. Solubility as a Function of Protein Structure and Solvent Components. *Bio/Technology* 1990; 8(4): 308–317.

Schmoekel H, Schense JC, Weber FE, Gratz KW, Gnagi D, Muller R, and Hubbell JA. Bone Healing in the Rat and Dog with Nonglycosylated BMP-2 Demonstrating Low Solubility in Fibrin Matrices. *J Orthop Res* 2004; 22(2): 376–381.

Shahrokh Z. Developing Pharmaceutical Protein Formulations: Assumptions and Analytical Tools. In: Shahrokh Z, Sluzky V, Cleland J, Shire S, Randolph T (eds.), Therapeutic Protein and Peptide Formulation and Delivery. *ACS Symposium Series 675* , 1997.

Shahrokh Z, Stratton PR, Eberlein GA, and Wang YJ. Approaches to Analysis of Aggregates and Demonstrating Mass Balance in Pharmaceutical Protein (Basic Fibroblast Growth Factor) Formulations. *J Pharm Sci* 1994; 83(12): 1645–1650.

Shaked Z, Stewart T, Thomson J, Thomson JW, Tafaro T, and Hershenson S. Pharmaceutical Composition of Recombinant beta-Interferon and Formulation Processes. Patent 2001: 00270799/EP-B1.

Shimizu S and Smith DJ. Preferential Hydration and the Exclusion of Cosolvents from Protein Surfaces. *J Chem Phys* 2004; 121(2): 1148–1154.

Shimpi S, Chauhan B, and Shimpi P. Cyclodextrins: Application in Different Routes of Drug Administration. *Acta Pharm* 2005; 55(2): 139–156.

Shire SJ, Shahrokh Z, and Liu J. Challenges in Development of High Protein Concentration Formulations. *J Pharm Sci* 2004; 93(6): 1390–1402.

Sommer A and Rifkin DB. Interaction of Heparin with Human Basic Fibroblast Growth Factor: Protection of the Angiogenic Protein from Proteolytic Degradation by a Glycosaminoglycan. *J Cell Physiol* 1989; 138(1): 215–220.

Stevenson C. Non-aqueous Solubilizing Systems for Parenteral Formulation Development—Proteins. Chapter 12 of this book.

Strickley R. Parenteral Formulations of Small Molecules Therapeutics in the United States. *PDA J Pharm Sci* 1999; 53(6): 324–349; 2000; 54(1): 69–96; 2000; 54(2): 152–169.

Tzannis ST, Hrushesky WJ, Wood PA, and Przybycien TM. Irreversible Inactivation of Interleukin 2 in a Pump-based Delivery Environment. *Proc Natl Acad Sci U S A* 1996; 93(11): 5460–5465.

van Reis R and Zydney A. Membrane Separations in Biotechnology. *Current Opinion in Biotech.* 2001; 12(2): 208–211.

Wang W. Instability, Stabilization, and Formulation of Liquid Protein Pharmaceuticals. *Intl J Pharm* 1999; 185: 129–188.

Zhang R and Hjerten S. A Micromethod for Concentration and Desalting Utilizing a Hollow Fiber with Special Reference to Capillary Electrophoresis. *Analyt Chem* 1997; 69(8): 1585–1592.

Zhu G. Stabilization and Controlled Release of Proteins Encapsulated in Poly(Lactide-co-Glycolide) Delivery Systems. Dissertation Thesis, Ohio State University, 1999, p. 199.

12

Non-Aqueous Systems for Formulation Development—Proteins

CYNTHIA L. STEVENSON

Nektar Therapeutics, San Carlos, CA

Introduction

Formulation of proteins and peptides has been primarily achieved through aqueous solutions or lyophilized cakes for reconstitution. Therefore, the incorporation of non-aqueous formulation techniques has been limited by the lack of general applicability. However, the advent of biotechnology, bioengineering and pharmaceutical delivery systems has increased the requirements for drug solubility and stability under a variety of conditions, and required scientists to pursue alternative strategies. For example, the exploration of drug delivery systems for macromolecules and novel applications for enzyme chemistry have driven the usage of alternate solvent systems.

Solubility, enzyme activity, chemical and conformational stability of pharmaceutically active proteins under non-aqueous conditions have been well characterized (Zaks and Klibanov, 1988a; Zaks and Klibanov 1988b; Houen, 1996). Many of the solvents utilized in the literature are not pharmaceutically acceptable, and much of this work has not been directly applied to non-aqueous pharmaceutical formulations. However, the fundamental science and elucidation of concepts important to successfully utilizing non-aqueous conditions are applicable from this literature base. Furthermore, prediction of activity, solubility, chemical and structural stability are not routine, and preformulation work must be done on a targeted basis.

This chapter reviews a) the characterization of proteins and peptides in a variety of non-aqueous or co-solvent conditions, both acceptable and unacceptable for pharmaceutical applications, b) the applicability of non-aqueous conditions for increasing solubility, stability and activity, and c) novel drug delivery and formulation process technology applications. This review focuses on non-aqueous solutions, suspensions and co-solvent systems that result in miscible conditions.

% EtOH (w/w)	Error (mVolt)	Error (pH)
0	0	0
16.2	5	0.08
33.2	−2	−0.03
52	−27	−0.46
73.4	−37	−0.63
85.4	−33	−0.56
100	−73	−1.23

Table 1. Calculated error due to liquid junction potential in water/ethanol solutions (Frant, 1995).

pH Measurements

Under aqueous conditions, the concentration of hydrogen ion in solution is usually expressed in terms of the hydrogen ion concentration or activity, or in terms of pH units. The pH of a non-aqueous solution or co-solvent solution is more difficult to measure, since pH only relates to purely aqueous conditions. The pH of a water miscible solution can be measured with an electrode, however interpretation of the pH value must be done with care.

First, the glass electrode may be off by as much as a pH unit due to the liquid junction potential (Table 1) (Frant, 1995). For example, a 52% solution of ethanol (EtOH) has a liquid junction potential error of −0.46 pH units (Frant, 1995).

Second, the addition of co-solvent with varying polarity will affect the solubility of the solute. For example, addition of EtOH increases the solubility of the un-ionized species, by decreasing the polarity of the solvent. Alcohol also decreases the dissociation of the solute, where solubility decreases as the pKa increases and the dissociation constant decreases (Martin et al., 1983).

Third, the dissociation constants of many organic solvents are not the same as that of water. For example, an aqueous solution is neutral at pH 7.0, but methanol (MeOH) and EtOH are neutral at 8.42 and 9.55, respectively (Frant, 1995). This means that alcohols are less dissociated than water at a pH value of 7.0.

Fourth, the pH scales for solvents are of a different range and breadth. The pH scale for water ranges from 0 to 14, while the pH scale for EtOH ranges from −4.2 to 14.9. In addition to the dissociation constants and the range of the pH scales, the effect of shielding by the solvent will vary. For example, MeOH has a dielectric constant of 32.7 compared with 78.3 for water at 25°C, indicating that MeOH will not shield separated charges as well as water.

Fifth, many co-solvent mixtures will have a combination of characteristics; however, these effects may not be linear with titration of organic content. Small amounts of MeOH in water will not affect the pH dramatically, but small amounts of water in MeOH will significantly change the pH [Popovych and Tomkins, 1981]. For EtOH, the required adjustment in the pH value is approximately 0.25 units at 70% EtOH. Some tables for adjusting pH readings are published, but are not comprehensive (Popovych and Tomkins, 1981; Frant, 1995).

Lastly, when working with macromolecules, the isoelectric point (pI) of the protein must be assessed to determine if the pH of the solution will have a large or small effect on the overall charge on the molecule. If the pH of the solution is far from the pI, the difference in the apparent pH and the actual pH may not have a large effect on the stability or conformation of the protein.

The pH measurements in the literature include a range of appropriately adjusted pH measurements, uncorrected pH readings and pH values pertaining to the "pH memory" of the lyophilized material prior to dissolution. Therefore, pH should be interpreted with some latitude, and minimally, serve a use to obtain reproducible experiments. One should keep in mind that little theoretical meaning can be attributed to a non-aqueous pH reading, and the definition of pH in co-solvent solutions may be better described as an apparent pH.

Solubility

Physical characteristics of organic solvents that are useful in understanding solubilization include hydrophobicity (log P), dielectric constant (ε), dipole moment (μ), and the Hildebrand solubility parameter (δ) (Table 2). Hydrophobic proteins avoid water, and self associate via van der Waals forces. These hydrophobic interactions are favored thermodynamically because of increased entropy of the water. If provided an alternate hydrophobic solvent (octanol), the partition coefficient can be measured. Solution polarity is determined by all the forces contributing to interactions between solvent and solution (coulombic, directional, inductive, dispersion and hydrogen bonding forces); however, we will focus on the contributions of permanent and induced dipoles. In large polar molecules with multiple functional groups, the separation of positively and negatively charged regions can be permanent (dipole moment) and/or inducible (dielectric constant). Finally, solubility parameters express the cohesion between like molecules (solute and solvent).

Bovine serum albumin (BSA) and lysozyme solubility are well documented in organic solvents, some of which are pharmaceutically acceptable (Table 3) (Houen, 1996). In general, good BSA solubility was observed in glycerol and dimethyl sulfoxide (DMSO), while poor solubility was observed in several alcohols. The solubility of lysozyme (pH 6.0) in 34 solvents and co-solvent systems was also characterized (Table 3) (Chin et al., 1994). To add complication, the solubility of lysozyme was difficult to measure, because solutions above 10 mg/ml lysozyme formed viscous gels. The protein solubility was then correlated with the characteristics previously outlined.

Solubility correlated well with hydrophobicity. Less correlation was observed with the dielectric constant and the solubility parameter, and no correlation with

Solvent	Dipole moment (μ)	Dielectric constant (ε)	Log P	Hildebrand solubility factor (δ)
Acetone	2.9	21.0	−0.23	19.70
Acetonitrile	3.4	35.9	−0.33	24.40
Benzyl Alcohol	1.7	11.9	–	–
1-Butanol	1.7	17.1	0.80	21.70
Chloroform	1.0	4.8	–	18.70
Decanol	–	7.9	4.00	–
Dimethyl Formamide	3.9	36.7	−1.00	24.80
Dimethyl Sulfoxide	4.0	46.5	−1.30	26.40
Dioxane	2.1	2.2	−1.10	20.50
Ethanol	1.7	24.6	−0.24	26.30
Ethyl Acetate	1.8	6.1	0.68	18.40
Ethylene Glycol	2.3	37.7	−1.80	32.90
Formamide	3.7	111.0	−2.10	39.30
Glycerol	–	42.5	−3.00	34.90
Heptanol	–	11.8	2.40	–

Hexane	–	1.9	3.50	–
Methanol	1.7	32.7	–0.76	29.60
Methylene Chloride	1.6	8.9	1.20	20.10
Nonane	–	2.0	5.10	–
1-Octanol	1.7	9.4	2.90	21.10
1,2-Propanediol	2.3	32.0	–1.40	28.00
1-Propanol	1.6	20.8	0.29	–
Pyridine	2.2	13.3	0.71	21.90
Tetradydrofuran	1.8	7.5	0.49	18.50
Tert Amyl Alcohol	1.7	5.8	1.30	–
Trifluroethanol	–	27.7	–2.10	–
Water	1.8	80.1	–	48.00

Table 2. Physical characteristics of organic solvents (Laane et al., 1987; Baron, 1991; Guinn et al., 1991; Volkin et al., 1991; Chin et al., 1994; Lide, 1995; Buchwald and Bodor, 1998).

Molecule	Solvent	Solubility (mg/ml)	Reference
Bovine Serum Albumin	Ethanol	0.01	(Houen, 1996)
	DMF	0.01	(Houen, 1996)
	Acetonitrile	0.02	(Houen, 1996)
	1-Propanol	0.03	(Houen, 1996)
	Methanol	0.8	(Bromberg and Klibanov, 1995)
	DMSO	5.12	(Houen, 1996)
	Glycerol	20	(Houen, 1996)
Cyclosporine A	Water	0.02	(Seeback et al., 1989)
	40% Glycofurol	0.8	(Seeback et al., 1989)
	35% PEG 400	0.8	(Seeback et al., 1989)
	30% PG	0.2	(Seeback et al., 1989)
	40% EtOH	3.6	(Seeback et al., 1989)
Cytochrome C	Ethanol	<0.005	(Bromberg and Klibanov, 1995)
	Methanol	1.4	(Bromberg and Klibanov, 1995)

Glucagon	Ethanol	1.1	(Bromberg and Klibanov, 1995)
Glucagon Like Peptide	DMSO	80	(Stevenson, 1999)
	Water	150	(Stevenson, 1999)
Growth Hormone	Methanol	ND	(Bromberg and Klibanov, 1995)
	Ethanol	2.4	(Bromberg and Klibanov, 1995)
Insulin	1-Octanol	0.005	(Matsuura et al., 1993)
	Ethanol	0.32	(Bromberg and Klibanov, 1995)
	Methanol	>5	(Bromberg and Klibanov, 1995)
Insulin:SDS (1:6)	1-Octanol	3	(Matsuura et al., 1993)
Leuprolide	DMSO	370	(Hall et al., 1999)
Lysozyme	Acetonitrile	0.03	(Houen, 1996)
	Isoamyl alcohol	0.08	(Houen, 1996)
	1-Octanol	<0.05	(Chin et al., 1994)
	EtOH	0.25	(Bromberg and Klibanov, 1995)
	DMF	0.61	(Houen, 1996)

(continued)

Molecule	Solvent	Solubility (mg/ml)	Reference
	MeOH	>10	(Chin et al., 1994)
	TFE	>10	(Chin et al., 1994)
	Ethylene glycol	>10	(Chin et al., 1994)
	DMSO	>100	(Houen, 1996)
	Glycerol	>100	(Houen, 1996)
Myoglobin	Ethanol	0.59	(Bromberg and Klibanov, 1995)
	Methanol	>5	(Bromberg and Klibanov, 1995)
Salmon Calcitonin	70% EtOH	50	(Stevenson and Tan, 2000)
	70% DMSO	50	(Stevenson and Tan, 2000)
	Water	100	(Stevenson and Tan, 2000)
Ribonuclease	Ethanol	<0.005	(Bromberg and Klibanov, 1995)
	Methanol	0.61	(Bromberg and Klibanov, 1995)

Table 3. Solubility of proteins in co-solvent and non-aqueous conditions.

dipole moment. Some success with polar protic solvents and polar aprotic solvents suggested that solubility can be enhanced by dissolution in an organic vehicle with a high dielectric constant (Chin et al., 1994; Houen, 1996). Increasing the dielectric constant may increase solubility, but it may also decrease stability. In general, no single parameter predicted lysozyme solubility, but protic polar solvents and hydrophilic solvents all reported solubilities above 10 mg/ml (Chin et al., 1994). For hydrophobic cyclosporine, decreasing the co-solvent polarity resulted in increased solubility, where EtOH, propylene glycol (PG), polyethylene glycol 400 (PEG 400), glycerin and glycofurol were studied (Ran et al., 2001).

Further enhancement of solubility in non-polar organic solvents can be achieved by addition of salts (Seeback et al., 1989; Rariy and Klibanov, 1999). Salts can act by non-specific electrostatic shielding or specific ion binding to the protein. Salts are characterized as kosmotropes (stabilizing water molecules through strong interactions) and chaotropes (disrupting water structure through weak interactions with water). Small ions with high charge density are kosmotropes and larger ions with diffuse charge density, weakly interacting with water, are chaotropes. An increase in protein solubility with the addition of salt is termed "salting in" and can be explained by the Debye-Huckel theory. Binding between the protein and the counter ion results in decreased electrostatic free energy of the protein (shielding) and increased solvent activity, resulting in increased solubility. The greatest effect is usually obtained from weakly hydrated ions, facilitating accumulation of water molecules and preferential hydration.

Conversely, "salting out" results from interfacial effects of strongly hydrated ions near the protein essentially removing water molecules and desolvating the surface. For example, the solubility of human insulin-like growth factor (hIGF) in 140 mM benzyl alcohol and 145 mM NaCl increased aggregation and decreased solubility (Fransson et al., 1997; Kim and Dordick, 1997). Usually, as the ionic strength increases, first a slight salting in effect is observed, followed by a salting out effect.

The Hofmeister lyotropic series can be used to select an anionic ($F^- > Cl^- > Br^- > I^- > NO_3^-$) or cationic ($Al^{+3} > Mg^{+2} > Ca^{+2} > Na^+ > K^+ > NH^{+4}$) species, where the precipitation ability is related to the hydration of the ion and its ability to separate water molecules from the hydrophilic regions of the molecule. Overall, anions hydrate more strongly than cations for the same ionic radius because they can approach closer to the hydrogen atoms when compared to the oxygen atoms, respectively (Grossfield et al., 2003). Anions are also usually more polarizable than cations due to their diffuse extra electrons (Kropman and Bakker, 2001).

Stoichiometric ratios of anionic detergents, such as sodium dodecyl sulfate (SDS), have been used to complex peptides and increase their partition coefficient into non-polar solvents by two- to four-fold (Powers et al., 1993). For example, insulin was complexed with SDS and resulted in a ten-fold increase in solubility in 1-octanol. The insulin remained in its native conformation with increased thermal stability (T_m), when compared with water (Powers et al., 1993; Manning et al., 1995). Insulin (pH 3.0) solubility in EtOH increased from 14 mg/ml to 830 mg/ml with the addition of SDS (Bromberg and Klibanov, 1995). Similar results were obtained with α-chymotrypsin in sodium bis(2-ethylhexyl) sulfosuccinate (Manning et al., 1995).

Solubility can also be enhanced by lyophilizing the proteins at a pH away from its pI (Zaks and Klibanov, 1988a; Schulze and Klibanov, 1991; Chin et al., 1994) prior to dissolution in a relatively polar organic vehicle. For example, lysozyme (pI 11.0) demonstrated a 1000-fold decrease in solubility in 56% acetonitrile/44% DMSO when reconstituted from pH 10.0, as compared with pH 2.0 (Chin et al., 1994). The solubility of several proteins was characterized as a function of pH of the lyophilized material (Bromberg and Klibanov, 1995). For example, lysozyme lyophilized at pH 3.0 and 6.0 yielded solubilities in EtOH of 250 μg/ml and 200 μg/ml, respectively. Insulin (pI 5.3) solubility increased from 160 μg/ml to 1100 μg/ml as the pH decreased from 7.4 to 3.0 (Bromberg and Klibanov, 1995). These findings suggest that pH values equidistantly above and below the pI should not be assumed to afford equivalent and/or linear increases in solubility, as the distribution of amino acid pKa's exposed to the surface will dictate these characteristics. The resultant increase in solubility can be generally attributed to increased charge repulsion, decreased aggregation and water retention on the protein (Houen, 1996).

Residual water content can also affect solubility; specifically, hydrophobic solvents can enhance water packing into a solvation layer. Hydrophilic solvents solubilize water, pulling it away from the protein surface, thus decreasing the protein hydration (Zaks and Klibanov, 1988a; Zaks and Klibanov, 1988b; Halling, 1990; Cowan, 1997). Elimination of the hydration shell can induce denaturation and inhibit enzymatic activity (Schulze and Klibanov, 1991). However, a balance is required since excess water can accelerate chemical degradation.

Lastly, structure and solubility of proteins are often related; specifically, conformational stability may affect unfolding, aggregation and precipitation processes (Xu et al., 1997). For example, increased β-sheet structure was observed to correlate with decreasing solubility for glucagon-like peptide, or insulinotropin (Kim et al., 1994; Senderoff, 1998). Disruption of β-sheet/aggregate states in peptides has been demonstrated utilizing DMSO (Kim et al., 1994; Tan et al., 1998). This increase in β-sheet conformation was correlated with a decrease in solubility for a range of peptides (Narita et al., 1993; Oh-Uchi et al., 1996). Specifically, increased β-sheet content and stability of penta- and decapeptides correlated well with a decrease in peptide solubility in organic solvents, such as DMSO, methylene chloride and dimethylformamide (DMF) (Oh-Uchi et al., 1996). Similarly, β-sheet conformation (1630 cm^{-1}) observed in Fourier Transform Infrared (FTIR) spectra of tri-, hepta- and octapeptides in methylene chloride decreased with increasing amounts of DMSO (Narita et al., 1989; Narita et al., 1993). DMSO was also observed to alleviate leuprolide β-sheet/aggregation and gelation, observed in aqueous solution (Tan et al., 1998).

Structure

Conformational stability and solubility are inter-related when minimizing irreversible unfolding and maintaining biological activity (Arakawa and Goddette, 1985; Huang et al., 1995). A folded protein usually contains most hydrophobic residues in the core, protected from the aqueous environment, and contains most hydrophilic residues on the surface. In some cases, a protein solubilized in a non-aqueous, less polar solvent will unfold. This process is initiated by the disruption of surface hydrogen bonds, and protein unfolding occurs in order to separate the hydrophilic residues from the hydrophobic solvent. Therefore, a non-aqueous solvent will reduce the free energy of the unfolded state by solvating the exposed hydrophobic residues (Dill, 1990).

Many of the unfolding processes are irreversible; however, some are reversible with the return of full biological activity. For example, proteins dissolved in DMSO refold upon dilution in water (Chang et al., 1991). Supercritical anti-solvent precipitation of insulin, lysozyme and trypsin from DMSO yielded partially unfolded intermediates, as characterized by spectroscopy (Yeo et al., 1994; Winters et al., 1996). However, these structures were reversible upon reconstitution in aqueous media, with recovery of biological activity (Yeo et al., 1994; Winters et al., 1996). Chymotrypsin also completely unfolded in DMSO, but regained activity upon rehydration (Zaks and Klibanov, 1988a; Yeo et al., 1994).

Furthermore, it has been proposed that DMSO and alcohols can be utilized to stabilize reversibly unfolded states (Arakawa and Goddette, 1985; Bhattacharjya and Balaram, 1997). For example, DMSO has been observed to stabilize a partially unfolded conformation of lysozyme (pH 3.0) (Bhattacharjya and Balaram, 1997). Lysozyme in 10% DMSO showed little change in structure by nuclear magnetic resonance (NMR) when compared with aqueous conditions. The lysozyme structure in 50% DMSO resembled an early kinetic intermediate observed in the refolding process. Conversely, a highly unfolded structure in 100% DMSO has been reported, probably due to the polar aprotic nature of the solvent (Jackson and Mantsch, 1991).

Alcohols (MeOH, EtOH, trifluoroethanol) have been used to both denature proteins and increase the structural stability of proteins and peptides. Stabilization of helices with alcohols is well characterized, where alcohols decrease hydrogen bonding with the solvent and increase intramolecular bonding between C=O and N-H groups, promoting stable secondary structures. Partially unfolded conformations of ubiquitin (pH 2.0), α-lactalbumin (pH 2.0) and monellin (pH 2.0) have been observed in 60% MeOH, 15%–50% trifluoroethanol (TFE) and 50% EtOH, respectively (Harding et al., 1991; Fan et al., 1993; Alexandrescu et al., 1994; Alonso and Daggett, 1995). The reversibility of these partially unfolded intermediates is unknown. However, denaturation of tumor necrosis factor α in TFE was irreversible with the formation of insoluble aggregates (Narhi et al., 1996). In another example, the T_m of interleukin receptor (IL-1R) (pH 7.4) increased at small ratios of EtOH, but at higher EtOH concentrations the

T_m was drastically reduced, indicating that titration studies are critical (Remmele et al., 1998).

Conversely, actin peptide (pH 4.0), apomyoglobin (pH 6.0) and melittin (pH 5.0) structures were stabilized in 80% TFE, 20% TFE and 100% MeOH, respectively (Shiraki et al., 1995; Sonnichsen et al., 1995; Bazzo et al., 1998). The solvent-induced structure in the actin peptide agreed well with that of the crystal structure for the actin molecule; however, the β-sheet regions were slightly under represented (Sonnichsen et al., 1995). Mellitin alcohol-induced structure was also similar to x-ray diffraction data (Bazzo et al., 1998).

It should be mentioned that crystal structure data are often obtained from co-solvent systems (Desai and Klibanov, 1995). For example, crystals of ubiquitin, papain and a heptapeptide were grown from 30% polyethylene glycol (PEG) 4000, 62% MeOH and a DMSO/isopropanol mixture, respectively (Kamphius et al., 1984; Karle et al., 1993; Love et al., 1997). The orthogonal and tetratogonal crystal forms of cyclosporine were prepared from 25% PEG 300 and acetone, respectively (Petcher et al., 1976; Kessler et al., 1985; Loosli et al., 1985; Verheyden et al., 1994a). Furthermore, cyclosporine and leuprolide form thermotropic liquid crystals when dried from EtOH and lyotropic liquid crystals when solubilized in PG, respectively (Tan et al., 1998; Lechuga-Ballesteros et al., 2003; Stevenson et al., 2003).

Solvents can also stabilize folded active conformations, depending on the driving forces (Ran et al., 2001; Stevenson et al., 2003). For example, cyclosporine dissolved in chloroform, resembles the tetragonal crystal structure (Stevenson et al., 2003). The crystalline cyclic peptide structure is described as having a twisted antiparallel β-sheet with a Type II β-turn at one end of the structure and a γ-loop at the other. Cyclosporine contains only four amide groups, which are all rotated inwards, stabilizing the secondary structure. When dissolved in organic solvents with varying polarity and hydrogen bonding donor (MeOH) or acceptor (acetonitrile) ability, two distinctly different conformations were obtained where the extent of hydrogen bonding (individual solvent effects on C=O and N-H groups) and β-sheet/β-turn structure varied. Cyclosporine is notably insoluble in water; however, when dissolved in 32% MeOH/32% acetonitrile/36% water, cyclosporine more closely mimicked the active conformation assumed under aqueous conditions with the backbone amides and carbonyls rotated outwards (Stevenson et al., 2003).

In addition to stabilizing or destabilizing protein structures, alcohols can also alter the ratio of secondary structures, and induce non-native conformations. For example, TFE can stabilize helical structure at the expense of β-sheet structure, as demonstrated with β-lactoglobulin, β-casein and the constant fragment of the immunoglobulin light chain (Jackson and Mantsch, 1992a; Shiraki et al., 1995). The native conformation of β-lactoglobulin (pH 2.0) is β-sheet rich, however upon titrating 15%–20% TFE, a highly cooperative transition from β-sheet to α-helix occurs (Shiraki et al., 1995). Interestingly, β-sheet rich concanavalin (pH 7.0), in which EtOH and MeOH both increased β-sheet structure, while TFE enhanced α-helical structure (Jackson and Mantsch, 1992a; Wang et al., 1992).

Therefore, the propensity of TFE to stabilize α-helices may be useful if a) the structural change increases the solubility or stability of the molecule and b) the intermediate is reversible.

The effect of several polyols on the T_m of thrombin and trypsin have also been reported (Boctor and Mehta, 1992). For example, the T_m of trypsin (pH 6.5) was observed to increase with increasing glycerol concentration (up to 50%) from 52°C to 60°C, resulting in a $\Delta T_m = 8.0°C$. Similar increases in thermal stability were observed for thrombin, were the $\Delta T_m = 9.8°C$ when comparing purely aqueous conditions with 50% glycerol. The effect of glycerol on myoglobin (pH 7.4) was more subtle, in which the overall tertiary structure was destabilized, but the overall α-helical secondary structure was stabilized (Barteri et al., 1996).

The thermal melting of lysozyme (pH 6.0) in 99% glycerol has also been characterized by Circular Dichroism (CD) (Knubovets et al., 1999a). The α-helix and β-turn content was essentially the same in both water and glycerol, with only a slight increase in the β-sheet content in glycerol. The tertiary structure of lysozyme in water and glycerol both exhibited a similar $T_m = 74 - 76°C$. However, the secondary structure of lysozyme denatured at very different temperatures: $T_m = 74°C$ in water and $T_m = 85°C$ in glycerol. This suggested that lysozyme secondary structure was stabilized by the presence of glycerol. To explore this concept, amide proton exchange experiments were monitored by ^1H NMR for 6 weeks. Hydrogen exchange rates and solvent accessibility provided a useful measurement of protein flexibility, and thus of structural stability. It was observed that amide protons exchanged at a much slower rate in glycerol than in water. The results suggested that lysozyme demonstrates increased stability in glycerol, and may have a slightly different unfolding mechanism in glycerol than in water (Knubovets et al., 1999a). Further studies showed increased secondary structure in ethylene glycol and MeOH, but not in DMSO or DMF (Knubovets et al., 1999b).

An additional method by which co-solvent systems stabilize proteins is by enhancing the hydrophobic interactions, altering the structure of water, reducing the interaction of the protein with other solutes and maintaining residual water in a protective monolayer around the protein (Ruelle and Kesselring, 1998; Wang 1999). Lysozyme (pH 8.2) refolding was monitored in varying ratios of glycerol and water (Rariy and Klibanov, 1997). Refolding yields in 90% glycerol and water were similar, where approximately 32% of the lysozyme refolded in 99% glycerol. Two trends were noted: a) lysozyme refolding decreased with decreasing water and b) lysozyme refolding in glycerol was time dependent, where refolding increased consistently over 40 hours. Refolding was observed in dilute solutions of DMSO, but not under concentrated conditions. Further studies in 60% ethylene glycol or 60% propylene glycol resulted in lysozyme refolding, however little refolding was noted in 60% TFE or 60% MeOH (Rariy and Klibanov, 1999).

Similarly, the effects of DMSO, glycerol and ethylene glycol on the conformation of lysozyme and cytochrome C were evaluated by FTIR (Huang et al., 1995). The results were protein specific: cytochrome C (pH 7.4) was little affected, but lysozyme showed large changes in structure in 30% co-solvent conditions. As the concentration of co-solvent increased, the α-helix structure (1657 cm^{-1}) increased. The change in secondary structure was attributed to an alteration in the structure of water molecules surrounding the protein. Similarly, the thermal stability of α-amylase (pH 6.5) was increased to 78°C in the presence of DMSO and PEG, again by a preferential hydration mechanism (Rajendran et al., 1995). For example, DMSO, a polar aprotic solvent, is capable of being a hydrogen bond acceptor, and can therefore form bonds with water, affecting the hydrogen bonding strength and the structure of water molecules around the protein. The effect of chloroform, water and MeOH on the helical conformation of pulmonary surfactant lipoprotein determined that only chloroform disrupted the secondary structure (Kovacs et al., 1995).

The addition of salts to co-solvents such as propylene glycol, ethylene glycol, glycerol and DMSO resulted in an increase in refolding of lysozyme (pH 8.2) (Rariy and Klibanov, 1999). The refolding increased 14-fold in 90% ethylene glycol with 1 M LiCl. Therefore, the addition of salts to non-aqueous conditions can increase folding and structural stability. Increasing the salt concentration above 3 M LiCl resulted in a dramatic decrease in refolding, probably due to salting out. Further studies with both anions and cations did not show a correlation between refolding and the Hofmeister lyotropic series. The observed increased stability with increasing salt concentration may be associated with increasing solubility, as previously discussed. Lysozyme refolding under all co-solvent conditions, including 100% water, was low. This low refolding yield was attributed to aggregation processes, where aggregation was reduced and solubility was increased in the presence of 1 M LiCl (Rariy and Klibanov, 1999).

Increasing the surface tension of the co-solvent solution has also been correlated with increased protein stability (Kita et al., 1994). Generally, solutes raising the surface tension of the co-solvent solution are usually partitioned away from the protein surface (Wang, 1999). This has been demonstrated with a variety of salts and related solutes (Kita et al., 1994). However, a decrease in surface tension is not always attributable to a loss of protein stability. For example, solvents like DMSO, 2-methyl-2,4-pentadiol and PEG (MW 200-1000) decreased the surface tension, but still induced preferential hydration and stabilization of lysozyme and BSA (Kita et al., 1994). Therefore, preferential hydration could not be linked to the surface free energy perturbation, suggesting that preferential hydration was governed by other factors, such as steric exclusion and solvophobic effects (Timasheff and Inoue, 1968; Kita et al., 1994; Wang, 1999).

Peptides are usually less adversely affected by non-aqueous conditions than proteins. The conformation of somatostatin (pH 1.25) has been characterized by NMR in water, DMSO, MeOH and ethylene glycol (Verheyden et al., 1990; Verheyden et al., 1991; Jackson and Mantsch, 1992b; Verheyden et al., 1994b). In all of the solvent conditions, the secondary structure was conserved, and was characterized to be a β-sheet structure with a β-turn. Further studies indicated that as the viscosity of the solvent system increased, the conformational flexibility decreased, stabilizing the structure. Conversely, significant changes in the conformation of myoglobin and concanavalin in DMSO were noted by FTIR spectroscopy (Jackson and Mantsch, 1991). For myoglobin, co-solvent ratios of 30% DMSO resulted in increased β-sheet structure and aggregation (1621 cm^{-1}). At very high DMSO concentrations, the DMSO competes with intramolecular C=O and N-H bonding, resulting in complete unfolding.

The structure of salmon calcitonin (sCT) has been characterized in a variety of co-solvent conditions (Motta et al., 1989; Meadows et al., 1991; Meyer et al., 1991; Motta et al., 1991; Arvinte and Drake, 1993). Salmon calcitonin in 90% TFE or 90% MeOH both formed α-helical structures with an N-terminal loop (Meadows et al., 1991; Meyer et al., 1991; Motta et al., 1991). Conversely, sCT in 90% DMSO exhibited an intermolecular β-sheet structure with two β-turns (Motta et al., 1989). These results indicate that sCT can adopt very different, reversible conformations in a variety of co-solvents. Similar conformational results for sCT (pH 5.0) were observed by FTIR in a series of stability studies at 37°C (Stevenson and Tan, 2000). Helical structures were observed in both 70% EtOH and 70% propylene glycol. A β-sheet structure was observed in 70% DMSO; however, a very loose α-helix was observed in 100% DMSO, probably caused by the aprotic nature of the solvent. In these studies, the conformation was correlated to the onset of gelation, where the co-solvents inducing an α-helical structure resulted in a decrease in gelation (Stevenson and Tan, 2000).

Growth hormone releasing factor (GRF) is known to form stable helical structures in alcohols and SDS micelles (Kloosterman et al., 1993; Stevenson et al., 1993a; Stevenson et al., 1993b; Stevenson et al., 1993c). The helical content of GRF (pH 10.0) was observed by CD to increase with increasing MeOH concentration (Stevenson et al., 1993a). Subsequently, the stability of the helical segment for GRF was characterized by a series of H/D exchange experiments by liquid chromatography tandem mass spectrometry (LC/MS/MS) (Stevenson et al., 1993c, Anderegg et al., 1994). Protons associated with protein and peptide structures will exchange at a varying rate depending on their location, secondary structure and solvent accessibility. For example, side chain protons exchange rapidly, however backbone amide protons in a stable secondary structure or compact tertiary structure will exchange slowly. With increasing ratios of MeOH, the amide protons in the helical segment of GRF exchanged more slowly, indicating that the structure was more stable. Furthermore, the increasing ratios of MeOH also resulted in decreasing rates of deamidation, due to helix stabilization (Stevenson et al., 1993a).

PEGs were also analyzed for their ability to affect structural stability (Kita et al., 1994; Vrkljan et al., 1994; Rajendran et al., 1995; Zhang et al., 1995; Remmele et al., 1998;). For example, PEG (300, 1000, 3350) all increased the T_m of IL-1R, where decreasing the PEG molecular weight increased the T_m, possibly because of the decreasing hydrophobicity with decreasing molecular weight (Remmele et al., 1998). Conversely, up to 15% PEG 4000 inhibited thermal aggregation of low molecular weight urokinase (LMW-UK), while PEG 300 had no effect (Vrkljan et al., 1994). In this case, the higher molecular weight PEG increased the thermal stability through preferential solute interaction, where the interaction increases with increasing molecular weight. Therefore, when using PEG, hydrophobicity and molecular weight affect stability and these effects may be dependent on temperature (Arakawa et al., 1991).

Subtilisin has been stabilized in DMF through site-directed mutagenesis (Martinez et al., 1992). First, a charged Asp^{248} on the surface of the protein was substituted with more hydrophobic residues, Asn, Ala or Leu, resulting in increased structural stability in 80% DMF. Second, Asn^{218} was substituted with Ser, altering internal hydrogen bonding interactions, resulting in stabilization in 40% and 80% DMF. When both Asp^{248} -> Asn^{248} and Asn^{218} -> Ser^{218}, the enzyme was 3.4 times more stable than the wild type in 80% DMF. A similar application of site-directed mutagenesis of subtilisin (Asp^{60} ->Asn^{60}, Glu^{103} -> Arg^{103} and Asn^{218} ->Ser^{218}, and resulted in 38 times more activity in 85% DMF (Chen and Arnold, 1991).

Stabilization observed with trace concentrations of solvents, less polar than water, may be attributed to reproducing a physiologic microenvironment more closely than purely aqueous conditions (Butler, 1979). Much of the work with alcohols aims to induce conformational changes similar to those adopted at an active site on a cell membrane. Often the hydrogen bonding potential and the dielectric constant of the co-solvent conditions are chosen to mimic that of a cell membrane (Jackson and Mantsch, 1992a). Some researchers have also explored the stability of proteins in oils or lipids (Yu et al., 1996; Lo and Rahman, 1998). When the thermal stability of ribonuclease and cytochrome C were characterized in dipalmitoylphosphatidylglycerol (DPPG), stabilization was observed at <4 mol% DPPG and destabilization was observed at > 4 mol% DPPG (Lo and Rahman, 1998). Suspension formulations of 40% GRF in Miglyol have demonstrated good stability by reversed-phase high performance liquid chromatography (RP-HPLC), size exclusion chromatography (SEC) and FTIR for up to 10 weeks at 25°C and 39°C (Yu et al., 1996). However, one must balance the requirements for solubility and structural stability with activity when choosing a solvent, and in this regard, a protein suspension may be more effective than a solution.

Suspensions

Much of the work done on proteins in non-aqueous conditions has been performed on suspensions (Zaks and Klibanov, 1984; Zaks and Klibanov, 1988a; Volkin et al., 1991; Desai and Klibanov, 1995; Yu et al., 1996). Overall, the same chemical and structural stability principles apply to proteins in solution and in suspension. However, the solubility of the protein is usually low in the solvent system. This suspension strategy allows low moisture, solid state stabilization of the protein structure in the particulate, avoiding irreversible unfolding. Suspensions also result in isolating the protein in the particulate away from the solvent system and decreasing protein flexibility (Zaks and Klibanov, 1984; Zaks and Klibanov, 1988a; Volkin et al., 1991). The strategy may be utilized to achieve higher concentration when the protein is limited by low solubility.

Suspensions of protein particulate can also be used to maintain the secondary and tertiary structure, which may not be possible with a solution. The secondary structure of lysozyme and subtilisin suspensions in acetonitrile, tetrahydrofuran and 1-propanol were characterized by FTIR (Griebenow and Klibanov, 1996). The secondary structure of lysozyme lyophilized powder and lysozyme suspended in 100% acetonitrile were found to be essentially equivalent. Furthermore, the structure appeared to be more stable in 100% acetonitrile than in 60% acetonitrile. The structure of bovine pancreatic trypsin inhibitor (BPTI) was characterized in several solvents using a hydrogen isotope exchange experiment with NMR detection (Desai and Klibanov, 1995). Suspensions of BPTI (pH 3.5) in acetonitrile, tetrahydrofuran, ethyl acetate or butanol yielded similar amide proton isotope exchange rates. When compared with the exchange rates for lyophilized powder, these results suggested that BPTI suspended in organics was partially unfolded, but no worse than the lyophilized powder. Similarly, NMR studies on the tyrosyl ring motion in α-lytic protease, electron spin resonance spectrometry and time-resolved fluorescence anisotropy measurements supported the restricted protein mobility in organic solvents (Guinn et al., 1991; Affleck et al., 1992; Burke et al., 1993; Broos et al., 1995).

Likewise, the structure of subtilisin (pH 3.0) suspended in varying ratios of acetonitrile and water demonstrated α-helical content similar to that in the lyophilized powder (Griebenow and Klibanov, 1996). Furthermore, the rate of transesterification reactions of subtilisin (pH 7.8) suspended in DMSO/acetonitrile, formamide/acetonitrile or formamide/dioxane were increased approximately 100-fold over aqueous conditions (Almarsson and Klibanov, 1996). Similar results were obtained for subtilisin (pH 7.8) in a tetrahydrofuran/1-propanol mixture (Affleck et al., 1992). These results can be attributed to the increased structural rigidity of the active conformation of the protein in the solid, and the denaturing characteristics of the solvent at the solvent-particulate interface. Preservation of this "molecular memory" or "molecular imprint" of the protein can also be used to stabilize structure and activity (Mishra et al., 1996; Rich and Dordick, 1997; Santos et al., 2001). Subtilisin was lyophilized from crown ethers, resulting in more native like structure, by FTIR, and increased enzyme activity in THF, acetonitrile and dioxane (Santos et al., 2001).

Overall, protein suspensions and solutions react similarly to the hydrophobicity of the solvent. For example, suspensions of ribonuclease, chymotrypsin and lysozyme in nonane, butanol or DMF demonstrated increased structural stability over aqueous conditions (Volkin et al., 1991). The solvents were categorized as non-polar (nonane), polar (butanol) or dipolar (DMF). Thermal stability increased with increasing hydrophobicity and apolarity. Similarly, chymotrypsin activity decreased with decreasing hydrophobicity, consistent with the hypothesis that hydrophilic solvents can more effectively remove bound water than can hydrophobic solvents (Zaks and Klibanov, 1988a). For example, the solubility of water in hydrophobic solvents (Log P < 2.0) was > 0.4 wt %, while the solubility of water in hydrophobic solvents (Log P > 4.0) was < 0.4

wt % (Laane et al., 1987). The moisture content of suspended α-chymotrypsin in octane and pyridine was 2.5% and 1.0% (w/w), respectively (Zaks and Klibanov, 1988a). These results are consistent with the more hydrophilic solvent effectively removing water from the protein.

Proteins suspended in organic solvents demonstrated increased resistance to thermal denaturation and chemical degradation, most likely because of the lack of protein flexibility and residual water content (Zaks and Klibanov, 1984; Volkin et al., 1991; Tuena de Gomez-Puyou and Gomez-Puyou, 1998). Residual water also affects protein solutions and suspensions similarly (Zaks and Klibanov, 1988a; Zaks and Klibanov, 1988b; Volkin et al., 1991; Affleck et al., 1992). For example, maintaining the bound layer of moisture increased the activity of subtilisin suspended in heptanol/tributyrin, and increased the thermal stability of ribonuclease suspended in nonane (Zaks and Klibanov, 1984; Volkin et al., 1991). Studies have shown that the presence of the organic solvent has little effect on the protein particulate or its residual bound water (Halling, 1990). However, non-polar solvents can increase the amount of water bound to the protein, while polar solvents can reduce the amount of bound water, possibly by replacing water with secondary hydration layers (Halling, 1990). Further studies investigated the hydration of lyophilized proteins suspended in non-aqueous conditions (McMinn et al., 1993). Moisture sorption curves were collected for lysozyme and chymotrypsin in di (n-butyl) ether, propanol and 1M propanol in benzene. A model was proposed in which the predicted moisture sorption isotherms provided good estimates for water sorption at low moisture levels.

Finally, Factor IX (pH 6.8) was suspended in perfluordecalin and perfluorotribultyamine for 6 months at 37°C with little loss of chemical stability (Knepp et al., 1998). The stability profile was similar to lyophilized powder stored at −80°C. Similar suspensions were prepared in methoxyflurane, octanol and PEG 400 with less promising stability. Halothane was observed to increase the stability of BSA, as characterized by differential scanning calorimetry (DSC) (Tanner et al., 1999). Furthermore, the conformation of lysozyme in tetrahydroethane and heptafluoropropane were characterized by raman spectroscopy (Quinn et al., 1999). These halogenated systems usually give the least interaction between the particulate and the solvent, and resulting in chemical and conformation stability similar to lyophilized powders.

Stability

One of the benefits of formulating under non-aqueous conditions is that hydrolytic degradation pathways can be minimized. For example, hydrolytic degradation of leuprolide in DMSO was limited when compared to aqueous conditions: leuprolide stored for 2 years at 37°C showed that 75%, 82% and 93% leuprolide remaining in water (pH 5.0), propylene glycol (PG) and DMSO, respectively (Hall et al., 1999; Stevenson et al., 1999). The increase in stability can be attributed to solubilization of leuprolide in an aprotic solvent, limiting the hydrogen source for degradation.

Metal-catalyzed oxidation in human relaxin was also inhibited by the addition of polyols (Li et al., 1996). When ethylene glycol or glycerol was added to relaxin in solution, the rate of methionine and histidine oxidation decreased (Li et al., 1996). Deuterium isotope experiments were conducted and concluded that the lack of difference in oxidation rate between glycerol and d_5-glcerol suggested that the stabilizing mechanism was not hydroxyl radical scavenging, but rather complexation of transition metal ions. Further studies on the metal-catalyzed oxidation of hGH also negated the radical scavenging mechanism, but observed a structural perturbation of the metal binding site (Hovorka et al., 2001). When hGH was oxidized in the presence of EtOH, ethylene glycol, TFE, propanol, PG, and butanol, the alcohols inhibited oxidation in a concentration dependent manner. No correlation was observed between an increase in helical content or solvent polarity. Similar to relaxin, the authors observed a transition metal complexation mechanism for ethylene glycol (Hovorka et al., 2001).

The deamidation rate in non-aqueous conditions has also been explored through changes in conformational stability using bovine growth hormone releasing factor (bGRF) (Stevenson et al., 1993a). The deamidation rate was observed to decrease with increasing MeOH concentration, due to an increase in α-helical stability. The resulting decreased solvent dielectric constant in increasing MeOH concentrations induced less hydrogen bonding with the solvent and more hydrogen bonding along the helix backbone. In a stable helical conformation, the backbone amide nitrogen is no longer in proximity to the asparagine side chain. Therefore, at high MeOH concentrations the helix structure is quite stable and does not readily unfold, allowing formation of the succinimide intermediate. Similar structure enhancing effects were observed for a β-hairpin peptide (Stotz et al., 2004).

Solvents with low dielectric constants have been utilized to stabilize peptide solutions (Brennan and Clarke, 1993). Specifically, the rate of asparagine deamidation for Val-Tyr-Pro-Asn-Gly-Ala (pH 7.4) in water, glycerol, EtOH and dioxane decreased with decreasing dielectric constant (Table 4). Theoretically, the decrease in the rate of deamidation may be due to destabilization of the deprotonated nitrogen anion in the peptide backbone responsible for attack on the asparagine side chain and formation of the succinimide intermediate. Similarly, increasing ratios of organic content decreased the deamidation rate of Boc-Asn-Gly-Gly-NH_2 due to a decrease in dielectric constant, where deamidation data was collected in water, MeOH, EtOH, dioxane, acetone and acetonitrile (Table 4) (Capasso et al., 1991).

However, no correlation was noted between the solvent dielectric and the degradation rate of salmon calcitonin (sCT) in water, DMSO, EtOH, glycerol and PG (Stevenson and Tan, 2000). The lack of correlation was probably due to the multiple degradation mechanisms reacting differently to the solvent dielectric.

The deamidation rate of the hexapeptide, Val-Tyr-Pro-Asn-Gly-Ala (pH 6.8), was also characterized in poly(vinyl alcohol) (PVA) and poly(vinyl pyrrolidone) (PVP) (Lai et al., 1999a; Lai et al., 1999b; Lai et al., 2000; Song et al., 2001). The rate of deamidation was observed to increase with increasing water content

Molecule	Solvent	Stability ($t_{1/2}$) at 37°C	Reference
Val-Tyr-Pro-Asn-Gly-Ala	100% Water, pH 7.4	29 hours	(Brennan and Clarke, 1993)
	40% Glycerol	60 hours	(Brennan and Clarke, 1993)
	80% Glycerol	142 hours	(Brennan and Clarke, 1993)
	15% Ethanol	59 hours	(Brennan and Clarke, 1993)
	30% Ethanol	101 hours	(Brennan and Clarke, 1993)
	40% Dioxane	243 hours	(Brennan and Clarke, 1993)
Boc-Asn-Gly-Gly-NH_2	100% Water, pH 7.5	8 days	(Capasso et al., 1991)
	50% Methanol	30 days	(Capasso et al., 1991)
	50% Ethanol	38 days	(Capasso et al., 1991)
	50% Acetone	48 days	(Capasso et al., 1991)
	50% Dioxane	65 days	(Capasso et al., 1991)
	50% Acetonitrile	206 days	(Capasso et al., 1991)
Leu^{27} bGRF(1-32)NH_2	100% Water, pH 10.0	8.5 hours	(Stevenson et al., 1993a)
	20% Methanol	12.4 hours	(Stevenson et al., 1993a)
	40% Methanol	17.0 hours	(Stevenson et al., 1993a)

Table 4. Stability of proteins in organic co-solvent conditions.

and decreasing glass transition temperature (T_g) (Lai et al., 1999a). For both PVA and PVP, deamidation was observed in the glassy state below the T_g, where the peptide was more stable in PVA. However, above the T_g in the rubbery state, the peptide was more stable in PVP, and the deamidation rate was much more sensitive to changes in PVA.

Subsequently, the authors characterized the deamidation rate in lyophilized formulations of PVP and glycerol and observed two effects: molecular mobility and solvent effects (Lai et al., 1999b). First, glycerol was used as a plasticizer to vary the T_g without changing the water content. At constant water content, increasing glycerol concentration decreased the T_g, increasing molecular mobility. Second, glycerol was used to obtain a constant T_g, while the water content was varied. Above the T_g, deamidation rates were independent of water content. Below the T_g, deamidation rates increased with increasing water content, suggesting a solvent effect. An increase in water content also shifted the ratio of degradation products, where less cyclic imide and more hydrolytic (Asp, isoAsp) products were observed.

Solution viscosity is also a measure of molecular mobility. The same hexapeptide, Val-Tyr-Pro-Asn-Gly-Ala (pH 10.0), with PVP and glycerol, was utilized to explore the effect of solvent polarity and viscosity on deamidation rate (Li et al., 2000; Li et al., 2002). The rate of deamidation, in solution, decreased with increasing PVP and glycerol concentrations. Overall, the deamidation rate constants decreased with increasing viscosity (0.7–13.0 cp), but showed no effect above 13 cp (Li et al., 2000). The rate constants also increased with increasing dielectric constant and decreasing refractive index. The pseudo first order rate constants were correlated with viscosity, dielectric constant and refractive index in a multi-dimensional free energy model (Li et al., 2000). The data suggested that a) a permanent dipole effect strongly stabilizes the transition state, as indicated by the dielectric constant, b) an induced dipole effect weakly stabilizes the reactant state, as indicated by the refractive index, and c) dynamic or frictional effects were of minimal importance, as indicated by the viscosity (Li et al., 2000).

Further characterization of solvent viscosity, at higher viscosities, in both solution and in hydrated solids was also explored within the hexapeptide Val-Tyr-Pro-Asn-Gly-Ala (pH 10.0) (Li et al., 2002). The effect of viscosity on the deamidation rate was determined in varying ratios of PVP and water. Glycerol was not used in these studies because of the added dielectric constant variable. The rate of deamidation decreased with increasing PVP content (0.1–0.8 w/w) or increasing PVP molecular weight (grades 17PF, 30 and 90F), both correlating with increasing solvent viscosity. The results suggested that the motion of the Asn side chain is a function of the macroscopic solvent viscosity (Li et al., 2002).

Finally, it was noted that during the deamidation studies in PVP and glycerol, mass balance was not observed (D'Souza et al., 2003). Samples containing Val-Tyr-Pro-Asn-Gly-Ala (pH 7.5) with PVP and glycerol were degraded until a plateau was reached at 30% peptide remaining. A PVP-peptide adduct was isolated, in which the pyrrolidone carbonyl of PVP and the N-terminus of the hexapeptide reacted to form a covalent adduct (D'Souza et al., 2003). Reactions between primary protein amines and carbonyls of reducing sugars are often referred to as the Maillard reaction. Ironically, moisture content is often reduced in formulations to increase stability, however these conditions may promote secondary reactions. These secondary reactions become more prevalent due to competition between water and peptide amine groups for carbonyls, so that the reaction is favored at low water content, and reduction in the dielectric constant of the reaction medium as water is removed, causing the peptide amine nitrogen to become more nucleophilic and thus more reactive.

Biological Activity

An important parameter for maintaining protein activity under non-aqueous conditions is the "pH memory" of the protein. The half-life of subtilisin in DMF was directly related to the pH of the aqueous solution from which the protein was lyophilized, where increased activity was observed as the pH was increased from 6.0 to 7.9 (Schulze and Klibanov, 1991). The "pH memory" of the protein from its last aqueous state can be attributed to the ionization state of the functional groups upon lyophilization, precipitation or other isolation methodology (Zaks and Klibanov, 1988a; Constantino et al., 1991; Guinn et al., 1991; Klibanov, 1997). Therefore, ionization of the protein, as well as solvent effects, must be considered.

In general, enzyme activity and thermostability increase as solvent polarity and hydrophobicity decrease. Enzymes in solvents with high molecular weight and low Hildebrand solubility factors (δ) or Log P result in higher activity (Hahn-Hagerdal, 1986; Adercruetz and Mattiasson, 1987; Laane et al., 1987; Cowan, 1997). When Log P < 2.0 enzymes are usually inactive, when Log P $= 2.0$–4.0 enzymes exhibit variable activity and when Log P > 4.0 enzymes exhibit good activity (Laane et al., 1987; Cowan, 1997). For example, *S. sufactaricus* malic enzyme activity in a series of alcohols indicates that activity is directly related to the hydrophobicity of the solvent (Table 5) (Guagliardi et al., 1989). A correlation has also been noted between proteins less susceptible to denaturation in organics and proteins with high thermal ($\sim$100°C) stability and resistance to thermal inactivation (Zaks and Klibanov, 1984; Adercruetz and Mattiasson, 1987; Cowan, 1997). Ribonuclease, chymotrypsin and lysozyme demonstrated increased thermostability when solubilized in non-aqueous solvents, as compared with water (Zaks and Klibanov, 1984; Volkin et al., 1991).

Similarly, enzyme activity has been correlated to solvent polarity. Oxidation of cinnamyl alcohol by horse liver alcohol dehydrogenase (LADH) (pH 7.5) was observed in anhydrous hexane, methylene chloride and acetonitrile (Guinn et al., 1991). The oxidation rates were observed to increase as the dielectric constant decreased (Table 5). Electron paramagnetic spectroscopy (EPR) and two active site directed spin labels were used to examine the effect of solvent dielectric on structural stability. As the dielectric constant of the solvent decreased, the spectra broadened, indicative of an increase in rigidity or stability.

Molecule	Solvent	Activity	Reference
α-Chymotrypsin	DMSO	14%	(Ko and Dalvit, 1992)
LADH	Hexane	11 nmoles/min g	(Guinn et al., 1991)
	Methylene Chloride	4.2 nmoles/min g	(Guinn et al., 1991)
	Acetonitrile	0 nmoles/min g	(Guinn et al., 1991)
Lysozyme	DMSO	88–100%	(Ko and Dalvit, 1992)
Pancreatic	Butanol	60 mmoles/min mg	(Zaks and Klibanov, 1984)
Lipase	Heptanol	75 mmoles/min mg	(Zaks and Klibanov, 1984)
	Decanol	80 mmoles/min mg	(Zaks and Klibanov, 1984)
Subtilisin	DMF (from pH 6.0)	$t_{1/2} = 48$ min at 45°C	(Schulze and Klibanov, 1991)
	DMF (from pH 7.9)	$t_{1/2} = 20$ hr at 45°C	(Schulze and Klibanov, 1991)
	Water (pH 7.8)	84 mmoles/hr mg	(Xu et al., 1997)
	Glycerol	5.4 mmoles/hr mg	(Xu et al., 1997)
	TFE	0.004 mmoles/hr mg	(Xu et al., 1997)
	90% DMSO	0.007 mmoles/hr mg	(Xu et al., 1997)
S. Sulfataricus	50% DMF	100%	(Guagliardi et al., 1989)
malic enzyme	15% EtOH	85%	(Guagliardi et al., 1989)
	50% EtOH	40%	(Guagliardi et al., 1989)
	50% MeOH	85%	(Guagliardi et al., 1989)
Trypsin	DMSO	73–88%	(Ko and Dalvit, 1992)

Table 5. Activity of proteins in organic and co-solvent conditions.

Solvent effects on activity must also be balanced with the requirement for some residual water to maintain activity. Residual water comprises a tightly bound solvation layer around the protein when suspending or solubilizing protein in organic media (Timasheff and Inoue, 1968; Zaks and Klibanov, 1988b; Khmelnitsky et al., 1991). Residual water content has also been correlated to increased solubility, structural stability and enzyme activity (Zaks and Klibanov, 1988a; Zaks and Klibanov, 1988b; Guinn et al., 1991; Huang et al., 1995; Cowan, 1997). Addition of water to LADH resulted in increased reaction rates, probably by modifying the electrostatic environment around the active site (Guinn et al., 1991). However, excess water beyond the solvation layer may result in a loss of activity and thermostability. For example, the half-life of porcine lipase decreased as the water content in tributyrin increased (Zaks and Klibanov, 1984).

Spectroscopic methods can be utilized to correlate changes in secondary structure with reduction in activity. Subtilisin (pH 7.8) solubilized in co-solvent solutions of glycerol, TFE or DMSO was characterized by FTIR (Griebenow and Klibanov, 1997; Xu et al., 1997). The structural information was then compared with the activity data (Table 5). Solutions in glycerol or water exhibited similar ratios of α-helix and β-sheet content, both correlating with good activity. TFE demonstrated an increase in α-helicity, and DMSO showed a decrease in both α-helix and β-sheet content, where both DMSO and TFE solutions showed little activity.

It is important to discriminate between denaturation and loss of activity, since loss of activity can occur without appreciable change in the tertiary structure (Cowan, 1997). Solvation in non-aqueous conditions can cause unfolding and denaturation, as previously discussed. Conversely, removal of bound water can leave the protein structurally intact, but without the critical water molecules in the active site that are required for activity (Cowan, 1997). Some water is necessary for the catalytic function of proteins and enzymes, but usually enzymes only require a monolayer of bound water for activity (Zaks and Klibanov, 1988a; Zaks and Klibanov, 1988b; Klibanov, 1997).

In some cases, proteins refold upon rehydration, or maintain activity in the organic solvent. Examples of the former include proteins that refold upon rehydration. Chymotrypsin (pH 7.8), dissolved in DMSO, showed dramatic unfolding and loss of activity (2). This was caused by the partitioning of bound water from the protein into the solvent, disrupting the hydration shell (Zaks and Klibanov, 1988a; Khmelnitsky et al., 1991). Upon dilution in water, chymotrypsin refolded and regained full activity (Zaks and Klibanov, 1988a). Subtilisin (pH 7.8) in tert-amyl alcohol or acetonitrile showed little activity, where the loss of bound water was correlated to enzyme activity in a time dependent manner (Schulze and Klibanov, 1991). Again, once the subtilisin was rehydrated from DMF, tert-amyl alcohol or acetonitrile, full activity was restored. Examples of the latter include full catalytic activity in organic conditions for yeast alcohol oxidase, mushroom polyphenol oxidase and LADH (Zaks and Klibanov, 1988b).

One theory suggests that the hydrophobic solvents enhance water packing into a solvation layer, while hydrophilic solvents solubilize water more equally through the solution (Zaks and Klibanov, 1988a; Zaks and Klibanov, 1988b). Hydrophobic solvents retard the bound water from partitioning away from the protein, preserving activity as previously discussed (preferential hydration), while hydrophilic solvents may more effectively strip the water molecules from the surface (Dordick, 1992; Gorman and Dordick, 1992; Reiersen and Rees, 2001; Yang et al., 2004). Co-solvent solutions of TFE were used to stabilize Type 2 β-turn structure in a short elastin peptide, Gly-Val-Gly-(Val-Pro-Gly-Val-Gly)$_3$. A model was proposed in which TFE molecules cluster to assist secondary structure formation by breaking down the interfacial water molecules, driven by an increase in entropy of the bulk water. Other studies have looked at the exchange of tritiated water bound to chymotrypsin, subtilisin and horseradish peroxidase. The exchange of T_20 from the proteins suspended in organic solvents (MeOH, DMF, THF, hexane) showed an increasing loss of bound T_20 with increasing solvent polarity (Gorman and Dordick, 1992). Solvent polarity, as compared with solvent hydrophobicity, was determined to provide the more meaningful correlation because water binds to polar and charged residues through mainly electrostatic forces. Increased solvent polarity weakens these electrostatic forces and enables water to be stripped from the protein.

Finally, researchers have also noted that the selection of non-aqueous solvents can affect enzyme activity in more subtle ways. Enzymes that bind with a dual pocket binding site (secondary alcohol dehydrogenases) possess enantiomeric conformations capable of binding (Cowan, 1997). It was observed that increasing the dielectric constant increased protein flexibility, which increased binding and catalysis. These conditions may also allow increased binding of the usually less favored enantiomer.

Overall, protein activity is affected by four major factors: a) the ground state, thermodynamic stabilization of the protein, b) active site flexibility response to solvent polarity and adsorbed water content, c) the impact of water stripping from the protein on the active conformation and d) direct solvent induced perturbation of the protein (Dordick, 1992).

Pharmaceutical Processing

Organic solvents are often used in the manufacture, purification, precipitation and crystallization of protein pharmaceuticals (Sukumar et al., 2005). Water miscible solvents such as ethanol, acetonitrile and propanol have been utilized to promote the stabilization of reversible denaturation and aggregation, with recovery of the native conformation. For example, insulin has been precipitated and/or crystallized in a wide variety of solvents (Brange and Lankjaer, 1993).

Solid state forms of macromolecules are a common method for preparing protein pharmaceuticals, due to stability advantages. Lyophilization is often utilized for proteins and peptides because it limits the moisture content and increases protein stability. Proteins are routinely lyophilized from aqueous solutions, but lyophilization from co-solvent conditions have also been documented (Santos et al., 2001).

Spray drying has become a more common drying method for macromolecules, again usually from an aqueous solution. However, cyclosporine has been successfully spray dried from a 95% ethanol solution, due to low solubility under aqueous conditions (Ko and Dalvit, 1992; Ran et al., 2001; Lechuga-Ballesteros et al., 2003; Stevenson et al., 2003).

A third macromolecular particle generation technique exposes the organic solvent protein solution during supercritical fluid aerosolization (Winters et al., 1997; Palakodaty and York, 1999; Sellers et al., 2001; Snavely et al., 2002; Bustami et al., 2003; Jovanovic et al., 2004). During the gas anti-solvent process, a dense gas (supercritical CO_2) is used as an anti-solvent to expand an organic solvent protein solution. As the protein solution expands, the dissolution power of the solvent decreases, precipitating the protein. The literature can be sub-divided into three modes of processing. The first mode is supercritical anti-solvent (SAS), and involves the addition of anti-solvent to the organic solution, containing protein, until precipitation occurs (Yeo et al., 1993; Yeo et al., 1994). The second mode is an aerosol solvent extraction system (ASES), and involves introducing aqueous or organic solutions through a capillary nozzle into the dense gas stream (Winters et al., 1997; Snavely et al., 2002; Bustami et al., 2003). The third mode uses a solution enhanced dispersion by supercritical fluid (SEDS), and involves an aqueous protein solution delivered through a coaxial nozzle (Palakodaty and York, 1999).

Supercritical gas anti-solvent (SAS) expansion of a DMSO solution of insulin with supercritical CO_2 produced precipitation of insulin with no difference in biological activity (Yeo et al., 1993; Winters et al., 1996). Infrared spectroscopy characterization of insulin, lysozyme and trypsin powders yielded partially unfolded intermediates (Yeo et al., 1994; Winters et al., 1997). Upon reconstitution of the powders in water, the structures were reversible, with recovery of biological activity (Chang et al., 1991; Winters et al., 1996). The powders were also placed on stability at −25°C to 60°C, where lysozyme showed no structural changes, by Raman spectroscopy, regardless of temperature (Winters et al., 1997). Trypsin and insulin showed structural unfolding at t=0, however this did not accelerate structural denaturation during storage. CD and bioactivity assays also confirmed that the partially unfolded intermediates did not hinder renaturation upon dissolution after storage. In most cases, the protein precipitate was amorphous; however, when catalase was precipitated from 90% ethanol, the particulate was crystalline (Debenedetti et al., 1993).

An alternate application of the SAS technology was to fractionally precipitate alkaline phosphatase, insulin, ribonuclease and trypsin mixtures from DMSO (Winters et al., 1999). Separation and purification of the proteins were achieved by their differing solubility in CO_2 expanded DMSO. Substantial biological activity was recovered for insulin, lysozyme, ribonuclease and trypsin; however alkaline phosphatase was irreversibly denatured.

Utilizing the ASES method, insulin solubilized in DMSO was precipitated from a solution of 1,1,1,3,3,3,-hexafluoro-2-propanol (HFIP) using supercritical CO_2 (Snavely et al., 2002). Insulin purity decreased from 99.1% to 97.9% and high molecular weight protein (HMWP) increased from 0.10% to 0.65%. No decrease in biological activity was observed, and CD and infrared spectra were essentially unchanged.

Protein encapsulation in polymers, resulting in controlled release depot biodegradable formulations, is beyond the scope of this chapter; however these formulations require organic solvent processing steps, such as methylene chloride or ethyl acetate (Cleland and Jones, 1996; Schwendeman et al., 1996; Johnson et al., 1997; Schwendeman et al., 1997; Carrasquillo et al., 1999). In order to optimize these processing steps, organic solvent solubility screens were correlated with the stability of PLGA encapsulated human growth hormone (hGH) and interferon-γ (IFN-γ) (Cleland and Jones, 1996). Additional excipients were screened for their ability to stabilize these proteins in organic solvents. Recovery of hGH emulsified in methylene chloride in phosphate buffer was 53.2%. The addition of PEG 3350 increased the recovery to 95.1%. Similarly, recovery of hGH from ethyl acetate was increased with PEG 3350, mannitol or trehalose. The excipients stabilized the proteins by preferential hydration and water substitution by providing a protective layer around the protein surface.

Similar results with trehalose were obtained in hGH dried biodegradeable polymer films (Carrasquillo et al., 1999). The structural integrity of hGH was monitored by FTIR. It was determined that the structure was significantly changed during the lyophilization step, but that prior suspension of the lyophilized powder in methylene chloride only introduced minor structural changes. The rationale, again, was that suspension of proteins in organic solvents limits the molecular mobility of the protein (Zaks and Klibanov, 1984; Zaks and Klibanov, 1988a; Volkin et al., 1991; Carrasquillo et al., 1999).

Finally, analysis of proteins during processing often utilizes organic solvents, and conformational changes become an additional retention mechanism when separating and eluting proteins using reversed phase chromatography (Katzenstein et al., 1986).

Formulation Applications

Pharmaceutically viable formulation applications include topical, pulmonary, depot and implantable drug delivery systems (Bhardwaj and Blanchard, 1996; Cleland and Jones, 1996; Schwendeman et al., 1996; Johnson et al., 1997; Schwendeman et al., 1997; Carrasquillo et al., 1999; Stevenson et al., 1999; Wright et al., 2001; Kikwai et al., 2004). Simple depot formulations, designed to decrease dissolution rates, have used non-aqueous conditions to achieve controlled release. Growth hormone has also been suspended in oil for depot injections (Yu et al., 1996).

Melanotan I is a tridecapeptide analog of α-MSH with melanotropic activity (Bhardwaj and Blanchard, 1996). It has been evaluated in clinical trials for its chemopreventative activity for sunlight-induced skin cancers (Bhardwaj and Blanchard, 1996). Melatonin I (pH 7.4) was formulated as a controlled release gel in Polaxamer 407 (P407), or combinations of P407 and PVP, methyl cellulose or hydroxymethyl cellulose.

Leuprolide, used for the palliative treatment of prostate cancer, was formulated at 370 mg/ml in DMSO within an osmotically-driven implantable pump lasting 1 year, where the choice of DMSO provided 2 year room temperature shelf life, plus 1 year of implant life at 37°C (Stevenson et al., 1999; Wright et al., 2001).

Cyclosporine has also been successfully solubilized in 10% EtOH and hydroflouroalkanes for delivery from pulmonary inhalers (Myrdal et al., 2004).

Finally, an anti-inflammatory peptide, Spantide II, blocks the inflammation associated with Substance P (Kikwai et al., 2004). The peptide was characterized in a topical formulation for the treatment of dermal inflammatory disorders (Kikwai et al., 2004). Preformulation studies indicated that Spantide II undergoes Lys-Pro diketopiperazine degradation. Formulation of Spantide II in PG, ethanol, N-methyl-2-pyrrolidone, ethanol/water and ethanol/ethyl oleate resulted in increased stability, where the Log K at 40°C were −0.93, −1.70, −1.81, −1.94 and −2.30, respectively.

Industrial Applications

Other industries have also utilized non-aqueous conditions for their biotechnological application, and this literature base can be applied to pharmaceutical applications. For example, industrial enzymes have been used for detergents, textiles, food production, dairy, animal feed, leather, bioremediation, paper, cosmetics and diagnostics (Matsuura et al., 1993; Dai and Klibanov, 1999; Laboret and Perraud, 1999; Stevenson, 1999; Loughlan and Hawkes, 2000; Gonzales-Navarro et al., 2001; Schmid et al., 2001; Secundo and Carrea, 2003; Kuntz et al., 2003; Sardessai and Bhosle, 2003; Gupta et al., 2004; Houde et al., 2004; Sardessai and Bhosle, 2004; Spreti et al., 2004).

Lipases have been utilized to hydrolyze triglycerides into diglycerides, monoglycerides, fatty acids and glycerol (Houde et al., 2004). Lipases are a subset of esterases. The detergent industry utilizes lipases to remove fat containing stains and formulates them at pH extremes in the presence of surfactants (Houde et al., 2004). The active site contains an Ser-His-Asp(Glu) catalytic triad that is shielded from the solvent by a flap (Gonzales-Navarro et al., 2001; Houde et al., 2004;). In an aqueous environment, the flap is closed. When bound to a lipid, or other interface, the flap is open and the active size is exposed to the solvent (Gonzales-Navarro et al., 2001). Researchers have utilized the concept of "molecular memory" to kinetically trap (freeze-drying) the enzyme conformation in its activated form and subsequently utilizing it in an organic media (Gonzales-Navarro et al., 2001). The benefits of utilizing an organic solvent include stabilizing the transition state and minimizing side reactions.

In addition to hydrolyzing carboxylic ester bonds, lipases can catalyze a variety of esterification reactions in non-aqueous media. Industrial applications of lipases have focused on their enantioselectivity, regioselectively and substrate specificity (Houde et al., 2004). For example, researchers have utilized acetone, ethanol and methanol to enhance lipase activity (Gupta et al., 2004). In the food industry, esterification of citronellol and geroniol with lipase was accomplished in n-hexane, where a series of solvents (correlated to Log P) and water content were also characterized (Laboret and Perraud, 1999).

An application of lipases in the pharmaceutical industry can be found in the manufacture of Ibuprofen (Houde et al., 2004). Ibuprofen is a racemic mixture, where the (S)-ibuprofen molecules is 160 times more potent that the (R)-ibuprofen one. Resolution of racemic ibuprofen can be achieved by esterification of (S)-ibuprofen with methanol or butanol, in organic media using lipase, leading to synthesis of the corresponding (S)-ester. This ester can be separated from (R)-ibuprofen and chemical transformed to (S)-ibuprofen.

Steroids are also of importance in the pharmaceutical industry. However, manufacture is challenged by poor solubility and transformation rates under aqueous conditions (Gupta et al., 2004). These challenges can be alleviated by solubilizing the steroid in organic solvent and using Organic Solvent Tolerant Bacteria (OSTB) for catalysis and biotransformations. OSTB are extremophilic microorganisms, isolated from coastal sediment, capable of thriving in organic solvents, specifically chloroform and butanol (Sardessai and Bhosle, 2003; Sardessai and Bhosle, 2004).

Similarly, chloroperoxidase has proven to be a versatile peroxidase enzyme for synthetic applications (Dai and Klibanov, 1999; Loughlan and Hawkes, 2000; Spreti et al., 2004). The industrial usage of chloroperoxidases under aqueous conditions have been hampered by poor stability (inactivation via peroxide) of the enzyme and poor solubility of the substrate (Spreti et al., 2004). However, suspension of the peroxidase in PEG 200 increased product yields (Dai and Klibanov, 1999; Spreti et al., 2004). Other researchers have determined the effect of DMSO, DMF, MeOH and acetonitrile on the chlorination activity of the enzyme (pH 2.8), where Log P < 0 was preferred (Loughlan and Hawkes, 2000).

Hydrolases have also been used in synthetic applications, primarily for the preparation of chiral building blocks (Secundo and Carrea, 2003). The usage of organic conditions allowed the solubilization of the substrate, increased thermal stability and the prevention of water dependent side reactions.

Conclusion

An overview of the utilization of non-aqueous solvents in macromolecular applications has been presented. A variety of organic solvents have been used to characterize and understand the parameters necessary for their applications to biotechnology. Although no single parameter is predictive of a final utility, the critical parameters that continue to surface in these studies are the dielectric constant, hydrophobicity, dipole moment, viscosity and solubility factor of the solvent. Furthermore, the pH memory, molecular memory and water content of the solvent play important roles in stabilizing these molecules. Not all of these studies are applicable for use in animals and pharmaceutically acceptable formulations in humans; however, many of the concepts translate well.

The advent of biotechnology, bioengineering and pharmaceutical delivery systems has increased the requirements for solubility and stability of macromolecules under a variety of versatile and unique conditions; and subsequently the use of non-aqueous solvents. Specific applications include a) isolation, purification, precipitation and crystallization of biopharmaceuticals, b) processing methods such as spray drying and microencapsulation, and c) formulation of proteins for delivery systems requiring high concentrations and prolonged stability, such as implants and depots.

List of Abbreviations

ASES....Aerosol Solvent Extraction System
BPTI....Bovine Pancreatic Trypsin Inhibitor
BSA....Bovine Serum Albumin
CD....Circular Dichroism
δ....Hildebrand Solubility Factor
DMF....Dimethyl Formamide
DMSO....Dimethyl Sulfoxide
DPPG....Dipalmitoylphosphatidylglycerol
DSC....Differential Scanning Calorimetry
ε....Dielectric Constant
EPR....Electron Paramagnetic Spectroscopy
EtOH....Ethanol
FTIR....Fourier Transform Infrared
GRF....Growth Hormone Releasing Factor
HFIP....1,1,1,3,3,3-hexafluoro-2-propanol
hIGF....Human Insulin Like Growth Factor
hGH....Human Growth Hormone
HMWP....High Molecular Weight Protein
INF....Interferon
LADH....Liver Alcohol Dehydrogenase
LC/MS/MS....Liquid Chromatography/Mass Spectrometry/Mass Spectrometry
LMW-UK....Low Molecular Weight Urokinase
Log D....Distribution coefficient
Log P....Partition Coefficient
μ....Dipole Moment
MeOH....Methanol
MSH....Melanocyte Stimulating Hormone
NMR....Nuclear Magnetic Resonance
OSTB....Organic Solvent Tolerant Bacteria
PEG....Polyethylene Glycol
PG....Propylene Glycol
pI....Isoelectric Point

PLGA........Poly-lactic-coglycolic acid
pKa........Dissociation Constant
PVA........Poly(vinyl alcohol)
PVP........Poly(vinyl pyrrolidone)
RP-HPLC........Reversed Phase High Performance Liquid Chromatrography
SAS........Supercritical Anti-solvent
sCT........Salmon Calcitonin
SDS........Sodium Dodecyl Sulfate
SEC........Size Exclusion Chromatography
SEDS........Solution Enhanced Dispersion by Supercritical Fluid
TFE........Trifluoroethanol
THF........Tetrahydrofuran
T_g........Glass Transition Temperature
T_m........Melting Temperature

Key Words

protein, peptide, non-aqueous, organic solvent

References

Adercruetz P and Mattiasson B. Aspects of Biocatalyst Stability in Organic Solvents. *Biocatalysis* 1987; 1: 99–108.

Affleck R, Xu ZF, Suzawa V, Focht K, Clark DS, and Dordick JS. Enzymatic Catalysis and Dynamics in Low-water Environments. *Proc Natl Acad Sci USA* 1992; 89: 1100–1104.

Alexandrescu AT, Ng YL, and Dobson CM. Characterization of a Trifluoroethanol Induced Partially Folded State of α-lactalbumin. *J Mol Biol* 1994; 235: 587–599.

Almarsson O and Klibanov AM. Remarkable Activation of Enzymes in Nonaqueous Media by Denaturing Organic Cosolvents. *Biotechnology and Bioengineering* 1996; 49: 87–92.

Alonso DOV and Daggett B. Molecular Dynamics Simulations of Protein Unfolding and Limited Refolding: Characterization of Partially Unfolded States of Ubiquitin in 60% Methanol and in Water. *J Mol Biol* 1995; 247: 501–520.

Anderegg RJ, Wagner DS, Stevenson CL, and Borchardt RT. The Mass Spectrometry of Helical Unfolding in Peptides. *J Am Soc Mass Spectrom* 1994; 5: 425–433.

Arakawa T and Goddette D. The Mechanism of Helical Transition of Proteins by Organic Solvents. *Arch Biochem Biophys* 1985; 240: 21–32.

Arakawa T, Kita Y, and Carpenter JF. Protein Solvent Interaction in Pharmaceutical Formulations. *Pharm Res* 1991; 8: 285–291.

Arvinte T and Drake AF. Comparative Study of Human and Salmon Calcitonin Secondary Structure in Solutions with Low Dielectric Constants. *J Biol Chem* 1993; 268: 6408–6414.

Barteri M, Guadiano MC, and Santucci R. Influence of Glycerol on the Structure and Stability of Ferric Horse Heart Myoglobin: a SAXS and Circular Dichroism Study. *Biochim Biophys Acta* 1996; 1295: 51–58.

Barton AFM. *Handbook of Solubility Parameters and Other Cohesion Parameters.* CRC Press, Ann Arbor, MI, 1991.

Bazzo R, Tappin MJ, Pastore A, Harvey TS, Carver JA, and Campbell ID. The Structure of Mellitin: A ^{1}H-NMR Study in Methanol. *Eur J Biochem* 1988; 173: 139–146.

Bhardwaj R and Blanchard J. Controlled-release Delivery System for the α-MSH Analog Melanotan-I using Poloxamer 407. *J Pharm Sci* 1996; 85: 915–919.

Bhattacharjya S and Balaram P. Effects of Organic Solvents on Protein Structures: Observations of a Structured Helical Core in Hen Egg-white Lysozyme in Dimethyl Sulfoxide. *Proteins: Structure, Function and Genetics* 1997; 29: 492–507.

Boctor AM and Mehta SC. Enhancement of the Stability of Thrombin by Polyols: Microcalorimetric Studies. *J Pharm Pharmacol* 1992; 44: 600–603.

Brange J and Langkjaer L. Insulin Structure and Stability. In Stability and Characterization of Protein and Peptide Drugs: Case Histories. Y. Wang, R. Pearlman, Eds. New York, NY: Plenum Press; 1993. 315–350.

Brennan TV and Clarke S. Spontaneous Degradation of Polypeptides at Aspartyl and Asparaginyl Residues: Effects of the Solvent Dielectric. *Protein Science* 1993; 2:331–338.

Bromberg LE and Klibanov AM. Transport of Proteins Dissolved in Organic Solvents Across Biomimetic Membranes. *Proc Natl Acad Sci USA*, 1995; 92:1262–1266.

Broos J, Bisser AJW, Engbersen JFJ, Verboom W, van Hoek A, and Reinhoudt DN. Flexibility of Enzymes Suspended in Organic Solvents Probed by Time-Resolved Fluorescence Anisotropy. Evidence that Enzyme Activity and Enantioselectivity are Directly Related to Enzyme Flexibility. *J Am Chem Soc* 1995; 117:12657–12663.

Buchwald P and Bodor N. Octanol-water Partition of Nonzwitterionic Peptides: Predictive Power of a Molecular Size-based Model. *Proteins: Structure, Function and Genetics* 1998; 30:86–99.

Burke PA, Griffin RG, and Klibanov AM. Solid-state Nuclear magnetic Resonance Investigation of Solvent Dependence of Tyrosyl Ring Motion in an Enzyme. *Biotechnology and Bioengineering* 1993; 42:87–94.

Bustami RT, Chan HK, Dehghani F, and Foster NR. Generation of Micro-particles of Proteins for Aerosol Delivery using High Pressure Modified Carbon Dioxide. *Pharmaceutical Research* 2000; 17:1360–1366.

Bustami RT, Chan HK, Sweeney T, Dehghani F, and Foster NR. Generation of Fine Powders of Recombinant Human Deoxyribonuclease using the Aerosol Solvent Extraction System. *Pharmaceutical Research* 2003; 20:2028–2035.

Butler LG. Enzymes in Non-aqueous Solvents. *Enzyme Microb Technol* 1979; 1:253–259.

Capasso S, Mazzarella L, and Zagari A. Deamidation via Cyclic Imide of Asparaginyl Peptide: Dependence of Salts, Buffers and Organic Solvents. *Pept Res* 1991; 4:234–238.

Carrasquillo KG, Costantino HR, Cordero RA, Hsu CC, and Griebenow K. On the Structural Preservation of Recombinant Human Growth Hormone in a Dried Film of a Synthetic Biodegradable Polymer. *J Pharm Sci* 1999; 88: 166–1173.

Chang N, Hen SJ, and Klibanov AM. Protein Separation and Purification in Neat Dimethyl Sulfoxide. *Biochem Biophys Res Commun* 1991; 176: 1462–1468.

Chen K and Arnold FH. Enzyme Engineering for Nonaqueous Solvents: Random Mutagenesis to Enhance Activity of Subtilisin E in Polar Organic Media. *Biotechnology* 1991; 9: 1073–1077.

Chin JT, Wheeler SL, and Klibanov AM. Protein Solubility in Organic Solvents. *Biotechnol Bioeng* 1994; 44: 140–145.

Cleland JL and Jones AJS. Stable Formulations of Recombinant Human Growth Hormone and Interferon-γ for Microencapsulation in Biodegradable Microspheres. *Pharm Res* 1996; 13: 1464–1475.

Constantino HR, Griebenow K, Langer R, and Klibanov AM. On the pH Memory of Lyophilized Compounds Containing Protein Functional Groups. *Biotechnol Bioeng* 1991; 53: 345–348.

Cowan DA. Thermophilic Proteins: Stability and Function in Aqueous and Organic Solvents. *Comp Biochem Physiol* 1997; 118: 429–438.

Dai L and Klibanov AM. Striking Activation of Oxidase Enzymes Suspended in Nonaqueous Media. *Proceedings-National Academy of Sciences USA Physical Sciences* 1999; 96: 9475–9478.

Debenedetti PG, Lim GB, and Prud'homme RK. Formation of Protein Microparticles by Antisolvent Precipitation. *EP Patent* 1993: No 542314.

Desai UR and Klibanov AM. Assessing the Structural Integrity of a Lyophilized Protein in Organic Solvents. *J Am Chem Soc* 1995; 117: 3940–3945.

Dill KA. Dominant Forces in Protein Folding. *Biochemistry* 1990; 29: 7133–7155.

Dordick JS. Designing Enzymes for use in Organic Solvents. *Biotechnology Process* 1992; 8:259–267.

D'Souza AJM, Schowen RL, Borchardt RT, Salsbury JS, Munson ER, and Topp EM. Reaction of a Peptide with Polyvinylpyrrolidone in the Solid State. *J Pharm Sci* 2003; 92: 585–593.

Fan P, Bracken C, and Baum J. Structural Characterization of Monellin in the Alcohol-denatured State by NMR: Evidence for β-sheet to α-helix Conversion. *Biochemistry* 1993; 32: 1573–1582.

Fransson J, Hallen D, and Florin-Robertsson E. Solvent Effects on the Solubility and Physical Stability of Human Insulin Like Growth Factor I. *Pharm Res* 1997; 14: 606–612.

Frant MS. How to Measure pH in Mixed and Nonaqueous Solutions. *Today's Chemist at Work* 1995; June:39–42.

Gonzales-Navarro H, Bano MC, and Abad C. The Closed-Open Model for Lipase Activation. Addressing Intermediate Active Forms of Fungal Enzymes by Trapping of Conformers in Water-Restricted Environments. *Biochemistry* 2001; 40: 3174–3183.

Gorman LAS and Dordick JS. Organic Solvents Strip Water off Enzymes. *Biotechnol Bioeng* 1992; 39: 392–397.

Griebenow K and Klibanov AM. On Protein Denaturation in Aqueous-organic Mixtures by not Pure Organic Solvents. *J Am Chem Soc* 1996; 118: 11698–11700.

Griebenow K and Klibanov AM. Can Conformational Changes be Responsible for Solvent and Excipient Effects on the Catalytic Behavior of Subtilisin Carlsberg in Organic Solvents? *Biotechnology and Bioengineering* 1997; 53: 351–362.

Grossfield A, Ren P, and Ponder JW. Ion Solvation Thermodynamics for Simulation with a Polarizable Force Field. *J Am Chem Soc* 2003; 125: 15671–15682.

Guagliardi A, Rossi M, and Bartolucci S. Enzyme-catalyzed Reactions in Organic Solvents. *Chimicaoggi* 1989; May, 31–36.

Guinn RM, Skerker PS, Kavanaugh P, and Clark DS. Activity and Flexibility of Alcohol Dehydrogenase in Organic Solvents. *Biotechnol Bioeng* 1991; 37: 303–308.

Gupta R, Gupta N, and Rathi P. Bacterial Lipases: An Overview of Production, Purification and Biochemical Properties. *Appl Microbiol Biotechnol* 2004; 64: 763–781.

Hahn-Hagerdal B. Water Activity: A Possible External Regulatory in Biotechnical Processes. *Enzyme Microb Technol* 1986; 8: 321–384.

Hall SC, Tan MM, Leonard JJ, and Stevenson CL. Characterization and Comparison of Leuprolide Degradation Profiles with Water and Dimethyl Sulfoxide. *J Peptide Res* 1999; 53: 432–441.

Halling PJ. High-affinity Binding of Water by Proteins is Similar in Air and in Organic Solvents. *Biochim Biophys Acta* 1990; 1040: 225–228.

Harding MM, Williams DH, and Woolfson DN. Characterization of a Partially Denatured State of a Protein by Two-dimensional NMR: Reduction of the Hydrophobic Interactions in Ubiquitin. *Biochemistry* 1991; 30: 3120–3128.

Houde A, Kademi A, and Leblanc D. Lipases and their Industrial Applications. *Appl Biochem Biotechnol* 2004; 118: 155–170.

Houen PG. The Solubility of Proteins in Organic Solvents. *Acta Chem Scand* 1996; 50: 68–70.

Hovorka SW, Hong J, Cleland JL, and Schoneich C. Metal-Catalyzed Oxidation of Human Growth Hormone: Modulation by Solvent-Induced Changes of Protein Conformation. *J Pharm Sci* 2001; 90: 58–69.

Huang P, Dong A, and Caughey WS. Effects of Dimethyl Sulfoxide, Glycerol and Ethylene Glycol on Secondary Structures of Cytochrome C and Lysozyme as Observed by Infrared Spectroscopy. *J Pharm Sci* 1995; 84: 387–392.

Jackson M and Mantsch HH. Beware of Proteins in DMSO. *Biochem Biophys Acta* 1991; 1078: 231–235.

Jackson M and Mantsch HH. Halogenated Alcohols as Solvents for Proteins: FTIR Spectroscopic Studies. *Biochim Biophys Acta* 1992a; 1118: 139–143.

Jackson M and Mantsch HH. Conformation of the Peptide Hormone Somatostatin in Aqueous Solution and Organic Solvents. *Vibrational Spectroscopy* 1992b; 3: 323–326.

Johnson OL, Jarowicz W, Cleland JL, Bailey L, Charnis M, Duenas E, Wu C, Shepard D, Magil S, Last T, Jones AJS, and Putney SD. The Stabilization and Encapsulation of Human Growth Hormone into Biodegradable Microspheres. *Pharm Res* 1997; 14: 730–735.

Jovanovic N, Bouchard A, Hofland GW, Witkamp GJ, Crommelin DJA, and Jiskoot W. Stabilization of Proteins in Dry Powder Formulations Using Supercritical Fluid Technology. *Pharm Res* 2004; 21: 1955–1969.

Kamphius IG, Kalk KH, Swartz BA, and Drenth J. Structure of Papain Refined at 1.65 Å Resolution. *J Mol Biol* 1984; 179: 233–256.

Karle IL, Flippen-Anderson JJ, Uma K, and Balaram P. Unfolding of an α-helix in Peptide Crystal by Solvation: Conformational Fragility in a Heptapeptide. *Biopolymers* 1993; 33: 827–837.

Katzenstein GE, Vrona SA, Wechler RJ, Steadman BL, Lewis RV, and Middaugh CR. Role of Conformational Changes in the Elution of Proteins from Reversed-phase HPLC Columns. *Proc Natl Acad Sci USA* 1986; 83: 4268–4272.

Kessler H, Loosli HR, and Oschkinat H. Assignment of the $^{1}H^{13}C^{15}N$-NMR Spectra of Cyclosporin A in $CDCl_3$ and C_6D_6 by a Combination of Homo and Heteronuclear Two-dimensional Techniques. *Helv Chim Acta* 1985; 68: 661–681.

Khmelnitsky YL, Belova AB, Levashov AV, and Mozhaev VV. Relationship between Surface Hyrophobicity of a Protein and its Stability agains Denaturation by Organic Solvents. *FEBS* 1991; 284: 267–269.

Kikwai L, Babu RJ, Kanikkannan N, and Singh M. Preformulation Stability of Spantide II, a Promising Topical Anti-inflammatory Agent for the Treatment of Psoriasis and Contact Dermatitis. *J Pharm Pharmacol* 2004; 56: 19–25.

Kim J and Dordick JS. Unusual Salt and Solvent Dependence of a Protease from an Extreme Halophile. *Biotechnol Bioeng* 1997; 55: 471–479.

Kim Y, Rose CA, Liu Y, Ozaki Y, Datta G, and Tu AT. FT-IR and Near-infrared FT-raman Studies on the Secondary Structure of Insulinotropin in the Solid State: α-helix to β-sheet Conversion Induced by Phenol and/or by High Shear Force. *J Pharm Sci* 1994; 83: 1175–1180.

Kita Y, Arakawa T, Lin TY, and Timasheff SN. Contribution of the Surface Free Energy Perturbation to Protein-solvent Interactions. *Biochemistry* 1994; 33: 15178–15189.

Klibanov AM. Enzymes that Work in Organic Solvents. *Chemtech* 1997; June, 354–359.

Kloosterman DA, Scahill TA, and Friedman AR. Secondary Structure of a Human Growth Hormone Release Factor Fragments (Leu^{27}-hGRF(15–32)NH_2) in Aqueous/SDS Micelle Environments. *Peptide Research* 1993; 6: 211–218.

Knepp VM, Muchnik A, Oldmark S, and Kalashnikova L. Stability of Nonaqueous Suspension Formulations of Plasma Derived Factor IX and Recombinant Human Alpha Interferon at Elevated Temperatures. *Pharm Res* 1998; 15: 1090–1095.

Knubovets T, Osterhout JJ, Connolly PF, and Klibanov AM. Structure, Thermostability and Conformational Flexibility of Hen Egg-white Lysozyme Dissolved in Glycerol. *Proc Natl Acad Sci USA* 1999a; 96: 1262–1267.

Knubovets T, Osterhout JJ, and Klibanov AM. Structure of Lysozyme Dissolved in Neat Organic Solvents as Assessed by NMR and CD Spectroscopies. *Biotechnol Bioeng* 1999b; 63: 242–248.

Ko SY and Dalvit D. Conformation of Cyclosporin A in Polar Solvents. *Int J Peptide Protein Res* 1992; 40: 380–382.

Kovacs H, Mark AE, Johansson J, and van Gunsteren WF. The Effect of Environment on Stability of an Integral Membrane Helix: Molecular Dynamics Simulations of Surfactant Protein C in Chloroform, Methanol and Water. *J Mol Biol* 1995; 247: 808–822.

Kropman MF and Bakker NJ. Dynamics of Water Molecules in Aqueous Solvation Shells. *Science* 2001; 291: 2118–2120.

Kuntz RL, Brown LR, Zappi ME, and French WT. Isopropanol and Acetone Induces Vinyl Chloride Degradation in Rhodoccocus Rhodochrous. *J Ind Microbiol Biotechnol* 2003; 30: 651–655.

Laane C, Boeren S, Vos K, and Veeger C. Rules for Optimized Biocatalysis in Organic Solvents. *Biotechnol Bioeng* 1987; 30: 81–87.

Laboret F and Perraud R. Lipase-catalyzed Production of Short-chain Acids Terpenyl Esters of Interest to the Food Industry. *Appl Biochem Biotechnol* 1999; 82: 185–198.

Lai MC, Hageman MJ, Schowen RL, Borchardt RT, and Topp EM. Chemical Stability of Peptides in Polymers. 1. Effect of water on Peptide Deamidation in Poly(vinyl alcohol) and Poly(vinyl pyrrolidone) Matrixes. *J Pharm Sci* 1999a; 88: 1073–1080.

Lai MC, Hageman MJ, Schowen RL, Borchardt RT, Laird BB, and Topp EM. Chemical Stability of Peptides in Polymers. 2. Discriminating between Solvent and Plasticizing Effects of Water on Peptide Deamidation in Poly(vinylpyrrolidone). *J Pharm Sci* 1999b; 88: 1081–1089.

Lai MC, Schowen RL, Borchardt RT, and Topp EM. Deamidation of a Model Hexapeptide in Poly(vinyl alcohol) Hydrogels and Xerogels. *J Peptide Res* 2000; 55: 93–101.

Lechuga-Ballesteros D, Abdul-Fattah A, Stevenson CL, and Bennett DB. Properties and Stability of a Liquid Crystal Form of Cyclosporine-The First Reported Naturally Occurring Peptide that exists as a Thermotropic Liquid Crystal. *J Pharm Sci* 2003; 92: 1821–1831.

Li R, D'Souza AJM, Laird BB, Showen RL, Borchardt RT, and Topp EM. Effects of Solution Polarity and Viscosity on Peptide Deamidation. *J Peptide Res* 2000; 56: 326–334.

Li R, Hageman MJ, and Topp EM. Effect of Viscosity on the Deamidation Rate of a Model Asn-hexapeptide. *J Peptide Res* 2002; 59: 211–220.

Li S, Patapoff TW, Nguyen TH, and Borchardt RT. Inhibitory Effect of Sugars and Polyols on the Metal-catalyzed Oxidation of Human Relaxin. *J Pharm Sci* 1996; 85: 868–872.

Lide DR. *Handbook of Chemistry and Physics, 75th Edition.* CRC Press, Ann Arbor, MI, 1995.

Lo YL and Rahman YE. Effects of Lipids on the Thermal stability and Conformational Changes of Proteins: Ribonuclease A and Cytochrome C. *Int J Pharm* 1998; 161: 137–148.

Loosli HR, Kessler H, Oschkinat H, Weber HP, Petcher T, and Wider A. The Conformation of Cyclosporin A in the Crystal and in Solution. *Helvetica Chimica Acta* 1985; 68: 682–703.

Loughlin WA and Hawkes DB. Effect of Organic Solvents on a Chloroperoxidase Biotransformation. *BioResour Technol* 2000; 71: 167–172.

Love SG, Muir TW, Ramage R, Shaw KT, Alexeev D, Sawyer L, Kelly SM, Price NC, Arnolds JE, Me MP, and Mayer RJ. Synthetic, Structural and Biological Studies of the Ubiquitin System: Synthesis and Crystal Structure on an Analogue Containing Unnatural Amino Acids. *Biochem J* 1997; 323: 727–734.

Manning MC, Matsuura JE, Kendrick BS, Meyer JD, Dormish JJ, Vrkljan M, Ruth JR, Carpenter JF, and Shefter E. Approaches for Increasing the Solution Stability of Proteins. *Biotechnol Bioeng* 1995; 48: 506–512.

Martin A, Sarbrick J, and Cammarata A. *Physical Pharmacy. Physical Chemical Principles in the Pharmaceutical Sciences.* Lea and Febiger, Philadelphia, PA, 1983.

Martinez P, Van Dam ME, Robinson AC, Chen K, and Arnold FH. Stabilization of Subtilisin E in Organic Solvents by Site-directed Mutagenesis. *Biotechnol Bioeng* 1992; 39: 141–147.

Matsuura J, Powers ME, Manning MC, and Shefter E. Structure and Stability of Insulin Dissolved in 1–octanol. *J Am Chem Soc* 1993; 115: 1261–1264.

McMinn JH, Sowa MJ, Charnick SB, and Paulaitis ME. The Hydration of Proteins in Nearly Anhydrous Organic Solvent Suspensions. *Biopolymers* 1993; 33: 1213–1224.

Meadows RP, Nikonowicz EP, Jones CR, Bastian JW, and Goresterin DG. Two Dimensional NMR and Structure Determination of Salmon Calcitonin in Methanol. *Biochemistry* 1991; 30: 1245–1247.

Meyer JP, Pelton JT, Holflack J, and Soudek V. Solution Structure of Salmon Calcitonin. *Biopolymers* 1991; 31: 233–241.

Mishra P, Griebenow K, and Klibanov AM. Structural Basis for the Molecular Memory of Imprinted Proteins in Anhydrous Media. *Biotechnol Bioeng* 1996; 52: 609–614.

Motta A, Morelli MAC, Goud N, and Temussi PA. Sequential 1H-NMR Assignment and Secondary Structure Determination of Salmon Calcitonin in Solution. *Biochemistry* 1989; 28: 7996–8002.

Motta A, Patore A, Goud NA, and Castiglione Morelli MA. Solution Conformation of Salmon Calcitonin in Sodium Dodecyl Sulfate Micelles as Determined by Two-dimensional NMR and Distance Geometry Calculations. *Biochemistry* 1991; 30: 10444–10450.

Myrdal PB, Karlage KL, Stein SW, Brown BA, and Haynes A. Optimized Dose Delivery of the Peptide Cyclosporine using Hydrofluoroalkane-based Metered Dose Inhalers. *J Pharm Sci* 2004; 93: 1054–1061.

Narhi LO, Philo JS, Li T, Zhang M, Samal B, and Arakawa T. Induction of α-helix in the β-sheet Protein Tumor Necrosis Factor-α: Thermal and Trifluoroethanol Induced Denaturation at Neutral pH., *Biochemistry* 1996; 35: 11447–11453.

Narita M, Lee JS, Hayashi S, and Hitomi M. The Estimation of the β-sheet-structure Stability of Protected Peptides in Organic Solvents. *Bull Chem Soc Jpn* 1993; 66: 489–493.

Narita M, Umeyama H, and Sasaki C. Weak Dependence of Octapeptide Solubility in Organic Solvents on the Amino Acid Sequence. *Int J Peptide Protein Res* 1989; 33: 340–344.

Oh-Uchi SK, Yang JY, Lee JS, Murakawa Y, and Narita M. Only Weak Dependence of the Protected Peptides Solubility in Organic Solvent on their Amino Acid Sequence. *Polym J* 1996; 28: 1033–1038.

Palakodaty S and York P. Phase Behavioral Effects on Particle Formation Processes using Supercritical Fluids. *Pharm Res* 1999; 16: 976–985.

Petcher TJ, Weber HP, and Ruegger A. Crystal and Molecular Structure on an Iodo-derivative of the Cyclic Undecapeptide Cyclosporin A. *Helv Chim Acta* 1976; 59: 1480–1489.

Popovych O and Tomkins RPT. *Nonaqueous Solution Chemistry*. John Wiley, New York, NY, 1981, pp. 221–224.

Powers ME, Matsuura J, Brassell J, Manning MC, and Shefter E. Enhanced Solubility of Proteins and Peptides in Nonpolar Solvents through Hydrophobic Ion Pairing. *Biopolymers* 1993; 33: 927–932.

Quinn EA, Forbes RT, Williams AC, Oliver MJ, McKenzie L, and Purewas TS. Protein Conformational Stability in the Hydroxyfluoroalkane Propellants Tetrafluoroethane and Heptafluoropropane Analyzed by Fourier Transform Raman Spectroscopy. *Int J Pharm* 1999; 186: 31–41.

Rajendran S, Radha C, and Prakash V. Mechanism of Solvent-induced Thermal Stabilization of α-amylase from Baccillus Amyloliquefaciens. *Int J Peptide Protein Res* 1995; 45: 122–128.

Ran Y, Zhao L, Xu Q, and Yalkowsky SH. Solubilization of Cyclosporin A. *AAPS PharmSciTech* 2001; 2:E2.

Rariy RV and Klibanov AM. Correct Protein Folding in Glycerol. *Proc Natl Acad Sci USA* 1997; 94: 13520–13523.

Rariy RV and Klibanov AM. Protein Refolding in Predominantly Organic Media Markedly Enhanced by Common Salts. *Biotechnol Bioeng* 1999; 62: 704–710.

Reiersen H and Rees AR. Trifluoroethanol may form a Solvent Matrix for Assisted Hydrophobic Interactions between Peptide Side Chains. *Protein Eng* 2001; 13: 739–743.

Remmele Jr RL, Nightlinger NS, Srinavasan S, and Gombotz WR. Interleukin-1 Receptor (IL-1R) Liquid Formulation Development using Differential Scanning Calorimetry. *Pharm Res* 1998; 15: 200–208.

Rich JO and Dordick JS. Controlling Subtilisin Activity and Selectivity in Organic Media by Imprinting with Nucleophilic Substrates. *J Am Chem Soc* 1997; 119: 3245–3252.

Ruelle P and Kesselring UW. The Hydrophobic Effect. 1. A Consequence of the Mobile Order in H-bonded Liquids. *J Pharm Sci* 1998; 87: 987–997.

Santos AM, Vidal M, Pacheco Y, Frontera J, Baez C, Omellas O, Barletta G, and Griebenow K. Effect of Crown Ethers on Structure, Stability, Activity and Enantioselectivity of Subtilisin Carlsberg in Organic Solvents. *Biotechnology and Bioengineering* 2001; 4:295–308.

Sardessai Y and Bhosle S. Isolation of an Organic-solvent-tolerant Cholesterol-transforming Bacillus Species, BC1, from Coastal Sediment. *Marine Biotechnol* 2003; 5: 116–118.

Sardessai Y and Bhosle S. Industrial Potential of Organic Solvent Tolerant Bacteria. *Biotechnol Prog* 2004; 20: 655–660.

Schmid A, Dordick JS, Hauer B, Kiener A, Wubbolts M, and Witholt B. Industrial Biocatalysis Today and Tomorrow. *Nature* 2001; 409: 258–268.

Schulze B and Klibanov AM. Inactivation and Stabilization of Subtilisins in Neat Organic Solvents. *Biotechnol Bioeng* 1991; 38:1001–1006.

Schwendeman SP, Cardamone M, Brandon MR, Klibanov AM, and Langer R. Stability of Proteins and their Delivery from Biodegradable Polymer Microspheres. In: Cohen S, Bernstein H (eds.), *Microparticulate Systems for Delivery of Proteins and Vaccines.* Marcel Dekker, New York, NY, 1996, pp. 1–49.

Schwendeman SP, Constantino HR, Gupta RK, and Langer R. Peptide, Protein and Vaccine Delivery from Implantable Polymeric Systems. Progress and Challenges. In: Park K (ed.), *Controlled Drug Delivery. Challenges and Strategies.* American Chemical Society, Washington DC, 1997, pp. 229–267.

Secundo F and Carrea G. Optimization of Hydrolase Efficiency in Organic Solvents. *Chemistry-A European Journal* 2003; 9:3194–3200.

Seeback D, Thaler A, and Beck AK. Solubilization of Peptides on Non-polar Organic Solvents by the Additions of Inorganic Salts: Facts and Implications. *Helvetica Chemica Acta* 1989; 72: 857–867.

Sellers SP, Clark GS, Sievers RE, and Carpenter JF. Dry Powders of Stable Protein Formulations from Aqueous Solutions Prepared using Supercritical CO_2-assisted Aersolization. *J Pharm Sci* 2001; 90: 785–797.

Senderoff RI, Kontor KM, Kreilgaard L, Chang JJ, Patel S, Krakavor J, Heffernan JK, Snell LB, and Rosenberg GB. Consideration of Conformations and Racemization during Process Development of Recombinant Glucagon-like Peptide-1. *J Pharm Sci* 1998; 87: 183–189.

Shiraki K, Nishikawa K, and Goto W. Trifluroethanol Induced Stabilization of the α-helical structure of β-lactoglobulin: Implications for Non-hierachical Protein Folding. *J Mol Biol* 1995; 245: 180–194.

Snavely W, Subramaniam B, Rajewski RA, and Defelippis MR. Micronization of Insulin from Halogenated Alcohol Solution using Supercritical Carbon Dioxide as an Antisolvent. *J Pharm Sci* 2002; 91: 2026–2039.

Song Y, Schowen RL, Borchardt RT, and Topp EM. Effect of 'pH' on the Rate of Asparagine Deamidation in Polymeric Formulations: 'pH'-Rate Profile. *J Pharm Sci* 2001; 90: 141–156.

Sonnichsen FD, VanEyk JE, Hodges JE, and Sykes BD. Effect of Trifluoroethanol on Protein Structure: an NMR and CD Study using a Synthetic Actin Peptide. *Biochemistry* 1995; 31: 8790–8798.

Spreti N, Germani R, Incani A, and Savelli G. Stabilization of Chloroperoxidase by Polyethylene Glycols in Aqueous Media: Kinetic Studies and Synthetic Applications. *Biotechnol Prog* 2004; 20: 96–101.

Stevenson CL. *Delivery of Proteins from Implantable Systems.* Macromolecular Drug Delivery Conference, Breckenridge, CO, 1999.

Stevenson CL, Anderegg RJ, and Borchardt RT. Probing Helical Content of Growth Hormone Releasing Factor Analogs Using Electrospray Mass Spectrometry. *J Am Soc Mass Spectrom* 1993c; 4: 646–651.

Stevenson CL, Donlan ME, Friedman AR, and Borchardt RT. Solution Conformation of Leu27 hGRF(1032)NH2 and its Deamidation Products by 2D NMR. *Int J Peptide Protein Res* 1993b; 42:24–32.

Stevenson CL, Friedman AR, Kubiak TM, Donlan ME, and Borchardt RT. Effect of Secondary Structure on the Rate of Deamidation of Several Growth Hormone Releasing Factor Analogs. *Int J Peptide Protein Res* 1993a; 42: 497–503.

Stevenson CL, Leonard JJ, and Hall SC. Effect of Peptide Concentration and Temperature on Leuprolide Stability in Dimethyl Sulfoxide. *Int J Pharm* 1999; 191: 115–129.

Stevenson CL and Tan MM. Solution Stability of Salmon Calcitonin at High Concentration for Delivery in an Implantable System. *J Peptide Res* 2000; 55: 129–139.

Stevenson CL, Tan MM, and Lechuga-Ballesteros D. Secondary Structure of Cyclosporine in a Spray-Dried Liquid Crystal by FTIR. *J Pharm Sci* 2003; 92: 1832–1843.

Stotz CE, Borchardt RT, Middaugh CR, Siahaan TJ, Vander Velde D, and Topp EM. Secondary Structure of a Dynamic Type I' β-hairpin Peptide. *J Peptide Res* 2004; 63: 371–382.

Sukumar M, Storms, SM, and De Felippis, MR. Non-native Intermediate Conformational States of Human Growth Hormone in the Presence of Organic Solvents. *Pharm Res* 2005; 22: 789–796.

Tan MM, Corley CA, and Stevenson CL. Effect of Gelation on the Chemical Stability and Conformation of Leuprolide. *Pharm Res* 1998; 15: 1442–1447.

Tanner JW, Eckenhoff RG, and Liebman PA. Halothan, an Inhalational Anesthetic Agent, Increases Folding Stability of Serum Albumin. *Biochim Biophys Acta* 1999; 1430: 46–56.

Timasheff SN and Inoue H. Preferential Binding of Solvent Components to Proteins in Mixed Water-Organic Solvent Systems. *Biochemistry* 1968; 7:2501–2513.

Tuena de Gomez-Puyou M, and Gomez-Puyou A. Enzymes in Low Water Systems. Crit. Rev. *Biochem Mol Biol* 1998; 33: 53–89.

Verheyden P, De Wolf E, Jaspers H, and Van Binst G. Comparing Conformations at Low Temperature and at High Viscosity. Conformational Study of Somatostatin and Two of its Analogues in Methanol and in Ethylene Glycol. *Int J Peptide Protein Res* 1994b; 44: 401–409.

Verheyden P, Francq W, Pepterman H, and Van Binst G. Conformational Study of a Somatostatin Analogue in DMSO/water by 2D NMR. *Biopolymers* 1990; 30: 855–860.

Verheyden PMF, Coy DH, and Van Binst G. Conformational Study of Somatostatin Analogs in Methanol at Low Temperature. *Magnetic Resonance in Chemistry* 1991; 29: 607–612.

Verheyden P, Jaspers H, Wolf E, and Van Binst G. Conformational Study of Cyclosporin A in Acetone at Low Temperature. *Int J Peptide Protein Res* 1994a; 44: 364–371.

Volkin DB, Staubli A, Langer R, and Klibanov AM. Enzyme Thermoinactivation in Anhydrous Organic Solvents. *Biotechnol Bioeng* 1991; 37: 843–853.

Vrkljan M, Foster TM, Powers ME, Henkin J, Porter WR, Staack J, Carpenter JF, and Manning MC. Thermal Stability of Low Molecular Weight Urokinase during Heat Treatment. II. Effect of Polymeric Additives. *Pharm Res* 1994; 11: 1004–1008.

Wang JM, Takeda A, Yang JT, and Wu CSC. Conformation of Concanavalin A and its Fragments in Aqueous Solution and Organic Solvent Water Mixtures. *J Protein Chem* 1992; 11: 157–164.

Wang W. Instability, Stabilization and Formulation of Liquid Protein Pharmaceuticals. *Int J Pharm* 1999; 185: 129–188.

Winters MA, Debenedetti PG, Carey J, Sparks HG, Sane SU, and Przybycien TM. *Pharm Res* 1997; 14: 1370–1378.

Winters MA, Frankel DZ, Debenedetti PG, Carey J, Devaney M, and Przybycien TM. Protein Purification with Vapor-Phase Carbon Dioxide. *Biotechnol Bioeng* 1999; 62: 247–258.

Winters MA, Knutson BL, Debenedetti PG, Sparks HG, Przybycien TM, Stevenson CL, and Prestrelski SJ. Precipitation of Proteins in Supercritical Carbon Dioxide. *J Pharm Sci* 1996; 85: 586–594.

Wright JC, Leonard ST, Stevenson CL, Beck J, Chen G, Leonard J, and Skowronski R. *J Control Release* 2001; 75: 1–10.

Xu K, Griebnow K, and Klibanov AM. Correlation between Catalytic Activity and Secondary Structure of Subtilisin Dissolved in Organic Solvents. *Biotechnol Bioeng* 1997; 56: 485–491.

Yang L, Dordick JS, and Garde S. Hydration of Enzyme in Nonaqueous Media is Consistent with Solvent Dependence of its Activity. *Biophys J* 2004; 87: 812–821.

Yeo SD, Lim GB, Debenedetti PG, and Bernstein H. Formation of Microparticulate Protein Powder using a Supercritical Fluid Antisolvent. *Biotechnol Bioeng* 1993; 41:341–346.

Yeo SD, Debenedetti PG, Patro S, and Przybycien TM. Secondary Structure Characterization of Microparticulate Insulin Powders. *J Pharm Sci* 1994; 83: 1651–1656.

Yu LX, Foster TP, Sarver RW, and Moseley WM. Preparation, Characterization and in vivo Evaluation of an Oil Suspension of a Bovine Growth Hormone Releasing Factor Analog. *J Pharm Sci* 1996; 85: 396–401.

Zaks A and Klibanov AM. Enzymatic Catalysis in Organic Media at 100°C. *Science* 1984; 224: 1249–1251.

Zaks A and Klibanov AM. Enzymatic Catalysis in Nonaqueous Solvents. *J Bio Chem* 1988a; 263: 3194–3201.

Zaks A and Klibanov AM. The Effect of Water on Enzyme Action in Organic Media. *J Biol Chem* 1988b; 263: 8017–8021.

Zhang J, Peng X, Jonas A, and Jonas J. NMR Study of the Cold, Heat and Pressure Unfolding of Ribonuclease A. *Biochemistry* 1995; 34: 8631–8641.

13

Pharmaceutical Solvents as Vehicles for Topical Dosage Forms

ADRIAN WILLIAMS

Professor of Pharmaceutics, School of Pharmacy, University of Reading, Reading, UK

Introduction

The selection of a vehicle can dramatically affect delivery and consequently efficacy of topical preparations. In terms of transdermal delivery, where delivering therapeutic agents for systemic effects is desired, solvents and co-solvent systems are widely used to improve both the amount and range of drugs that can be administered at therapeutic levels through the skin. Vehicles used in transdermal systems, such as patches, have recently been reviewed (Williams, 2003). In contrast, the focus of this chapter is on the use of solvents in topical dosage forms, i.e. preparations intended for a local or regional effect on the skin.

Topical formulations rarely contain a simple vehicle comprising a single solvent, except for some aqueous lotions, polar gels or soaks. Even with such simple formulations, gelling agents, thickeners, perfumes or other excipients are often included. More typically, mixed solvent systems are employed and multiphasic systems are often used to provide efficacy from the vehicle (such as anti-inflammatory, cleansing or emollient activity) in addition to therapeutic effects of the active pharmaceutical ingredient. However, in describing the selection of solvents as vehicles for topical dosage forms, it is impractical to consider the multitude of mixed solvent systems. Likewise, the literature describes numerous "novel" solvents and solvent systems that have been used in research investigations. This review considers the properties and uses as well as the advantages and disadvantages of solvents typically used for topical formulations. Prior to this, a brief review of the structure and function of the membrane to which the solvent will be applied as well as background relevant to solvent selection is provided.

Structure and Function of Human Skin

Human skin is a highly efficient, self-repairing barrier which permits terrestrial life by regulating heat and water loss from the body while preventing the ingress of noxious chemicals or micro-organisms. It is the largest organ of the human body providing around 10% of the body mass of an average person and an area of approximately 1.8 m^2. The tissue can be examined at various levels of complexity. Simplistically, skin can be regarded merely as a physical barrier with more sophistication introduced by considering the various skin layers, providing barriers in parallel. We can introduce barriers in series by considering drug transport through pores in the tissue. Degrees of complexity also exist when examining basic structures and functions of the membrane. In some extreme cases it may be that topical drug delivery is limited by metabolic activity within the membrane. Alternatively, immunological responses may prevent the clinical use of a formulation that has proven to be optimal during *in vitro* studies. A further complication is introduced in clinical situations where topical delivery is intended to treat diseased states; here the barrier nature of the membrane may be compromised or, in some scaling conditions such as psoriasis, drug uptake into the tissue can be dramatically altered. This brief overview considers primarily the structure and function of healthy human skin but also alludes to issues that arise when selecting solvents for topical formulations for skin disorders. A fuller description of skin structure and function in health and disease is given in the literature (Williams, 2003).

Healthy Skin Structure and Function

Due to experimental and ethical concerns, development of topical formulations tends to utilise skin *ex vivo* (*in vitro*) which inherently reduces some of the above noted complexity—regeneration stops, immune responses cease and metabolic activity is usually greatly reduced or lost in these studies. However, it should always be borne in mind that data obtained from excised skin may not translate directly to the *in vivo* situation. In particular, applying exogenous chemicals, such as solvents in a formulation, can induce immunological, histological and metabolic changes that may not be evident from *in vitro* studies.

For the purpose of topical formulation design and application, we can examine the structure and function of human skin categorised into four main layers; 1—the innermost subcutaneous fat layer (hypodermis), 2—the overlying dermis, 3—the viable epidermis and 4—the outermost layer of the tissue (a non-viable epidermal layer) the stratum corneum (Figure 1).

The Subcutaneous Fat Layer

The subcutaneous fat layer, or hypodermis, bridges between the overlying dermis and the underlying body constituents. In most areas of the body, this layer is relatively thick, (in the order of several mm) but there are areas in which the subcutaneous fat layer is absent, such as the eyelids. This layer of adipose tissue

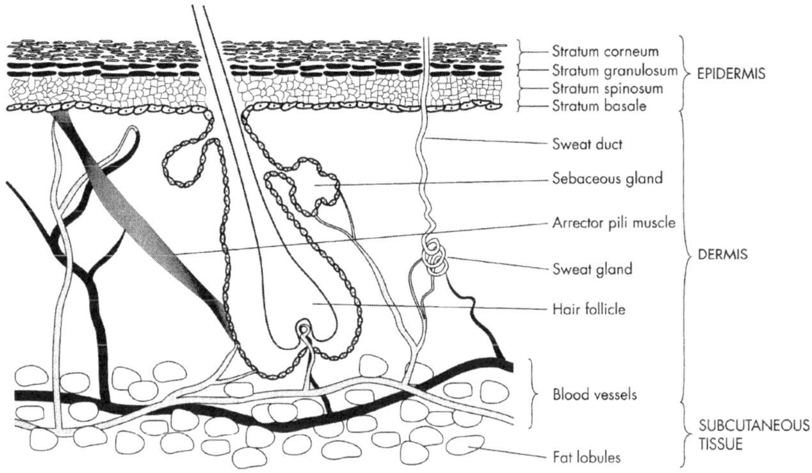

Figure 1. Diagrammatical representation of the structure of human skin (from Williams, AC, 2003, with permission).

principally serves to insulate the body and to provide mechanical protection against physical shock and carries the principle blood vessels and nerves to the skin. In terms of topical formulations and administration, the fatty layer is usually discounted as a controlling factor.

The Dermis

The dermis is generally 3–5 mm thick and is the major component of human skin. It is composed of a network of connective tissue, predominantly collagen fibrils providing support and elastic tissue providing flexibility, embedded in a mucopolysaccharide gel (Wilkes et al., 1973). In terms of topical drug delivery, this layer is often viewed as essentially gelled water thus providing a minimal barrier to delivery of most polar drugs, although the dermal barrier may be significant when delivering highly lipophilic molecules. The dermis has numerous structures embedded within it: blood and lymphatic vessels, nerve endings, pilosebaceous units (hair follicles and sebaceous glands), and sweat glands (eccrine and apocrine).

The extensive vasculature of the skin is essential for regulation of body temperature while also delivering oxygen and nutrients to the tissue and removing toxins and waste products. The rich blood flow, around 0.05 $mLmin^{-1}$ per mg of skin, is very efficient for the removal of molecules at the dermis-epidermis interface, ensuring that dermal concentrations of most permeants are very low. The lymphatic system also reaches to the dermo-epidermal layer and while it is important in regulating interstitial pressure, for facilitating immunological responses to microbial assault and for waste removal, the lymphatic vessels may also remove permeated molecules from the dermis. Cross and Roberts (1993) showed that while dermal blood flow affected the clearance of relatively small solutes, such as lidocaine, lymphatic flow was a significant determinant for the clearance of larger molecules such as interferon.

There are three main appendages found on the surface of human skin that originate in the dermis. Hair follicles are found over the entire surface of the skin except for the load-bearing areas of the body (soles of feet, palms of hands) and the lips. The sebacceous gland associated with the hair follicle secretes sebum, composed of free fatty acids, waxes and triglycerides, which lubricates the skin surface and helps to maintain the surface pH of around 5. Typically, there are 60–100 hair follicles per cm^2 of skin. Eccrine (or sweat) glands and apocrine glands also originate in the dermal tissue. Eccrine glands are found over most of the body surface, typically at a density of 100–200 per cm^2 of skin. Secreting sweat, a dilute salt solution at a pH of around 5, these glands are stimulated in response to heat and emotional stress. The apocrine glands are located near the dermo-epidermal layer but are limited to specific areas of the skin including the axillae, nipples and ano-genital regions. The lipoidal and "milk" protein secretions are primarily responsible for imparting the odour of "sweat".

With topical formulations, the appendages (hair follicles, sweat ducts) may offer a potential route by which molecules could enter the lower layers of the skin without having to enter the "intact" barrier provided by the stratum corneum. However, the fractional area offered by these shunt routes is so small that the predominant pathway for molecules to enter the tissue remains across the bulk of the skin surface.

The Epidermis

The epidermis is itself a complex multiply layered membrane which varies in thickness from around 0.06 mm on the eyelids to around 0.8 mm on the load-bearing palms and soles of the feet. The epidermis contains no blood vessels and hence nutrients and waste products must diffuse across the dermo-epidermal layer to maintain tissue integrity. Likewise, molecules permeating across the epidermis must cross the dermo-epidermal layer in order to be cleared into the systemic circulation.

The epidermis contains five histologically distinct layers which, from the inside to the outside, are the stratum germinativum, stratum spinosum, stratum granulosum, stratum lucidum and the stratum corneum (Figure 2). The stratum corneum, comprising anucleate (dead) cells, provides the main barrier to topical drug delivery and hence is considered in further detail here.

The Stratum Corneum (Horny Layer). Typically, the stratum corneum comprises only 10 to 15 cell layers and is around 10 μm thick when dry, although it may swell to several times this thickness when wet (see section 4.1). As with the viable epidermis, the stratum corneum is thickest on the palms and soles and is thinnest on the lips. This thin membrane, consisting of dead, anucleate, keratinised cells embedded in a lipid matrix allows for survival of terrestrial animals without desiccation. The stratum corneum serves to regulate water loss from the body while preventing the entry of harmful materials including microorganisms. The stratum corneum has been represented as a "brick and mortar" model (Michaels

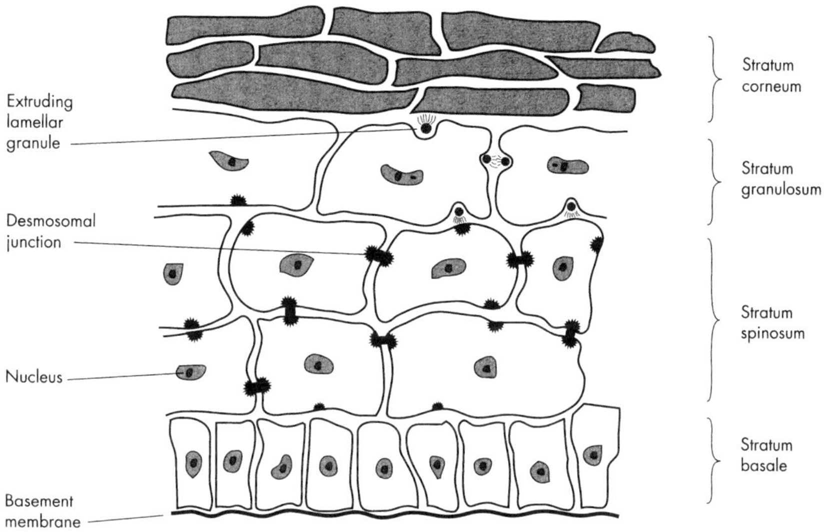

Figure 2. A representation of human epidermal cell layers and differentiation (from Williams, AC, 2003, with permission).

et al., 1975, Elias, 1981a) in which the keratinised cells are embedded in a mortar of lipid bilayers.

The barrier nature of the stratum corneum depends critically on its unique constituents; 75–80% is protein, 5–15% is lipid with 5–10% unidentified on a dry weight basis (Wilkes et al., 1973). The protein is primarily located within the keratinocytes and is predominantly alpha-keratin (around 70%) with some beta-keratin (approximately 10%) and a proteinaceous cell envelope (around 5%). Enzymes and other protein account for approximately 15% of the protein component. The cell envelope protein is highly insoluble and is very resistant to chemical attack. Human stratum corneum contains a unique mixture of lipids and, for most topically applied drugs, the continuous multiply bilayered lipid component of the stratum corneum is key in regulating drug permeation through the tissue. While the lipid content of the stratum corneum varies between individuals and with body site, the major components of the domain include ceramides, fatty acids, cholesterol, cholesterol sulphate and sterol/wax esters (see Table 1). The stratum corneum lipids are arranged in multiple bilayers, but in contrast to all other lipid bilayers in the body, phospholipids are largely absent. When considering solvent selection for topical formulations, their influence on stratum corneum lipids is an important factor. For example, if acetone is deposited onto skin *in vivo*, volatilisation of the solvent leaves a white residue from lipids (and other materials) that has been extracted from the membrane. Such extraction can clearly compromise the barrier function of the tissue.

Other components of the stratum corneum can also be compromised by solvent selection—water plays a key role in maintaining stratum corneum barrier integrity and Natural Moisturising Factor (NMF) is a highly efficient humectant

Lipid constituent	Skin site		
	Abdominal[a]	Plantar[b]	Unspecified[c]
Ceramides	18	35	41
Glucosylceramides	Trace		0.0
Fatty acids	19	19	9.1
Cholesterol	14	29	27
Cholesteryl sulphate	1.5	1.8	1.9
Sterol/wax esters	5.4	6.5	10
Di- and triglycerides	25	3.5	0.0
Squalene	4.8	0.2	
n-Alkanes	6.1	1.7	
Phospholipids	4.9	3.2	

Table 1. Content of lipid types within human stratum corneum, expressed as a percentage of the total lipid content.

[a]Lampe et al., (1983), [b]Melnik et al., (1989), [c]Wertz and Downing (1989).

Note: The presence of alkanes may be erroneous as there are no reported metabolic pathways for the production of alkanes within the skin, and it appears likely that their presence may arise from external contamination. In addition, Lampe and co-workers emphasised that the high triglyceride content of their stratum corneum sample was probably due to contamination by triglyceride rich subcutaneous lipids.

synthesised and hence located within the stratum corneum. Additionally micro-organisms are present on the skin surface and solvent could alter the microfloral balance on the tissue.

Diseased/Compromised Skin and Solvent Selection

Numerous physiological factors in healthy and diseased skin can affect topical drug delivery and consequently formulation design. For example, neonatal skin is more permeable than adult skin and site to site variations in permeability are well known with genital tissue more permeable than that on the head and neck with arm and leg skin even less permeable. However, it is with compromised skin, found in many conditions where topical dosage forms are applied, where solvent selection can be problematic.

Numerous disorders result in an eruption of the skin surface. Clearly in such cases, the barrier properties of the stratum corneum are compromised, allowing easier passage of drugs (and solvent) into and through the skin. Likewise, the erupted skin surface will allow increased water loss from the body. Psoriasis is a chronic recurring non-infectious scaling skin condition characterised by erythematous plaques covered with silvery scales. For topical therapy the loss of skin barrier integrity has been shown to be valuable for targeting drugs to the required site of action while minimising side effects (Anigbogu et al., 1996). Lichenoid eruptions are characterised by intensely itchy flat-topped papules while eczema is a further non-infectious eruptive condition, in which blistering occurs. Contact dermatitis can result from a direct irritant action of a substance on the skin (irritant contact dermatitis) or further exposure, following previous sensitisation of the skin, from a contact allergen (allergic contact dermatitis). Irritant dermatitis is the more common of the two manifestations, and can be caused by many chemicals, solvents and detergents; sodium lauryl sulphate was used to induce irritant dermatitis before *in vivo* percutaneous absorption of several drugs was assessed (Wilhelm et al., 1991). Clearly having implications for therapy, hydrocortisone absorption was shown to increase nearly three-fold through the affected site, with a two-fold increase seen for indomethacin. Likewise, solvent permeation from an applied formulation will tend to increase where the barrier is compromised.

General Principles for Solvent Selection

Many factors can influence the choice of solvent for topical preparations, ranging from chemical considerations such as drug:solvent compatibility or solvent stability to packaging, storage and handling issues. Some critical concerns are discussed below.

Thermodynamics

In order to ensure that an active pharmaceutical ingredient passes from a topical dosage form into the skin, formulators must generate a vehicle that will release the drug. Indeed, if suspended, then the drug must initially dissolve in the vehicle prior to release. Subsequently, the drug must partition into the outer layer of the skin, the stratum corneum, before it diffuses into the deeper skin strata. Thus, in selecting a vehicular solvent, the physicochemical properties of the drug (partition coefficient, solubility in the solvent) must be considered.

Numerous relationships have been described for drug release from topical preparations, and release may follow, for example, zero- or first-order kinetics, or may obey the Higuchi relationship depending whether the formulation is a simple suspension or semi-solid. As a general principle, drug release relates to the concentration of drug in the vehicle. More accurately, thermodynamic activity of the therapeutic agent in the vehicle will provide the driving force for penetration into the stratum corneum. Consequently, a formulation with the drug at saturation will maximise release and penetration compared with delivery from sub-saturated systems. Solvent selection or the use of mixed solvent systems can thus provide this optimal thermodynamic driving force.

Vehicle Effects on the Skin

While the vehicle can modify drug release from the formulation and partitioning into the stratum corneum, some solvents are also well known to affect the integrity of the stratum corneum as a barrier. For example, organic solvents such as ethanol are able to solubilise and extract lipids from the inter-cellular multiple bilayers of the stratum corneum. More caustic solvents, such as acids and phenols, can break the desmosomal junctions between the corneocytes whereas polar solvents such as dimethylsulphoxide (DMSO) can alter the intracellular keratin conformation while also introducing disorder into the lipid domains. In addition to affecting the skin barrier integrity, solvents have been used to alter partitioning into the membrane; pyrrolidones have been used which can partition rapidly into stratum corneum where they consequently alter partitioning of the drug into the membrane. Even water can affect drug delivery from topical formulations as the tissue hydrates whereupon drug delivery usually increases. Ideally, solvents for topical formulations are inert but clearly this is not the case in practice hence formulators look for a controlled and predictable effect on the skin.

Biological Factors

Beyond considering the solvent *per se*, formulators must also consider the biology and state of the skin, and whether transient or sustained delivery is desired. Generally, semisolid formulations are selected for increased residence on the skin, and liquid formulations for a rapid short-term input of permeant into the skin. In both the clinical and cosmetic domains, skin type can affect the choice of formulation base in that usually for normal to oily skin types, gels are preferred but for normal to dry skin types lotions are usually selected and for dry skin, creams are often the preferred base. As well as skin type, the skin site to be treated can affect vehicle selection. For example, for hairy areas lotions, gels or sprays are usually preferable as these spread better whereas for intertriginous areas, creams or lotions are usually employed.

However, it is mainly clinical rationale as to which formulation type (and hence vehicle) is selected for topical therapy. Depending upon the lesion type of the condition to be treated then:

- For a wet, vesicular or weeping lesion, a "wet", usually aqueous based, formulation is generally preferred (cream, lotion, gel).
- For a dry, thickened scaly lesion, a "dry", usually fatty, formulation is preferred (ointments, pastes)

Beyond these simple considerations, clinicians tend not to apply liniments, which may be alcoholic or oily solutions, to broken skin. Evaporation from aqueous solutions (lotions) cools and soothes the skin which can be valuable when treating inflamed skin sites. Alcohol can increase the soothing effect, but should only be used on intact skin. Paints and tinctures generally contain volatile solvents (e.g. acetone or ether) designed to deposit a layer of the active agent onto the skin. Clearly solvent selection is important in clinical efficacy but also impacts upon patient compliance since patients tend to favour creams rather than gels or ointments.

Solvents Used As Vehicles

Numerous solvents have been employed in topical formulations and researchers often select solvent vehicles to probe scientific principles rather than for commercial formulations. Thus the literature describes many solvents that are inappropriate for clinical use. Here, typical solvents that are widely used in commercial preparations, and which have thus received regulatory approval are considered, and are grouped rather arbitrarily into "water", "alcohols", "glycols", "oils and waxes" and "other solvents" though clearly some materials could fit into several of the groupings. A brief summary of some of the main examples of solvents from these classes is given in Table 2. Further details, uses, advantages and disadvantages of a broader range of these solvents are given below. The following is by no means exhaustive, but provides an overview of the more commonly employed solvents for typical dosage forms such as creams, ointments, emulsions etc.

Solvent	Principle advantages	Principle disadvantages
Water	Safe, low irritation potential	Not compatible with lipophilic drugs
Alcohols	Widely used, compatible with many drugs. May get supersaturation	Can modify the barrier nature of the stratum corneum
Ethanol	Listed as inactive ingredient by FDA	Concentration dependent effects on skin barrier and drug delivery. Readily absorbed through skin
Isopropyl alcohol	Listed as inactive ingredient by FDA	Can affect skin barrier
Benzyl alcohol	Commonly used in creams and emulsions	Usually used as co-solvent with other solvents
Lanolin alcohols	Historical use	Potential allergic responses
Fatty alcohols	Widely used as emulsifying agents	Penetration enhancing effects have been reported

Solvent	Principle advantages	Principle disadvantages
Glycols	Very commonly used, good range of solubilising effects	Occasional adverse reactions reported. Act synergistically with other solvents to enhance permeation
Propylene glycol	Widely used and at high concentrations. Listed as inactive ingredient by FDA	Some adverse reactions reported. Readily absorbed through skin
Polyethylene glycols	Widely used at relatively high concentrations. Not readily absorbed through skin	Some contact dermatitis reports
Oils and waxes	Widely used, cost-effective, well accepted	May hinder delivery of polar drugs
Mineral oils	Used in various preparations, including baby lotions. Can occlude skin surface hence emollient	Can hinder polar drug delivery
Paraffins	Widely used, well tolerated, occlusive	Polycyclic aromatic hydrocarbon impurities can sensitise skin
Other solvents	Varied materials for specific applications	Not as widely used as above materials
Isopropyl myristate	Compatible with many drugs	Possible (mild) sensitiser. Readily absorbed through skin
Oleic acid	Has been used as a component of arachis oil	Potent penetration enhancer

Table 2. Summary of some typical solvents used in topical formulations.

Water

A natural concentration gradient across human skin exists; the inner dermal layer is essentially fully hydrated whereas the outer stratum corneum water content depends on external factors such as relative humidity and temperature; typically, stratum corneum water content is around 20–30% of the tissue dry weight. Of this, approximately 25–35% is associated with some structural component of the tissue (i.e. "bound") with the remainder "free" (Walkley, 1972). Increasing the water content of the barrier stratum corneum is well known to

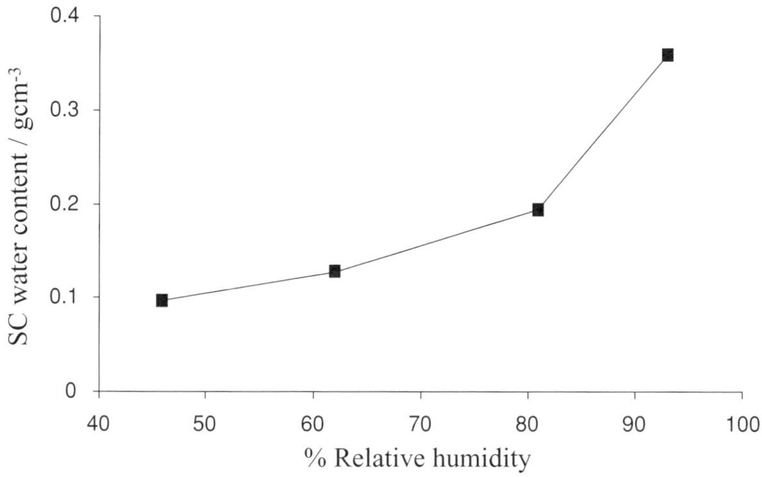

Figure 3. The water content of human stratum corneum as a function of the environmental relative humidity, plotted from data provided by Blank et al. (1984).

affect transport of topically applied medicaments through the membrane. Indeed, this is a strategy often used clinically where occlusive dressings are used to cover topically applied preparations. Occlusion prevents transepidermal water loss and can lead to stratum corneum water contents approaching 400% (of the dry tissue weight). The ability of stratum corneum to imbibe water from topical formulations can be illustrated using data from Blank et al. (1984), who determined water content of the tissue at various relative humidities at 31°C (Figure 3).

In general, increasing stratum corneum hydration tends to increase transdermal delivery of both hydrophilic (e.g. alkanols, Behl et al., 1980) and lipophilic (e.g. steroids, McKenzie and Stoughton, 1962) permeants. The increase in drug delivery with occlusion can be dramatic; McKenzie and Stoughton (1962) showed that covering a steroid application on the skin led to a 100-fold increase in the amount absorbed compared to that from a non-occluded application. However, Bucks and Maibach (1999) caution against such generalisations stating that occlusion does not necessarily increase percutaneous absorption and that transdermal delivery of hydrophilic compounds may not be enhanced by occlusion. Further, they caution that occlusion can cause some local skin irritation with clear implications for design and manufacture of topical formulations.

While water content of the stratum corneum affects permeability of the tissue, hydration also impacts various physical properties of the membrane such as tensile strength and elasticity, modifies the microenvironment for microorganisms on the tissue surface, alters the thermal conductivity of the tissue and also affects skin appearance. Further, increasing hydration also alters the thickness of the stratum corneum as shown in Figure 4, again taken from data provided by Blank et al. (1984).

The above serves to highlight an important principle in solvent selection for topical preparations. Even the (apparently) most benign and acceptable solvent

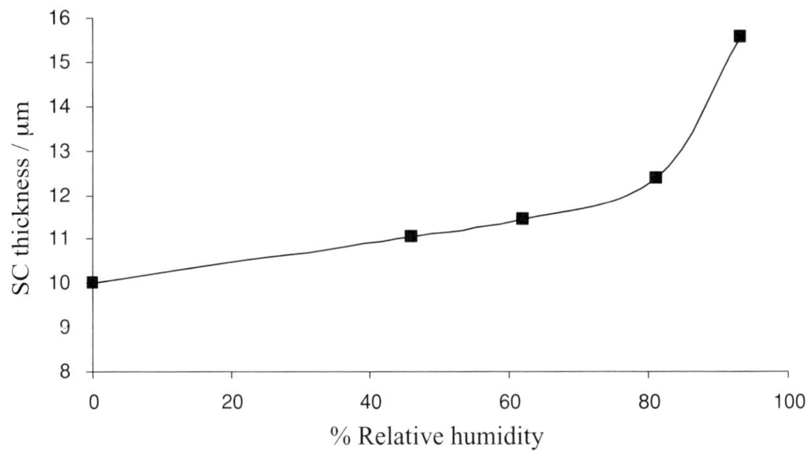

Figure 4. Thickness of human stratum corneum as a function of the environmental relative humidity, plotted from data provided by Blank et al. (1984).

for application to the skin can have significant impact upon the structure and function of the membrane. Consequently, solvent selection aims to minimise the effects on the skin while delivering the therapeutic agent to the tissue.

Alcohols

Various alcohols are currently used in topical preparations. **Ethanol** (ethyl alcohol) is often selected as a solvent or co-solvent but, as a small molecule, it can itself readily permeate through human skin with a reported steady state flux through the membrane of approximately 1 mg/cm^2/h (Berner et al., 1989). Ethanol as a vehicle has been shown to influence the delivery of many drugs into and through the skin, including steroids and salicylate ions. Various actions on topical formulations and skin have been described for the solvent. As a rapidly permeating solvent, or by vaporisation, ethanol can leave behind a higher drug concentration in the formulation as it leaves the system, hence increasing the thermodynamic activity of the drug in the residual formulation so increasing delivery. Alternatively, the solvent is capable of extracting some lipids from the stratum corneum, so modifying the barrier properties of the tissue.

Ethanol exhibits interesting concentration dependent effects on delivery through the skin. In ethanol/water systems, delivery of salicylate ions (Kurihara-Bergstrom et al., 1990), estradiol (Megrab et al., 1995a) and zidovudine increased to an ethanol fraction of around 0.63. Increasing the ethanol content beyond this point decreased delivery of these compounds, probably because, at higher ethanol levels, dehydration of the biological membrane reduced permeation across the tissue.

Permeation of ethanol into the stratum corneum can alter the solubility properties of the tissue with a consequent improvement of drug partitioning into the membrane (Megrab et al., 1995a). A further potential mechanism of action, arising as a consequence of rapid ethanol permeation across skin, is that a "solvent drag" effect may carry permeant into the tissue as the ethanol traverses, although such a mechanism was discounted for morphine hydrochloride permeation from ethanol and menthol containing formulations (Morimoto et al., 2002).

Ethanol is listed (as alcohol) in the FDA inactive ingredient guide (January 1996). At that time, some 15 approved drug products used ethanol in a topical solution, at compositions ranging from 33 to 88% of the formulation. This solvent was also present in topical gels (at 52%) and in lotions (71–80.5%). Likewise, dehydrated alcohol was listed as present between 55 and 77% in topical solutions and at between 20 and ~95% in topical gel formulations. Similarly, denatured alcohol was evident at 75 to ~97% in gels and between 44 and 60% in solutions. Thus, while it has been shown that ethanol may alter the nature of human stratum corneum, it is a widely used and accepted solvent for topical dosage forms.

Isopropyl alcohol (2-propanol) is also commonly listed in the inactive ingredient guide for topical lotions (2.7 to 78%) and solutions (4 to ~50%). Indeed, this alcohol is widely used at around 70% as an antimicrobial agent and has been shown to be effective for preoperative skin preparation (Keblish et al., 2005). However, isopropyl alcohol can also traverse human skin in significant amounts; ten healthy adult volunteers applied a commercially available isopropyl alcohol-containing hand rub every 10 min over a 4 h period. Blood levels were recorded in 9 of the 10 volunteers at levels ranging from 0.5 to1.8 mg/l (Turner et al., 2004).

Isopropyl alcohol can clearly have an effect on the skin barrier. Thus, it has recently been shown to increase delivery of capsaicinoids in human skin *in vivo* compared with delivery from mineral oil or propylene glycol (Pershing et al., 2004). Importantly, isopropyl alcohol poisoning has been reported in neonates and infants as a result of absorption through the skin. Because of this neonatal toxicity, isopropyl alcohol is excluded from antiseptics used for neonates, but it remains widely used in house cleaning and other products. Tragically, a neonatal death from a skin burn caused by his mother using a disposable antiseptic wipe has recently been reported (Brayer et al., 2004) and highlights the difficulties in selecting safe solvents for skin ranging from neonates to adults, in both health and disease.

Other alcohols are used as co-solvents or for other purposes in topical formulations, such as **benzyl alcohol** (α-hydroxytoluene) in many emulsions and creams as well as in some lotions (typically 0.2 to 2.5%). Fatty alcohols are often included as emulsifying agents, including **cetearyl alcohol** (usually around 2 to 12%), **cetyl alcohol** (up to ~10%) and **lanolin alcohols** (typically at 2 to 3%). However, allergic responses have been reported for lanolins and other wool alcohols and so these agents are largely being removed from newer formulations. Again, and as with most components of topical formulations, literature suggests that fatty alcohols (or alkanols) can act on the stratum corneum barrier to increase drug delivery. Thus, structure activity relationships have been drawn for fatty alcohol effects on drug delivery using melatonin permeating through porcine and human skin *in vitro* (Andega et al., 2001). When comparing activities for saturated fatty alcohols, ranging from octanol to myristyl alcohol, a parabolic relationship was found with a maximum effect for decanol. The alcohol structure was shown to be important with increased delivery noted when adding up to two unsaturated bonds into the alcohols, but activity fell where three double bonds were introduced.

Glycols

Glycols are probably the most commonly selected solvents for topical dosage forms and of these, **propylene glycol** (PG) is the most popular. Considering its chemical structure, it is not surprising that this solvent also readily penetrates into human stratum corneum, and its effects on topical drug delivery are probably similar to those suggested above for ethanol. Permeation of the solvent through the tissue could alter the thermodynamic activity of the vehicle which would in turn modify the driving force for diffusion and there may be some minor disturbance to the intercellular lipid packing within the stratum corneum bilayers on treatment with PG. Probably more important for topical formulations, the solvent may partition into the tissue and change the solvent properties of the stratum corneum so facilitating uptake of the drug into skin. This was elegantly demonstrated by Jacobi and co-workers (2005) who used laser scanning microscopy to probe the long term reservoir effects of PG, water and paraffin oil on fluorescent dyes in the stratum corneum. The results showed that the glycol maintained a reservoir of the dyes for up to 5 days in both inter- and intra-cellular domains with similar findings when using water as a solvent but with paraffin oil the reservoir effect was lost after 2 days. Propylene glycol is also widely used as a solvent or co-solvent for delivering penetration enhancers to the skin; while the glycol may have a minor enhancing effect itself, it often works synergistically with other enhancers, probably by improving partitioning of the agents into the stratum corneum (Williams and Barry, 1989).

Propylene glycol is extensively used in numerous topical formulations as illustrated in Table 3, abstracted from the FDA Inactive Ingredient Guide (1996). Over 230 drug products at that time contained the solvent, and at levels up to 99.99% of the formulation.

Occasional adverse reactions to PG have been reported. Recently, allergic contact dermatitis was reported from a patient exposed to a propylene glycol containing ultrasonic gel (Horiguchi et al., 2005). Similar findings have been reported from ECG electrodes (Connolly and Buckley, 2004) and also from topical preparations containing propylene glycol such as Efudix cream (Farrar et al., 2003).

Polyethylene glycols (PEGs) have also found use in topical preparations, though are generally used in emulsions and ointments. Molecular weights ranging from 200 up to 8000 have been selected and included at levels typically up to about 60%. Smaller polyethylene glycols have also been selected as solvents and

Topical dosage form	Number of new drug applications	Concentration range of propylene glycol in the dosage form/%
Aerosol	1	
Cream, augmented	1	
Emulsion, aerosol foam	3	5.376
Emulsion, cream	124	0.2—67.43
Gel	9	3.0—98.09
Lotion	24	2.0—50.0
Lotion, augmented	1	
Ointment	26	0.12—38.0
Ointment, augmented	2	10.0
Shampoo	1	
Solution	39	3.0—99.99
Sponge	2	
Suspension, shampoo	1	
Swab	3	25.0

Table 3. Classes of topical formulations described in the FDA Inactive Ingredient Guide (1996) containing propylene glycol with the number of new drug applications and their propylene glycol concentration range.

co-solvents for solution formulations; PEG 300 was used up to around 30% while solutions containing 9 to 50% PEG 400 have been approved. While less widely used than propylene glycol, the larger molecular weight of PEGs indicates that they are not readily absorbed through intact skin, though clearly where the skin barrier is compromised (as in diseased states for which a topical preparation may be applied) then absorption is possible. Thus, while PEGs have been reported as not sensitising normal skin, they are associated with contact dermatitis and systemic toxicity in burn patients (Lanigan and Yamarik, 2001).

Ethylene glycol and **hexylene glycol** have also been used in topical ointments and emulsions, though are not widely favoured.

Oils and Waxes

Mineral oil (liquid paraffin) and **light mineral oil** (light liquid paraffin) are widely used components of topical formulations such as emulsions, creams, lotions and ointments as well as other preparations such as baby lotions, sunscreens and cosmetics, and they provide cost effective solvents. Typically incorporated at up to ~15% in lotions but at levels up to 95% in ointments, mineral oils comprise a complex mixture of branched alkanes (also termed "paraffinics") and alkylated saturated ring compounds ("napthenics"). These materials are naturally occlusive, providing an additional layer on the stratum corneum surface that can reduce transepidermal water loss, so hydrating the tissue and potentially allowing greater drug delivery as described above for water. However, oils can also provide an additional skin surface barrier to penetration of polar molecules which must traverse this additional barrier, but could provide a reservoir for lipophilic materials on the skin surface when the bulk of the formulation has been removed. Further, mineral oils are also well known moisturising agents, again partly through skin surface occlusion and hence are a valuable emollient for topical formulations.

As mineral oils have been so widely used in medicinal and cosmetic preparations, extensive safety data exists showing very few adverse reactions. Using repeat insult patch tests, Mahagaokar (1996) found that a mineral oil gel did not produce sensitisation in human skin. Indeed, mineral oil has even been used as an over-the-counter orally administered laxative, illustrating its safety (though it does have an unpleasant taste!). Mineral oil can be valuable for delivering anhydrous agents, or for preparing medicinal agents at or near saturation so increasing thermodynamic activity and hence delivery. Wang et al. (2005) recently showed physostigmine delivery was greater from mineral oil than propylene glycol due to solubility differences of the drug in the vehicles. Clearly such an effect relates to the physicochemical properties of the medicinal agent as propylene glycol was superior to mineral oil for delivering capsaicinoids (Pershing et al., 2004). **Coconut oil** is less widely used in absorbable ointment bases, but also features in the FDA inactive ingredient guide, **palm kernel oil** is used as an emollient and ointment base and **almond oil** has been used as a base in ear drops.

In addition to nut oils, **vegetable oils** are common components of topical formulations. Typically fatty oils, the primary components are triglycerides but vegetable oils also contain varying amounts of waxes, free fatty acids and other glycerides. Vegetable oils are typically used in ointments, primarily where their emollient actions are desirable. However, vegetable oils vary in composition and, importantly, are readily oxidised requiring careful storage conditions (cool, dark) and/or the inclusion of anti-oxidants such as butylated hydroxyanisol. Different grades of vegetable oils can be obtained; *virgin oils* are obtained from particular grades of raw materials by a mechanical process such as cold expression. *Refined oils* are obtained following expression or solvent extraction and a refining step such as alkali refining followed by bleaching and deodorisation. *Hydrogenated oils* take either the virgin or the refined oils and hydrogenate to convert unsaturated fatty acids to saturated species, resulting in semi-solid or solid products.

Vegetable oils have long been used in topical formulations and hence many are widely regarded as safe. For example, **castor oil** from the castor seed has been used medicinally for over 6000 years. As a solvent, castor oil is surprisingly polar since over 90% of its fatty acid content is ricinoleic acid, a C_{18} fatty acid with a hydroxyl moiety at C_{12}. Castor oil is used orally as a laxative, but in topical formulations it acts as an emollient and it is a component (50%) of Zinc and Castor Oil Ointment BP, used as a barrier preparation in, for example, diaper rash. Hydrogenated castor oil is also used to provide stiffness to creams and emulsions, and is found in cosmetics as well as pharmaceutical preparations. Other derivatives of castor oil have also been used in commercial topical formulations, such as polyoxyethylene castor oil (polyoxyl castor oil); these materials act as emulsifying, solubilising and wetting agents. **Olive oil** also has a long history of medicinal use, and is regarded as a healthy dietary oil because it contains a high proportion of mono-unsaturated fat and polyphenols. Interestingly, the predominant mono-unsaturated fat is oleic acid, a well established skin permeation enhancer (see section 4.5). Again, olive oil tends to oxidise rapidly if stored inappropriately. Other vegetable oils used in topical formulations include **refined maize oil**, **refined soya oil**, **cottonseed oil**, **palm oil** and **vegetable oil** which may be a mixture of oils derived from various plants.

Soft and **hard paraffins** (variously termed paraffin wax, white soft paraffin, yellow soft paraffin, petrolatum, petroleum jelly) are also common components of primarily ointments, emulsions and creams, and are used up to 99.98% in ointments. These paraffins are often used as stiffening agents and are emollient for skin disorders; they also find use in sterile wound dressings. Through their widespread use, very few adverse reactions have been reported, though a series of studies by Dooms-Goossens and co-workers (1983a,b,c) showed that polycyclic aromatic hydrocarbon impurities can sensitise the skin. Consequently, highly purified white soft paraffin tends to be preferred to yellow soft paraffin.

Numerous **waxes** also find use largely in ointments and creams as stiffening agents. **Beeswaxes** are used in white and yellow ointments, and enable water to be incorporated to produce water-in-oil emulsions. However, some hypersensitivity reactions to beeswaxes have been reported. **Emulsifying wax** is produced primarily from cetostearyl alcohol and can be used to generate oil-in-water emulsions which are non-greasy and which readily absorb into the skin. **Cetyl esters wax** and **microcrystalline wax** are also used as stiffening agents in various ointments and creams.

Other Solvents

Several other solvents are used extensively in topical formulations. **Isopropyl myristate** (IPM) could be classified as an oily ingredient, which can be grouped within a series of isopropyl fatty acid esters such as **isopropyl linoleate** and **isopropyl palmitate**. IPM has been included in emulsions and gels at around 10%

and at 2% in lotions. It penetrates readily into the skin, and can affect partitioning and permeation of other compounds into and through the membrane. For example, Brinkmann and Muller-Goymann (2005) showed that IPM integrated into the lipophilic regions of the stratum corneum lipid matrix, an effect that was enhanced when IPM was used as a co-solvent with propylene glycol. IPM is generally accepted to be non- or only very mildly sensitising to the skin and so is a useful solvent for topical dosage forms (Uter et al., 2004).

Oleic acid is also used in some topical formulations such as White Liniment BP. However, oleic acid has been shown to effectively enhance delivery of numerous chemicals to and through human skin, and indeed is widely regarded as one of the most potent penetration enhancers. For example, oleic acid increased the flux of salicylic acid 28-fold and 5-flourouracil flux 56-fold through human skin membranes *in vitro* (Goodman and Barry, 1989). The fatty acid is effective at relatively low concentrations (typically less than 10%) and can work synergistically when included in formulations with other solvents such as propylene glycol or dimethyl isosorbide. Considerable efforts have been directed at investigating the interactions of oleic acid with human skin constituents. It is clear from numerous literature reports that oleic acid interacts with and modifies the lipid domains of the stratum corneum, as would be expected for a long chain fatty acid with a *cis*-configuration. Spectroscopic investigations using deuterated oleic acid in human stratum corneum indicates that oleic acid at higher concentration can also exist as a separate phase (or as 'pools') within the bilayer lipids (Ongpipattanakul et al., 1991). More recently, electron microscopic studies have shown that a discreet lipid domain is induced within stratum corneum bilayer lipids on exposure to oleic acid (Tanojo et al., 1997). The formation of such pools would provide permeability defects within the bilayer lipids thus facilitating permeation of hydrophilic permeants through the membrane and increasing partitioning of lipophilic drugs into the oleic acid phase.

Silicones are polymers comprising alternate atoms of silicon and oxygen, with organic groups attached to the silicon atoms. Various grades are available and so they have many uses as resins, waxes and rubbers, including use as adhesives for topical application. **Dimethicones** (Dimeticone in Europe) are fluid silicones with the organic element a methyl group. These fluid silicones are water repellent and have been used to protect the skin from water soluble irritants. Thus they have been used to protect against bed sores and diaper rash. However, though rare, adverse reactions to silicones have been reported, notably when used for joint or breast implants.

Many other excipients are incorporated into topical preparations, including preservatives, fragrance materials and surfactants. Though not typically employed as solvents *per se*, these materials may have solubilising properties, such as **sodium lauryl sulphate** or **terpene** fragrances. Again, many of these materials can interact with the stratum corneum to modify drug delivery.

Solvents Not Widely Used for Topical Preparations

While the above describes some of the more commonly used solvents for topical formulations, and emphasises those that appear in the FDA inactive ingredient guide (1996), the listing does not describe some of the solvents that are widely used in research studies of topical delivery. **Dimethylsulphoxide** (DMSO) and similar solvents such as **dimethylacetamide** (DMA) and **dimethylformamide** (DMF) are popular in these studies, but present problems. All three of these solvents penetrate well into human skin and can be potentially toxic; dimethylformamide is a widely used industrial solvent and cases have reported testicular cancer from exposure, though this may have been due to the solvent enhancing skin uptake of heavy metal carcinogens. Both DMA and DMF are Class 2 solvents according to the FDA Q3C Guidance for Industry (2003), and as such are to be limited in pharmaceutical products because of their inherent toxicity.

Dimethylsulphoxide is a "universal solvent" but can cause itching erythema and uticaria when applied to skin. However, it has been used medicinally in bladder instillations (at 50% in water) for interstitial cystitis, and is a vehicle for idoxuridine for herpes infections (though is of little value). Together with **acetone**, DMSO is a class 3 solvent under the FDA guidance, not known as a human health hazard at levels normally expected in pharmaceuticals and which can thus be limited by appropriate GMP practices.

Conclusions

From the above it is readily apparent that solvent selection for topical dosage forms is far from trivial. It is thus impractical to simply state which solvent should be selected for a given drug. The complexity of solvent selection is illustrated through the example decision tree given in Figure 5. In this example, the first factor to be considered is the diseased state and nature of the skin barrier to which the formulation will be applied. This is often overlooked in formulation science where physicochemical principles tend to be the primary concern.

It is important to note that all solvents can affect (one or more of) the skin barrier, the formulation thermodynamics and/or release and can promote delivery by interacting with the membrane; even the most "inert" of solvents, water, can dramatically affect the stratum corneum barrier and so alter drug delivery to and through the tissue. Further, the biology of the tissue can also be modified by solvent application—microbiological, immunological and histological changes have been reported for most solvents when applied to human skin. Add to this the further level of complexity in that most formulations employ mixed/co-solvent systems and that many solvents have complex synergistic influences (such as isopropyl myristate with propylene glycol) then selection of a solvent can be fraught with difficulties. However, accepting that the vehicle will

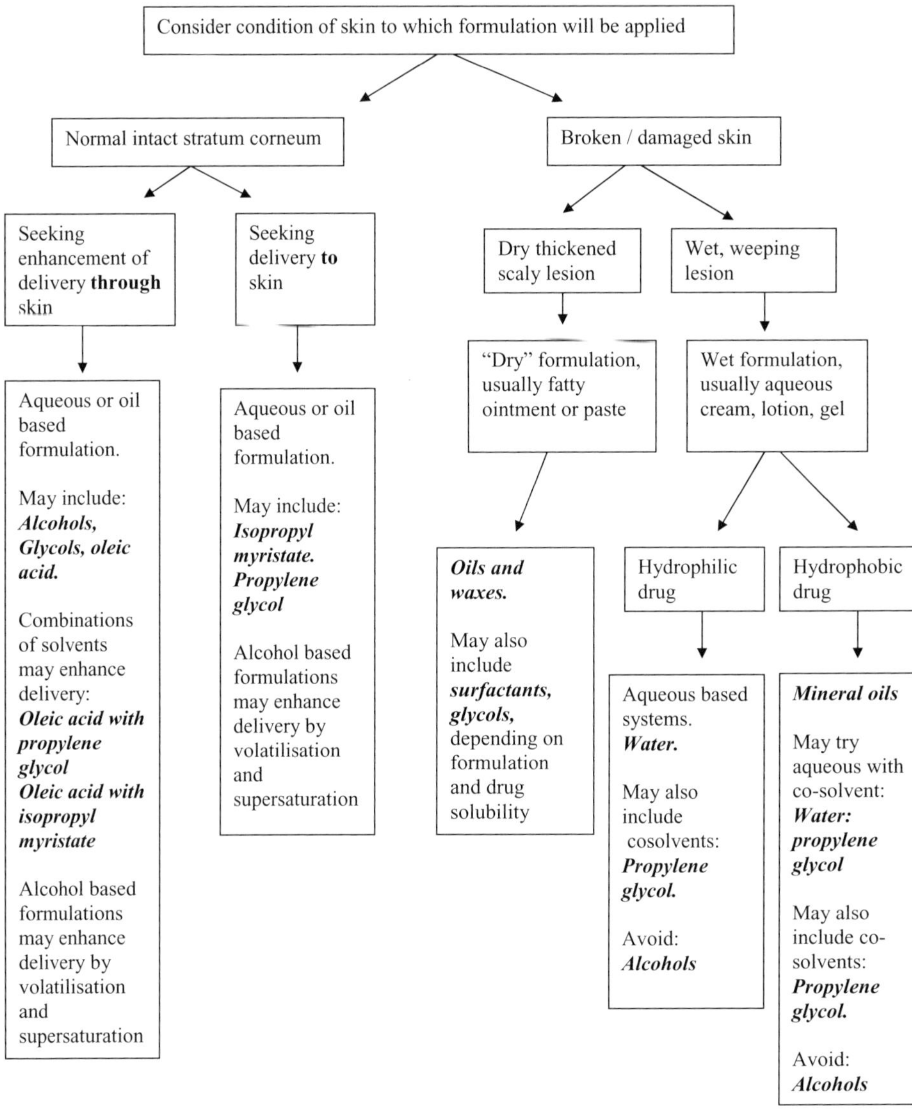

Figure 5. An example of a decision tree for solvent selection in topical formulations. The tree is not exhaustive but illustrates some of the factors to consider when selecting a solvent.

have some influence on the barrier, then increased drug delivery by appropriate selection of the solvent, such as ensuring saturation of the active pharmaceutical ingredient so maximising thermodynamic activity and hence delivery or choosing a vehicle to occlude the skin surface, is feasible. However, it must be borne in mind that vehicle selection will also depend on the disease state to be treated.

References

Anigbogu ANC, Williams AC, and Barry BW. Permeation Characteristics of 8-methoxypsoralen through Human Skin; Relevance to Clinical Treatment. *J Pharm Pharmacol* 1996; 48: 357–366.

Behl CR, Flynn GL, Kurihara T, Harper N, Smith W, Higuchi WI, Ho NFH, and Pearson CL. Hydration and Percutaneous Absorption. I. Influence of Hydration on Alkanol Permeation through Hairless Mouse Skin. *J Invest Dermatol* 1980; 75: 346–352.

Berner B, Mazzenga GC, Otte JH, Steffens RJ, Juang R-H, and Ebert CD. Ethanol: Water Mutually Enhanced Transdermal Therapeutic System II: Skin Permeation of Ethanol and Nitroglycerin. *J Pharm Sci* 1989; 78: 402–407.

Blank IH, Moloney J, Emslie AG, Simon I, and Apt C. The Diffusion of Water Across the Stratum Corneum as a Function of its Water Content. *J Invest Dermatol* 1984; 82: 188–194.

Brayer C, Micheau P, Bony C, Tauzin L, Pilorget H, Samperiz S, and Alessandri JL. Neonatal Accidental Burn by Isopropyl Alcohol. *Archives de Pediatrie* 2004; 11: 932–935.

Brinkmann I and Muller-Goymann CC. An Attempt to Clarify the Influence of Glycerol, Propylene Glycol, Isopropyl Myristate and a Combination of Propylene Glycol and Isopropyl Myristate on Human Stratum Corneum. *Pharmazie* 2005; 60: 215–220.

Bucks D and Maibach HI. Occlusion does not Uniformly Enhance Penetration in vivo. Chapter 4. In: Bronaugh RL and Maibach HI (eds.). *Percutaneous Absorption; Drugs—Cosmetics—Mechanisms—Methodology (3rd Edition).* Marcel Dekker Inc., New York, 1999.

Connolly M and Buckley DA. Contact Dermatitis from Propylene Glycol in ECG Electrodes, Complicated by Medicament Allergy. *Contact Dermatitis* 2004; 50: 42.

Cross SE and Roberts MS. Subcutaneous Absorption Kinetics of Interferon and other Solutes. *J Pharm Pharmacol* 1993; 45: 606–609.

Dooms-Goossens A and Degreef H. Contact Allergy to Petrolatums. 1. Sensitizing Capacity of Different Brands of Yellow and White Petrolatums, *Contact Dermatitis* 1983a; 9: 175–185.

Dooms-Goossens A and Degreef H. Contact Allergy to Petrolatums. 2. Attempts to Identify the Nature of the Allergens, *Contact Dermatitis* 1983a; 9: 247–256.

Dooms-Goossens A and Dooms M. Contact Allergy to Petrolatums. 3. Allergenicity Prediction and Pharmacopoeial Requirements, *Contact Dermatitis* 1983a; 9: 352–359.

Elias, PM. Epidermal Lipids, Membranes and Keratinization. *Int J Dermatol* 1981; 20: 1–9.

Farrar CW, Bell HK, and King CM. Allergic Contact Dermatitis from Propylene Glycol in Efudix (R) Cream. *Contact Dermatitis* 2003; 48: 345–345.

Goodman M and Barry BW. Lipid-protein-Partitioning Theory (LPP) Theory of Skin Enhancer Activity: Finite dose Technique. *Int J Pharm* 1989; 57: 29–40.

Horiguchi Y, Honda T, Fujii S, Matsushima S, and Osaki Y. A Case of Allergic Contact Dermatitis from Propylene Alycol in an Ultrasonic Gel, Sensitized at a Leakage Skin Injury Due to Transcatheter Arterial Chemoembolization for Hepatocellular Carcinoma. *Int J Dermatol* 2005; 44: 681–683.

Keblish DJ, Zurakowski D, Wilson MG, and Chiodo CP. Preoperative Skin Preparation of the Foot and Ankle: Bristles and Alcohol are Better. *Bone and Joint Surgery-American Volume*, 2005; 87A: 986–992.

Kurihara-Bergstrom T, Knutson K, De Noble LJ, and Goates CY. Percutaneous Absorption Enhancement of an Ionic Molecule by Ethanol-Water Systems in Human Skin. *Pharm Res* 1990; 7: 762–766.

Jacobi U, Waibler E, Sterry W, and Lademann J. In Vivo Determination of the Long-Term Reservoir of the Horny Layer using Laser Scanning Microscopy. *Laser Physics* 2005; 15: 565–569.

Lampe MA, Burligame AL, Whitney J, Williams ML, Brown BE, Roitman E, and Elias PM. Human Stratum Corneum Lipids; Characterisation and Regional Variations. *J Lipid Res* 1983; 24: 120–130.

Lampe MA, Williams ML, and Elias PM. Human Epidermal Lipids; Characterization and Modulations During Differentiation. *J Lipid Res* 1983; 24: 131–140.

Lanigan RS and Yamarik TA. Final Report on the Safety Assessment of PEG-6,-8, and-20 Sorbitan Beeswax. *Int J Toxicol* 2001; 20: 27–38.

Mahagaokar, S. Evaluation of the Skin Sensitization Potential of White Mineral Oil, Petrolatum, and Gelled White Oil. *J Toxicol Cut Ocular Toxicol* 1996; 15: 315–323.

McKenzie AW and Stoughton RB. Method for Comparing Percutaneous Absorption of Steroids. *Arch Dermatol* 1962; 86: 608–610.

Megrab NA, Williams AC, and Barry BW. Oestradiol Permeation Across Human Skin, Silastic and Snake Skin Membranes: the Effects of Ethanol/Water cosolvent Systems. *Int J Pharm* 1995; 116: 101–112.

Melnik BC, Hollmann J, Erler E, Verhoeven B, and Plewig G. Microanalytical Screening of all Major Stratum Corneum Lipids by Sequential High Performance Thin-Layer Chromatography. *J Invest Dermatol* 1989; 92: 231–234.

Michaels AS, Chandrasekaran SK, and Shaw JE. Drug Permeation Through Human skin: Theory and In Vitro Experimental Measurement. *AIChE J* 1975; 21: 985–996.

Morimoto Y, Wada Y, Seki T, and Sugibayashi K. In Vitro Skin Permeation of Morphine Hydrochloride During the Finite Application of Penetration-Enhancing System Containing Water, Ethanol and l-menthol. *Biol Pharm Bull* 2002; 25: 134–136.

Ongpipattanakul B, Burnette RR, Potts RO, and Francoeur ML. Evidence that Oleic Acid Exists in a Separate Phase within Stratum Corneum Lipids, *Pharm Res* 1991; 7: 350–354

Pershing LK, Reilly CA, Corlett JL, and Crouch DJ. Effects of Vehicle on the Uptake and Elimination Kinetics of Capsaicinoids in Human Skin in Vivo. *Toxicol Appl Pharmacol* 2004; 200: 73–81.

Tanojo H, BosvanGeest A, Bouwstra JA, Junginger HE, and Bodde HE. *In Vitro* Human Skin Barrier Perturbation by Oleic Acid: Thermal Analysis and Freeze Fracture Electron Microscopy Studies, *Thermochim Acta* 1997; 293: 77–85.

Thomas NS and Panchagnula R. Transdermal Delivery of Zidovudine: Effect of Vehicles on Permeation Across Rat Skin and Their Mechanism of Action, *Pharm Sci* 2003; 18, 71–79.

Turner P and Saeed Band Kelsey MC. Dermal Absorption of Isopropyl Alcohol from a Commercial Hand Rub: Implications for its Use in Hand Decontamination. *J Hosp Infection* 2004; 56: 287–290.

Uter W, Schnuch A, Geier J, and Lessmann H. Isopropyl Myristate Recommended for Aimed Rather than Routine Patch Testing. *Contact Dermatitis* 2004; 50: 242–244.

Walkley K. Bound Water in Stratum Corneum Measured by Differential Scanning Calorimetry. *J Invest Dermatol* 1972; 59: 225–227.

Wang MY, Yang YY, and Heng PWS. Skin Permeation of Physostigmine from Fatty Acids-Based Formulations: Evaluating the Choice of Solvent. *Int J Pharm* 2005; 290: 25–36.

Wertz PW and Downing DT. Stratum Corneum: Biological and Biochemical Considerations. In: Hadgraft J and Guy RH (eds.). *Transdermal Drug Delivery: Developmental Issues and Research Initiatives.* Marcel Dekker, Inc., New York 1989; pp. 1–22.

Wilhelm K, Surber C, and Maibach HI. Effect of Sodium Lauryl Sulfate-Induced Skin Irritation on In Vivo Percutaneous Absorption of Four Drugs. *J Invest Dermatol* 1991; 97: 927–932.

Wilkes GL, Brown IA, and Wildnauer RH. The Biomechanical Properties of Skin. *CRC Crit Rev Bioeng* 1973; 453–495.

Williams AC. *Transdermal and Topical Drug Delivery; From Theory to Clinical Practice.* London, Pharmaceutical Press; 2003.

Williams AC and Barry BW. Urea Analogues in Propylene Glycol as Penetration Enhancers in Human Skin. *Int J Pharm* 1989; 56: 43–50.

14

Pharmaceutical Solvents for Pulmonary Drug Delivery

ERIK MOGALIAN AND PAUL BRIAN MYRDAL

College of Pharmacy, University of Arizona, Tucson, AZ, USA

Introduction

It must first be recognized that formulating compounds and delivering them as aerosols is complex. Not only does it involve the formulation of a stable solution or suspension in a medium (propellant) that is not as well characterized as other systems, but the resultant system is also subject to performance limitations. In order to efficiently reach the lung, the formulation must be atomized into particles having aerodynamic sizes between approximately 1 and 5 microns. Due to these particle size constraints, as well as inhalation toxicology concerns, the range of possible excipients to choose from during the formulation phase is substantially reduced. Additionally, limiting the concentration of excipients in a formulation is crucial for maintaining adequate aerosol performance. Thus, given the complexity of this relationship, formulating aerosols is a challenging endeavor.

Although complex, the successful formulation of drugs for pulmonary delivery provides a valuable therapeutic route. Upon introduction of the metered dose inhaler (MDI), medical treatment of lung diseases changed significantly. Since that time, MDIs have become the most effective means of controlling symptoms of lung diseases such as asthma and chronic obstructive pulmonary disorder (COPD). More recently, formulation modifications were merited when chlorofluorocarbon (CFC) propellants were linked to the depletion of the ozone layer (Molina and Rowland, 1974). With the successful transition to new propellant systems, MDIs are still well accepted and highly utilized by patients across the globe today. Looking forward, the effectiveness, ease of use, and relatively low cost of aerosol preparations in combination with modifications in delivery technology and formulation sciences, will likely expand the treatment of diseases previously untreated via the respiratory tract. The approval of inhaled insulin in early 2006 is a paramount example of the potential for delivering proteins and peptides via the pulmonary route.

A Brief History: From CFCs to HFAs

The first metered dose inhaler, Medihaler Epi™ was introduced in 1956 by Riker Laboratories (3M Pharmaceuticals) for the management of asthma and COPD. The delivery device has been well accepted since that time, as evidenced by the annual production of over a half-billion units (McDonald and Martin, 2000). Since the introduction of the MDI, β-adrenergic agonists, anticholinergics, corticosteroids, and cromolyn compounds have been the staple of management for the commonly occurring lung diseases, asthma and COPD.

Historically, the MDIs have utilized CFC propellants to provide the energy for atomizing the formulation upon actuation. Some characteristics that made CFC propellants attractive for pharmaceutical aerosols were their limited toxicity, inertness and suitable vapor pressures (Smyth, 2003). CFCs were not only readily used in MDIs, but were also highly utilized in a myriad of household aerosol sprays, air conditioners (as refrigerants), fire extinguishers, and also for industrial manufacturing of foams and insulations, including NASA's application of insulation to space shuttle rocket boosters (NASA, website). Despite the significant advances that CFC propellants enabled, they were found to be contributing to depletion of the ozone layer and to the greenhouse effect (Molina and Rowland, 1974).

Due to the environmental ramifications of CFC use, the Montreal Protocol was devised, and then ratified in 1989, initiating the phase-out of CFC propellants, including those used in MDIs (Figures 1a-c). As of 2002, the Montreal Protocol has been ratified by 183 countries (UN Environmental Program, 1996). However, because pharmaceutical inhalers are considered life saving for many asthmatic and COPD patients, they were exempted from the protocol pending availability of suitable alternatives (FDA, 21CFR(2), 2002).

As a result of the Montreal Protocol, significant research and resources were invested for the development of alternative non-CFC containing products, namely dry powder inhalers and nebulized solutions, in addition to other suitable propellants to replace CFCs for use in MDIs. Two candidates for CFC replacement were identified, HFA 134a (Figure 1d) and HFA 227 (Figure 1e). These hydrofluoroalkanes lack the ozone depleting characteristics of their predecessors; however, they still contribute to the greenhouse effect, albeit to a lesser degree than their CFC counterparts, as displayed in Table 1 (Smith, 1995). Additionally, the half-life of these HFA propellants in the atmosphere is a fraction of that of the CFCs they would ultimately replace (McDonald and Martin, 2000).

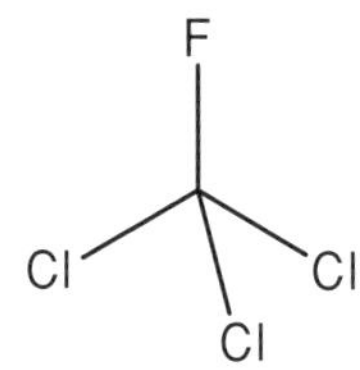

Figure 1a. CFC 11, trichlorofluoromethane.

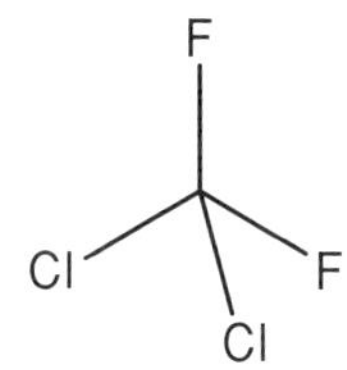

Figure 1b. CFC 12, dichlorodifluoromethane.

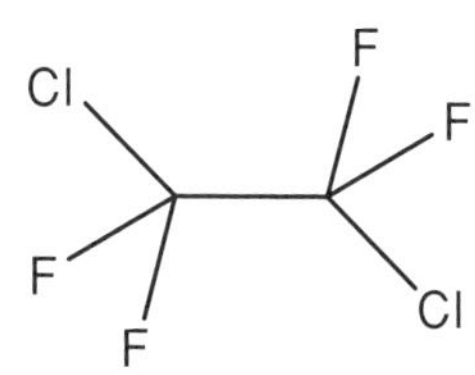

Figure 1c. CFC 114, 1,2-dichlorotetrafluoroethane.

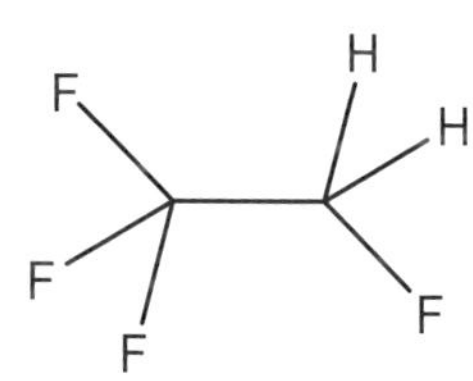

Figure 1d. HFA 134a, 1,1,1,2-tetrafluoroethane.

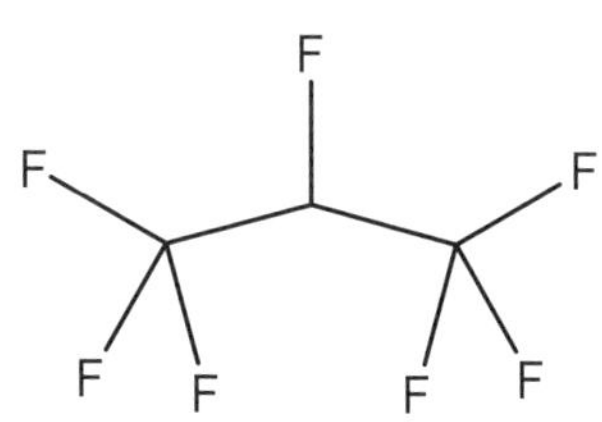

Figure 1e. HFA 227, 1,1,1,2,3,3,3-heptafluoropropane.

Choosing a Propellant: Characteristics of HFA 134a and 227

Several authors have described the physiochemical properties and differences observed between CFCs and HFAs (Smyth, 2003, Pischtiak et al., 2000, Vervaet and Byron, 1999, Tiwari et al., 1998). As Table 2 exhibits, the two HFA

Propellant	Ozone depletion potential*	Atmospheric life (years)	Global warming potential*
CFC 11	1	60	1
CFC 12	1	125	3
CFC 114	0.7	200	3.9
HFA 134a	0	16	0.3
HFA 227	0	33	0.7

Table 1. Environmental impact of pharmaceutical propellants.

* Relative to CFC 11. [7]

propellants that are currently utilized have boiling points and vapor pressures comparable to CFC 12 (dichlorodifluoromethane), the chief propellant used to obtain sufficient vapor pressure when used in blends (McDonald and Martin, 2000). However, HFA 134a and 227 do not display the same solvency characteristics of the CFCs (Vervaet and Byron, 1999). Presumably, this is due to the lack of polarizability of the fluorinated hydrocarbons as compared to the partially chloro-substituted CFCs (Smyth, 2003). This decrease in polarizability relative to CFC propellants could help explain some solubility differences of solutes in HFA based systems, despite their increased polarity over CFCs. Another major difference between the propellants is the hydrogen(s) on the HFAs, resulting in an increased dipole moment relative to CFC propellants which are completely chloro- and fluoro-substituted. As a result of this dipole, the highly electropositive hydrogen(s) appear to make the environment much less amenable to nonpolar solutes, while potentially enabling a degree of hydrogen bonding.

Although the above characteristics may begin to explain the difference in observed propellant-excipient/drug interactions, it is arguably academic, as CFC propellants are not a propellant option for future therapeutics. Thus, when formulating MDIs, there are only two propellants currently available, HFA 134a and HFA 227 (pictured in Figures 1d and 1e, respectively).

Propellant	Liquid density (g/ml)	Dipole moment (debye)	Boiling point (°C)	Vapor pressure (psig @ 20 °C)
CFC 11	1.49	0.46	23.7	−1.8
CFC 12	1.33	0.51	−29.8	67.6
CFC 114	1.47	0.50	3.6	11.9
HFA 134a	1.21	2.06	−26.3	68.4
HFA 227	1.41	0.93	−16.5	56

Table 2. Physiochemical properties of pharmaceutical propellants [10, 11].

Differences in the physical properties of HFAs, although minor, may warrant using 134a versus 227, or vice versa for a given formulation. Purity profiles of both propellants show a very low degree of impurity (both > 99.9% pure) (Solvay, 227, 134a Prod. Information), and do not significantly impact the choice of propellant. Compared to CFC propellants, both HFAs have relatively low boiling points (as seen in Table 2) which afford sufficient vapor pressure at lower temperatures without compromising efficiency (Stein and Stefely, 2003; Hoye et al., 2005). Additionally, they are completely miscible in one another and vapor pressure upon mixing behaves ideally, thus they may be blended in different proportions to obtain a specific vapor pressure or density (Williams et al., 1998). There is no toxicological advantage to either propellant, nor is there a degradation concern of one over the other, under relatively normal storage conditions (Solvay, 227, 134a Prod. Information). That said, some potential differences which may persuade a formulator mainly relate to chemical structure and resultant properties. HFA 227 has a $\log K_{ow}$ of 2.05 versus 1.06 for 134a, and as such, water has nearly four-fold increased solubility in HFA 134a versus 227 (2200 and 610 ppm, respectively). Of note, both HFA 134a and 227 have significantly greater water uptake as compared to the aforementioned CFC propellants (all ~120ppm), likely due to the relatively increased polarity (Solvay, 227, 134a Prod. Information). Thus, when formulating a suspension of a compound, physical stability as a function of water shall require addressing. Likewise, if in a solution formulation, the compound of interest is water labile, HFA 227 may offer advantages, though the formulation may still be susceptible to water migration. Williams and Hu (2000) confirmed these findings experimentally, though did not obtain the same magnitude of difference. They also showed that depending on the drug, emitted particle size could change, and likewise the fine particle fraction (fraction of aerosol less than 4.7 microns). Additionally, they indicated container lining and storage temperature for impacting increased water content.

Excipients for Metered Dose Inhalers

The Montreal Protocol, mandating the discontinuation of CFC propellants, gave the pharmaceutical industry a chance to reevaluate technical attributes of the MDI system. Due to the different physical characteristics of the propellants, excipients previously used in CFC based systems do not behave the same in HFA propellants and therefore also needed to be reinvestigated.

Surfactants were one excipient family which required reevaluation in HFAs. Surfactants are used in MDIs for several reasons: seal lubrication, emulsification, dispersion, solubilization, and as a preservative. Surfactants used in CFC formulations such as lecithin, sorbitan trioleate, soya lecithin, and oleic acid are highly soluble in CFC propellants (particularly CFC 11), however in HFA propellants, the solubility of these surfactants is relatively poor. Several authors point out that surfactant polarity, indicated by their respective hydrophilic-lipophilic balance (HLB) correlates with the incompatibility of the aforementioned surfactants

Excipient[a]	Product	Function	Maximum approved concentration[b]
Sorbitan trioleate (Span 85)	Aerobid, Alupent, Intal, Maxair, Tilade	Surfactant Dispersion Suspension Solubilization agent	0.069%
Soya lecithin	Atrovent, Combivent	Dispersion	0.28%
Lecithin	Flovent, Serevent	Dispersion Solubilization	0.00025%
Oleic acid	Beclovent, Proventil, Proventil HFA, Vanceril, Ventolin, Xopenex HFA	Dispersion Emulsification	0.267%
Cetylpyridinium chloride	Asthmahaler Mist, Bronkaid Mist	Preservative Surfactant	

Table 3. Commonly used surfactants in previously approved products.

[a]All excipient tables (3, 4, 5) were produced using US FDA Center for Drug Evaluation and Research Inactive Ingredient List for Approved Drug Products, updated 10/31/2005, along with product monographs for respective listed products.

[b]Maximum concentration in a product previously approved by the United States Food and Drug Administration.

in the more polar HFA environment (Vervaet and Byron, 1999; Ridder et al., 2005). Table 3 lists previously approved surfactants utilized in MDIs along with their maximum concentration approved by the United States Food and Drug Administration (U.S. FDA).

Co-solvents in metered dose inhalers were commonly used in CFC formulations to aid in drug solubilization. In HFA formulations, co-solvents continue this same function, but have additional benefits in the new systems, such as solubilization of other excipients. Vervaet and Byron discuss water solubility in the various propellants addressed here, where the addition of ethanol to the HFA system considerably increases the solubility of water (Vervaet and Byron, 1999). Likewise, ethanol was found to increase the solubility of several surfactants in HFA (Vervaet and Byron, 1999; Stein and Stefely, 2003). Suspension formulations using this technique (surfactant plus ethanol) must be made with caution however, as ethanol can also increase the solubility of the drug substance, potentially causing increased particle growth via Ostwald ripening. Nonetheless, without the use of ethanol as a co-solvent, several currently accepted surfactants would be virtually unusable in HFA systems.

Dosing Efficiency

Although co-solvent use is a successful method of improving drug/excipient solubilization in aerosol systems, there are limitations to this technique. Simply increasing the drug concentration in a formulation does not guarantee an increase in the total amount of respirable drug delivered.

The addition of non- or semi-volatile excipients (including the active drug substance) has multiple effects on the delivery process. These systems will have a reduced vapor pressure, which affect the atomization of the formulation. All else equal, there is less force generating aerosol droplets of a given surface tension, viscosity, etc., thus larger initial droplets. Second and potentially more important, these larger (and fewer) aerosol droplets take longer to dry as the total surface area is decreased, compared to a higher number of smaller aerosolized droplets (analogous to micronizing solid drug particles to increase solubility). The result of these two dynamic processes, albeit occurring simultaneously, results in larger final particle diameter. An increased proportion of larger particles directly correlates to a decrease in respirable fraction/mass (decreased fine particle fraction/mass), thus decreased efficiency (Stein and Myrdal, 2006).

A prime example of this relationship was obtained from Gupta et al. (2003) where a linear increase in the solubility of beclomethasone dipropionate was observed with an increase in ethanol concentration from 0 to 20% (Gupta et al., 2003). Unfortunately, respirable fraction displayed an accompanying decrease (Myrdal et al., 2004a; Mogalian et al., 2005). Examining the sum of this relationship reveals diminishing increases in respirable deposition once exceeding 10% ethanol in the system (see Figure 2). Therefore, when increasing solubility of a compound via co-solvent (or other excipients) to achieve an appropriate dose or dose per actuation, delivery of the formulation may be simultaneously compromised, thus limited net gain is achieved. In some instances however, this relationship could be advantageous. If formulating an aerosol that displays a mass median aerodynamic diameter (MMAD) smaller than desired, as in the case of dilute (low dose) solution formulations, addition of a less volatile excipient such as water or propylene glycol, could help increase the particle size to a more desirable range of aerodynamic diameter.

Table 4 lists non-surfactant excipients that have been previously approved for use in MDIs by the U.S. FDA. The maximum concentration listed in Table 4 (and other related tables) was obtained from this source; however, the excipients in given products were obtained from their respective drug monographs. Given the independent sources of this information, it is impossible to determine if the cited concentration correlates to a particular product. Citing ethanol as an example, note that although the maximum approved concentration is 34.5%, it would greatly decrease the performance of a MDI, and likely no newer formulation

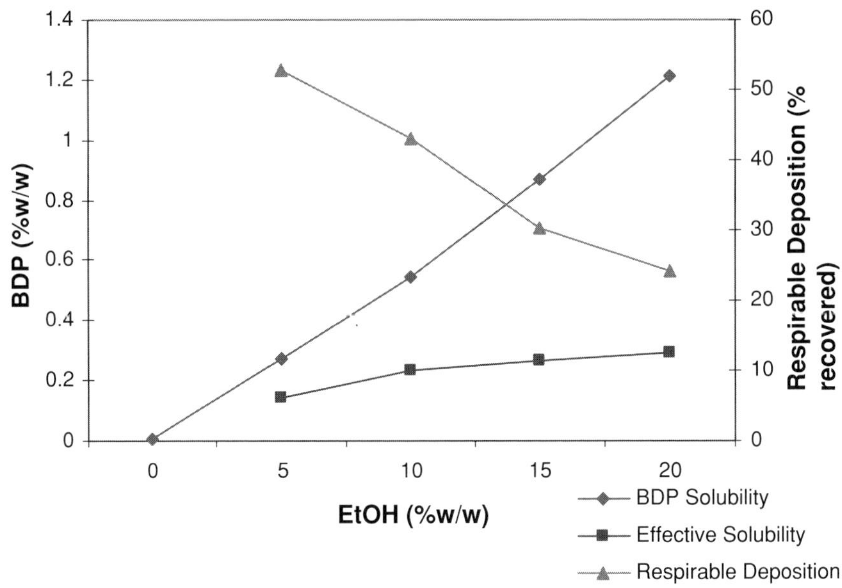

Figure 2. Gupta et al. [18] shows a linear increase in drug solubility with the addition of ethanol; however, there is an accompanying decrease in aerosol performance. Due to decreased efficiency, limited gains in respirable mass are observed when adding more than 10% ethanol (BDP = beclomethasone dipropionate).

would contain ethanol in that quantity. More typically, ethanol concentrations from 1-20% w/w are found in the literature.

Considerations for Nebulized Solutions

In formulating solutions for nebulization, more common techniques are available for solubilizing compounds. Unlike MDIs, nebulization solutions are aerosolized via an external energy source (ultrasonic or jet spray) in place of propellants. Thus, solutions for nebulization can be formulated as routine aqueous systems with additional considerations being made for the pulmonary toxicity of the excipients, volatility (with regard to final particle size), and flavor. Citric acid, sodium citrate, and saccharin sodium are three flavoring agents which have been used in nebulization solutions, the former two of these also acting as buffers to control pH within the formulation.

In addition to the buffers previously mentioned, other agents have also been employed to adjust the pH of nebulization solutions for drug stability and/or solubility reasons (Steckel et al., 2003). These agents include hydrochloric acid, sulfuric acid, and sodium hydroxide, and all have been used in marketed products, as seen in Table 5.

Microbial studies in MDIs showed that CFC propellant blends tended to be bactericidal against the commonly occurring bacterial contaminant *Staphylococcus aureus*. The same was found to be true for HFA 134a, however

Excipient[a]	Product	Function	Maximum approved concentration[b]
Ethanol/ Dehydrated Alcohol/ Alcohol	Azmacort, Isuprel, Primatene Mist, Tornalate, Qvar, Atrovent HFA, Proventil HFA, Xopenex HFA, Aerospan HFA, Alvesco	Co-solvent	34.5%
Water	Atrovent HFA	Co-solvent	
Menthol	Aerobid, Tornalate	Flavoring agent	0.05%
Saccharin sodium		Flavoring agent	0.045%
Saccharin	Tornalate	Flavoring agent	0.112%
Citric acid (anhydrous)	Atrovent HFA	Flavoring agent	0.00022%
Hydrochloric acid		pH adjustment	1.72%
Nitric acid		pH adjustment	1.67%
Ascorbic acid	Primatene Mist, Isuprel, Tornalate	Antioxidant	1.02%

Table 4. Common co-solvents and miscellaneous excipients in previously approved products.

[a]All excipient tables (3, 4, 5) were produced using US FDA Center for Drug Evaluation and Research Inactive Ingredient List for Approved Drug Products, updated 10/31/2005, along with product monographs for respective listed products.

[b]Maximum concentration in a product previously approved by the United States Food and Drug Administration.

HFA 227 was found to be bacteriostatic, which is at least sufficient to stop proliferation of *S. aureus* in the MDIs (McDonald and Martin, 2000). However, for nebulization solutions, propellants are not used and thus antimicrobial properties of these propellants do not apply; sterility must therefore be obtained by other measures. Table 5 displays several antimicrobial preservatives used in nebulization solution.

Excipient[a]	Product	Function	Maximum approved concentration[b]
Alcohol (ethanol)	Tornalate	Co-solvent	25%
Glycerin	Isuprel	Co-solvent Humectant Preservative Tonicity agent	7.3%
Propylene glycol	Tornalate	Co-solvent Preservative	25%
Methylparaben		Preservative	0.07%
Propylparaben		Preservative	0.037%
Chlorobutanol	Isuprel	Preservative	0.5%
Sodium meta bisulfite	Isuprel	Preservative	1%
Sodium bisulfite		Preservative	0.32%
Sodium sulfite		Preservative	0.1%
Sodium sulfate (anhydrous)		Tonicity agent	0.025%
Thymol		Preservative	0.01%
Benzalkonium chloride	Alupent, Proventil, Ventolin	Preservative Wetting agent Solubilizing agent	20%
Sodium chloride	Airet, Proventil, Isuprel, Xopenex, Atrovent, Duovent	Tonicity	3.16%
Sodium citrate/ Citric acid	Airet, Isuprel, Tornalate	Buffering agent Chelating agent Flavoring agent	0.6%/0.44%
Edetate sodium/ Edetate disodium	Airet, Alupent, Atrovent	Chelating agent	0.02%/0.03%

Excipient[a]	Product	Function	Maximum approved concentration[b]
Saccharin sodium		Flavoring agent	
Hydrochloric acid	Airet, Atrovent, Duovent	pH adjustment	3.5%
Sulfuric acid	Proventil, Ventolin, Xopenex	pH adjustment	12.5%
Sodium hydroxide	Tornalate	pH adjustment	8%
Ascorbic acid		Antioxidant	1.02%
Water			

Table 5. Previously approved excipients for inhaled solutions for nebulization.

[a]All excipient tables (3, 4, 5) were produced using US FDA Center for Drug Evaluation and Research Inactive Ingredient List for Approved Drug Products, updated 10/31/2005, along with product monographs for respective listed products.

[b]Maximum concentration in a product previously approved by the United States Food and Drug Administration.

Fortunately, many of the frequently used antimicrobial excipients are also used to enhance solubility. Glycerin and propylene glycol, for example, are both commonly used co-solvents but also exert antimicrobial properties through increased osmotic pressure. Similarly, benzalkonium chloride is a cationic surfactant that is commonly used as a wetting agent, emulsifying agent, as well as an antimicrobial. It exhibits antimicrobial effects through surface activity and is most effective against gram positive bacteria such as *S. aureus* via interaction with the cell wall. However, because it is a quaternary ammonium compound, it may not be compatible for use with coexisting anionic species, whether they be the drug itself, or other excipients in the formulation (Sigma-Aldrich, Prod. Information).

Novel Solubilization, Suspension, and Delivery Techniques

With the change to HFA propellants, several pharmaceutical companies saw an incentive to develop formulations, while companies with established inhalation products sought to protect their business (Stein and Stefely, 2003). Regardless, research in the aerosol field is far from limited to updating MDIs. New biological targets are regularly being discovered, which gives rise to new therapies and, ultimately, new formulation challenges. This section will address some of the newer trends in inhalation drug therapy, and some of the challenges of formulation.

Recently, biologically active proteins and peptides have received a great deal of interest. Although promising, inhalation therapy with proteins and peptide drugs has proved to be difficult. Due to the larger molecular size of these therapeutic proteins, and the importance of sterics to their activity, the high stress of aerosolization has proven to be of significance (Mumenthaler et al., 1994).

Williams and Liu investigated the delivery of the protein bovine serum albumin (BSA) via MDI as a suspension, using HFA 134a and combinations of ethanol with different surfactants and were able to obtain >1mg BSA/g formulation with respirable fractions up to 50% (Williams and Liu, 1998). Myrdal et al. examined cyclosporine as a model peptide in a formulation containing HFA 134a or 227 and low amounts of ethanol (2004b). They were able to deliver 500 μg/actuation with adequate particle size (MMAD roughly 2μm) and respirable fraction, while also showing formulation stability over a two year period. Additionally, Taljanski et al. demonstrated success dosing aerosolized cyclosporine A in rats utilizing the surfactant Cremophor® EL (Taljanski et al., 1997). This micellar solution resulted in increased pulmonary bioavailability over the regular solution (ethanol co-solvent) and suspension (saline based) in the study, thereby supporting the use of Cremophor® EL as a permeability enhancer for this drug, though the solution was not delivered by MDI (Taljanski et al., 1997).

Another recently developed delivery technology includes the use of phospholipids in HFA MDI formulations. Dellamary et al. describe their method of formulating an HFA based suspension of phospholipids, water and drug, using lecithin as a stabilizing agent (Dellamary et al., 2000). This formulation technique was initially employed to contain suspensions of cromolyn sodium, albuterol sulfate, or formoterol fumarate, and was later successfully tested using human immune globulin (IgG) to initiate an immune response in mice, via the respiratory tract (Bot et al., 2000).

Sustained release of aerosolized drugs has been investigated to some degree. Several approaches have been considered, including, liposomes, microspheres, prodrug formation, cyclodextrins, and *in vivo* precipitation. Many of these excipients are thought to demonstrate sustained-release properties via the same mechanism by which they display solvency characteristics. Liposomes, microspheres, and cyclodextrins were hypothesized to display sustained-release characteristics via entrapment or complexation. Zeng et al. describes a liposomal inhalation formulation containing sodium cromoglycate, an anti-asthmatic drug, where blood samples showed detectable levels of drug in the blood at 25 hours with a maximum concentration (C_{max}) seven times less than the regular solution formulation. The same solution formulation also resulted in undetectable concentrations in half of the time that the liposomal formulation took to reach undetectability, suggesting some degree of sustained-release (Zeng et al., 1995). Cyclodextrin formulations had received interest after showing slowed absorption rates of albuterol through pulmonary epithelial tissue *in vitro* when complexing with hydroxypropyl-β-cyclodextrin (HP-β-CD) (Wall et al., 1994). However, Wall et al. found that rat models showed no sustained-release properties with the same formulation. A phenomenon which may account for this has been previously documented, in which cholesterol would disrupt a drug complex and preferentially take the place of the drug, due to higher affinity of cholesterol for HP-β-CD (Frinjlink et al., 1991).

Conclusions

Pulmonary drug delivery has proven useful for decades due to its tremendous advantage in treating diseases of the lung. Recently, pulmonary drug delivery has benefited from advances in technological (device) design, enabling generation of consistent aerosol delivery while also more effectively controlling emitted particle size. A large impact has also taken place in aerosol drug formulation. Doing away with environmentally detrimental CFC propellants forced reformulation and, consequently, gave an opportunity to improve upon existing formulations. With this came an improved understanding of HFA propellants and how they interacted differently with excipients which were formerly well characterized.

Despite the success of reformulating previously marketed products, research still needs to be conducted evaluating the basic physical properties of HFA propellants for most efficient future application. This includes further evaluation of solubilization properties and prediction for use with both small and large molecule drugs, as well as newer excipients including novel co-solvents, surfactants, lipids, polymers, and cyclodextrins. Moreover, new biologically active targets or diseases previously untreated via the pulmonary route will lead to new drugs and inevitably additional formulation challenges.

List of Abbreviations

CFC........chlorofluorocarbon
COPD........chronic obstructive pulmonary disorder
HFA........hydrofluoroalkane
MMAD........mass median aerodynamic diameter
MDI........metered dose inhaler

References

Benzalkonium Chloride Product Information. Sigma-Aldrich Inc., http://www.sigmaaldrich.com/sigma/product%20information%20sheet/b1383pis.pdf (accessed 5/2006)

Bot AI, Tarara TE, Smith DJ, Bot SR, Woods CM, and Weers JG. Novel lipid-based hollow-porous microparticles as a platform for immunoglobulin delivery to the respiratory tract. *Pharm Res* 2000; 17(3): 275–283.

Dellamary LA, Tarara TE, Smith DJ, Woelk CH, Adractas A, Costello ML, Gill H, and Weers JG. Hollow Porous Particles in Metered Dose Inhalers. it Pharm Res 2000; 17: 168–174.

FDA, 21 CFR (2), Use of Ozone-Depleting Substances; Essential-Use Determinations, Federal Register, 2002.

Frinjlink HW, Eissens AC, Hefting NR, Poelstra K, Lerk CF, and Meijer DKF. The Effects of Parenterally Administered Cyclodextrins on Cholesterol Levels in the Rat. *Pharm Res* 1991; 8: 9–16.

Gupta A, Stein SW, and Myrdal PB. Balancing Ethanol Cosolvent Concentration with Product Performance in 134a-based Pressurized Metered Dose Inhalers. *J Aerosol Med* 2003; 16(2): 167–174.

Hoye WL, Mogalian E, and Myrdal PB. Effects of Extreme Temperatures on Drug Delivery of Albuterol Sulfate Hydrofluoroalkane Inhalation Aerosols. *Am J Health-Sys Pharm* 2005, 62: 2271–2277.

McDonald KJ and Martin GP. Transition to CFC-free Metered Dose Inhalers—into the New Millennium. *Int J Pharm* 2000; 201: 89–107.

Mogalian E and Myrdal PB. Application of USP Inlet Extensions to the TSI Impactor System 3306/3320 Using HFA 227 Based Solution Metered Dose Inhalers. *Drug Devel Ind Pharm* 2005; 31: 977–985.

Molina MJ and Rowland FS. Stratospheric Sink for Chlorofluoromethanes: Chlorine Atom-catalyzed Destruction of Ozone. *Nature* 1974; 249: 810–812.

Mumenthaler M, Hsu CC, and Pearlman R. Feasibility study on Spray-drying Protein Pharmaceuticals: Recombinant Human Growth Hormone and Tissue-type Plasminogen Activator. *Pharm Res* 1994; 11(1): 12–20.

Myrdal PB, Karlage KL, Stein SW, Brown BA, and Haynes A. Optimized Dose Delivery of the Peptide Cyclosporine Using Hydrofluoroalkane-Based Metered Dose Inhalers. *J Pharm Sci* 2004b; 93(4): 1054–1061.

Myrdal PB, Stein SW, Mogalian E, Hoye W, and Gupta A. Comparison of the TSI Model 3306 Impactor Inlet with the Andersen Cascade Impactor: Solution Metered Dose Inhalers. *Drug Devel Ind Pharm* 2004a; 30(8): 859–868.

Pischtiak AH, Pittroff M, and Schwarze. Characteristics, supply and use of the ydrofluorocarbons HFA 227 and HFA 134a for medical aerosols in the past, present and future—Manufacturers perspectives. *Drug Delivery to the Lungs XI.* 2000; 1–5.

Ridder KB, Davies-Cutting CJ, and Kellaway IW. Surfactant solubility and aggregate orientation in hydrofluoroalkanes. *Int J Pharm* 2005; 295: 57–65.

Smith IJ. The Challenge of Reformulation. *J Aerosol Med* 1995; 8(1): S19–S27.

Smyth HDC. The Influence of Formulation Variables on the Performance of Alternative Propellant-driven Metered Dose Inhalers. *Adv Drug Deliv Rev* 2003; 55: 807–828.

Solkane 227 pharma and Solkane 134a pharma Product Information. Solvay Fluor GmbH. www.solvay-fluor.com.

Space Shuttle Program Petition for HCFC 141b Exemption Allowance. NASA http://www.nasa.gov/pdf/45331main_hcfc5_002.pdf (accessed 5/2006)

Steckel H, Eskandar F, and Witthohn K. Effect of Excipients on the Stability and Aerosol Performance of Nebulized Aviscumine. *J Aerosol Med* 2003; 16(4): 417–432.

Stein SW and Myrdal PB. The Relative Influence of Atomization and Evaporation on Metered Dose Inhaler Drug Delivery Efficiency. *Aerosol Science and Technology* 2006; 40: 335–347.

Stein SW and Stefely JS. Reinventing Metered Dose Inhalers: From poorly Efficient CFC MDIs to Highly Efficient HFA MDIs. *Drug Deliv Tech* 2003; 3(1): 46–51.

Taljanski W, Pierzynowski SG, Lundin DPD, Westrom BR, Eirefelt S, Podnesly J, Dahlback M, Siwinska-Golebiowska H, and Karlsson BW. Pulmonary Delivery of Intratracheally Instilled and Aerosolized Cyclosporine A to Young and Adult Rats. *Drug Metabolism and Disposition* 1997; 25(8): 917–920.

Tiwari D, Goldman D, Malick WA, and Madan PL. Formulation and Evaluation of Albuterol Metered Dose Inhalers Containing Tetrafluoroethane (P134a), a non-CFC Propellant. *Pharm Dev Technol* 1998; 3(2): 163–174.

U.N. Handbook for the International Treaties for the Protection of the Ozone Layer, United Nations Environment Programme, Kenya, 1996.

Vervaet C and Byron PR. Drug-Surfactant-Propellant Interactions in HFA Formulations. *Int J Pharm* 1999; 186: 13–30.

Wall DA, Marcello J, Pierdomenico D, and Farid A. Administration as Hydroxypropyl β-cyclodextrin Omplexes does not Slow Rates of Pulmonary Drug Absorption in Rat. *STP Pharma Scienes* 1994; 4(1): 63–68.

Williams III RO and Hu C. Moisture Uptake and its Influence on Pressurized Metered-dose Inhalers. *Pharm Devel Technol* 2000; 5(2): 153–162.

Williams III RO and Liu J. Formulation of a Protein with Propellant HFA 134a for Aerosol Delivery. *Eur J Pharm Sci* 1998; 7: 137–144.

Williams III RO, Repka M, and Liu J. Influence of Propellant Composition on Drug Delivery from a Pressurized Metered-dose Inhaler. *Drug Devel Ind Pharm* 1998; 24(8): 763–770.

Zeng XM, Martin GP, and Marriott C. The Controlled Delivery of Drugs to the Lung. *International Journal of Pharmaceutics* 1995; 124: 149–164.

Index

A

Abraxaner®[12], 334
acceptable daily intake (ADI), 237
Accutane®, 278
acetaminophen, 270
acid ionization constant, defined as, 33
Active pharmaceutical ingredients (API), 78, 236
  role of crystallization in, 53–54
activity, 30–31
acylovir sodium, 317
ADME/TOX screens, 113
ADMET Predictor™, 224
Advil®, 277
aerosol solvent extraction system (ASES) mode, 384
Agenerase® oral solution, 245, 260, 302
Agenerase® soft gelatin capsules, 271, 296, 302
aggregate-based inhibition, 124
alfacalcidol, 271, 284
Alkeran®, 318
allopurinol sodium, 317
alprostadil (PGE1), 324, 326
ALZAmer® depot platform, 334
Amitiza™ soft gelatin capsules, 279, 291
amorphous solids, 16–17
  solubility of, 19
amphotericin B, 329
  injections, 321
amprenavir, 271, 302
amsacrine hydrochloride, 318–319
Aptivus® soft gelatin capsules, 282, 298
aurothioglucose, 331
autoionization, of water, 31–32

Avodart®, 276, 293
azacitidine, 331

B

base ionization constant, defined as, 34
benzoic acid, 40
betamethazone acetate+betamethasone sodium phosphate, 331
bexarotene, 272
bile acids, molecular structure of, 153
composition in human intestine, 154, 155
bile salts, 153
Critical Micellar Concentration (CMC) of, 152
kinetics of solubilization of phospholipids by, 154
synthesis by hepatocytes, 154
Biopharmaceutical Classification System (BCS), 137
application, 222–223
BCS Class II drug compounds, solid formulations of, 167
Class 2 compounds, 151
defined, 222
biorelevance
pH, bile salts and proteins for, 202–206
biorelevant media, 152, 160, 166–167, 169
biorelevant DSPAMPA model, 206
Bjerrum difference plots, 43
Boltzmann constant, 20
bovine growth hormone releasing factor (bGRF), 377
bovine serum albumin (BSA), 122, 205, 361, 364
brick dust category, of compounds, 234
Brønsted-Lowry acid model, 33
buffer capacity, 36
buprenorphine, 43
butylated hydroxy toluene (BHT), 291

C

Caco-2 cell culture model, 181, 186, 189, 191, 204, 229
calcitrol, 273, 291, 301
Cambridge Structural Database (CSD), 65
capric and caprylic mono- and diglycerides, 293
capsules, gelatin, soft and hard, 269–287
lipid-containing solubilizing vehicles, 291–299
solubilizing vehicles in oral formulations, 265–266
water-soluble solubilizing vehicles, 287–290

carbamazapine
  intrinsic dissolution rates of, 57–59
  chemical structure of, 65
  nucleation rate of, 64
  single-solvent polymorph screen for, 95, 96
  solvent-mediated transformation of, 77
carmustine, 319
cell culture models
  for permeability screening, 183
cetylpyridinium chloride, 432
ChemCards™ technology, 120
chemical stability, of a drug, 260–261
chemiluminescent nitrogen detectors (CLND), 119
ChiSys®, 334
chlordiazepoxide, 319
chlorofluorocarbon (CFC) propellants, 428–431
chloroperoxidase, 387
chlorpheniramine maleate, 270
chymotrypsin, 369
CiL® 200 mg Kapselin, 293
Clausius-Clapeyron equation, of solubility, 6
clofazimine, 284
clomethiazole edisilate, 285
co-solvent systems, 144
colfosceril palmitate (dipalmitoylphosphatidylcholine (DPPC)), 331
corn oil, 292
cortisone acetate, 332
coulombic interaction, 14
cremophor, 245–246
Cremophor® EL, 260, 262, 296, 310, 320, 438
Cremophor® RH40, 260, 264, 287
Cremophor® RH60, 264, 296
critical micellar concentration (CMC), 152, 320
crystal growth
  effect of molecular recognition phenomena on, 68
  imporatnce of desolvation step in, 68
  kinetics and stages in, 67, 68
  volume-diffusion controlled, 67
crystal structure prediction, 90
crystallization
  interplay between phenomena governing, 54
  kinetics and energetics of solvation and desolvation events during, 65
  processes involved in, 60–69
  role of solvent in, 54
  solvent systems for, 53

cyclodextrins, 47, 226, 267–268, 322–323, 350, 438
  regulatory status of, 243
  types, 242
  use, 242–243
cyclosporin A, 273, 275, 303, 321, 370, 386
CYP450 2D6 inhibition assay, 112
cytarabine, 329

D

d-*α*-tocopherol polyethylene glycol 1000 succinate (TPGS), 260, 264, 266, 295
danazol
  impact of food intake, 168
  solubility of, 165
daunorubicin, 328
DCS (developmental classification system), 223–224
Debye forces, 15
Debye-Hückel theory, 30, 367
Deoxycholic acid, as secondary bile acids, 156
Depakene® soft gelatin capsules, 283, 292
Depo-Medrol® Sterile Aqueous Suspension, 330
Developmental Classification System (DCS), 223–224
dexamethasone acetate, 333
dexamethasone palmitate, 326
dextromethorphan, 270
diazepam, 319, 326
diethanolamine, 39
differential scanning calorimetry (DSC), 140
digoxin, 275, 300, 319
dimethylacetamide (DMA), 235
dimethylformamide (DMF), 368
dimethylsulfoxide (DMSO), 140, 318, 361, 368, 371–372, 376–377, 381–382
dipalmitoylphosphatidylglycerol (DPPG), 374
Diprivan®, 323
dissolution phenomena, 20–23
DMSO, solubility issues
  effects of concentration, 113–115
  effects of freeze-thaw cycles, water uptake and crystallization, 116–117
  effect of storage at low concentration, 115
  impact of poor solubility on assay results, 112
  solubility prediction, 117, 118
  strategies to enhance solubility, 118–119
  strategies to overcome, 118–120
  stock solution, haystack solutions, 228

dolasetron mesylate monohydrate, 316
DotFoil™ concept, 119
doxercalciferol, 276
doxorubicin, 316, 325, 328
doxycycline hyclate, 316
dronabinol, 276, 292
drug absorption models
    advantages and disadvantages of, 183
    impact of bile salts in GI tract on, 188–190
    limitations of, 186–190
drug candidates
    elimination of poorly soluble compounds, 141
    goal of screening in, 188
    oral absorption potential of, 180
drug classes, 223, 229
dutasteride, 276
dynamic light scattering (DLS) plate reader, for aggregates, 124

E

efavirenz, 285, 301–302
elixirs, 300–301, 301–304
enantiotropic systems, 145
enthalpy and entropy of fusion, 6–7
enthalpy of solution, ideal state, 6
Entocort® EC capsules, 261
entropy of mixing, ideal state, 6
Epadel®, 277, 288, 293
ethanol, 260, 296, 318, 435
ethanolamine, 39
ethyl icosapentate, 277, 288, 293
etoposide, 277
evaporative light scattering detectors (ELSD), 119

F

FaSSIF (fasted state simulated intestinal fluid), composition of, 162
FeSSIF (fed state simulated intestinal fluid), composition of, 162, 164, 165, 207
fed stomach
    composition of fluids in, 157
    digestion of lipid, 158–160
    milk for simulation, 167
    nutritional drinks for simulating, 163, 164

felbinac ethyl ester, 326
Fenofibrat AZU® 200 mg Kapselin, 293
fenofibrate, 285
Fenogal, 293
Fick's second law of diffusion, 20
flavopiridol, 47
flurbiprofen, 43
flurbiprofen axetil, 326
Fortovase® soft gelatin capsules, 281, 294
Fourier Transform Infrared (FTIR) spectra, 368, 374–375
free energy-reaction progress diagram, 69, 70
Freeze thaw cycles
    impact of water absorption in, 116
    impact on precipitation, 116

G

gastric emptying rate, 157
gastric lipase, 158
gastrointestinal tract, conditions in
    biorelevant lipids and surfactants, 152–154
GastroPlus™, 224, 228
Gelucire®44/14, 260, 291
Gelucire®50/13, 264
GRAS (generally regarded as safe) materials, 236
Gengraf® soft gelatin capsules, 275, 298–299
Geodon®, 322
Gibbs free energy curves, 56
Gibbs phase rule, 41
Glakay® soft gelatin capsules, 280, 288, 294
glycerin, 318
glyceryl monooleate, 296
greaseballs category, of compounds, 234
growth hormone releasing factor (GRF), 373
Guidance for Industry—Drug Product, Chemistry, Manufacturing and Controls Information draft, 236
Guidance for Industry—Nonclinical Studies for the Safety Evaluation of Pharmaceutical Excipients draft, 237

H

Hanks' Balanced Salt Solution (HBSS), 187
Hectorol®, 258, 276
heminevrin capsules, 273, 285, 291
Henderson-Hasselbalch equation, 36

hexamethylene tetramine
growth kinetics of, 68
high molecular weight protein (HMWP), 385
Hildebrand solubility equation, 7, 11–12, 15, 42
Hildebrand solubility parameter model, 12
Hixson-Crowell model, of dissolution, 21
Hofmeister lyotropic series, 367, 372
HTS and bioassays quality
effects of insolubility of compounds in DMSO, 114, 115
HTS Hit rates
effect of solubility on, 120, 121
human growth hormone (hGH), 385
human insulin-like growth factor (hIGF), 367
hydrate crystals, formation, 58
hydrate-anhydrate systems
phase diagrams in, 60
physical stability of, 58
hydrates, 18
hydrate screening, 93, 94
hydration/dehydration equation, 58
hydrocortisone acetate, 333
hydrofluoroalkane (HFA) propellants, 428–431
hydrogen bonding effects, 311
hydrogenated oil, 296
hydrophilic-lipophilic balance (HLB), 431
hydrophobic effect, 312
hydrophobic interaction chromatography, 342
2-hydroxypropyl-ß-cyclodextrin (HPßCD), 226–228, 259, 260, 322, 438
hydroxpropylmethylcellulose (HPMC), 240

I

IAM (immobilized artificial membrane), 180
ibuprofen, 277, 387
IDD®-P solubilization technology, 261
ideal solutions
defined, 5
forces on, 5
mixing process of, 5–6
predicting solubility in, 7–9
steps in dissolution process, 5–7
immobilized artificial membrane (IAM), 180
in vitro absorption models
additives as solubility enhancers, 191–201
in vitro–in vivo correlations (IVIVC), 166

In Silico polymorph screening, 78
In situ salt screening method, 144
in well sonication, 118
indomethacin farnesil, 288
indomethacin, 277
Infree® soft gelatin capsules, 277, 288, 296
Intestinal fluids, composition of
  during fasted and fed state, 156–160
ionic equilibria
  and solubility, 43–47
  of acidic and basic substances, 33–34
  of buffer systems, 35–39
  of conjugate acids and bases, 34–35
ionic strength
  and protein solubility, 346–349
  of a solution, 30
isotretinoin, 278
itraconazole, 226, 235, 278, 301, 324
IVIVC using dissolution, 168

J

Juvela®, 283, 289
Juvela® N soft gelatin capsules, 295

K

Kaletra®, 261, 279, 297
Keesom forces, 14–15
Keldjal nitrogen measurements, 346
Kirchoff equation, 7
Krebs Bicarbonate Ringer's Solution (KBR), 187

L

Labrafil® M-2125CS, 264, 287
Labrasol®, 260, 264, 287
Lamprene® soft gelatin capsules, 298
Lanoxin®, 275
lecithin, 432
leuprolide, 386
library screening, 224–228
lipid digestion, and emulsification
  pancreatic colipase role in, 158–160
Lipirex, 293

liposomes, 325
London dispersion forces, 12, 15
lopinavir, 279, 303
lorazepam, 319
low molecular weight urokinase (LMW-UK), 373
lubiprostone, 279, 291
Lucifer yellow, 202
Lupronr® Depot, 334
lyso-phospholipids, 152

M

^{14}C-mannitol, 202
Marinol®, 276
maximum absorbable dose (MAD), 137
Medihaler Epi™, 428
medroxyprogesterone acetate, 333
melanotan, 385
melphalan hydrochloride, 318
membrane based models
  for drug premeability screening, 181, 182
menaquinone-4, 326
menatetrenone, 280, 288
methanesulfonic acid, 322
methylene chloride, 368
methylprednisolone acetate, 333
micelle-forming amphiphiles, 152
microdialysis, osmotic pressure dependent, 343
mitomycin, 324
monotropic systems, 145
Montreal Protocol, 428
morphine, 313
MTT test, for cell viability evaluation, 202

N

N-methyl-2-pyrrolidone (NMP), 318
NanoStore technology, 119
natural moisturising factor (NMF), 407
Neoral®, 273
Neoral® soft gelatin capsules, 298–299
nephelometric method, 125
new chemical entity (NCE), 111
nimodipine, 280
Nimotop®, 280

nitroglycerin, 319
non-ideal solutions
  intermolecular interactions on, 14–15
  properties of, 9–14
non-ionic surfactants, 206
Norvir® soft gelatin capsules, 281, 297, 302–303
Noyes-Whitney equation, 167
  of dissolution rate, 20, 22, 234
nucleation equation, 62
nucleation rate
  factors governing, 66, 67
  determination and dependence on solubility, 63, 64
  effect of molecular recognition phenomena on, 62–67

O

oleic acid, 260, 268, 292–293, 296, 420, 432
One-Alpha® soft gelatin capsules, 293
organic crystals, phase transitions of, 76
Organic Solvent Tolerant Bacteria (OSTB), 387
Ostwald-Freundlich equation, 325
Ostwald's law of stages, 63, 75, 76
Ostwald's ripening, 70

P

P-glycoprotein transporter, 258
paclitaxel, 320, 321
PAMPA assay (parallel artificial membrane permeability assay), 180, 229
  Double-Sink PAMPA (DS-PAMPA), 206
pancreatic phospholipase A2, 158
parenteral formulation development
  adjustment of pH and salt formation, 313–315
  design, 310–313
  dispersed systems, 323–334
  flowchart for solubilization, 314
  inclusion complexes, 322–323
  use of cosolvents, 315–320
  use of surfactants, 320–321
paricalcitrol, 280
peanut oil, 292
pentobarbital sodium, 319
peppermint oil, 268, 290
perflenapent, 326

perflubron-based emulsion, 327
perfluorodecalin, 327
perfluorotripropylamine mixture, 327
permeability
  and impact of bile salts, 204
  and impact of pH gradient on, 203, 204
permeability screening, experiment design, 206–208
permeation assessment, 181
  biological models for, 180
  cell-based systems for, 182
  Immobilised artificial membrane (IAM), 181
  liposomes for, 181
  PAMPA approach, 181, 182
  tissue-based systems for, 185
permeation models, for drugs
  non-specific binding, 186, 187
  poor aqueous solubility, 186
  solvent systems of, 191–200
permitted daily exposure (PDE), 237
pH measurements
  adjustments, 233, 313–315
  of aqueous solutions, 39–48, 346
  of non-aqueous solutions, 360–361
pH stability profile, of a drug substance, 37
pH-solubility profile, 39, 43–44, 140, 141
pH-stat, 142
preformulation
  and solubility determination methods, 138
pharmaceutical salts
  solubility determination for, 142
phase transformations process, solvent-mediated, 75–77
phenobarbital elixir, 280, 301
phenobarbital sodium, 319
phenytoin sodium, 317, 319
phosphate salt solubility, 143
Polaxamer 407 (P407), 386
polyelectrolytes, 349–350
polyethylene glycol, 202, 318
polyethylene glycol 400 (PEG 400), 244, 260, 262, 287–290, 367
polyethyleneglycol ether, 318
polymorph screening
  solvents used and their properties, 80–89
polymorphic system
  of enantiotropically related polymorphs, 71

polymorphism, defined, 16
polymorphs
transition temperature determination for, 145
crystalline solids, 53
free energy relationships between, 55
screening approaches, 78–79
selection strategy, 91
solvent effects on, 70–75
prediction-based screening, 78–90
porcine bile extract, 162, 163
potentiometric titration method, 43
progesterone, 281
Prometrium® soft gelatin capsules, 281, 292
propofol, 323, 327
propylene glycol, 260, 296, 318
prostaglandin E1, 326
protein binding, impact on drug transport, 205
proteins: non-aqeous pharmaceutical formulations
biological activity, 380–383
formulation applications, 385–386
industrial applications, 386–387
pH measurements, 360–361
pharmaceutical processing, 383–385
physical characteristics of organic solvents, 362–363
solubility, 361–368
stability, 376–380
structure, 369–374
suspensions, 374–376
protein solubility
defined, 342
methods for measuring, 342–346
methods for prediction, 342
solubility principles, 346–348
strategies for enhacing solubility, 348–351
pseudoephedrine, 270
pseudopolymorphism, 53
pulmonary drug delivery formulations
chlorofluorocarbon (CFC) propellants, 428–429
choosing a propellant, 429–431
dosing efficiency, 433–434
excipients, 431–433, 435–437
hydrofluoroalkane (HFA) propellants, 428–429
trends, 437–439
PXRD (Powder X-ray diffraction), 140, 144

R

Raman spectroscopy, analytical technique, 146
ranitidine, 262
Raoult's law, 58
Rapamune®, 282
Restandol 40 mg soft gelatin capsules, 292–293
ritonavir, 258, 261, 279, 281, 302–303
  stable polymorph screen of, 93
Robitussin® Cold, Cough & Flu liquid-filled capsules, 270, 295
Rocaltrol®, 273, 291

S

salmon calcitonin (sCT), 373
salts solubility, 142–144
Sandimmune® soft gelatin capsules, 274, 298–299
saquinavir, 281
Selbex®, 289
self-emulsifying drug delivery systems (SEDDS), 264, 269, 287, 297
sesame oil, 292
shake-flask method, based on phase solubility technique
  steps invovled in, 126, 138–140
sirolimus, 282, 303
size exclusion chromatography (SEC), 374
skin
  dermis layer, 405–406
  epidermis layer, 406
  stratum corneum layer, 406–408
  subcutaneous fat layer, 404–405
sodium chloride injection, 322
sodium dodecyl sulfate (SDS), 367
Softigenr® 767, 264, 287
solid-state stability
  and free energy relationships, 54–60
solubility
  defined, 1
  dissolution phenomena, 20–23
  hydrogen bonding effects, 15
  and impact of bile salt concentration, 165
  of a solid, 40–41
  of metastable polymorph, 17–18

solubility (*Cont.*)
of stable crystal form, 17
of substances in solid-state forms, 16–19
parameters for solvents, 13
pH measurements of aqueous solutions, 39–48
thermodynamics of, 1, 4–16
units for expression of, 2–3
solubility determination
for polymorphs and solvates, 144, 145
role of pH-solubility profile in, 142
in pharmaceutical preformulation, 138
solubility enhancement approaches, 191–201
solubility prediction, 126
solubilizing excipients and mixtures
cyclodextrins, 267–268
microemulsion oral formulations, 269
surfactant molecules, 262–267
water-soluble organic solvents, 262
solute-solute interaction energies, 11
solutions
concept of percentage, 3
molarity of, 2
mole fraction of, 3
normality of, 2
saturated, 2
supersaturated and supersaturation, 2, 60–62, 99, 100
unsaturated, 2
volume fraction of, 3
solvated crystals analyses, 65
solvates
and non-solvates free energy relationships between, 56–58
solvatomorphs, 18
solvent-mediated polymorphic transformation
effect of impurities on, 92
effective in achieving stable drug polymorph, 76, 77
solvent-solvent interaction energies, 11
sorbitan trioleate, 432
soya lecithin, 432
soybean oil, 296
Span 80, 287
Sporanox®, 278
Stock-Einstein equation, 20

stomach, fasted state
volume of gastric fluid in, 156, 157
strong acid, defined, 33
strong base, defined, 34
Sudafed®, 270
sulfathiazole
impact on polymorphic outcome, 73, 74
sulfobutyl ether β-cyclodextrin (SBE-ßCD), 322
supercritical anti-solvent (SAS) mode, 384
supersaturation, 60–62, 98, 99, 100
Sustiva®, 285
Symyx Discovery Tools Solubility and Liquid Formulations Workflow, 146

T

Targretin®, 272
Taxol®, 320
teprenone, 289
testosterone undecanoate, 286
tetrahydrofuran (THF), 235
tetrahydrofurfuryl alcohol, 318
theophylline hydrate/anhydrate phase diagram, 60
thermal analysis techniques, 140
thermodynamic equilibrium constant, of a substance, 30
thermodynamics, of solubility
chemical potential, 4
equilibrium solubility, 4
in ideal solutions, 5–9
in regular solutions, 9–14
intermolecular interactions in non-ideal solutions, 14–15
temperature effects, 15–16
Threlfall's analysis, for isolation of polymorphs, 72
tipranavir, 282
tissue factor pathway inhibitor (TFPI), 350
α-tocopherol, 293
tocopherol nicotinate, 283, 289
tolerable daily intake (TDI), 237
topical preparations, solvent selection principles
biological factors, 410–411
thermodynamics, 409
vehicle effects on skin, 410

topical formulations, solvents
  dimethicones, 420
  glycols, 416–417
  isopropyl linoleate, 419
  isopropyl myristate (IPM), 419
  isopropyl palmitate, 419
  oils and waxes, 417–419
  oleic acid, 420
  silicones, 420
  sodium lauryl sulphate, 420
  water, 412–414
Transcutol®, 262
transepithelial electrical resistance (TEER), 191
tretinoin, 283
triamcinolone diacetate, 332
triamcinolone hexacetonide, 332
triglide™, 261
triglycerides, 263, 268, 291
turbidimetric method, 126
Tween 20, 264
Tween 80, 261, 264, 287, 320
Tylenol®, 270

U

USP test fluid, 161

V

valproic acid, 283, 286
van der Waals forces, 11, 14, 311, 323, 361
van't Hoff equation, of solution, 8, 15, 42
van't Hoff plot, of solubility of naphthalene, 9
Van't Hoff-type plot, intrinsic dissolution rates, 57
Vesanoid®, 283
Vfend®, 322
vitamin A, 327
vitamin D2, 327
vitamin E, 327
vitamin E TPGS, 202, 246
vitamin K_1, 327
VNP40101M (1,2-bis(methylsulfonyl)-1-(2-chloroethyl)-2-[(2-methylamino) carbonyl] hydrazine), 318
voriconazole, 322, 324

W

Walden rule, 8
warfarin sodium, 317
water activity, a_w, 58

Y

Yalkowsky models, of fusion, 8

Z

Zemplarl®, 280
ziprasidone mesylate, 322, 324

Printed in the United States of America

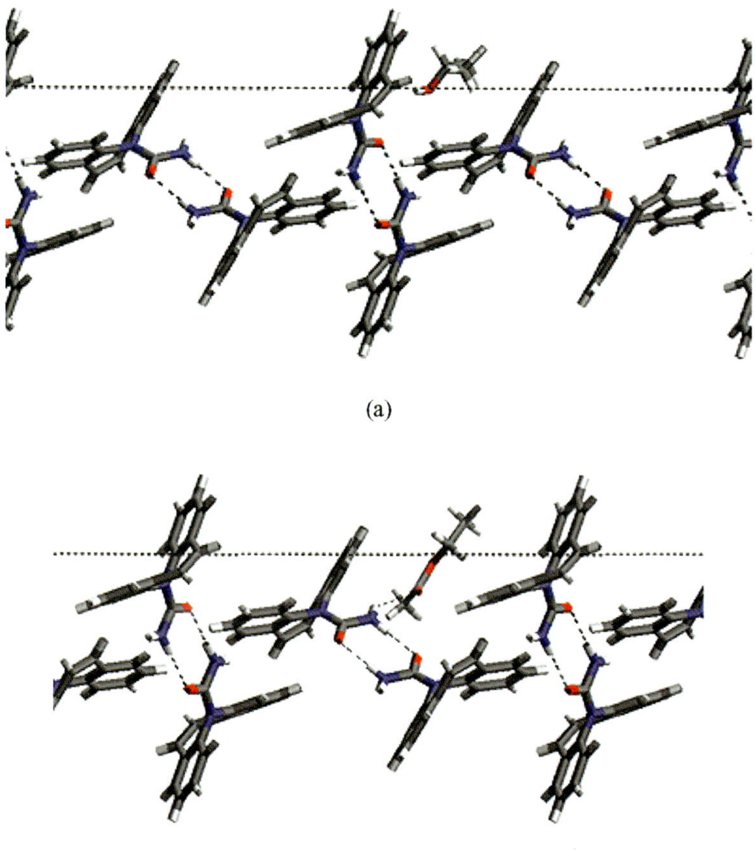

(a)

(b)

Chapter 3 - Figure 13. Solvent interaction at the (011) crystal face of carbamazepine form III. (a) 2-propanol and (b) ethyl acetate. Reprinted from Kelly, 2003 with permission.

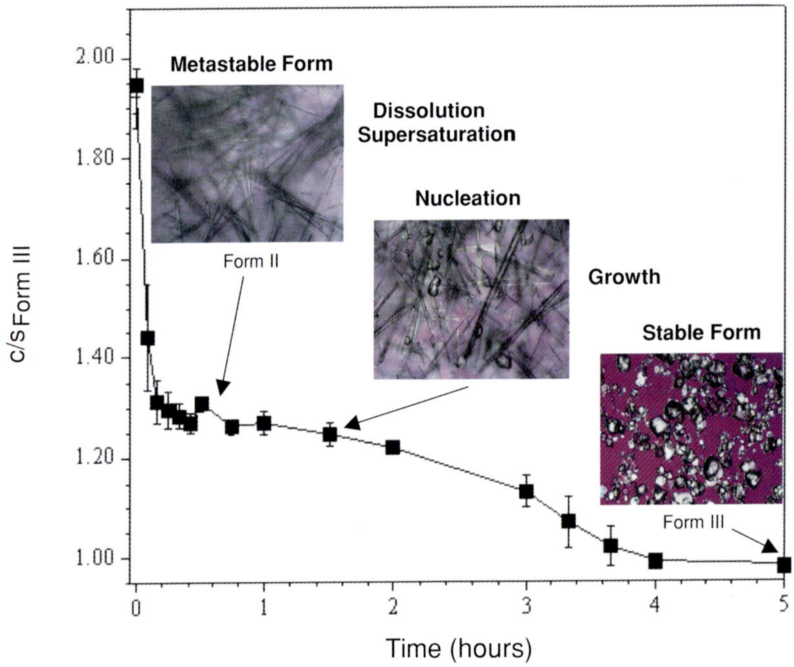

Chapter 3 - Figure 15. Solvent-mediated transformation of carbamazepine polymorphs in ethyl acetate at 25 °C and initial $c/s_{FormIII} = 2.0$. Reprinted from Kelly, 2003 with permission.

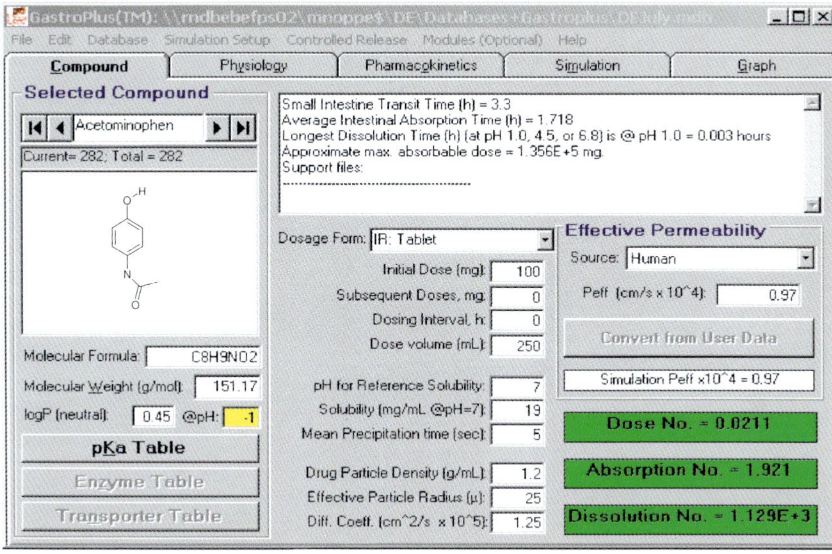

A

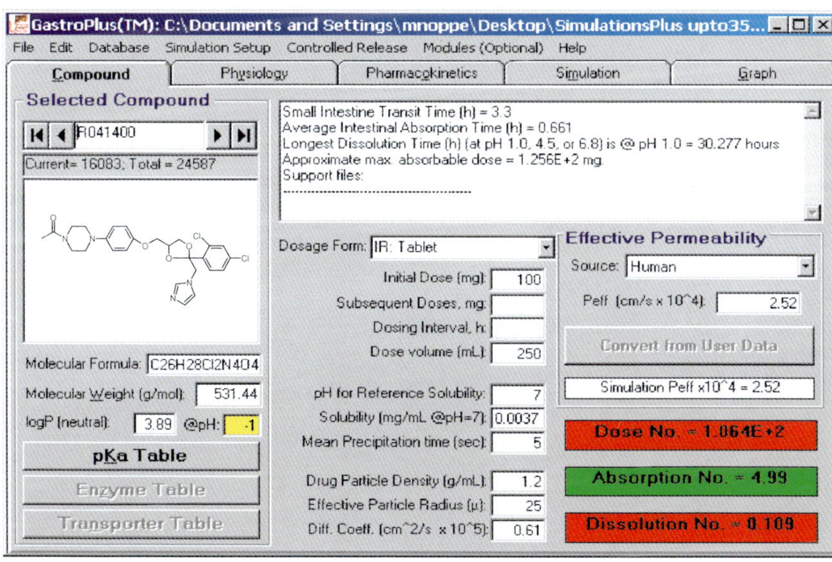

B

Chapter 8 - Figure 2. Estimation of DCS class based on the GastroPlus software of SimulationPlus. Based on the dimensionless Dose, Absorption and Dissolution Numbers, drugs can be defined as DCS Class I (A), II (B), III (C) and IV (D) based solely on structural information. The program gives predicted values for solubility, pKa and effective human intestinal permeability in order to suggest a fraction absorbed into the portal vein.

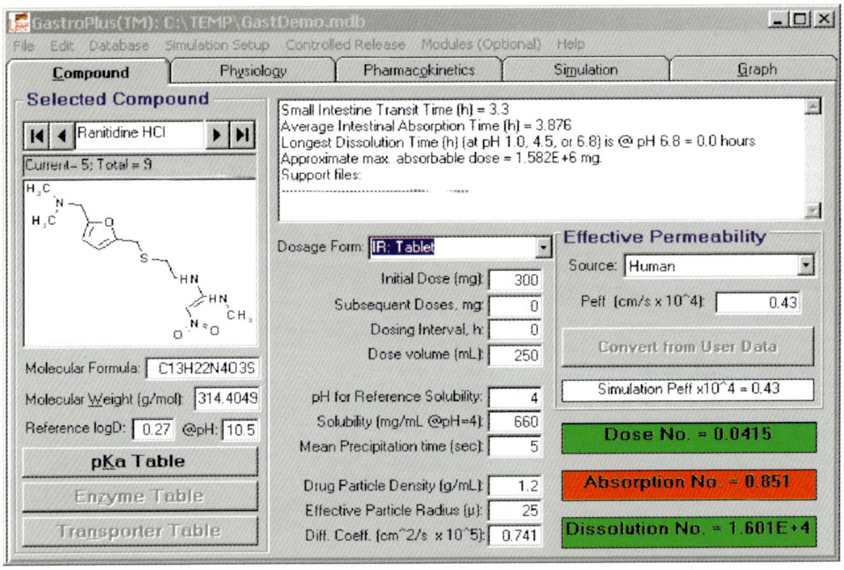

C

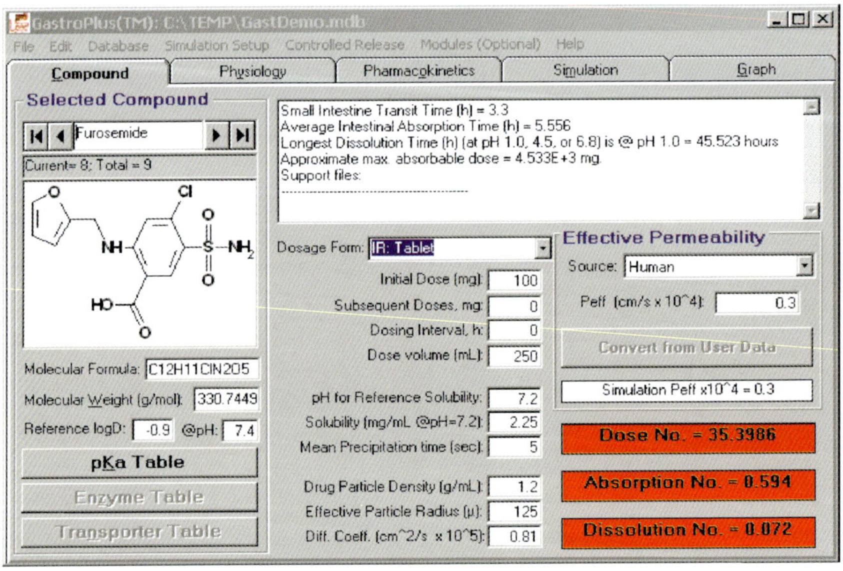

D

Chapter 8 - Figure 2. (*Continued*)

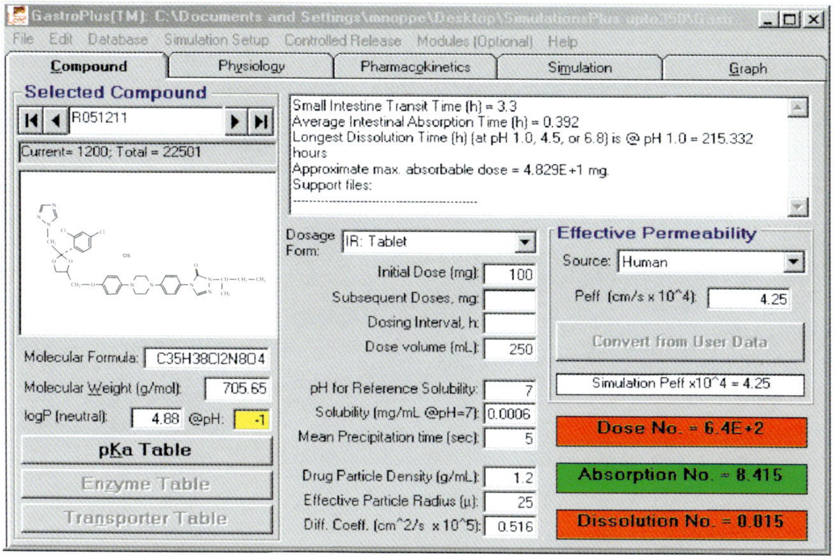

A

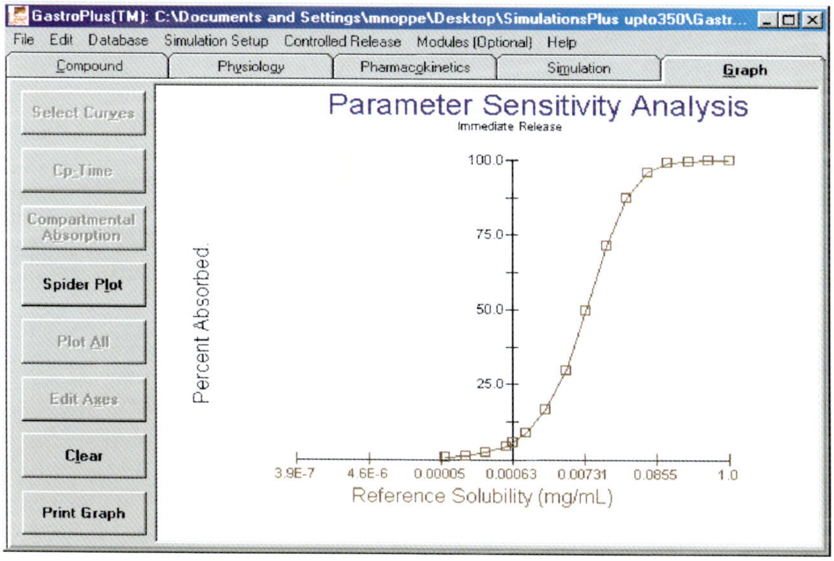

B

Chapter 8 - Figure 3. Predicted DCS classification of Itraconazole (A) as well as the effect of solubility on fraction absorbed (B). The Spider plot suggest good oral bioavailability at solubility values above 100 µg/mL. The solubility of itraconazole at neutral pH is estimated at 1 ng/mL. Itraconazole can be solubilized in 2-hydroxypropyl-β-cyclodextrin to levels in excess of 10 mg/mL which suggests Class I behavior (C).

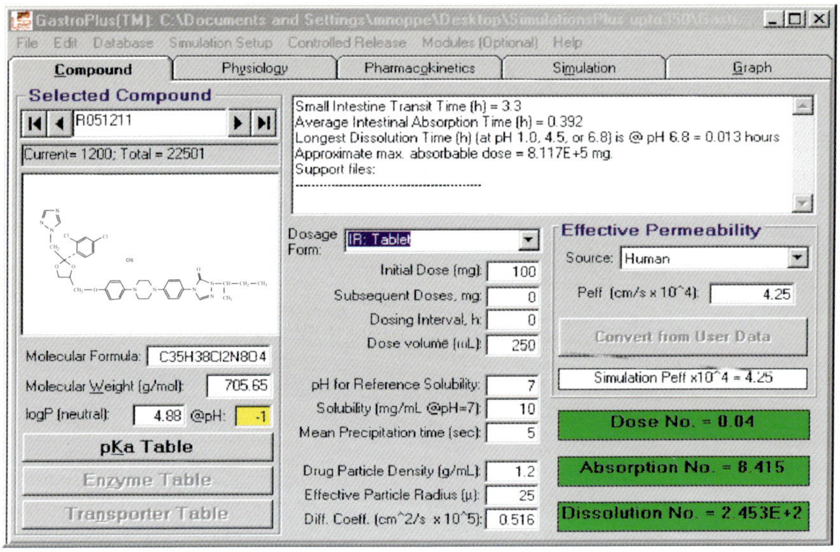

C

Chapter 8 - Figure 3. (*Continued*)

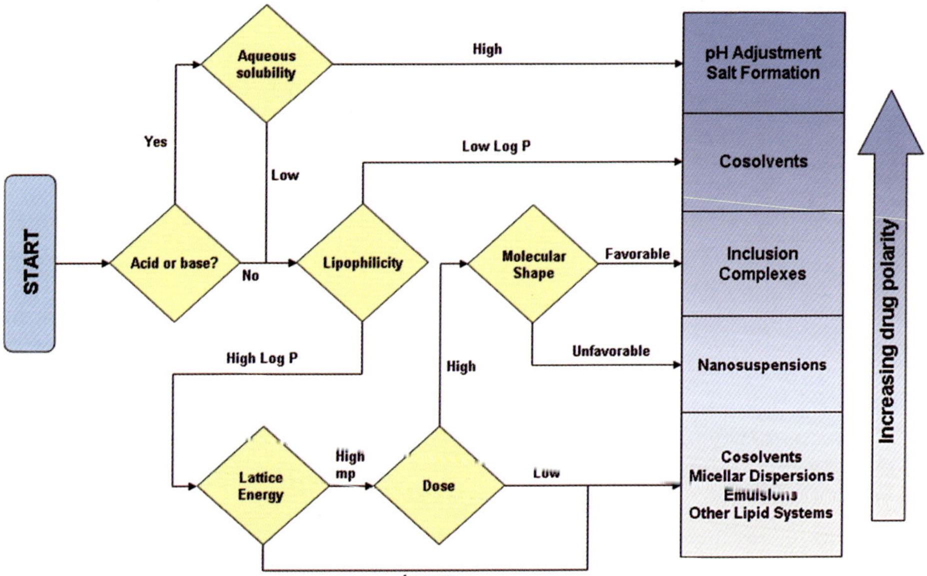

Chapter 10 - Figure 2. Flowchart for solubilization of parenteral drugs.

Printed in the United States of America